AF895580

Environmental Science and Engineering

Series Editors

Ulrich Förstner, Buchholz, Germany

Wim H. Rulkens, Department of Environmental Technology, Wageningen, The Netherlands

The ultimate goal of this series is to contribute to the protection of our environment, which calls for both profound research and the ongoing development of solutions and measurements by experts in the field. Accordingly, the series promotes not only a deeper understanding of environmental processes and the evaluation of management strategies, but also design and technology aimed at improving environmental quality. Books focusing on the former are published in the subseries Environmental Science, those focusing on the latter in the subseries Environmental Engineering.

Nitish Kumar · Clemens Walther ·
Dharmendra K. Gupta
Editors

Chromium in Plants and Environment

Springer

Editors
Nitish Kumar
Department of Biotechnology
Central University of South Bihar
Gaya, Bihar, India

Dharmendra K. Gupta
Ministry of Environment, Forest
and Climate Change
New Delhi, Delhi, India

Clemens Walther
Institut für Radioökologie Und
Strahlenschutz
Gottfried Wilhelm Leibniz Universität
Hannover, Germany

ISSN 1863-5520 ISSN 1863-5539 (electronic)
Environmental Science and Engineering
ISBN 978-3-031-44028-1 ISBN 978-3-031-44029-8 (eBook)
https://doi.org/10.1007/978-3-031-44029-8

© The Editor(s) (if applicable) and The Author(s), under exclusive license to Springer Nature Switzerland AG 2023

This work is subject to copyright. All rights are solely and exclusively licensed by the Publisher, whether the whole or part of the material is concerned, specifically the rights of translation, reprinting, reuse of illustrations, recitation, broadcasting, reproduction on microfilms or in any other physical way, and transmission or information storage and retrieval, electronic adaptation, computer software, or by similar or dissimilar methodology now known or hereafter developed.
The use of general descriptive names, registered names, trademarks, service marks, etc. in this publication does not imply, even in the absence of a specific statement, that such names are exempt from the relevant protective laws and regulations and therefore free for general use.
The publisher, the authors, and the editors are safe to assume that the advice and information in this book are believed to be true and accurate at the date of publication. Neither the publisher nor the authors or the editors give a warranty, expressed or implied, with respect to the material contained herein or for any errors or omissions that may have been made. The publisher remains neutral with regard to jurisdictional claims in published maps and institutional affiliations.

This Springer imprint is published by the registered company Springer Nature Switzerland AG
The registered company address is: Gewerbestrasse 11, 6330 Cham, Switzerland

Paper in this product is recyclable.

Preface

Chromium was discovered by the French chemist Nicolas-Louis Vauquelin in the year 1797; its symbol is Cr and atomic number is 24. Cr is a steely-grey, lustrous, hard and brittle transition metal and is valued for its high corrosion resistance and hardness. Cr metal and Cr(III) ions are considered as non-toxic, and hexavalent chromium Cr(VI) is toxic and carcinogenic. Naturally occurring Cr is composed of four stable isotopes, i.e. 50Cr, 52Cr, 53Cr and 54Cr, with 52Cr being the most abundant (83.789% natural abundance).

Cr is regarded as a severe environmental pollutant due to its extensive industrial use. Cr contamination of soil and water has drawn attention in the past decades, due to Cr contamination of groundwater and food sources, but it still poses a serious risk to human health on a global scale. The speciation of Cr largely determines its toxicity, mobility and bioavailability. Cr's behaviour in soil, soil–plant transfer and accumulation in various plant parts depends on the chemical form, type of plant and physicochemical characteristics of the soil. Cr speciation and behaviour in soil are significantly influenced by the soil microbial population. Plants can prevent Cr from entering the cell through external rejection of organic acids, amino acids and proteins that combine with heavy metals. Accumulation of Cr in plants inevitably disrupts homeostasis, damages cell structure and affects the dynamic balance of the antioxidant enzyme system.

This book *Chromium in Plants and Environment* presents the most recent research on chromium's chemistry, sources, acute and long-term impacts on the environment, human health and remediation measures. This book is offering a thorough examination of chromium exposure, toxicity and toxicity prevention so that readers can evaluate the hazards associated with it efficiently. The book compiles the most recent research on how chromium affects the environment, human health and remediation. Recent methods in chromium detoxification, speciation and molecular mechanisms are included, and it offers all the knowledge required for efficient risk assessment, prevention and countermeasure.

The editors (Dr. Nitish Kumar, Prof. Clemens Walther and Dr. Dharmendra K. Gupta) personally thank the authors for contributing with their valuable time, knowledge and eagerness to bring this book into this shape.

Patna, India — Dr. Nitish Kumar
Hannover, Germany — Prof. (Dr.) Clemens Walther
New Delhi, India — Dr. Dharmendra K. Gupta

Contents

Chapter 1
Health Impacts of Cr Contamination in Soil

Ratna Dwi Puji Astuti and Annisa Utami Rauf

Abstract Chromium (Cr) has been widely used for many purposes and products. The elevation of Cr concentration in the environment is mainly related to anthropogenic sources. From a long time ago, chromium trivalent was known as an essential element but in a recent study, chromium can be only acknowledged as a pharmacologically active substance and not an essential substance/element. Under particular conditions, chromium may be oxidized to Cr(III) or Cr(VI) and vice versa. Increased soil Cr content may be hazardous to terrestrial organisms and humans. Chromium is included as a top-priority chemical substance because it may cause toxic, mutagenic and carcinogenic effects on humans. Humans can expose to Cr via direct dermal contact, ingestion of food or accidental ingestion of soil, and inhalation of particulate in the air. Plants exposed to Cr may experience negative impacts on growth, development, and photosynthesis. The excessive deposition of Cr in soil has caused growth retardation and germination issues in a number of food crops, which has significant implications for the food supply chain and raises the probability of serious health effects on humans and animals. Epidemiologic studies have linked chronic and acute dietary Cr exposures with several adverse health effects and symptoms like lung cancer, skin lesions, neurological diseases, and problems in reproductive system. Advanced research strategies and technologies to reduce Cr contamination in the environment and diminish human health risks are needed.

Keywords Chromium · Trace elements · Soil · Health effects · Humans

R. D. P. Astuti (✉)
Department of Environmental Health, Faculty of Public Health, Universitas Airlangga, Surabaya 60115, Indonesia
e-mail: ratna.dwi@fkm.unair.ac.id

A. U. Rauf
Department of Health Behavior, Environment and Social Medicine, Faculty of Medicine, Public Health and Nursing, Universitas Gadjah Mada, Yogyakarta 55281, Indonesia

© The Author(s), under exclusive license to Springer Nature Switzerland AG 2023

N. Kumar et al. (eds.), *Chromium in Plants and Environment*, Environmental Science and Engineering, https://doi.org/10.1007/978-3-031-44029-8_1

1.1 Introduction

Potentially toxic elements (PTEs) contamination in our environment is one of the environmental problems nowadays due to its potential to cause health problems to humans and living organisms, as well as ecological degradation (Tchounwou et al. 2012; Mishra and Bharagava 2016; Ali et al. 2019; Kiran and Sharma 2022; Mitra et al. 2022). This element can become toxic and/or carcinogenic through accumulation in living organisms. One of the PTEs that potentially generate environmental problems is chromium (Cr). The United States Agency for Toxic Substances and Disease Registry ranks this PTE among the most hazardous substances (U.S ATSDR). First found by the scientist named Vauquelin in 1797 in lead ore (crocoite) (Baruthio 1992). Until now, chromium has been widely used for many purposes and products. Chromium (Cr) is naturally found as chromite ore (Cr) in the trivalent form (Cr(III)). Until now, this ore is utilized for manufacturing Cr pigments, chromic acid, dichromates, monochromates, and Cr metal. Cr is mostly utilized in the leather tanning, chrome plating, paint pigment, and wood treatment sectors. The most significant industrial source of chromium in the atmosphere is metallurgical, chemical, and heat-resistant application sectors (ATSDR 2012). Hexavalent chromium (Cr(VI) and trivalent chromium (Cr(III)) are two oxidation states of chromium that occur naturally (Sun and Costa 2022). The toxicity of Cr is based on the states where hexavalent chromium is more hazardous than trivalent chromium (Sun and Costa 2022). A long time ago, several researchers still proposed Cr(III) is commonly used as essential dietary nutrient at low dose (Anderson 1997; Krejpcio 2001; Heimbach and Anderson 2005; Pechova and Pavlata 2007). This statement was based on (1) studies related to low concentration exposure of chromium in rats' diet, (2) studies with total parental nutrition diets given to subjects, (3) studies related to absorption of chromium, however, those evidences are still problematic (Vincent 2017).

The benefit of Cr supplementation is not only for humans but also for animals. There are numerous benefits of Cr, including improved insulin activity (Chen et al. 2017), diminished inflammation and oxidative stress (Preuss et al. 1997; Tuzcu et al. 2011), and reduced body weight and improved blood lipid profile (Nachtigal et al. 2005; Kuryl et al. 2008). Untea et al. (2017) previously demonstrated that chromium picolinate (Cr(pic)) supplementation had effects on fat metabolism and protein, as well as increased amino acids in pork meat. In contrast to animals, the negative effect of Cr occurred to plants due to the accumulation of Cr such as diminished nutrients and the number of tomato fruit (Moral et al. 1995).

In contrast to previous studies, a novel study by Vincent (2017) proposed that chromium can only be included as an active pharmacological substance not a vital element/substance for a living organism. This statement is supported by the European Food Safety Authority (EFSA) (2014) assertion that there is no evidence of the health advantages of chromium consumption in healthy individuals. Other than that, there are no international acceptable plasma Cr values or ranges for the general population (Chen et al. 2017). Meanwhile, prolonged exposure to high levels of Cr can trigger genotoxic and cytotoxic effects, ultimately leading to immune system problems (Shrivastava et al. 2002). Others showed that Cr exposure may limit absorption of Zn or Fe from the gastrointestinal tract which can induce mineral deficiency and anemia (Lukaski 1999). The sixth most abundant element in the earth's crust, Cr, is one of the most dangerous inorganic soil pollutants (ATSDR 2012). Its average concentration is 125 mg/kg. (Sun et al. 2015a). Cr may enter the environment by air, water, soil, and the food chain (Mishra et al. 2019). Soil is one of the environmental media in which inorganic contaminants, such as Cr, can be found. It can serve as both a source and a sink for contaminants in terrestrial ecosystems.

Cr accumulation can easily infiltrate the soil, influencing the soil biota and humans. Nonetheless, the effects of the Cr accumulation in soil on terrestrial organisms are much more extensive because Cr accumulates in the food chain. Food crops are the most important nutrition source for human (Hefferon 2015; Dobermann et al. 2022). The nutritional and toxic substances of food crops will depend on the medium of planting. Plants can absorb Cr concentrated in the soil and accumulate it in the tissues. The accumulation of Cr in soil may contribute to bioaccumulation of Cr in plants. Fruit and vegetable consumption is the primary source of PTE exposure, accounting for 90% of metal intake, with inhalation of contaminated dust and direct skin contact accounting for the remaining 10% (Mawari et al. 2022). Hence, exposure of Cr to humans via ingestion or oral routes is significant for identifying the health risk in the future.

Cr will be absorbed into the body through inhalation, ingestion, and skin contact. Aside from skin contact, the principal routes of Cr exposure in humans are ingestion and inhalation (Langård and Costa 2015). Cr(VI) absorption is quicker than Cr(III) absorption through the digestive and respiration tract. Since food demand is increasing over time, food security and safety issues have become a significant public health concern in terms of human health. Transport of Cr from soil to plant is a crucial stage in the transmission of Cr between trophic levels in our food chain. From pesticide applications, wastewater disposal, sewage sludge applications, industrial effluents, and municipal waste, the food chain is regularly replenished with numerous compounds, including non-essential and essential trace elements such as chromium. Cr may be absorbed into plants from polluted soils and transported up the food chain to animals. Thus, Cr accumulation in soil and its relationship with the food chain are a concern due to human health impacts that may occur in the future. This section explains how soil contamination with Cr will affect human health via the food chain.

1.2 Fate and Source of Chromium in Terrestrial Ecosystems

The trace element such as chromium (Cr) can be found in volcanic dust and rocky soil (Sharma et al. 2020a). Chromium content in rocks is 100 mg/kg on average (Ma and Hooda 2010). Basaltic igneous rocks have an average Cr concentration of 200 mg/kg, ultramafic rocks 1800 mg/kg, sandstone 35 mg/kg, granitic rocks 20 mg/kg, shales and clays 120 mg/kg, and limestone 10 mg/kg (Ma and Hooda 2010). As a consequence of its high redox potential, complex valence shell chemistry, and electron chemistry, chromium may rapidly change from one oxidation phase to another (Prado et al. 2016; Shahid et al. 2017; Sharma et al. 2020b). In nature, Cr exists in multiple oxidation states ranging from valence −2 to +6. (ATSDR 2012; Shahid et al. 2017). The most common and stable oxidation phases of Cr in nature are 0 (elemental metal), +3 (trivalent), and +6 (hexavalent) (ATSDR 2008). Trivalent chromium (Cr(III)) exists as $FeOCr_2O_3$ (chromite), whereas hexavalent chromium (Cr(VI)) exists as CrO_4^{2-} (chromate) or $Cr_2O_7{}^{2-}$ (dichromate), both of which are toxic to organisms (Ertani et al. 2017). Since Cr(III) is highly immobile and has a strong bond with silicates, oxides, and organic matter, thus it is difficult to absorb by plants (Saha et al. 2017). Mobility of Cr is lower than that Cd and Ni. That relative mobility characteristic can be described by transfer coefficient (Sauerbeck 1987). The greater transfer coefficient indicates greater PTE mobility from soil to plant. According to Saha et al. (2017), heavy metals such as Co, Cr, Hg, Pb, Sn, As, Be, and F have low transfer coefficient value (0.01–0.1). Cr(VI) or hexavalent chromium is soluble and toxic to organisms (Astuti et al. 2023). This substance is highly toxic and is only immobile under an alkaline environment in the soil (Shanker et al. 2005; Sun et al. 2015b; Astuti et al. 2023).

The presence of Cr(VI) is associated with oxygen as CrO_4^{2-} or chromate and $Cr_2O_7^{2-}$ or dichromate oxyanions (Shanker 2019). Cr in the soil is representing the combination of trivalent chromium Cr(III) and hexavalent chromium Cr(VI) content. Naturally, chromium can transform into different oxidation states. Cr change can occur by reduction, precipitation, oxidation, dissolution, and sorption, among others (Kimbrough et al. 1999). In the soil, hexavalent chromium can change into trivalent chromium and vice versa (Apte et al. 2006). The transformation process of chromium can be influenced by various natural conditions in nature, such as changing of pH, Eh, temperature, and the presence of organic matter, oxygen and manganese oxide (Apte et al. 2006; Ma and Hooda 2010). Dissolved oxygen and manganese oxides are the oxidants in soil that facilitate the transformation of Cr (MnO_2) (Oliveira 2012; Shanker 2019). While Cr reduction can be influenced by the presence of iron, vanadium, sulphydes, and organic matters (Oliveira 2012). When Cr(VI) enters the soil environment with sewage sludge or polluted irrigation water, it rapidly converts to Cr(III) within one or two days in the presence of organic matter and acidic conditions (Saha et al. 2017).

Because trivalent chromium is less mobile than hexavalent chromium, the conversion from Cr(VI) to Cr(III) reduces Cr accumulation in plant tissue. The development of organic compounds in acidic conditions will increase the solubility of Cr(III) (Ma and Hooda 2010). Under aerobic conditions, Cr(III) can be oxidized to Cr(VI), most likely due to the presence of manganese oxide. Due to its high levels in the environment (soil and water) as a result of multiple anthropogenic and natural activities, Cr pollution in the environment has garnered significant attention (Quantin et al. 2008; Ashraf et al. 2017). The accumulation of Cr in plants will affect to plants' growth and photosynthesis process. Exposure to chromium compound is immensely harmful for

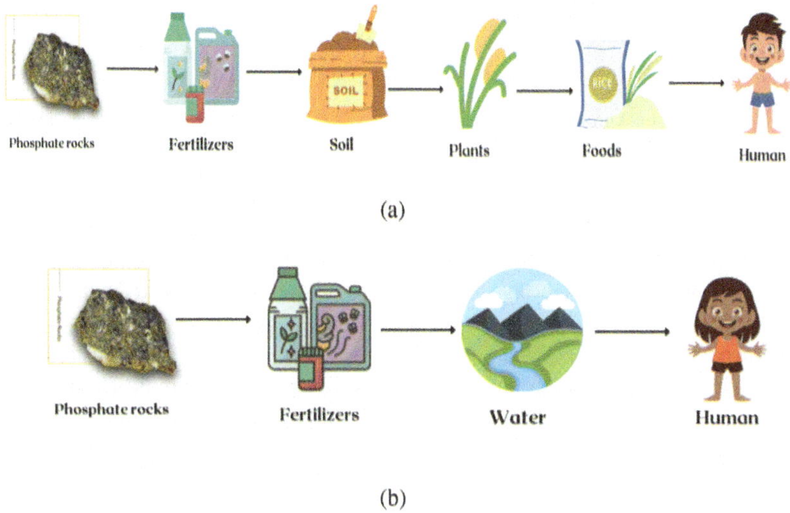

Fig. 1.1 Transfer Cr from fertilizers to human body

various plants. A 100 μM/Kg dry weight of Cr is hazardous to the majority of plant species (Davies et al. 2002). Chromium will enter the environment through surface runoff from industrial and mining areas, the leaching process from landfills, fossil fuel combustion, agricultural activities (e.g., fertilizer and pesticide applications), cement plant emissions, mineral leaching, and waste incineration. Textile industry, leather tanneries, electroplating industries, and steel industries are the common source of Cr (Ali et al. 2019). Fertilizers also contain Cr significantly (Krüger et al. 2017). Phosphate fertilizers derived from phosphate rocks may contribute to the global movement of Cr in the environment and have a substantial impact on the rise in human health risk and ecological risk. There are two major routes for transfer chromium from fertilizers to human (Fig. 1.1).

Cr has been widely used in many industries, primarily in chrome plating, paint pigment, wood treatment, and the leather tanning industry. Thus, anthropogenic sources of Cr have led to widespread pollution in the environment and have elevated biomobility and bioavailability. In addition to human-made sources, natural sources including the weathering of parent rocks also contribute to Cr pollution. Cr occurs naturally as chromite or $FeCr_2O_4$ in serpentine and ultramafic rocks, as well as in other metals including bentorite, $Ca_6(Cr, Al)_2(SO_4)_3$, crocoite, $PbCrO_4$, vauquelinite, $CuPb_2CrO_4PO_4OH$, and tarapacaite, K_2CrO_4 (Babula et al. 2008). Some natural, background, and average Cr concentration are presented in Table 1.1. The levels of Cr in the earth's crust ranges from 0.1 to 0.3 mg/kg (Shahid et al. 2017). In normal settings, Cr content ranges from 10 to 50 mg/kg; however, Cr buildup in agricultural soil can reach as high as 350 mg/kg (Ertani et al. 2017). Pathways of Cr transfer from soil to the human body are shown in Fig. 1.2. Human can expose to Cr through ingestion (food crops and geophagy or pica), direct contact with skin, and inhalation (wind-borne particulate).

Table 1.1 Natural, background, and average Cr concentration in the soil reported by several authors

Parameter	Concentration (mg/kg)	Place	References
Natural concentration	5–3000	India	Shanker et al. (2005), Shanker (2019)
Natural concentration	4.89–106	Korea	Kim et al. (2010)
Natural concentration	2–60	World	World Health Organization (1998)
Natural concentration	10–50	–	Shahid et al. (2017)
Background concentration	50–600	–	Ma and Hooda (2010)
Average concentration	93	Indonesia	Mallongi et al. (2021)
Average concentration	100	Jamaica	Mandal and Voutchkov (2011)
Average concentration	20.71–123.76	Turkey	Isıklı et al. (2003)
Average concentration	59.5	Poland	Kabata-Pendias (2011)
Average concentration	22	Sweden	Eriksson (2001)
Average concentration	58	Japan	Takeda et al. 2004)
Average concentration	25.6	Thailand	Zarcinas et al. (2004)

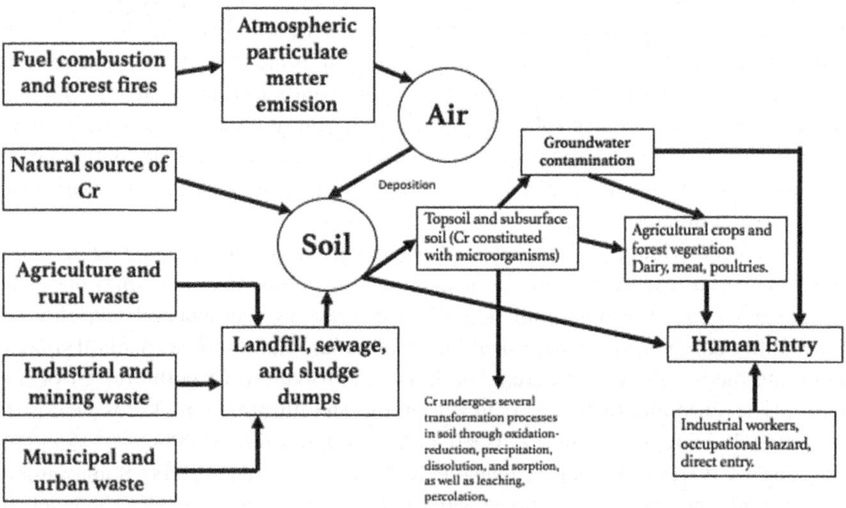

Fig. 1.2 Pathways of Cr transportation from soil to human body

1.3 Chromium in Terrestrial Ecosystems

1.3.1 Chromium Contamination in Plants

Contamination of the environment by Cr has become a major concern. Cr(III) occurs naturally as an essential nutrient in the environment, whereas Cr(VI) is a byproduct of industrial activities. (Sangwan et al. 2014). Low toxicity of Cr(III) associated with its immobility and insolubility. Moreover, both bioavailability as well as biomobility is increased in soil as well as in water. Plants suffer through unstable environmental conditions, biochemical accumulation and production of free radicals as they cannot escape from unfavourable environmental conditions. In hexavalent state, chromium is attributed to the cell impairment and cell damage (Barceloux and Barceloux 1999; Wise et al. 2019; Rauf et al. 2021). In plants, the accumulation of Cr comes from rock weathering, chemical fertilizers, various industrial effluents, fly ash disposal, soil and groundwater contamination. Soil conditions, pesticide residues, water quality, organic matter, pH and geological conditions are potential sources of PTEs (Covarrubias et al. 2018; Mallongi et al. 2019; Astuti et al. 2021a).

The absorption step is the initial interaction between Cr and a plant. Due to the similarities in composition between Cr(VI) and phosphate and sulfate ions, this reaction often occurs in the roots. Through the plasma membrane, root cells will transport phosphate or sulfate ions (de Oliveira et al. 2016; Sharma et al. 2020a). Previous study in *Phaseolus vulgaris* showed the highest bioaccumulation of Cr located in roots (Oruko Ongon'g et al. 2020). Cr(VI) accumulation can reduce the seed germination rate, growth, cell damage, pigment degradation, changes in enzymatic function and nutritional balance. In Fig. 1.3, the uptake of Cr(VI) impairs the beginning of lateral root primordia and influences nutrient absorption.

In developing countries with limited understanding of sustainable agriculture, high levels of Cr in media and crops will become a long-chain problem. Several studies reported the effect of Cr toxicity in plant growth that influences their essential metabolic processes. When there is a deficiency and an increase in the activity of the chlorophyllase enzyme, there is a decrease in the chlorophyll content of plants (Sharma et al. 2020a, b). In agronomic plants, the concentration of Cr in nutrient solution ranges from 0.5 to 5.0 mg/ml and from 5 to 100 mg/g in soil. The typical content of Cr in plants, according to Oliveira, is less than 1 μg/g (Oliveira 2012). Hexavalent chromium can deactivate numerous proteins that bind or displace metal ions from the active core of the protein. In addition, the generation of reactive oxygen species (ROS) during redox processes will activate antioxidant signaling and initiate cellular oxidative stress. This mechanism induces adverse alterations in plant morphology and physiology (Stambulska et al. 2018; DesMarias and Costa 2019; Cooper et al. 2022), caused a stunted growth in shoot and leaf. Edible vegetables with higher Cr(VI) content in the leafy parts of the plant represent a high health risk to consumers, causing harm to children who require primary nutrition from vegetables. Usually vegetables absorb inorganic and organic Cr ions during the growth period which are released through the fertilizers application (Oruko Ongon'g et al. 2020).

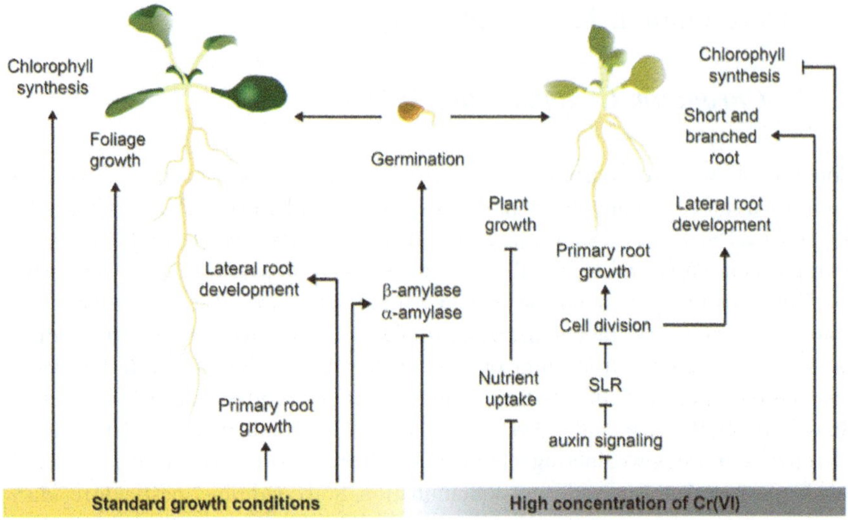

Fig. 1.3 Growth repression imposed by Cr(VI) in plants. Adopted from López-Bucio et al. (2022)

Using the bioaccumulation factor (BAF) index, one may assess the capacity of plants to absorb metals from the soil. This ratio is an assessment for determining metal concentration levels, showing the potential consequences of metal accumulation in the vegetables or fruit that will be consumed by humans or animals (Kováčik et al. 2012). If the BAF value is higher than one, the plant is a hyperaccumulator and phytoremediation actions are needed, whereas a value less than one means the plant is an excluder and not a good candidate for removal due to the low uptake of metal contents. The BAF index is calculated by measuring the concentration of metals in the roots, stems, and leaves of various plant parts (Agarwal et al. 2019; Mishra and Pandey 2019). The formula for calculating BAF showed in Eq. 1.1.

$$\text{BAF} = \frac{\text{Metal (plant part)}}{\text{Metal (substrate)}} \tag{1.1}$$

According to a study conducted in Bangladesh, the levels of Cr in Spinach, Red Amaranth, Jute Leaf, Bottle Gourd, and Mustard Green surpassed the levels permitted by the Food and Agriculture Organization (FAO) and the Chinese Ministry of Health. This is owing to the accumulation and emission of effluents containing 30–40% Cr in the Buriganga River, which will rise due to the proximity of a battery manufacturing to the river (Miclean et al. 2018; Islam et al. 2020). Similarly, Cr has a high concentration in all parts of the plant as shown in Sidnh and Punjab, India, where increasing the value of BAF (>1) was recorded. In fruit, the high accumulation of Cr found in grapes in Cairo and Fayoum were surpassed the allowable limit of 0.10 mg/kg, which may lead to adverse health effects if consumed for a long time (Amer et al. 2019). Exposure to Cr(VI) reduced the root length and number of root hairs in *Zea mays*. The brown colour of this plant has also been attributed to less cell division (Xu et al. 2020).

Agriculture is the primary source of pollution and Cr presence. Transporting nutrients in agricultural crops will hinder plant growth. Some hyperaccumulating plants are utilized in the bioremediation procedure. Consistent monitoring of water quality, crops, soil, and sophisticated agricultural practices can reduce Cr content (Astuti et al. 2021b; Kapoor et al. 2022; Ali et al. 2022). To prevent the environmental contamination of Cr, the released and discharged of metals from anthropogenic activities must be reduced. Unfortunately, the use of coal and crude oil is one of the main factors for the high accumulation of metals in environmental media, especially in developing countries with low environmental awareness. Contamination of groundwater media is one of the causes of metal contamination, especially Cr, which is generated from agricultural and pond effluents. This element is often found in fly ash from cement dust and fine dust from vehicle exhaust and road surfaces. Government and society must implement cross-sectoral collaboration, communication, and environmental quality management to avoid the onset of Cr (Stambulska et al. 2018; Astuti et al. 2022). An ecofriendly strategy in the agricultural sector is highly desirable. Consumers are advised to clean and separate dirt from fruit or vegetables so that cross contamination can be avoided and the health risk will be lower.

1.3.2 Chromium Contamination in Animals

Trivalent chromium decreases heat stress, which induces lipid peroxidation, and enhances food metabolism and the action of insulin and cortisol hormones in animals, protecting them from the negative effects of heat stress. The minor effect of Cr exposure on rats presented the occurrence of motor coordination dysfunction. It was determined that the brain injury was not very dangerous. However, the detection of time-dependent changes poses a risk of increased server toxicity with prolonged exposure (Hegazy et al. 2021). During its reduction, the toxicity of Cr(VI) releases reactive oxygen species (ROS), which can cause protein, lipid, and DNA protein damage. In chronic animal investigations, the Reference dosage for Cr(III) is 1 mg/kg/d (Barceloux and Barceloux 1999). Based on one year of Sprague-Dawley rat drinking water research, the RfD for Cr(VI) is 5 g/kg/d. In Punjab, Cr levels in soil and forages were related to Cr buildup in the blood plasma of goats and sheep due to feeding practices (Khan et al. 2020).

The index of bioconcentration factor (BCF) is the most important component in determining exposure level. In a separate study on mice, exposure to Cr boosted the activity of the transaminase enzyme in all organs. It began by activating fatty acid synthase activity in hepatic tissue, boosting isocitrate dehydrogenase activity in the liver and skeletal muscle, and changing malate dehydrogenase activity in all tissues (Shil and Pal 2019). As depicted in Fig. 1.4, the presence of Cr in the environment originates from many sources and has detrimental effects on living organisms.

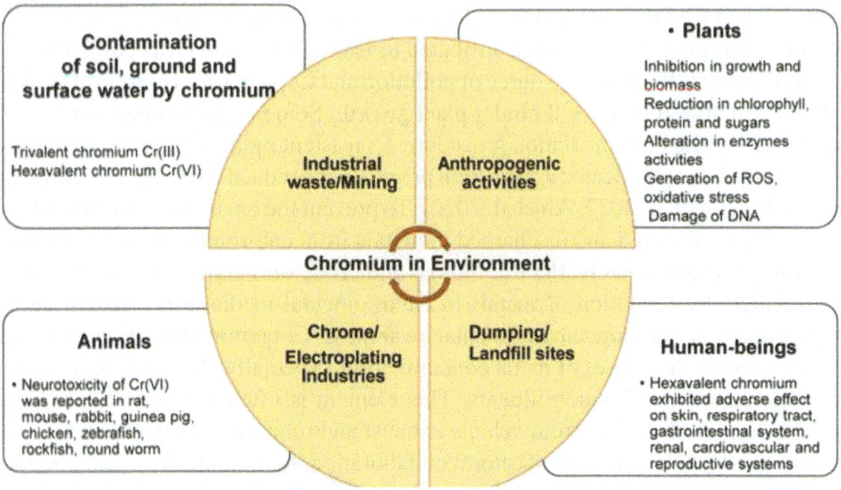

Fig. 1.4 The sources and interaction of Cr in environment. Adopted from Kapoor et al. (2022)

A study in crucian carp (*Carassius auratus*) reported the higher uptake rate for Cr(VI) in fish head comparing to fish muscle. Dynamic processes of metals play an important role during uptake, distribution and elimination. In aquatic exposure, the first organ to acquire Cr is the gills (VI). After absorption, Cr(VI) can be transported to various tissues, including the liver, kidney, and intestines, before being deposited in muscle (Mohamed et al. 2020; Yin et al. 2021; Astuti et al. 2022). As a vital organ for storage and detoxification, liver Cr levels are usually elevated after purification. The fish's head may act as a short-term buffer for Cr^{6+}, which is possibly connected with the high lead bone. The antioxidant responses of S. Schlegelii were affected by dietary Cr exposure. The expression of metallothionein (MT) in crustacea or aquatic animals is a sensitive biomarker for assessing the toxic effects of metal exposure, including Cr(VI). The gene expression of MT was significantly increased about 120 ascorbic acids (AsA) (Mohamed et al. 2020).

1.3.3 Chromium in the Human Body

Inducing severe clinical symptoms, allergies, and chronic diseases, Cr-contaminated foods and other exposures may pose concerns to human health. The previous study found the association of Cr(VI) with allergies and skin problems in human, specifically in workers (Bregnbak et al. 2016; Chou et al. 2016; Rauf et al. 2020a). Redness, itching, and inflammation of the skin were observed among industrial employees exposed to Cr in the cement manufacturing region of Taiwan. Additionally, some workers are diagnosed with dermatitis, ulcers, eczema, and an elevated risk of skin cancer (Wang et al. 2011; Chou et al. 2016). Workers exposed to Cr usually work in the electroplating, coating and printing process sectors. Apart from skin, workers also experience increased lung cancer due to the daily inhalation of Cr(VI). According to the Agency for Toxic Substances and Disease Registry (ATSDR), chromate tank employees in a chrome plating factory with inadequate ventilation reported the neurological effects of breathed Cr(VI) (ATSDR 2012). Cr(III) is absorbed orally at 1%, whereas Cr(VI) is absorbed primarily by intestinal bacteria in the duodenum.

a. **Absorption**

According to a number of studies, Cr(VI) compounds are more quickly absorbed in the lungs than Cr(III) compounds, due in part to their ability to penetrate cell membranes. Short-term exposure of 0.005–0.23 mg Cr(III) lignosulfonate dust to workers is still detectable in their urine at the end of their shift. Previous research in rats exposed to Cr demonstrated that particle size is the primary determinant of pulmonary clearance in both valence states, with Cr(VI) being more mobile and rapidly delivered to the circulation than Cr(III) (ATSDR 2012). Children are exposed to Cr when they eat food or drink water. Adults are exposed to Cr when they smoke cigarettes, use agricultural fertilizers, or eat food.

The entry of Cr into the body is associated with environmental conditions of air, soil and groundwater. Farmers who work around 3–6 h and above every day will be exposed to Cr through dermal exposure due to contact with agricultural soil (Rauf et al. 2020b; Motas et al. 2021; Ali et al. 2022). In addition, dermal contact with chromium can occur when bathing, although the risk is very small and negligible (Mallongi et al. 2022). Within a few hours, the body absorbs about 5% of the Cr(III) chloride. Several studies demonstrated the poor water solubility of chromates through the gastrointestinal tract, whereas more water-soluble chromates are absorbed into the blood (ATSDR 2016). Previous research has shown that ingesting Cr(VI) compounds causes nasal and pharyngeal discomfort in humans. Symptoms may include a runny nose, sneezing, coughing, itching, and a burning sensation. Exposure to low concentrations of Cr does not cause adverse effects directly in workers, but some of them will experience allergies when inhaling Cr compounds, which are associated with asthma characteristics like wheezing and shortness of breath. Maximum hexavalent chromium absorption was detected at 6 h, with a reduction at 72 h (Alvarez et al. 2021).

b. **Distribution**

The circulation of Cr(III) in the human body is affected by age, gender, and occupation. Cr may be transferred to fetuses while breast-feeding. A previous study found the association of woman worker and the breast milk quality in Murcia Region (Spain) (Motas et al. 2021). The highest levels of Cr and Mn were detected in the breast milk of women living near an agricultural region. Continuous exposure from infant's daily consumption and children age increase the intake of heavy metals and higher than those recommended. The distribution of Cr through the blood will carry over to the organs and affect the liver and kidney. Transferrin is responsible for Cr(III) transport and metabolism as a serum protein. It has the same size and charge as ferric ions (Edwards et al. 2020). The generation of free radicals can damage the protein molecules, lipids and nucleic acids. The molecular mechanism includes mutation of bases, strand breakage, DNA aberrations, RNA polymerase activity, and alterations in gene expression (Kapoor et al. 2022).

c. **Metabolism**

Cr(III) absorption into cells is 500 times lower than Cr(VI), and only 5% of Cr is absorbed by the body. In chromium-treated cell membranes, chromium increases the activity of insulin receptor tyrosine kinase. If chromium is transported to tissues, it could bond with both large and small molecules. An animal study showed the metabolic fate of $CrCl_3$ and CrO_4^{2-} collected from rat liver and their blood after the oral intake. Significant results of Cr(III) were observed in rat blood which has high binding activity for transferrin in plasma, whereas Cr(VI) is permeable to red blood cells and bound to hemoglobin. Bands from different chromium sources significantly differ in the cytosol of hepatocytes, Cr of $CrCl_3$ is an example of a primary bond with a high molecular weight protein (Feng 2007).

d. **Elimination**

Mutagenic carcinogens are strongly linked to DNA-reactive mutagenic mechanisms in Cr's chronic and carcinogenic effects (VI). In animal and human, elimination of Cr in the body can occur in urinary and sweat events. The nature of heavy metals, which makes decomposition difficult and time-consuming, will cause Cr to accumulate in the body for quite some time. Specific decomposition will be left in the body, especially Cr ions with a fairly high mobility in blood cells. This was proven through the previous studies on workers who were exposed to Cr and could still excrete it several hours later at work hours (Kapoor et al. 2022).

1.4 Effects of Chromium Exposure on Human Health

As an element with a high level of toxicity that is categorized as a heavy metal, Cr has a strong correlation with chronic diseases and is associated with bad environmental conditions. Consumed fruit or vegetables will accumulate Cr in the human body. Poor waste management, residual fertilizer carried by water into rice fields, and dry deposition of vehicle-generated particle dust on plant leaves would enhance the Cr accumulation in the soil. The use of fertilizers containing many trace metals will raise the risk of carcinogenic and non-carcinogenic substances, especially when consumed daily. Several cases of Cr exposure that disrupt the food supply chain and human health are presented in Table 1.2.

Based on the list of disease in Table 1.1, the carcinogenic effects in the form of cancer and skin disorders in workers dominate the negative effects of Cr exposure. As a result of consuming agricultural goods or coming into contact with high Cr levels, agricultural and industrial workers are exposed to adverse long-term impacts and perhaps elevated carcinogenic hazards that pose a future cancer hazard. Women and children are vulnerable populations due to their physical condition and higher

Table 1.2 Diseases and cases related to Cr exposure in human

S. No	Diseases	Symptoms	Country/locations	References
1	Dermatitis and eczema in cement workers	Itching, burning sensation and redness	Taiwan	Chou et al. (2016)
2	Psoriasis	Redness, itching and rash on the skin. High levels of Cr and other metals in scalp hair, blood and urine	Pakistan	Afridi et al. (2011)
3	Neuropsychological problems	Poor performance test, reaction and lower IQ	Spain	Caparros-Gonzalez et al. (2019)
4	Occupational contact dermatitis	Irritation and allergies from wet cement, skin permeation that causes sensation	Australia	Wong et al. (2015)
5	Chromium allergy	Itching, redness and chromium allergies	Denmark	Thyssen et al. (2009)
6	Kidney injury	Higher plasma blood and urinary chromium	Mexico	Cárdenas-gonzález et al. (2016)
7	Lung cancer	Small cell lung and breathing problem	Slovakia	Halasova et al. (2010)
8	Lung injury and cancer development	Suffering to breath and lung injury	USA	Beaver et al. (2009)
9	Cancer and non-cancer effects	Learning disabilities, decrease intelligence and growth disturbance	Poland	Mainka and Fantke (2022)
10	Tumor	Unstable metabolism and genomic instability	–	Wang et al. (2019)
11	Allergic contact dermatitis and irritant contact dermatitis	Psychosomatic condition, physical trauma, redness	–	Packham (2007)

intake of heavy metals, especially Cr (Glorennec et al. 2012; Caparros-Gonzalez et al. 2019; Motas et al. 2021).

References

Afridi HI, Kazi TG, Kazi N et al (2011) Evaluation of cadmium, chromium, nickel, and zinc in biological samples of psoriasis patients living in Pakistani cement factory area. Biol Trace Elem Res 142:284–301

Agarwal P, Sarkar M, Chakraborty B, Banerjee T (2019) Phytoremediation of air pollutants. In: Phytomanagement of polluted sites. Elsevier, pp 221–241

Ali H, Khan E, Ilahi I (2019) Environmental chemistry and ecotoxicology of hazardous heavy metals: environmental persistence, toxicity, and bioaccumulation. J Chem 2019:1–14

Ali W, Zhang H, Mao K et al (2022) Chromium contamination in paddy soil-rice systems and associated human health risks in Pakistan. Sci Total Environ 826:153910

Alvarez CC, Bravo Gómez ME, Hernández Zavala A (2021) Hexavalent chromium: regulation and health effects. J Trace Elem Med Biol 65:126729

Amer MM, Sabry BA, Marrez DA et al (2019) Exposure assessment of heavy metal residues in some Egyptian fruits. Toxicol Reports 6:538–543

Anderson RA (1997) Chromium as an essential nutrient for humans. Regul Toxicol Pharmacol 26:S35–S41

Apte A, Tare V, Bose P (2006) Extent of oxidation of Cr(III)–Cr(VI) under various conditions pertaining to natural environment. J Hazard Mater 128:164–174

Ashraf A, Bibi I, Niazi NK et al (2017) Chromium(VI) sorption efficiency of acid-activated banana peel over organo-montmorillonite in aqueous solutions. Int J Phytoremediation 19:605–613

Astuti RDP, Mallongi A, Amiruddin R et al (2023) Hexavalent chromium contamination in groundwater and its implication to human health: a Monte Carlo model approach in Indonesia. Sustain Water Resour Manag 9:22

Astuti RDP, Mallongi A, Choi K et al (2022) Health risks from multiroute exposure of potentially toxic elements in a coastal community: a probabilistic risk approach in Pangkep Regency, Indonesia. Geomat Nat Hazards Risk 13

Astuti RDP, Mallongi A, Rauf AU (2021a) Natural enrichment of chromium and nickel in the soil surrounds the karst watershed. Glob J Environ Sci Manag 7:1–18

Astuti RDP, Mallongi A, Rauf AU (2021b) Risk identification of Hg and Pb in soil: a case study from Pangkep Regency, Indonesia. Soil Sci Annu 72:1–15

ATSDR (2008) Case studies in environmental medicine chromium toxicity

ATSDR (2012) Toxicological profile for chromium. Atlanta, Georgia

ATSDR (2016) The priority list of hazardous substances that will be the subject of toxicological profiles. Atlanta

Babula P, Adam V, Opatrilova R et al (2008) Uncommon heavy metals, metalloids and their plant toxicity: a review. Environ Chem Lett 6:189–213

Barceloux DG, Barceloux D (1999) Chromium. J Toxicol Clin Toxicol 37:173–194

Baruthio F (1992) Toxic effects of chromium and its compounds. Biol Trace Elem Res 32:145–153

Beaver LM, Stemmy EJ, Schwartz AM et al (2009) Lung inflammation, injury, and proliferative response after repetitive particulate hexavalent chromium exposure. Environ Health Perspect 117:1896–1902

Bregnbak D, Thyssen JP, Jellesen MS et al (2016) Experimental skin deposition of chromium on the hands following handling of samples of leather and metal. Contact Dermatitis 75:89–95

Caparros-Gonzalez RA, Giménez-Asensio MJ, González-Alzaga B et al (2019) Childhood chromium exposure and neuropsychological development in children living in two polluted areas in southern Spain. Environ Pollut 252:1550–1560

Cárdenas-gonzález M, Osorio-yáñez C, Gaspar-ramírez O, Pavkovi M (2016) Environmental exposure to arsenic and chromium in children is associated with kidney injury molecule-1. Environ Res 1–10

Chen S, Jin X, Shan Z et al (2017) Inverse association of plasma chromium levels with newly diagnosed type 2 diabetes: a case-control study. Nutrients 9:294

Chou TC, Wang PC, Wu JD, Sheu SC (2016) Chromium-induced skin damage among Taiwanese cement workers. Toxicol Ind Health 32:1745–1751

Cooper KL, Liu R, Zhou X (2022) Particulate arsenic trioxide induces higher DNA damage and reactive oxygen species than soluble arsenite in lung epithelial cells. Toxicol Appl Pharmacol 457:116320

Covarrubias SA, Flores de la Torre JA, Maldonado Vega M et al (2018) Spatial variability of heavy metals in soils and sediments of "La Zacatecana" Lagoon, Mexico. Appl Environ Soil Sci 2018:1–8

Davies FT, Puryear JD, Newton RJ et al (2002) Mycorrhizal fungi increase chromium uptake by sunflower plants: influence on tissue mineral concentration, growth, and gas exchange. J Plant Nutr 25:2389–2407

de Oliveira LM, Gress J, De J et al (2016) Sulfate and chromate increased each other's uptake and translocation in As-hyperaccumulator Pteris vittata. Chemosphere 147:36–43

DesMarias TL, Costa M (2019) Mechanisms of chromium-induced toxicity. Curr Opin Toxicol 14:1–7

Dobermann A, Bruulsema T, Cakmak I et al (2022) Responsible plant nutrition: a new paradigm to support food system transformation. Glob Food Sec 33:100636

Edwards KC, Kim H, Vincent JB (2020) Release of trivalent chromium from serum transferrin is sufficiently rapid to be physiologically relevant. J Inorg Biochem 202:110901

Eriksson J (2001) Concentrations of 61 trace elements in sewage sludge, farmyard manure, mineral fertiliser, precipitation and in oil and crops

Ertani A, Mietto A, Borin M, Nardi S (2017) Chromium in agricultural soils and crops: a review. Water, Air, Soil Pollut 228:190

European Food Safety Authority (EFSA) (2014) Scientific opinion on dietary reference values for chromium. EFSA J 12:3845

Feng W (2007) The transport of chromium(III) in the body: Implications for function. In: The nutritional biochemistry of chromium(III). Elsevier, pp 121–137

Glorennec P, Lucas J-P, Mandin C, Le Bot B (2012) French children's exposure to metals via ingestion of indoor dust, outdoor playground dust and soil: contamination data. Environ Int 45:129–134

Halasova E, Adamkov M, Matakova T et al (2010) Lung cancer incidence and survival in chromium exposed individuals with respect to expression of anti-apoptotic protein survivin and tumor suppressor P53 protein. Eur J Med Res 15:55

Hefferon K (2015) Nutritionally enhanced food crops; progress and perspectives. Int J Mol Sci 16:3895–3914

Hegazy R, Mansour D, Salama A et al (2021) Exposure to intranasal chromium triggers dose and time-dependent behavioral and neurotoxicological defects in rats. Ecotoxicol Environ Saf 216:112220

Heimbach JT, Anderson RA (2005) Chromium: recent studies regarding nutritional roles and safety. Nutr Today 40:189–195

Isıklı B, Demir TA, Ürer SM et al (2003) Effects of chromium exposure from a cement factory. Environ Res 91:113–118

Islam MD, Hasan MM, Rahaman A et al (2020) Translocation and bioaccumulation of trace metals from industrial effluent to locally grown vegetables and assessment of human health risk in Bangladesh. SN Appl Sci 2:1315

Kabata-Pendias A (2011) Trace elements in soils and plants. CRC Press, Boca Raton, FL, Fourth

Kapoor RT, Bani Mfarrej MF, Alam P et al (2022) Accumulation of chromium in plants and its repercussion in animals and humans. Environ Pollut 301:119044

Khan ZI, Ahmad K, Siddique S et al (2020) A study on the transfer of chromium from meadows to grazing livestock: an assessment of health risk. Environ Sci Pollut Res 27:26694–26701

Kim RY, Sung JK, Lee JY et al (2010) Chromium distribution in Korean soils: a review. Korean J Soil Sci Fertil 43:296–303

Kimbrough DE, Cohen Y, Winer AM et al (1999) A critical assessment of chromium in the environment. Crit Rev Environ Sci Technol 29:1–46
Kiran BR, Sharma R (2022) Effect of heavy metals: an overview. Mater Today Proc 51:880–885
Kováčik J, Grúz J, Klejdus B et al (2012) Accumulation of metals and selected nutritional parameters in the field-grown chamomile anthodia. Food Chem 131:55–62
Krejpcio Z (2001) Essentiality of chromium for human nutrition and health. Polish J Environ Stud 10:399–404
Krüger O, Fiedler F, Adam C et al (2017) Determination of chromium(VI) in primary and secondary fertilizer and their respective precursors. Chemosphere 182:48–53
Kuryl T, Debski B, Martinik K (2008) The effect of microelements supplementation on beta-oxidation activity in healthy and type 1 diabetic rats. Cent Eur J Public Health 16:205–208
Langård S, Costa M (2015) Chromium. In: Nordberg GF, Fowler BA, Nordberg M (eds) Handbook on the toxicology of metals, 2nd edn. Academic Press, London, pp 717–742
López-Bucio JS, Ravelo-Ortega G, López-Bucio J (2022) Chromium in plant growth and development: toxicity, tolerance and hormesis. Environ Pollut 312:120084
Lukaski HC (1999) Chromium as a supplement. Annu Rev Nutr 19:279–302
Ma Y, Hooda PS (2010) Chromium, nickel and cobalt. In: Trace elements in soils. Wiley, Chichester, UK, pp 461–479
Mainka A, Fantke P (2022) Preschool children health impacts from indoor exposure to PM2.5 and metals. Environ Int 160:107062
Mallongi A, Astuti RDP, Amiruddin R et al (2021) Identification source and human health risk assessment of potentially toxic metal in soil samples around Karst Watershed of Pangkajene, Indonesia. Environ Nanotechnol Monit Manag 17:100634
Mallongi A, Rauf A utami, Daud A et al (2022) Health risk assessment of potentially toxic elements in Maros karst groundwater: a monte carlo simulation approach. Geomat Nat Hazards Risk
Mallongi A, Stang MS et al (2019) Risks assessment of silica contamination on the communities living surround the cement industry, Pangkep Indonesia. Indian J Public Heal Res Dev 10:1619
Mandal A, Voutchkov M (2011) Heavy metals in soils around the cement factory in Rockfort, Kingston, Jamaica. Int J Geosci 02:48–54
Mawari G, Kumar N, Sarkar S et al (2022) Heavy metal accumulation in fruits and vegetables and human health risk assessment: findings from Maharashtra, India. Environ Health Insights 16:117863022211191
Miclean M, Cadar O, Levei L et al (2018) Metal contents and potential health risk assessment of crops grown in a former mining district (Romania). J Environ Sci Heal Part B 53:595–601
Mishra S, Bharagava RN (2016) Toxic and genotoxic effects of hexavalent chromium in environment and its bioremediation strategies. J Environ Sci Health Part C Environ Carcinog Ecotoxicol Rev 34:1–32
Mishra S, Bharagava RN, More N et al (2019) Heavy metal contamination: an alarming threat to environment and human health. Springer, Singapore
Mishra T, Pandey VC (2019) Phytoremediation of red mud deposits through natural succession. In: Phytomanagement of polluted sites. Elsevier, pp 409–424
Mitra S, Chakraborty AJ, Tareq AM et al (2022) Impact of heavy metals on the environment and human health: novel therapeutic insights to counter the toxicity. J King Saud Univ Sci 34:101865
Mohamed AAR, El-Houseiny W, ELMurr AE et al (2020) Effect of hexavalent chromium exposure on the liver and kidney tissues related to the expression of CYP450 and GST genes of *Oreochromis niloticus* fish: role of curcumin supplemented diet. Ecotoxicol Environ Saf 188:109890
Moral R, Pedreno JN, Gomez I, Mataix J (1995) Effects of chromium on the nutrient element content and morphology of tomato. J Plant Nutr 18:815–822
Motas M, Jiménez S, Oliva J et al (2021) Heavy metals and trace elements in human breast milk from industrial/mining and agricultural zones of Southeastern Spain. Int J Environ Res Public Health 18:9289

Nachtigal MC, Patterson RE, Stratton KL et al (2005) Dietary supplements and weight control in a middle-age population. J Altern Complement Med 11:909–915
Oliveira H (2012) Chromium as an environmental pollutant: insights on induced plant toxicity. J Bot 2012:1–8
Oruko Ongon'g R, Edokpayi JN, Msagati TAM et al (2020) The potential health risk associated with edible vegetables grown on Cr(VI) polluted soils. Int J Environ Res Public Health 17:470
Packham C (2007) Investigating a suspected occupational skin problem. Expert Rev Dermatol 2:681–683
Pechova A, Pavlata L (2007) Chromium as an essential nutrient: a review. Vet Med (praha) 52:1–18
Prado C, Chocobar Ponce S, Pagano E et al (2016) Differential physiological responses of two Salvinia species to hexavalent chromium at a glance. Aquat Toxicol 175:213–221
Preuss HG, Grojec PL, Lieberman S, Anderson RA (1997) Effects of different chromium compounds on blood pressure and lipid peroxidation in spontaneously hypertensive rats. Clin Nephrol 47:325–330
Quantin C, Ettler V, Garnier J, Šebek O (2008) Sources and extractibility of chromium and nickel in soil profiles developed on Czech serpentinites. Comptes Rendus Geosci 340:872–882
Rauf AU, Mallongi A, Astuti RDP (2020a) Heavy metal contributions on human skin disease near cement plant: a systematic review. Open Access Maced J Med Sci 8:117–122
Rauf AU, Mallongi A, Astuti RDP (2020b) Mercury and chromium distribution in soil near Maros karst ecosystem. Carpathian J Earth Environ Sci 15:453–460
Rauf AU, Mallongi A, Daud A et al (2021) Ecological risk assessment of hexavalent chromium and silicon dioxide in well water in Maros Regency, Indonesia. Gac Sanit 35:S4–S8
Saha JK, Selladurai R, Coumar MV et al. (2017) Soil pollution—An emerging threat to agriculture. Springer Singapore, Singapore
Sangwan P, Kumar V, Joshi UN (2014) Effect of chromium(VI) toxicity on enzymes of nitrogen metabolism in clusterbean (*Cyamopsis tetragonoloba* L.). Enzyme Res 2014:1–9
Sauerbeck D (1987) Effects of agricultural practices on the physical, chemical and biological properties of soils: part II—Use of Sewage Sludge and agricultural wastes. In: Barth H, L'Hermite P (eds) Scientific basis for soil protection in the European community, 1st edn. Springer, Netherlands, Dordrecht, pp 181–210
Shahid M, Shamshad S, Rafiq M et al (2017) Chromium speciation, bioavailability, uptake, toxicity and detoxification in soil-plant system: a review. Chemosphere 178:513–533
Shanker A, Cervantes C, Lozatavera H, Avudainayagam S (2005) Chromium toxicity in plants. Environ Int 31:739–753
Shanker AK (2019) Chromium: environmental pollution, health effects and mode of action. In: Encyclopedia of environmental health. Elsevier, pp 624–633
Sharma A, Kapoor D, Wang J et al (2020a) Chromium bioaccumulation and its impacts on plants: an overview. Plants 9:1–17
Sharma A, Kapoor D, Wang J et al (2020b) Chromium bioaccumulation and its impacts on plants: an overview. Plants (Basel, Switzerland) 9
Shil K, Pal S (2019) Metabolic and morphological disorientations in the liver and skeletal muscle of mice exposed to hexavalent chromium. Comp Clin Path 28:1729–1741
Shrivastava R, Upreti RK, Seth PK, Chaturvedi UC (2002) Effects of chromium on the immune system. FEMS Immunol Med Microbiol 34:1–7
Stambulska UY, Bayliak MM, Lushchak VI (2018) Chromium(VI) toxicity in legume plants: modulation effects of rhizobial symbiosis. Biomed Res Int 2018:1–13
Sun H, Brocato J, Costa M (2015a) Oral chromium exposure and toxicity compliance with ethics guidelines human and animal rights and informed consent HHS public access. Curr Env Heal Rep 2:295–303
Sun H, Brocato J, Costa M (2015b) Oral chromium exposure and toxicity. Curr Environ Heal Reports 2:295–303
Sun H, Costa M (2022) Chromium. In: Handbook on the toxicology of metals. Elsevier, pp 197–220

Takeda A, Kimura K, Yamasaki S (2004) Analysis of 57 elements in Japanese soils, with special reference to soil group and agricultural use. Geoderma 119:291–307

Tchounwou PB, Yedjou CG, Patlolla AK, Sutton DJ (2012) Heavy metal toxicity and the environment. In: Luch A (ed) Molecular clinical and environmental toxicology. Springer, Basel, Switzerland, pp 133–164

Thyssen JP, Jensen P, Carlsen BC et al (2009) The prevalence of chromium allergy in Denmark is currently increasing as a result of leather exposure. Br J Dermatol 161:1288–1293

Tuzcu M, Sahin N, Orhan C et al (2011) Impact of chromium histidinate on high fat diet induced obesity in rats. Nutr Metab (lond) 8:28

Untea AE, Varzaru I, Panaite TD et al (2017) Effects of chromium supplementation on growth, nutrient digestibility and meat quality of growing pigs. S Afr J Anim Sci 47:332

Vincent JB (2017) New evidence against chromium as an essential trace element. J Nutr 147:2212–2219

Wang BJ, Wu JD, Sheu SC et al (2011) Occupational hand dermatitis among cement workers in Taiwan. J Formos Med Assoc 110:775–779

Wang Z, Lin HP, Li Y et al (2019) Chronic hexavalent chromium exposure induces cancer stem cell-like property and tumorigenesis by increasing c-Myc expression. Toxicol Sci 172:252–264

Wise JTF, Shi X, Zhang Z (2019) Toxicology of chromium(VI). In: Encyclopedia of environmental health. Elsevier, pp 1–8

Wong CC, Gamboni SE, Palmer AM, Nixon RL (2015) Occupational allergic contact dermatitis to chromium from cement: estimating the size of the problem in Australia. Australas J Dermatol 56:290–293

World Health Organization (WHO) (1998) Environmental health criteria 61 Chromium. Environ Heal Criteria

Xu T, Nan F, Jiang X et al (2020) Effect of soil pH on the transport, fractionation, and oxidation of chromium(III). Ecotoxicol Environ Saf 195:110459

Yin J, Zhang F, Wang L et al (2021) A kinetic study on accumulation and depuration of hexavalent chromium in crucian carp (*Carassius auratus*) reveals the potential health risk of fish head consumption. Food Control 130:108291

Zarcinas BA, Pongsakul P, McLaughlin MJ, Cozens G (2004) Heavy metals in soils and crops in Southeast Asia 2. Thailand Environ Geochem Health 26:359–371

Chapter 2
Speciation, Chemistry, Geogenic Formation and Dispersion of Chromium in Groundwater

Elisavet Amanatidou

Abstract The chromium origin in the environment can be geogenic, anthropogenic, or both. The most common forms of chromium in soil and water are various species of Cr(III) or Cr(VI). Cr(III) has low solubility at most environmental conditions, and rarely exceeds the maximum permissible drinking water concentration, while it is an essential nutrient trace element for plant and animal metabolism. Unlike Cr(III), Cr(VI) is very soluble, strong oxidizer, unstable in the presence of reducing agents (electron donors), toxic and carcinogenic. The limit of chromium concentration in drinking water according to the International Health Organization and the EU is 50 μg/L referred as total chromium. Key parameters that determine the distribution of various chromium species in the natural environment are: the geochemical environment, pH and the redox potential (Eh). The conversion of Cr(III) to Cr(VI) is possible in alkaline and oxidizing ($Eh > 0$) conditions. There are a few agents in the natural environment capable of oxidizing chromium. Mn oxides (Mn(III), Mn(IV) oxy-hydroxides and Mn(IV) oxides such as pyrolusite), are the only naturally occurring minerals capable of oxidizing Cr(III) to Cr(VI). Inhibition of Cr(III) oxidation can occur due to competitive adsorption of some cations (La, Al, Mn(II)) on Mn oxide surface. Precipitation of $Cr(OH)_3 \cdot nH_2O$ and $Al(OH)_3$ formed on Mn oxide surface can also inhibit Cr(III) oxidation. A portion of Cr(VI) formed by oxidation of Cr(III) is released back into solution while a part of it remains adsorbed on the MnOx surface and associated to the matrix. The release of hexavalent chromium into aqueous phase can be enhanced by competitive anion adsorption (e.g., phosphates) increasing hexavalent chromium concentration: phosphates directly remove chromates by competing Cr(VI) for the adsorption sites or indirectly by increasing the pH of solution. Cr(VI) is reduced in soil and water to Cr(III), in the presence of reductants such as S^{2-}, V^{2+}, Fe^{2+}, HNO_2, HSO_3^- and some organic species. Cr(VI) reduction in water with soil and sediment is very rapid while the reduced chromium is resistant to reoxidation. The inhibition of Cr(III) oxidation and Cr(VI) reduction can reduce the pollution caused by Cr(VI), only by natural attenuation, without any

E. Amanatidou (✉)
Department of Chemical Engineering, School of Engineering, University of Western Macedonia, Macedonia, Greece
e-mail: eamanatidou@uowm.gr

© The Author(s), under exclusive license to Springer Nature Switzerland AG 2023

N. Kumar et al. (eds.), *Chromium in Plants and Environment*, Environmental Science and Engineering, https://doi.org/10.1007/978-3-031-44029-8_2

human intervention. The deeper levels of an aquifer have a clearly different water composition from the shallow levels. At greater depths the dominant form is Cr(VI), at intermediate depths the dominant form is again Cr(VI) but in lower concentrations. On the other hand, the shallow wells have Cr(III) as the dominant form, in coexistence with nitrates as a result of fertilization and increased water reflow, through the unsaturated zone during irrigation. The differences in water quality, as a function of depth, are very often due to the increased pumping.

Keywords Cr(VI) natural occurrence · Chromium chemistry · Chromium oxidation · Natural attenuation

2.1 Introduction

In the natural environment Cr is found mainly in two oxidation states, Cr(III) and Cr(VI), being of either geogenic or anthropogenic origin, or both. Cr(III), is the most common form of chromium. It has relatively low toxicity, is very insoluble and relatively immobile under moderate alkaline to slightly acidic conditions (Richard and Bourg 1991; Palmer and Puls 1994). Cr(III) is an essential nutrient for glucose and fat metabolism and for synthesis of amino- and nucleic acid in plants and animals. Its accumulation can generate serious troubles and diseases (nausea, skin ulcerations, lung cancer). When its concentration reaches 0.1 mg/g of body weight, it can ultimately become lethal (Richard and Bourg 1991). Cr(VI) is relatively mobile in the environment and is highly toxic and mutagenic (Bianchi et al. 1984; Beyersmann et al. 1984; Bonatti et al. 1976; Paschin et al. 1983), teratogenic (Abbasi and Soni 2007), and carcinogenic (WHO 1990; Stern 2010). Geogenic Cr has been related to ultramafic rocks, naturally enriched in Cr(III). Few Cr(VI) minerals have been found worldwide, but the ability of manganese dioxide to oxidize Cr(III) to Cr(VI) is well known. In the presence of an oxidative agent, such as manganes dioxide, Cr(III) can be oxidized to Cr(VI) which is a threat to groundwater quality. On the other hand, hexavalent chromium, can be reduced back to the less toxic trivalent form in an acidic environment and in presence of reducing compounds (electron donors). The electron donors, that are commonly found in soils, include aqueous Fe(II), ferrous iron minerals, reduced sulfur, and soil organic matter. However, the redox chemistry of chromium, manganese and iron are closely related and can easily react with each other (Bartlett and James 1983; Eary and Rai 1987; Fendorf and Zasoski 1992; Gonzalez et al. 2005).

The anthropogenic origin of chromium in air, water, and soil may be due to activities such as agricultural, industrial, groundwater overexploitation, and urbanization which have been shown to affect water quality in a region (Machiwal and Jha 2015). Chromium in water bodies is considered as a major environmental issue around the world. Overall, groundwater is at high risk for hexavalent chromium contamination as a result of industrial activities, natural geology and, potentially, land use (Hausladen et al. 2018).

The sources of anthropogenic Cr contamination of water bodies include the disposal of industrial effluents from leather tanning, electroplating, metal finishing, pigments, dyes, paints and ceramics industries (Khalil et al. 2021). Wastewaters from industries, discharged or escaped, can increase chromium levels in ground and surface waters, either in the trivalent or hexavalent form depending on solubility.

Air chromium emissions represent the sum of the hexavalent and trivalent species. The latter emissions are present in air as particulate matter (PM) and are subjected to PM emission control regulations. However, they are not considered hazardous in contrast to hexavalent chromium emissions. Cr(VI) air pollution is emitted from chromic acid anodizing, hard chromium plating processes and coatings spraying, usually primers, that contain pigments rich in chromates.

In soils and waters, the use of phosphate fertilizers, which contain hexavalent chromium, contributes to its increasing presence (Perraki et al. 2021). In addition, chromic acid is used as a fungicide and insecticide. Chromium in soil can also be increased by the deposition of ash produced during the combustion of coal for electricity production (Jacobs and Testa 2004).

Due to the high toxicity of Cr(VI), wastewaters containing hexavalent chromium must be treated. Conventional methods used for Cr(VI) removal are: chemical precipitation, electrocoagulation, adsorption and filtration, membrane separation, ion exchange and electrodialysis. However, the latter methods have significant drawbacks such as low efficiency, high operating and maintenance costs, and sludge generation, which causes disposal problems or produces a secondary pollutant, limiting their overall applicability (Alemu et al. 2018). The adsorption and reduction of Cr(VI)–Cr(III), with subsequent precipitation of Cr(III) as hydrated Cr(III) oxide, are simple in operation, and efficient methods for the removal of Cr(VI) (Khalil et al. 2021; Gheju and Balcu 2011; Gheju et al. 2008; Barrera-Díaz et al. 2012; Hu and Noubactep 2019). In recent years, the development of nanotechnology enables the use of nano-adsorbents to enhance the chemical reactivity between the adsorbed Cr(VI) and the reductant adsorbent. The development of iron-based nanomaterials (NMs) contribute significantly to the effective removal of Cr(VI) from waters predominantly via adsorption or conversion to Cr(III). For this purpose, chemically stable, energy effective, and recyclable NMs with reduced operating and processing costs can be used (Maitlo et al. 2019; Farooqi et al. 2021). Moreover, hexavalent chromium reduction by discarded or low-cost iron-based reducing materials, is under research regarding the creation of a compact, pump-and-treat (P&T) filter for Cr(VI) removal from groundwater (Samiotis et al. 2023). P&T belongs to the most applied technologies for groundwater remediation. It generally involves the extraction of polluted ground water and its on-site or off-site treatment applying appropriate techniques in designed plants. The treated water can be re-injected into the aquifer or discharged to a surface receiver, in relation to the applicable quality standards (EPA 2005). Although it is feasible to remove high levels of chromium from the subsurface applying P&T methods, as concentrations decrease it becomes more difficult to remove the remaining chromium. Under certain circumstances, expensive remedial measures may not even be necessary, i.e. if natural attenuation occurs (Palmer and Puls 1994; Bortone et al. 2020).

In recent year's research has focused on the impact of pumping on the hexavalent chromium concentration and on groundwater quality generally. Many studies showed that depending on the geological background, the water in deeper aquifer levels has a clearly different composition from the water in the shallow aquifer levels. In the deeper aquifer levels, the longer water residence time (old water) and rock-water interaction create hydrochemical conditions which allow the release of naturally occurring Cr(VI) into aqueous solution. Thus, dissolved chromium, essentially present in its hexavalent form at the intermediate and deep levels of an aquifer, has its highest concentrations in the deeper aquifer levels (Bourotte et al. 2009; Chenini et al. 2010; Bexfield and Jurgens 2014; Mavromatidou et al. 2021; Guo et al. 2020).

In natural waters, the range of chromium concentrations is quite large. However, natural concentrations as high as 4 μmol/L have been observed, exceeding the drinking water maximum acceptable level of 50 μg/l (approx. 1 μmol/l) for total chromium, recommended for drinking water by the Commission of European Communities, the World Health Organization or the U.S. Environmental Protection Agency. Regarding hexavalent chromium, limit of 5 μg/L is established in Italy and 10 μg/L is proposed in California (Chrysochoou et al. 2016). These high concentrations of dissolved chromium are usually associated with the very soluble chromate species. The Cr(VI) to Cr_{total} ratios are higher than 0.90 indicating that more than 90% of total chromium exists as hexavalent chromium in natural waters (Robertson 1975; Mavromatidou et al. 2021; Yan et al. 2022).

Elevated Cr(VI) concentrations have been reported in many places around the world, due to the geology of the area, such as for example in USA (Ball and Izbicki 2004; Gonzalez et al. 2005; Coyte et al. 2020), in Italy (Tiwari et al. 2019; Fantoni et al. 2002), Brazil (Bourotte et al. 2009), in Mexico (Robles-Camacho and Armienta, 2000) in China (He and Wu 2019; Guo et al. 2020), in Cyprus (Zissimos et al. 2021), in Greece (Dermatas et al. 2012; Economou-Eliopoulos et al. 2013; Koilakos 2017; Vasileiou et al. 2019; Stamos et al. 2019; Mavromatidou et al. 2021; Perraki et al. 2021) and elsewhere.

The widespread natural waters contamination with Cr(VI) around the world makes it necessary to fully understand the oxidation mechanisms of Cr(III) in nature both to determine the risk of hexavalent chromium formation and to develop and apply suitable techniques to prevent or reduce its formation.

2.2 Chromium Chemistry

Chromium belongs to the transition metals, it is the first chemical element of the VIB group of the periodic table, it has an atomic number $Z = 24$ and a relative atomic mass of 51.9961. Its name comes from the Greek word *χρώμα* (*chroma*, color) and is due to the wide variety of colors displayed by its compounds. Chrome metal is silver-gray in color, shiny, hard and brittle, and does not oxidize easily. It has a melting temperature of 1907 °C and a boiling point of 2671 °C (Lide 2006). The principal Chromium ore is *chromite* ($FeCr_2O_4$). The oxidation states of chromium vary from −2 to +6. The oxidation states −2, −1, 0 and +1 occur in synthetic organic compounds of chromium (Motzer 2005).

The compounds of Cr(II) are very similar to those of Fe(II). The hydrated ion of Cr(II) in solution and in salts is blue in color. The negative standard potential (E^0) of the Cr(III)/Cr(II) metal ion pair indicates that Cr(II) is readily oxidized to Cr(III), and that Cr(II) species are not stable in aqueous solutions even in the absence of oxygen or any other oxidizing agent (Kotaś and Stasicka 2000). In the presence of oxygen, it oxidizes to Cr(III) much more easily than Fe(II).

The main oxidation states of chromium are three:

(a) Cr^0, appearing as metallic Cr
(b) Cr(III), trivalent chromium and
(c) Cr(VI), the hexavalent chromium appearing as CrO_4^{-2} and $Cr_2O_7^{-2}$.

In the environment, chromium is found in only two stable oxidation states, the trivalent and the hexavalent form (Peterson et al. 1996), depending on the pH, the prevailing redox potential, and the chemical and physical properties of soil (Rai et al. 1989; Richard and Bourg 1991; Kotaś and Stasicka 2000). In most cases the presence of Cr(III) in the environment prevails over that of Cr(VI) (Kotaś and Stasicka 2000), although in groundwater the form of chromium is mainly Cr(VI) (Gonzalez et al. 2005).

Cr(III) is relatively insoluble and immobile at alkaline and oxidizing environments, in contrast to Cr(VI), which is mostly soluble and mobile at these conditions (Cranston and Murray 1978; Jacobs and Testa 2004; Sharma et al. 2008). Therefore, Cr(VI) is the form mainly found in natural waters (Rai et al. 1987; Richard and Bourg 1991; Sperling et al. 1992; Kotaś and Stasicka 2000), particularly in water systems interacting with ultramafic rocks (Vasileiou et al. 2019).

2.2.1 *Trivalent Chromium, Cr(III)*

Trivalent chromium species have low solubility in most environmental conditions. The solubility of $Cr(OH)_3$ is above the drinking water limit (10^{-6} M) between pH 6 and 12 (Rai et al. 1989). Cr(III) generally forms insoluble chromic oxide (Cr_2O_3) at pH 5–13,5 and at an electroreduction potential (Eh) of +0.8 to −0.75 V. At pH slightly lower than 5 and under the influence of low redox potential (Eh), Cr_2O_3 dissolves and forms soluble $CrOH^{2+}$ cations. In aqueous environments under low Eh conditions, the main trivalent chromium species are the Cr^{+3} and $CrOH^{2+}$ cations (Richard and Bourg 1991; Motzer 2005; Hlihor et al. 2009). In a slightly acidic to alkaline environment (pH values 6–8, typical pH values of natural water reservoirs), Cr(III) occurs as amorphous chromium hydroxide, $Cr(OH)_3^0$, which can quickly precipitate. Amorphous $Cr(OH)_3$ can crystallize as hydrated $Cr(OH)_3 \cdot 3H_2O$ or as Cr_2O_3, (eskolaite) under different conditions (Tumolo et al. 2020; Swayambunathan et al. 1989). Cr(III) tends to be extremely insoluble (<20 μg/l) between pH 7 and pH 10. At pH 8 the minimum solubility (about 1 μg/l) of Cr(III) occurs forming the

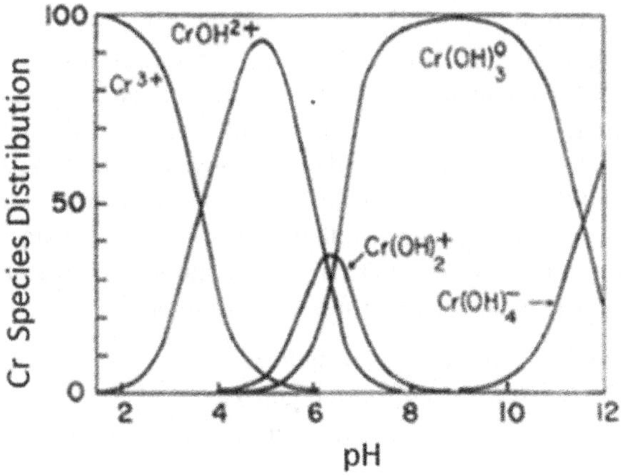

Fig. 2.1 Calculated inorganic chromium (III) speciation as a function of pH (solution in equilibrium with $Cr(OH)_3(s)$. Adopted from Richard and Bourg (1991)

insoluble (Ksp = 6.7 × 10^{-31}) amorphous hydroxide $Cr(OH)_3$ (Fig. 2.1) (Rai et al. 1987; Hlihor et al. 2009; Gorny et al. 2016). In extreme reducing and pH conditions (Eh < 0.0 and pH > 12.0) soluble chromium hydroxide anions, $Cr(OH)_4^-$, are formed (Hem 1977; Richard and Bourg 1991; Hlihor et al. 2009).

In the Cr(III)–H_2O system, Cr(III) exists predominantly as Cr^{+3} below pH 3.5. With increasing pH, hydrolysis of Cr^{+3} yields $CrOH^{2+}$, $Cr(OH)_2^{+1}$, $Cr(OH)_3^0$ and $Cr(OH)_4^-$ (Rai, et al. 1987; Hem 1977; Richard and Bourg 1991; Palmer and Puls 1994; Tumolo et al. 2020) according to Eqs. 2.1–2.4 (Figs. 2.1 and 2.2).

$$Cr^{+3} + H_2O \leftrightarrow CrOH^{2+} + H^+ \quad (2.1)$$

$$CrOH^{2+} + H_2O \leftrightarrow Cr(OH)_2^+ + H^+ \quad (2.2)$$

$$Cr(OH)_2^+ + H_2O \leftrightarrow Cr(OH)_3 + H^+ \quad (2.3)$$

$$Cr(OH)_3 + H_2O \leftrightarrow Cr(OH)_4^- + H^+ \quad (2.4)$$

As can be seen from the above reactions, soluble cations and anions are produced from the insoluble $Cr(OH)_3$. However, in groundwater, under prevailing natural conditions, Cr(III) is relatively insoluble (Fig. 2.1) and rarely occurs in concentrations exceeding the maximum permissible drinking water concentration of 1 pmol/l or 50 ppb (Motzer 2005).

In the aqueous environment trivalent chromium Cr(III) is considered a strong Lewis acid and has a tendency to form complexes of hexacoordinate octahedral structure, (with six substituents) such as water, hydroxyl, ammonium, urea, ethylenediamine and with natural or synthetic organic compounds. In addition, trivalent chromium creates bonds with other substituents such as SO_4^{-2}, NO_3^{-1}, CO_3^{-2}. The

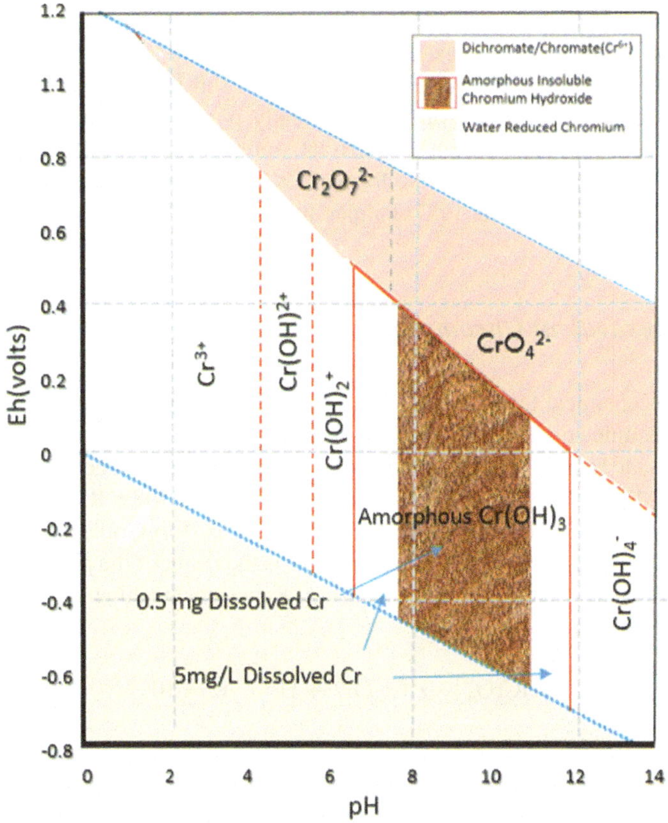

Fig. 2.2 Eh–pH diagram for the chromium–water–oxygen system. Adopted from Hem (1977), Richard and Bourg (1991), Hlihor et al. (2009)

complexes formed by trivalent chromium with substituents in the form of discrete molecules or ions other than OH^-, increase its solubility.

In the absence of complexing agents (except H_2O and OH^-), it appears as a complex $Cr(H_2O)_6^{3+}$ (hexa-aquachromium (III)) and as its hydrolysis products (Eqs. 2.5–2.7). $Cr(H_2O)_6^{3+}$ is a moderately strong acid (pK ~ 4) and its deprotonated forms, denoted for simplicity as $Cr(OH)^+_{2(aq)}$, $Cr(OH)^{2+}_{(aq)}$ and $Cr(OH)_{3(aq)}$ dominate successively between pH values 4–10 (Rai et al. 1987; Kotaś and Stasicka 2000).

$$Cr(H_2O)_6^{3+} + H_2O \leftrightarrow Cr(OH)(H_2O)_5^{+2} + H_3O^+ \tag{2.5}$$

$$Cr(OH)(H_2O)_5^{+2} + H_2O \leftrightarrow Cr(OH)_2(H_2O)_4^{+} + H_3O^+ \tag{2.6}$$

$$Cr(OH)_2(H_2O)_4^{+} + H_2O \leftrightarrow Cr(OH)_{3(aq)} + H_3O^+ \tag{2.7}$$

When the donor atoms are united in polymer systems such as humic acids, then the trivalent chromium complexes are considered to have low mobility (Kotaś and Stasicka 2000). But if the complexation with such substituents can be ignored, in the redox conditions and pH values that usually prevail in natural systems, then Cr is removed from the solution as a $Cr(OH)_3$ precipitate. If Fe(III) is present, Cr precipitate as a solid solution in the form of $Cr_x\ Fe_{1-x}\ (OH)_3$ (where x is the mole fraction of Cr) (Sass and Rai 1987; Palmer and Puls 1994).

2.2.2 *Hexavalent Chromium, Cr(VI)*

The various chemical forms of Cr(VI) are mainly encountered under oxidizing (Eh > 0) and alkaline conditions (pH > 6.0). In aqueous solutions, the dominant Cr(VI) species include $HCrO_4^-$ (for pH values < 6.5, when the Cr(Vl) concentrations are low, up to 10^{-2} M) and CrO_4^{-2} (for pH values > 6.5) (Rai et al. 1989; Richard and Bourg 1991; Palmer and Puls 1994; Koilakos 2017). Cr(VI) also forms other soluble species, such as $HCr_2O_7^{-1}$ and $Cr_2O_7^{-2}$. The dichromate is the result of the polymerization of the monomeric hydrogen chromate ions (Eq. 2.8) to form the dimer, $Cr_2O_7^{-2}$ (Palmer and Puls 1994).

$$2HCrO_4^{-1} \rightarrow Cr_2O_7^{-2} + H_2O \quad K = 10^{2,2} \tag{2.8}$$

However, their formation requires Cr(VI) concentrations greater than $10^{-2,1}$ M which are not found commonly in natural waters (Rai et al. 1989). But it is possible that Cr(VI) concentrations on the surface of a treatment medium are high enough to favor the appearance of dichromates (James et al. 1983).

Unlike Cr(III), Cr(VI) is very soluble and in acidic solutions presents a very high positive redox potential (Fig. 2.2), which shows that it is a strong oxidant and unstable in the presence of electron donors (Rai et al. 1989). Under normal conditions, the zone of Cr(VI) chemical stability, which takes place in the system $Cr–H_2O–O$ (Fig. 2.2), is much smaller than that of Cr(III). In this zone of stability, Cr(VI) generally forms soluble chromic anions (CrO_4^{2-}) at a pH of about 6.0–14.0 and Eh approximately −0.1 to +0.9 V (Brookins 1988; Hlihor et al. 2009; Motzer 2005).

Although Cr(VI) is soluble at most environmental conditions, it can form salts with divalent cations such as Sr^{2+}, Zn^{2+}, Pb^{2+}, and Cu^{2+} and form precipitates, which can be a continuous source of chromate to the environment (Kimbrough et al. 1999).

H_2CrO_4 belongs to the strong acids and for pH > 1 the unprotonated forms prevail (Eqs. 2.9 and 2.10).

$$H_2CrO_4 \rightarrow H^+ + HCrO_4^{-1} \quad k_{a1} = 10^{-0,75} \tag{2.9}$$

$$HCrO_4^{-2} \rightarrow H^+ + CrO_4^{-2} \quad k_{a1} = 10^{-6,45} \tag{2.10}$$

Given the small radius of the Cr(VI) ion (0.0325–0.052 nm), it easily replaces the SO_4^{-2} anion, whose ionic radius is similar: 0.029–0.034 nm (Motzer 2005). This is important because CrO_4^{2-} anions can replace SO_4^{2-} anions and appear chromates as an isoform replacement for sulfate ions in sediments (Motzer 2005).

Cr(VI) in soil and water is formed from Cr(III), which is adsorbed by Mn oxides and oxidized. Cr(III) typically accumulates on the surface of iron oxides and clay minerals. This process is rapid and about 90% of the added chromium is sorbed by clay minerals and iron oxides in 24 h. The adsorption of Cr(III) increases with pH and the organic matter content of soils. It decreases when other competitive inorganic cations or dissolved organic ligands are present in the solution (Bartlett and James 1983; Richard and Bourg 1991; Koilakos 2017). Oxidation is achieved by Mn(III) and Mn(IV) oxy-hydroxides and by Mn(IV) oxides (such as pyrolusite). The oxidation efficiency increases with pH (Eary and Rai 1987; Hug et al. 1997; Motzer 2005) but not necessarily the hexavalent chromium production rate. Cr(III) concentration, pH, and the amount of initial available surface are the parameters controlling the degree of oxidation. The competition for adsorption sites between Cr(III) and Mn(II) (produced from Mn(III) and Mn(IV) reduction) limited Cr(III) oxidation (Fendorf and Zasoski 1992). Reaction rates for the oxidation of Cr(III) to Cr(VI) are relatively slow with half-life ranging from 0.58 to 37.2 year (Saleh et al. 1989).

Amacher and Baker (1979) presented the following general reaction for Cr(III) oxidation by Mn(IV) oxide:

$$Cr(III) + 1,5\delta - MnO_2 + H_2O \rightarrow HCrO_4^- + 1,5Mn^{2+} + H^+ \qquad (2.11)$$

A similar equation for this reaction was determined experimentally by Palmer and Puls (1994).

As already mentioned, under oxidizing conditions, hexavalent chromium can exist in soil as CrO_4^{-2} and $HCrO_4^-$ (James et al. 1997). In deeper soils where anaerobic conditions exist, chromium (VI) will be reduced to chromium (III) by the S^{-2} and Fe^{+2} ions present in the soil. Organic forms of chromium (III), such as humic acid complexes, are more readily oxidizable than insoluble oxides. The barrier for the Cr(III) oxidation process, by the immobile on soil surfaces manganese dioxide, is the lack of mobile chromium (III) availability due to its adsorption and complexation with soil materials. Thus, a large amount of chromium in soil will be not oxidized to chromium (VI), even in the presence of manganese dioxide and favorable pH conditions (James et al. 1997).

In general, organic compounds reduce Cr(VI)–Cr(III). Soil organic matter rapidly adsorbs and reduces Cr(VI)–Cr(III). The latter remains mobile only if its concentration exceeds the sorption and reduction capacity of the soil (Richard and Bourg 1991). Reduction of Cr(VI) by organic matter and other electron donors (e.g., Fe(II) and sulfides) is favored at lower pH. Both oxidation and reduction are inhibited under more alkaline conditions (James et al. 1997). The reduction of hexavalent chromium in water with soil and sediment is very rapid with half-life ranged from instantaneous to 53 days under anaerobic or reducing conditions and 15 min to 21.5 days for aerobic conditions (Saleh et al. 1989). Richard and Bourg (1991) note that simple amino, humic and fulvic acids produce a reduction intermediate Cr(V) which is converted to Cr(III) in a few days. Reduction of Cr(VI) can also result from reaction with Fe(II). This generally involves a three-step process (Sedlak and Chan 1997):

$$Fe^{2+} + Cr(VI) \rightarrow Fe^{3+} + Cr(V) \tag{2.12}$$

$$Fe^{2+} + Cr(V) \rightarrow Fe^{3+} + Cr(IV) \tag{2.13}$$

$$Fe^{2+} + Cr(IV)+ \rightarrow Fe^{3+} + Cr(III) \tag{2.14}$$

In surface waters, the ratio of Cr(III) to Cr(VI) varies widely; relatively high concentrations of Cr(VI) can be found locally, as a result of anthropogenic activities.

2.3 Natural Occurrence of Cr(VI) in Ground Water

The aqueous geochemistry of chromium has been reviewed mainly with reference to the work of Rai et al. (1986, 1988). The natural water hexavalent chromium content is generally low although rocks show a wide range of Cr(III) concentrations, but the range of Cr(VI) concentrations is quite large.

The following compounds control the content of dissolved chromium in natural waters: the sparingly soluble forms of trivalent chromium, $Cr(OH)_{3(s)}$ and (Fe, Cr)$(OH)_{3(s)}$; the soluble forms of this, $CrOH^{2+}$, $Cr(OH)_2^{+1}$ and $Cr(OH)_4^-$ as well as some organic and few inorganic complexes.

The distribution of chromium in the natural environment is controlled by the redox reactions that convert Cr(III) to Cr(VI). These reactions are almost negligible by dissolved oxygen while they are very efficient in the presence of solid MnO_2 (Gorny et al. 2016). Anthropogenic activities (industrial processes) and Mn oxides in nature can oxidize Cr(III) to Cr(VI) and release chromium in its toxic form into natural systems (Fendorf 1995; Sparks 1998). The oxidation of Cr(III) to Cr(VI) is of great importance in assessing the risk posed by chromium to the geoenvironment, as it converts the apparently harmless Cr(III) to the toxic Cr(VI).

Because of the very high redox potential required to convert Cr(III) to Cr(VI), there are few oxidizing agents in the natural environment capable of oxidizing chromium (Rai et al. 1989). Mn oxides are the only naturally occurring minerals capable of oxidizing Cr(III) to Cr(VI), while V^{2+}, Fe^{2+}, S^{2-}, HNO_2, HSO_3^- and some organic species can reduce Cr(VI) (Fendorf 1995; Sparks 1998; Kimbrough et al. 1999). Since water is one of the reactants involved in many chromium oxidation reactions, along with Cr(III) and the oxidizing agent, the conversion from Cr(III) to Cr(VI) rarely occurs in dry systems.

Other possible oxidizing compounds of Cr(III) are disinfectants often used to disinfect drinking water before it is released into the distribution network, such as sodium hypochlorite (NaOCl). Although these compounds are not present in groundwater, they are likely to cause secondary contamination of drinking water with hexavalent chromium, in chlorination tanks and in the distribution network (Lee and Hering 2005). Nevertheless, hexavalent chromium pollution rarely results from the chlorination process, due to the low solubility of trivalent chromium in water.

The oxidation of Cr(III) by Mn oxides has been extensively studied over the past decades (Eary and Rai 1987; Rai et al. 1988; Sparks 1998; Landrot et al. 2012). Recent investigations have shown that groundwater and surface water suffer from hexavalent chromium pollution resulting from the oxidation of trivalent chromium, which is released by the dissolution of chromite ($FeCr_2O_4$), from birnessite (Ndung'u et al. 2010; Zhong et al. 2015).

Although many different manganese oxides have been shown to oxidize Cr(III), the most common oxides in nature have mixed valences containing both Mn(III) and Mn(IV) (Fendorf et al. 1992; Silvester et al. 1995; Fritsch and Navrotsky 1996).

The oxidation degree and the Cr(VI) production rate is highly dependent on the minerals surface charge, (PZC, Point of Zero Charge), surface area, Mn(IV) and Mn(III) content, the density of active Mn(IV) sites as well as pH and competing ions, which are able to change the distribution of Cr(III) and Cr(VI) between the liquid and solid phases (Gorny et al. 2016). The reactivity of Mn oxides is closely related to their degree of crystallinity, which affects the surface (Weaver and Hochella 2003; Kim et al. 2002). Also, large intracrystalline distances or defects in the crystal increase the available surface area and can act as active adsorption centers. Most naturally occurring Mn oxides are poorly crystalline with a layered structure (Post 1999). According to Weaver and Hochella (2003), the following hierarchy represents the ability of different minerals to oxidize Cr(III): birnessite > hausmannite >> romanechite > cryptomelane >> manganite > pyrolusite >> lithiophorite. The above work is in aliment with Kim et al. (2002) regarding the birnessite activity, but they concluded a different hierarchy: birnessite >> todorokite > lithiophorite > pyrolusite. These results suggest that the differences in the manganese oxides structure, mainly related to their amorphous character, determine the minerals oxidation activity. In both studies, the least crystalline manganese oxide mineral, i.e. birnessite, presents the greater activity.

The oxidation of Cr(III) to Cr(VI) on the surface of manganese oxide is done by successive reactions. In general, heterogeneous redox reactions of this type arise through a complex mechanism (Amacher and Baker 1979; Rai et al. 1986; Eary and Rai 1987; Richard and Bourg 1991; Manceau et al. 1992; Gorny et al. 2016). Three general stages can occur in sequence:

(i) adsorption of Cr (III) on the surface of the mineral and formation of surface complexes Cr(III)–MnO_2 and thus
(ii) electron transfer between Cr(III) and Mn(IV)/Mn(III), as well as rearrangements of the oxygen ligands, leading to Cr(VI) and Mn(II) formation, followed by
(iii) desorption of reaction products into the aqueous solution.

The kinetics and mechanisms of these three steps are not yet fully understood although they have been studied by many researchers (Manceau et al. 1992; Johnson and Xyla 1991; Fendorf 1995; Silvester et al. 1995; Banerjee and Nesbitt 1999; Nico and Zasoski 2000; Silvester et al. 1995; Eary and Rai 1987; Kim and Moon 1998).

A general stoichiometric equation for the oxidation of Cr(III) as proposed by Amacher and Baker (1979), is presented in Eq. 2.11. Fendorf and Zasoski (1992) observed very rapid initial rates of oxidation at pH = 5 which are followed by a brusque termination of oxidation. The oxidation reaction according to Fendorf and Zasoski (1992) and Fendorf (1995) for pH ~ 5, in which the largest fraction of soluble Cr(III) is present as the first hydrolysis product, $CrOH^{2+}$, is presented in Eq. 3.1:

$$Cr(OH)^{+2} + 1.5\delta MnO_2 \rightarrow HCrO_4^{-1} + 1.5Mn^{+2} \quad (2.15)$$

While according to Eary and Rai (1987) the oxidation reaction is as follows:

$$Cr(OH)^{2+} + 3\beta - MnO_{2(s)} + 3H_2O \leftrightarrow HCrO_4^{-1} + 3MnOOH_{(s)} + 3H^+ \quad (2.16)$$

The solid $MnOOH_{(s)}$ then decomposes to Mn^{+2}. Cr(III) in solution interacts with the MnOx surface. Some amount of Cr(III) is adsorbed on the MnOx surface, while some is oxidized. The Cr(III) oxidation rate, related to the amount and the surface area of Mn oxides is initially rapid and slows down significantly after 20–60 min (Eary and Rai 1987; Richard and Bourg 1991; Landrot et al. 2012).

The main factor affecting the oxidation of Cr(III) is electrostatic interaction between the surface of Mn and Cr(III) which depends on the change of valence (charge) in the surface bound forms of manganese (surface bound Mn species). A schematic illustration of the oxidation process is shown in the Fig. 2.3 (Nico and Zasoski 2000; Fischel et al. 2015).

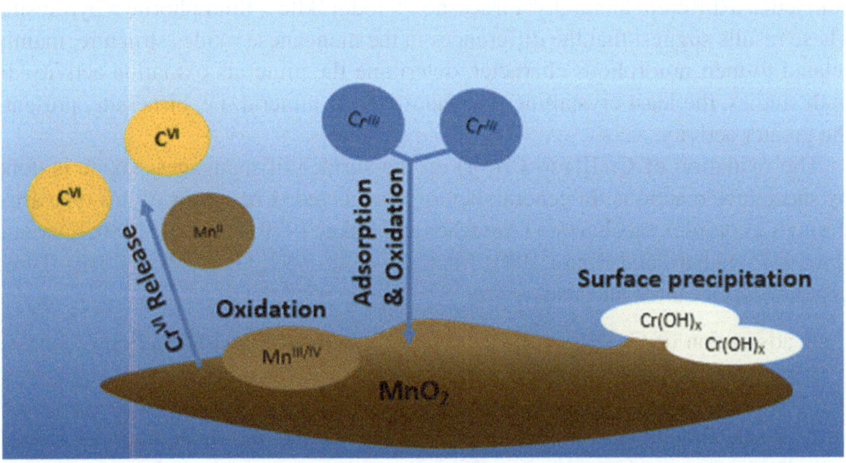

Fig. 2.3 Schematic illustration of the Cr(III) oxidation process. Adopted from Fischel et al. (2015)

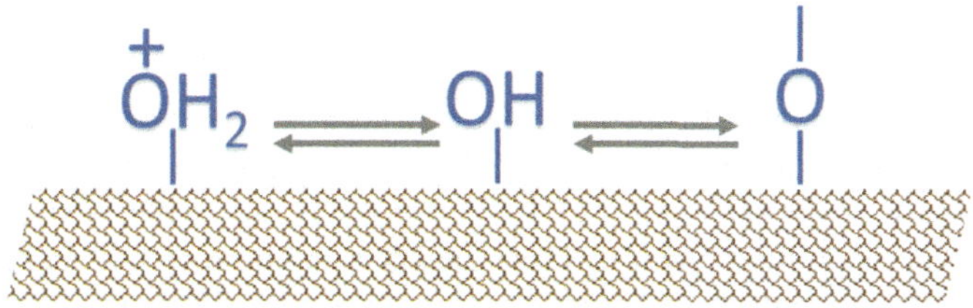

Fig. 2.4 Mn oxide surface with protonated and deprotonated hydroxyls

The surface of Mn oxide in aqueous solution (rock in contact with water) is covered by a large number of hydroxyl groups which can accept or release H^+ into the solution. The number of protonated or deprotonated OH^- groups is different depending on the prevailing solution conditions (e.g., pH), and thus the surface of the Mn oxide appears positively or negatively charged. Depending on the prevailing charge it can attract and adsorb positively or negatively charged ions (Figs. 2.3 and 2.4).

According to the literature, as the pH increases, the Mn surface in the mineral becomes more negative increasing the electrostatic interaction between the surface of Mn and Cr(III) and attracting further positively charged Cr(III), which exhibits a typical cationic adsorption behavior (Palmer and Puls 1994; Fendorf 1995). Consequently, more Cr(III) is adsorbed as an inner-sphere complex (Landrot et al. 2012) on the Mn surface leading initially to higher Cr(VI) production rates (Rai et al. 1987). However, not all Cr(III) is oxidized because Mn(IV) and Cr(III) share the same structural environment. Therefore, some amount of Cr(III) may be adsorbed on the manganese surface and remain unoxidized. Furthermore, as the pH of the solution increases, more hydroxyl ions are introduced into the reaction, enhancing the formation of a Cr hydroxide precipitate on the Mn mineral surface. Thus, higher pH ultimately reduces the amount of adsorbed Cr(III) which reduces Cr(VI) production (Fendorf and Zasoski 1992; Fischel et al. 2015). Furthermore, Cr(III) adsorption is reduced when competing ions are present (Fendorf 1995).

Not all of the Cr(VI) formed by oxidation of Cr(III) is released back into solution as a portion may remain adsorbed on the MnOx surface (matrix associated). The total concentration of Cr(Vl) formed is the sum of the aqueous Cr(Vl) and the matrix associated Cr(Vl) (Palmer and Puls 1994). The interaction of the solid-liquid phase can regulate the chromium content in the water. Cr(VI), as chromate anions, exhibits a typical anionic adsorption behavior and forms mainly outer-sphere (Fig. 2.3) complexes with Mn oxide (Landrot et al. 2012). At low concentrations, decreasing pH, adsorption of Cr(VI) increases, whatever the adsorbent, (Bartlett and James 1979; Rai et al. 1988; Richard and Bourg 1991). This result suggests that Cr(VI) adsorption is favored on adsorbents which are positively charged at low to neutral pH (Richard and Bourg 1991). Competitive adsorption can release chromates into aqueous phase. According to many researchers, it appears that Cr(VI) and phosphate are competing for the same adsorption sites. The phosphate removes chromate by both directly competing for the adsorption sites in the soil and indirectly (in some cases) by increasing the pH (Bartlett and Kimble 1976; James and Bartlett 1983; Palmer and Puls 1994).

In general, oxidation of trivalent chromium can occur over a pH range of 3–10.1 (Eary and Rai 1987). However, according to Fendorf and Zasoski (1992), when the pH increased above 3.5 the oxidation of Cr(III) decreased. In more detail, above pH = 3.5, the increased availability of hydroxyl ions enhances the formation of Cr hydroxides leading to passivation of the manganese oxide surface (Fendorf and Zasoski 1992) and inhibition of Cr(III) oxidation due to precipitation of $Cr(OH)_3.nH_2O$ (Fendorf 1995). This was confirmed by Feng et al. (2006), who found that the amount of Cr(III) oxidized by birnessite at pH between 2 and 3.5 was nearly constant, but that Cr(III) oxidation decreased as pH increased beyond 3.5 (Fischel et al. 2015). In general, oxic and alkaline conditions enhance Cr(VI) mobility. These conditions are not favorable for Cr(VI) sorption and reduction reactions that result in Cr(III) production and its immobilization throuth precipitation (Chrysochoou et al. 2016).

Chromium hydroxide precipitate inhibits oxidation in two ways. On the one hand, through the creation of a redox-stable reservoir of soluble, oxidizable Cr(III), and on the other hand due to the existence of a physical barrier between the dissolved chromium and the manganese surface. More specifically, the reduced mobility of the chromium hydroxide that precipitates on the surface of the manganese oxide has the effect of reducing the possibility of its transport and oxidation. Also, the chromium hydroxide precipitate is interposed between the manganese oxide surface and the Cr(III) ions so that the oxidation reaction cannot take place (Fendorf 1995). Fendorf (1995) investigated the extent to which competitive adsorption of La, Al, Mn(II) and Cr(III) occurs in the oxidation to form Cr(VI). It has been shown that aluminum effectively inhibit the oxidation of Cr(III) for pH values greater than 4, in which a precipitate, $Al(OH)_3$, is formed. When the aluminum hydroxide precipitate is extensive enough to completely cover the manganese oxide surface, oxidation of Cr(III) is almost negligible (Amacher and Baker 1979; Fendorf et al. 1992; Fendorf 1995). According to Amacher and Baker (1979) the $Mn(II)_{(aq)}$ ions released from the redox reactions are adsorbed on the surface of δ-MnO_2 and they also compete with the adsorption of Cr(III), and thus inhibit further oxidation of this (Fendorf et al. 1992).

As shown above, hexavalent form of chromium can be transformed by several natural reductants to the less toxic trivalent form, which precipitates as insoluble and immobile hydroxide. Therefore, in certain sites, pollution caused by Cr(VI) may be reduced without human intervention by natural attenuation, which is a viable option depending on both the aquifer and the polluted plume under investigation. Both oxidation and reduction of chromium are occurring simultaneously within the subsurface as part of a geochemical cycle: manganese dioxides, present the soil, oxidize Cr(III), while some reductants, such as soil organic carbon or Fe(II) and Fe(II)-bearing minerals, can reduce Cr(VI)–Cr(III). At pH between about 5 and 12, Cr(III) formed precipitates as $Cr(OH)_3$ or as part of a solid solution with Fe(III), inhibiting further Cr(III) oxidation and keeping Cr(III) concentrations below 1 pmol/l (0.05 mg/l).

Sorption to sediments and Cr(VI) sequestration under some conditions as well as chromates precipitation can also be an important attenuation mechanism. Precipitation of chromates Cr(VI) has been shown to be associated with calcium carbonate precipitates (Landrot et al. 2012). Studies indicate that reduced chromium is resistant to reoxidation (Palmer and Puls 1994; Tumolo et al. 2020; Fendorf 1995). The identification of hexavalent chromium natural attenuation mechanisms in groundwater was also studied by Lamar et al. (2017).

2.3.1 Aquifer Depth Effect on Chromium Concentration

Age and quality stratification of groundwater in an aquifer is well known. Depending on the geological background, the water in the deeper aquifer levels has a clearly different composition from the water in the shallow aquifer levels (Bourotte et al. 2009; Bexfield and Jurgens 2014; Chenini et al. 2010; Izbicki et al. 2015; Li et al. 2015; Dragon and Gorski 2015; Mavromatidou et al., 2021). Apart from the geological background, the longer water residence time (Robertson 1991; Böhlke et al. 2007; Bourotte et al. 2009; Smedley et al. 2018) and the various reactions (such as ion exchange which occur in the deeper aquifer levels) contributes to the different water composition.

Three hydrochemical zones can describe the water quality stratification in an aquifer (Bourotte et al. 2009):

(a) the Shallow Zone (up to a depth of about 30 m), exhibiting Na^+–Ca^{2+}–Cl^-–NO_3^- facies
(b) the Intermediate Zone (up to a depth of about 30–70 m), showing Ca^{2+}–HCO_3^- facies
(c) the Deep Zone (for depth greater than 70 m), showing Na^+–Ca^{2+}–HCO_3^- facies.

The transmission of recharging water as well as of contaminants from the land surface to groundwater is controlled by the soil surface above the water table of an aquifer, called vadose or unsaturated zone.

Basins with longer groundwater residence time have higher Cr(VI) concentrations (concentrations are inversely related to recharge rates) (Robertson 1991). For example, in an area of Brazil at greater depths of the aquifer Cr(VI) was the dominant form and was detected in concentrations up to 120 μg/l, in shallow wells with young water the dominant form was Cr(III) while in the intermediate depths again the dominant form is Cr(VI) but in clearly lower concentrations (up to 25 μg/l) (Bourotte et al. 2009).

Apart from the Cr(VI) concentration there are other variations with aquifer depth in various ions, with a clear increase in the Na/Cl molar ratio and a decrease in the Ca^{+2}/HCO_3^- ratio in deeper aquifer levels. These variations were attributed to ion exchange between water and rocks and the existence of competing anions favoring Cr(VI) desorption from the rock (Bourotte et al. 2009; Chenini et al. 2010; Dragon and Gorski 2015). This may be the most likely phenomenon occurring in the deep aquifer, as the increase in pH (from the top to the base of the aquifer), due to cation exchange reactions involving Na^+, Ca^{2+} and Mg^{2+}, and by the dissolution of carbonate minerals to the limit of their solubility, is likely to cause Cr(VI) anions to desorb to the water table. The Na^+ increase accompanied by a decrease in Ca^{2+} and Mg^{2+} concentration as well as the Ca^{2+}/HCO_3^- ratio indicates the existence of cation exchange reactions in the deeper parts of the aquifer (Bourotte et al. 2009; Yan et al. 2022; Smedley et al. 2018).

Corresponding variations such as increased concentration of geogenic As at greater depths and increased concentration of NO_3^- at shallow zones have been reported for Mexico, Greece and elsewhere (Bexfield and Jurgens 2014; Mavromatidou et al. 2021). In this case, the differences in water quality depending on the depth were noticeable due to the increased pumping.

Many research has been conducted on the effect of pumping on groundwater quality, with several of them focusing on its effect on the hexavalent chromium concentration. Hence, pumping and intensive use of groundwater has recently been recognized as an anthropogenic activity with an effect on Cr(VI) formation (Robertson 1975). The over-pumping of the aquifer for e.g. irrigation, water supply, industrial use, in general causes a change in water quality (Fig. 2.5) which can even occur before a decrease in the aquifer water level is noticed (Llamas et al. 2003). Also, the seasonal change in water quality is known, depending on whether "new" or "old" water is pumped, which is affected by increased pumping (Bexfield and Jurgens 2014; Mavromatidou et al. 2021). According to Hausladen et al. (2018) and Izbicki et al. (2015), land use is likely to be a high risk for hexavalent chromium contamination of California groundwater, along with the industrial activities and natural geology. According to the same researchers, irrigation combined with groundwater pumping increases the tendency for downward movement of components through soil and underlying sediments, e.g. nitrates, and thus may be an additional link between agricultural activities and groundwater Cr(VI) concentrations. Irrigation has an increased contribution to groundwater recharge relative to precipitation (Mills et al. 2011; Bertoldi et al. 1991). According to Davisson et al. (1993), up to half of the groundwater in some areas can be attributed to irrigation return water as indicated by $d^{18}O_{H2O}$ values and elevated NO_3^- concentrations. Thus, the simplest link between irrigation and Cr(VI) concentrations in groundwater is the faster transport of the produced Cr(VI) in the unsaturated zone sediments to the shallow aquifer zone, due to increased returned water. As has been suggested from Hausladen et al. (2018) and Manning et al. (2015), within the unsaturated zone, interaction between Cr(III) minerals and Mn(III/IV) oxides results in the formation of Cr(VI). Extensive changes in the aquifer discharge–charge cycle leads to Cr(VI) formation as has been suggested from Hausladen and Fendorf (2017). For example, in the California region, it has been reported that during increased pumping in summer, mainly new water with high NO_3^- and uranium values is pumped in contrast to winter. Correspondingly, in a region of Mexico, old water with high As values is pumped in winter, while in summer this is less likely to happen (Bexfield and Jurgens 2014). So, in drilling of local aquifers with small water potential, even reasonable pumping has a great effect on the water level, while in large aquifers, over-pumping leads to the mixing

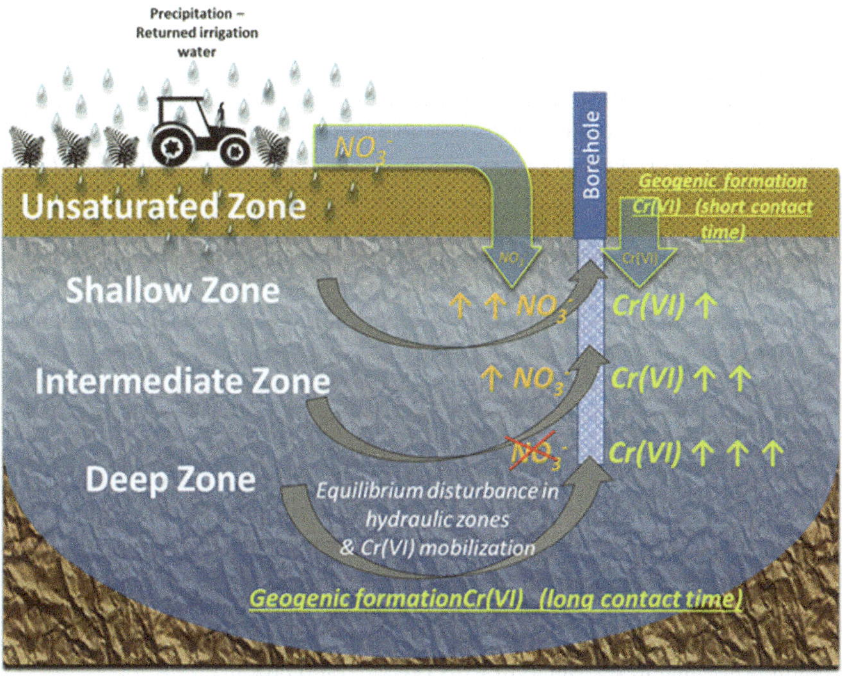

Fig. 2.5 Hexavalent chromium presence in relation to hydrochemical zones

of the various water layers or the pumping from lower layers (leading to increasing concentrations of Cr, As, etc.).

Therefore, in the shallow reservoirs and zone of an aquifer, due to fertilization and irrigation water reflow through the unsaturated zone, high nitrate and low hexavalent chromium concentrations may coexist. In this zone, the presence of Cr(VI) is attributed to its formation in the ophiolite sediments of the vadose zone and its low concentrations are a result of the short contact time with infiltrating water (Fig. 2.5) and whitout achieving redox equilibrium that lead to low Cr(VI)/Cr_{total} ratio values. In the deep reservoirs and zone of an aquifer, soluble iron and manganese may coexist along with high Cr(VI) concentrations, in absence or low presence of nitrate ions (Fig. 2.5), indicating geogenic formation of Cr(VI). In this zone, the high Cr(VI) concentrations are a result of the long contact time of ophiolite sediments with water, the relative redox equilibrium and the resultant higher Cr(VI)/C_{rtotal} ratio values. Pumping from the deeper zone of an aquifer (Fig. 2.5), influences the Cr(VI) release in water, leading to higher (>50 μg/l) Cr(VI) concentrations (Hausladen et al. 2018; Mills et al. 2011; Mavromatidou et al. 2021).

2.4 Concluding Remarks

- Mn oxides are the only naturally occurring minerals that can oxidize Cr(III) to Cr(VI) due to the very high redox potential that it is required.
- Reduction Cr(VI)–Cr(III) can be achieved by V^{2+}, Fe^{2+}, S^{2-}, HNO_2, HSO_3^- and some organic species.
- The conversion of Cr(III) to Cr(VI) rarely occurs in dry systems, since water is one of the reactants involved in many chromium oxidation reactions, along with Cr(III) and the oxidizing agent.
- The oxidation degree and the Cr(VI) production rate are highly dependent on the minerals amorphous character, surface charge, (PZC, Point of Zero Charge), surface area, Mn(IV) and Mn(III) content, density of active Mn(IV) sites, pH and competing ions, which are able to change the distribution of Cr(III) and Cr(VI) between the liquid and solid phases.
- Oxic and alkaline conditions enhance Cr(VI) mobility. These conditions are not favorable for Cr(VI) sorption and reduction reactions that result in Cr(III) production and its immobilization throuth precipitation.
- Increasing nitrate and low hexavalent chromium concentrations may coexist in shallow zone of an aquifer as a result of fertilization and water reflow through the unsaturated zone during irrigation. In this zone, short contact time, no redox equilibrium and lower values of Cr(VI)/C_{rtotal} ratio prevail.
- In the deep zone of an aquifer soluble iron and manganese may coexist along with high Cr(VI) concentrations, in absence or low presence of nitrate ions, indicating geogenic formation of Cr(VI).
- In an aquifer, Cr(VI) pollution may be reduced without human intervention by natural attenuation.

References

Abbasi SA, Soni R (2007) Teratogenic effects of chromium (vi) in environment as evidenced by the impact on larvae of amphibian *Rana tigrina*: implications in the environmental management of chromium. Int J Environ Stud 23:131–137

Alemu A, Lemma B, Gabbiye N, Alula MT, Destaf MT (2018) Removal of chromium (VI) from aqueous solution using vesicular basalt: A potential low-cost wastewater treatment system. Heliyon 4(7):e00682.

Amacher C, Baker DE (1979) Report redox reactions involving chromium, plutonium, and manganese in soils. Final report DE-AS08-77DPO4515. Institute for Research on Land and Water Resources, Pennsylvania State University

Ball JW, Izbick JA (2004) Occurrence of hexavalent chromium in ground water in the western Mojave Desert California. Appl Geochem 19:1123–1135

Banerjee D, Nesbitt HW (1999) Oxidation of aqueous Cr(III) at Birnessite surfaces: constraints on reaction mechanism. Geochim Cosmochim Acta 53:1671–1687

Barrera-Díaz CE, Lugo-Lugo V, Bilyeu B (2012) A review of chemical, electrochemical and biological methods for aqueous Cr(VI) reduction. J Hazard Mater 223–224:1–12

Bartlett RJ, James B (1979) Behavior of Chromium in Soils: III. Oxidation1. J Environ Qual 8. https://doi.org/10.2134/jeq1979.00472425000800010008x
Bertoldi GL, Johnston RH, Evenson KD (1991) Ground water in the Central Valley, California-A summary report. USGS. Professional Paper 401 (A)
Bartlett RJ, Kimble JM (1976) Behavior of Chromium in Soils: 1. Trivalent Forms II. Hexavalent Forms. J Environ Qual 5:379–386
Bexfield LM, Jurgens BC (2014) Effects of seasonal operation on the quality of water produced by public-supply wells. Ground Water 52:10–24
Beyersmann D, Koster A, Buttner B, Flessel P (1984) Model reactions of chromium compounds with mammalian and bacterial cells. Toxicol Environ Chem 8:279–286
Bianchi V, Zantedeschi A, Montaldi A, Majone F (1984) Trivalent chromium is neither cytotoxic nor mutagenic in permeabilized hamster fibroblasts. Toxicol Lett 23:51–59
Böhlke JK, O'Connell ME, Prestegaard KL (2007) Ground water stratification and delivery of nitrate to an incised stream under varying flow conditions. J Environ Qual 36:664–680
Bonatti S, Meini M, Abbondandolo A (1976) Genetic effects of Potassium Chromate in Schizosaccharomyces pornbe. Mutat Res 38:147–149
Bortone I, Erto A, Nardo AD, Santonastaso GF, Chianese S, Musmarra D (2020) Pump-and-treat configurations with vertical and horizontal wells to remediate an aquifer contaminated by hexavalent chromium. J Contam Hydrol 235:103725
Bourotte C, Bertolo R, Almodovar M, Hirata R (2009) Natural occurrence of hexavalent chromium in a sedimentary aquifer in Urânia, State of São Paulo, Brazil. Earth Sci 81:227–242
Brookins DG (1988) Eh-pH Diagrams for geochemistry. Springer-Verlag, New York, USA
Chenini I, Farhat B, Ben Mammou AB (2010) Identification of major sources controlling groundwater chemistry from a multilayered aquifer system. Chem Speciat Bioavailab 22:183–189
Chrysochoou M, Theologou E, Bompoti N, Dermatas D, Panagiotakis I (2016) Occurrence, origin and transformation processes of geogenic chromium in soils and sediments. Curr. Pollut. Reports. 2:224–235
Coyte RM, McKinley KL, Jiang S, Karr J, Dwyer GS, Keyworth AJ, Davis CC, Kondash AJ, Vengosh A (2020) Occurrence and distribution of hexavalent chromium in groundwater from North Carolina, USA. Sci Total Environ 711
Cranston RE, Murray JW (1978) The determination of chromium species in natural waters. Anal Chim Acta 99(2):275–282
Davisson ML, Criss RE, Campbell KR (1993) Preliminary report on the stable isotope imaging and characterization of surface and ground water resources in the southern Sacramento Valley. LLNL. 303, UCRL-ID-115393
Dermatas D, Vatseris C, Panagiotakis I, Chrysochoou M (2012) Potential contribution of geogenic chromium in groundwater contamination of a Greek Heavily Industrialized Area. Chem Eng Trans 28
Dragon K, Gorski J (2015) Identification of groundwater chemistry origins in a regional aquifer system (Wielkopolska region, Poland). Environ Earth Sci 73:2153–2167
Eary LE, Rai D (1987) Kinetics of chromium(III) oxidation to chromium(VI) by reaction with manganese dioxide. Environ Sci Technol 21:1187–1193
Economou-Eliopoulos M, Megremi I, Atsarou C, Theodoratou C, Vasilatos C (2013) Spatial evolution of the chromium contamination in soils from the assopos to thiva basin and C. Evia (Greece) and potential source(s): anthropogenic versus natural processes. Geoscience 3:140–158
Environmental Protection Agency U.S. (2005) Cost-effective design of pump and treat systems. EPA 542-R-05-008
Fantoni D, Brozzo G, Canepa M, Cipolli F, Marini L, Ottonello G, Vetuschi Zuccolini M (2002) Natural hexavalent chromium in groundwaters interacting with ophiolitic rocks. Environ Geol 42:871–882
Farooqi ZH, Akram MW, Begum R, Wu W, Irfan A (2021) Inorganic nanoparticles for reduction of hexavalent chromium: physicochemical aspects. J Hazard Mater 402:123535
Fendorf SE (1995) Surface reactions of chromium in soils and waters. Geoderma 67:55–71

Fendorf SE, Fendorf M, Sparks DL, Gronsky R (1992) Inhibitory mechanisms of Cr(lll) oxidation by δ-Mn02. J Colloid Interface Sci 153:37–54

Fendorf SE, Zasoski RJ (1992) Chromium (Ill) Oxidation by δ-MnO2. 1. Characterization. Environ Sci Technol 26:79–85

Feng XH, Zhai LM, Tan WF, Zhao W, Liu F, He JZ (2006) The controlling effect of pH on oxidation of Cr(III) by manganese oxide minerals. J Colloid Interface Sci 298:258–266

Fischel JS, Fischel MH, Sparks DL (2015) Advances in understanding reactivity of manganese oxides with arsenic and chromium in environmental systems manganese oxide reactivity with arsenic. ACS Symp Ser 1197:1–27

Fritsch S, Navrotsky A (1996) Thermodynamic properties of manganese oxides. J Am Ceram Soc 79:1761–1768

Gheju M, Balcu I (2011) Removal of chromium from Cr(VI) polluted wastewaters by reduction with scrap iron and subsequent precipitation of resulted cations. J Hazard Mater 196:131–138

Gheju M, Iovi A, Balcu I (2008) Hexavalent chromium reduction with scrap iron in continuous-flow system. Part 1: effect of feed solution pH. J Hazard Mater 153:655–662

Gonzalez AR, Ndung'u K, Flegal AR (2005) Natural occurrence of hexavalent chromium in the Aromas Red Sands aquifer, California. Environ Sci Technol 39:5505–5511

Gorny J, Billon G, Noiriel C, Dumoulin D, Lesven L, Madé B (2016) Chromium behavior in aquatic environments: a review. Environ Rev 24:503–516

Guo H, Chen Y, Hu H, Zhao K, Li H, Yan S, Xiu W, Coyte RM, Vengosh A (2020) High hexavalent chromium concentration in groundwater from a deep aquifer in the Baiyangdian Basin of the North China plain. Environ Sci Technol 54:10068–10077

Hausladen DM, Alexander-Ozinskas A, McClain C, Fendorf S (2018) Hexavalent chromium sources and distribution in California groundwater. Environ Sci Technol 52:8242–8251

Hausladen DM, Fendorf S (2017) Hexavalent chromium generation within naturally structured soils and sediments. Environ Sci Technol 51:2058–2067

He S, Wu J (2019) Hydrogeochemical characteristics, groundwater quality, and health risks from hexavalent chromium and nitrate in groundwater of Huanhe Formation in Wuqi County, Northwest China. Expo Health 11:125–137

Hem JD (1977) Reactions of metal ions at surfaces of hydrous iron oxide. Geochim Cosmochim Acta 41:527–538

Hlihor RM, Apostol LC, Vasile LP, Smaranda C, Robu BM, Caliman FA, Gavrilescu MARIA (2009) Overview on chromium ocurrence in the environment and its remediation. Buletinul Institutului Politehnic din Iasi, Chimie si Inginerie Chimicá 60:67–83

Hu R, Noubactep C (2019) Redirecting research on Fe0 for environmental remediation: the search for synergy. Int J Environ Res Public Health 16:4465

Hug SJ, Buerge IJ, Weidler PG (1997) Transformations of chromium in the environment. Analusis Mag 25(7):M12–M15

International Agency for Research on Cancer (1990) Chromium, nickel and welding. IARC monographs on the evaluation of carcinogenic risks to humans 49

Izbicki JA, Wright MT, Seymour WA, McCleskey RB, Fram MS, Belitz K, Esser BK (2015) Cr(VI) occurrence and geochemistry in water from public-supply wells in California. Appl Geochem 63:203–217

Jacobs J, Testa SM (2004) Overview of Chromium(VI) in the environment. Chapter In book: Chromium(VI) handbook. CRC Press

James BG, Bartlett RJ (1983) Behavior of chromium in soils: VII. Adsorption and reduction of hexavalent forms. J Environ Qual 12:177–181

James RB, Petura CJ, Vitale JR, Mussoline RG (1997) Oxidation-reduction chemistry of chromium: Relevance to the regulation and remediation of chromate-contaminated soils. Soil Sediment Contam 6(6):569–580

Johnson CA, Xyla AG (1991) The oxidation of chromium(III)–chromium(VI) on the surface of manganite (γ-MnOOH). Geochim Cosmochim Acta 55:2861–2866

Kim JG, Dixon JB, Chusuei CC, Deng Y (2002) Oxidation of Chromium(III)–(VI) by manganese oxides. Soil Sci Soc Am J 66:306–315

Kim JG, Moon HS (1998) Oxidation of chromium +3 to chromium +6 by a series of synthesized birnessites (δ-MnO2): kinetics and oxidation capacity. Clay Sci 10:363–374

Kimbrough DE, Cohen Y, Winer AM, Creelman L, Mabuni C (1999) A critical assessment of chromium in the environment. Crit Rev Environ Sci Technol 29:1–46

Khalil U, Shakoor MB, Ali S, Ahmad SR, Rizwan M, Alsahli AA, Alyemeni MN (2021) Selective removal of hexavalent chromium from wastewater by rice husk: kinetic, isotherm and spectroscopic investigation. Water 13(3):263

Koilakos DI (2017) Aspects of hexavalent chromium pollution of thebes plain aquifer, boeotia, greece. Water 9(8):611

Kotas J, Stasicka Z (2000) Chromium occurrence in the environment and methods of its speciation. Environ Pollu 107:263–283

Lamar M, Lyons J, Baker K, Bowen I, Burgesser T, Olsen R, Smith N, Whiting K (2017) Identification of natural attenuation mechanisms of hexavalent chromium in groundwater through geochemical and matrix diffusion evaluations. Bioremediation and Sustainable Environmental Technologies, The Fourth International Symposium – Miami, Florida

Landrot G, Ginder-Vogel M, Livi K, Fitts JP, Sparks DL (2012) Chromium(III) oxidation by three poorly-crystalline manganese (IV) oxides. 1. Chromium(III)-oxidizing capacity. Environ Sci Technol 46:11594–11600

Lee G, Hering JG (2005) Oxidative dissolution of chromium (III) hydroxide at pH 9, 3, and 2 with product inhibition at pH 2. Environ Sci Technol 39:4921–4928

Li W, Wang MY, Liu LY, Yan Y (2015) Assessment of long-term evolution of groundwater hydrochemical characteristics using multiple approaches: a case study in Cangzhou, Northern China. Water 7:1109–1128

Lide DR (ed) (2006) CRC handbook of chemistry and physics, internet version. Taylor and Francis, Boca Raton, FL. http://www.hbcpnetbase.com

Llamas MR, Custodio E (2003) Intensive use of groundwater. Challenges and opportunities. Balkema, Rotterdam, The Netherlands

Machiwal D, Jha MK (2015) Identifying sources of groundwater contamination in a hard-rock aquifer system using multivariate statistical analyses and GIS-based geostatistical modeling techniques. J Hydrol Reg Stud 4:80–110

Maitlo HA, Kim KH, Kumar V, Kim, Park SJW (2019) Nanomaterials-based treatment options for chromium in aqueous environments. Environ Int 130:10478

Manceau A, Charlet L, Boisset MC, Didier B, Spadini L (1992) Sorption and speciation of heavy metals on hydrous Fe and Mn oxides. From microscopic to macroscopic. Appl Clay Sci 7:201–223

Manning AH, Mills CT, Morrison JM, Ball LB (2015) Insights into controls on hexavalent chromium in groundwater provided by environmental tracers, Sacramento Valley, California, USA. Appl Geochem 62:186–199

Mavromatidou C, Samiotis G, Batsi A, Amanatidou (2021) Correlation of hexavalent chromium concentration to groundwater hydrochemical zones chemistry. Groundw Sustain Dev 15

Mills TC, Morrison MJ, Goldhaber BM, Ellefsen JK (2011) Chromium(VI) generation in vadose zone soils and alluvial sediments of the southwestern Sacramento Valley, California: A potential source of geogenic Cr(VI) to groundwater. Appl Geochem 26:1488–1501. https://doi.org/10.1016/j.apgeochem.2011.05.023

Monks PS, Ravishankara AR, Schneidemesser EV, Sommariva R (2021) Opinion: papers that shaped tropospheric chemistry. Atmosheric Chem Phys Dis

Motzer WE (2005) Chemistry, geochemistry, and geology of chromium and chromium compounds. In: Chromium handbook. Guertin J, Jacobs JA, Avakian CP (Eds), Boca Raton, London, New York, Washington DC 23–91

Ndung'u K, Friedrich S, Gonzalez AR, Flegal AR (2010) Chromium oxidation by manganese (hydr)oxides in a California aquifer. Appl Geochem 25:377–381

Nico PS, Zasoski RJ (2000) Importance of Mn (III) availability on the rate of Cr(III) oxidation on δ-MnO2. Environ Sci Technol 34:3363–3367
Palmer CD, Puis RW (1994) Natural attenuation of hexavalent chromium in groundwater and soils. EPA Ground Water Issue
Paschin YV, Kozachenko VI, Sal'nikova LE (1983) Differential mutagenic response at the HGPRT locus in V79 and CHO Chinese hamster cells after treatment with chromate. Mutat Res Lett 122:361–365
Perraki M, Vasileiou E, Bartzas G (2021) Tracing the origin of chromium in groundwater: current and new perspectives. Curr Opin Environ Sci Health 22:100267
Peterson ML, Brown Jr GE, Parks GA (1996) Direct XAFS evidence for heterogeneous redox reaction at the aqueous chromium/magnetite interface. Coll Surf Phy Eng Asp 107:77–88
Post JE (1999) Manganese oxide minerals: crystal structures and economic and environmental significance. Proc Natl Acad Sci USA. 96(7):3447–3454. https://doi.org/10.1073/pnas.96.7.3447. PMID: 10097056; PMCID: PMC34287.
Rai D, Zachara JM (1986) Geochemical behavior of chromium species. Technical Report EA-4544
Rai D, Sass BM, Moore DA (1987) Chromitim(III) hydrolysis constants and solubility of chromium(III) hydroxide. Inorg Chem 26:345–349
Rai D, Zachara JM (1988) Chromium reactions in geologic materials. Technical Report EA-5741
Rai D, Zachara JM, Eary LE, Ainsworth CC, Amonette JE, Cowan CE, Szelmeczka RW, Resch CT, Schmidt RL, Girvin DC, Smith SC (1988) Chromium reactions in geological materials. Interim Report EPRI EA-5741. E.P.R.I., Pale Alto, Calif
Rai D, Eary LE, Zachara JM (1989) Environmental chemistry of chromium. Sci Total Environ 86:15–23
Richard FC, Bourg ACM (1991) Aqueous geochemistry of chromium: a review. Wat Res 25:807–816
Robertson FN (1991) Geochemistry of ground water in alluvial basins of Arizona and adjacent parts of Nevada, New Mexico, and California. USGS Publications Warehouse, Professional Paper 1406-C
Robertson FN (1975) Hexavalent Chromium in the Ground Water in Paradise Valley, Arizona. Groundwater 13:516–527. https://doi.org/10.1111/j.1745-6584.1975.tb03621.x
Robles-Camacho J, Armienta MA (2000) Natural chromium contamination of groundwater at León Valley, México. J Geochem Explor 68:167–181
Samiotis G, Stimoniaris A, Ristanis I, Kemmou L, Mavromatidou C, Amanatidou E (2023) Novel applications of iron in water and wastewater treatment. Resources 12:39. https://doi.org/10.3390/resources12030039
Saleh FY, Parkerton TF, Lewis RV, Huang JH, Dickson KL (1989) Kinetics of chromium transformations in the environment. Sci Total Environ 86:25–41
Sass BM, Rai D (1987) Solubility of amorphous chromium(III)–iron(III) hydroxide solid solutions. Inorg Chem 26:2228–2232
Sedlak DL, Chan PG (1997) Reduction of hexavalent chromium by ferrous iron. Geochim Cosmochim Acta 61(11):2185–2192
Sharma SK, Petrusevski B, Amy G (2008) Chromium removal from water: a review. J Water Supply: Res Technol 57(8):541–553
Silvester E, Charlet L, Manceau A (1995) Mechanism of chromium(III) oxidation by Na-Buserite. J Phys Chem 99:16662–16669
Smedley PL, Shand P, Butcher AS (2018) Age and quality stratification of groundwater in the Triassic Sherwood Sandstone aquifer of South Yorkshire and the East Midlands, UK. Appl Geochem 97:109–122
Smith JV (1999) Geology, mineralogy, and human welfare. Proc Natl Acad Sci USA 96:3348–3349
Sparks DL (1998) Soil physical chemistry, 2nd edn. CRC Press
Sperling M, Xu S, Welz B (1992) Determination of chromium (III) and chromium (VI) in water using flow injection on-line preconcentration with selective adsorption on activated alumina and flame atomic absorption spectrometric detection. Anal Chem 64(24):3101–3108

Stamatis G, Alexakis D, Gamvroula D, Migiros G (2011) Groundwater quality assessment in Oropos-Kalamos basin, Attica, Greece. Environ Earth Sci 64:973–988

Stamos A, Samiotis G, Pekridis G, Tsioptsias C, Amanatidou E (2019) Natural presence of hexavalent chromium in spring waters of South-West Mountain Vermion, Greece. In: 16th international conference on environmental science and technology

Stephen MT, Javobs J (2004) Overview of chromium(VI) in the environment. In: Chromium(VI) handbook, pp 1–21

Stern AH (2010) A quantitative assessment of the carcinogenicity of hexavalent chromium by the oral route and its relevance to human exposure. Environ Res 110:798–807

Swayambunathan V, Liao YX, Meisel D (1989) Stages in the evolution of colloidal chromium(III) oxide. Langmuir 5(6):1423–1427

Testa SM, Jacobs J (2004) Overview of chromium(VI) in the environment. In: Chromium(VI) handbook. CRC Press

Tiwari AK, Orioli S, De Maio M (2019) Assessment of groundwater geochemistry and diffusion of hexavalent chromium contamination in an industrial town of Italy. J Contam Hydrol 225:103503

Tumolo M, Ancona V, De Paola D, Losacco D, Campanale C, Massarelli C, Felice VU (2020) Chromium pollution in european water, sources, health risk, and remediation strategies: An overview. Int J Environ Res Public Health 17:5438 https://doi.org/10.3390/ijerph17155438

Vasileiou E, Papazotos P, Dimitrakopoulos D, Perraki M (2019) Expounding the origin of chromium in groundwater of the Sarigkiol basin, Western Macedonia, Greece: a cohesive statistical approach and hydrochemical study. Environ Monit Assess 191:509

Weaver RM, Hochella MF (2003) The reactivity of seven Mn-oxides with $Cr^{3+}{}_{aq}$: a comparative analysis of a complex, environmentally important redox reaction. Am Min 88:2016–2027

WHO-International Agency for Research on Cancer (1990) Chromium, Nickel and Welding. IARC monographs on the evaluation of carcinogenic risks to humans. 49

Yan S, Guo H, Yin J, Hu H, Cui D, Gao B (2022) Genesis of high hexavalent chromium groundwater in deep aquifers from loess plateau of Northern Shaanxi, China. Water Res 216:118323

Zhong L, Yang J, Liu L, Xing B (2015) Oxidation of Cr(III) on birnessite surfaces: the effect of goethite and kaolinite. J Environ Sci (China) 37:8–14

Zissimos AM, Christoforou IC, Christofi C, Rigas M, Georgiadou EC, Christou A (2021) Occurrence and distribution of hexavalent chromium in ground and surface waters in cyprus. Bull Environ Contamin Toxicol 106:428–434

Chapter 3
Chromium: Sources, Speciation, Toxicity, and Chemistry

John B. Vincent

Abstract In the environment, chromium is found almost exclusively in the trivalent, 3^+, and hexavalent, 6^+, oxidation states. Chromium^{3+} is stable in the presence of water and air; however, chromium^{6+} as chromate is kinetically stable, but thermodynamically unstable. Chromium has a range of effects in living organisms. In plants, as will be the focus of the following chapters of this work, chromium is a toxin as either Cr^{3+} or Cr^{6+}. In animals, the status has been highly debated since the element as the trivalent ion was first proposed as an essential element about seventy years ago; however, the element is no longer considered to be essential but may have beneficial pharmacological effects. The modes of Cr^{3+} transport in animals has recently been elucidated and may provide insight on how the metal ion can be transported in plants. Chromate is toxic and carcinogenic. A focus of the chapter will be on techniques to characterize chromium in mammals in terms of structure and potential function to suggest how the methods cold be extended to plants.

Keywords Chromium · Insulin sensitivity · Trace element

3.1 Introduction

Chromium is known in eleven oxidation states from 4 to 6^+, although essentially only two are found in the environment, Cr^{3+} and Cr^{6+}. Chromium is the 7th most abundant element on the earth but only the 21st most abundant element in the earth's crust. In the presence of air and moisture, Cr^{3+} is the most stable oxidation state. Cr^{3+} in the environment occurs as coordination complexes with a wide variety of primarily oxygen- and nitrogen-based ligands. As Cr^{6+}, chromium is normally found in the environment as chromate or dichromate. Chromate and dichromate are thermodynamically unstable but kinetically stable. In other words, in the absence of an appropriate reductant, these forms of Cr^{6+} can sit in the environment, despite their oxidizing potential. However, they are readily reduced releasing their oxidizing power when an

J. B. Vincent (✉)
Department of Chemistry and Biochemistry, The University of Alabama, Tuscaloosa, AL, USA
e-mail: jvincent@ua.edu

© The Author(s), under exclusive license to Springer Nature Switzerland AG 2023
N. Kumar et al. (eds.), *Chromium in Plants and Environment*, Environmental Science and Engineering, https://doi.org/10.1007/978-3-031-44029-8_3

appropriate reductant is encountered. The natural occurrence of chromium is heterogeneous in rock, and its occurrence in the environment is made more heterogeneous by anthropogenic activity. The major applications of chromium are in stainless steel (>11% Cr), where introduction of chromium increases corrosion resistance, chrome plating, and pigments. Chromium is also important in the tanning industry, where Cr^{3+} crosslinks collagen fibers in tanned leather stabilizing the leather, and in the preservation of wood, where added chromate resists infiltration by insects and fungi. These applications have led and continue to lead to the introduction of significant amounts of toxic Cr^{6+} into the environment.

The toxicology and biochemistry of both Cr^{3+} and Cr^{6+} have extremely controversial histories. The public is perhaps more familiar with toxicology, chemistry, and biochemistry of Cr^{6+} because of its carcinogenicity and toxicity, which was the focus of the Hollywood movie *Erin Brockovich* in 2000, starring Julia Roberts. The d^0 Cr^{6+} ion is most commonly encountered as the intensely colored chromate (CrO_4^{2-}) and dichromate ($Cr_2O_7^{2-}$) anions, which are interconvertible as a function of pH in water. Chromate occurs at basic pH's where the dianion has a distinctive yellow color, which led to its use as the insoluble salt $PbCrO_4$ as the yellow pigment in the paint used for highway lines. Below pH 6, chromate and orange dichromate are in equilibrium; acidic solutions of dichromate are potent oxidants. The Cr^{6+} center in both chromate and dichromate possesses tetrahedral geometry (Fig. 3.1). Their intense colors arise from charge transfer bands. Mixed ligands complexes of Cr^{6+} with oxides and halides or oxides and amines are also well known, as are peroxo complexes. As the Cr^{6+} center is diamagnetic, it does not give rise to electron spin resonance (ESR) spectra, while nuclear magnetic resonance (NMR) studies of complexes with oxo, peroxo, and halo ligands are of limited utility (Fig. 3.2).

Coordination complexes of Cr^{3+}, the chromic ion, are almost always octahedral (Fig. 3.1). As a result, the chromic center has a d^3 electron configuration with three unpaired electrons giving an electron spin of 3/2. This is responsible for the kinetic inertness of complexes of Cr^{3+}, where at room temperature the ligand exchange half-times are often in the range of hours. While the hexaaquo ion of Cr^{3+}. $[Cr(H_2O)_6]^{3+}$, is purple in color in aqueous solution, aqueous solutions of the cation are acidic. At neutral or basic pH's, the ion is susceptible to oligomerization generating hydroxo-or-oxo bridged species. The most commonly used source of Cr^{3+} in the laboratory is green crystalline $CrCl_3 \cdot 6H_2O$, which is actually *trans*-$[Cr(H_2O)_4Cl_2]Cl \cdot 2H_2O$ (Fig. 3.1). Dissolving this salt in water initially generates the green cation $[Cr(H_2O)_4Cl_2]^+$, which hydrolyzers with the replacement of one (and with time both chlorides depending on concentration) with water. The chromic ion has a large charge to size ratio, so that it is a hard Lewis acid and prefers oxygen-based ligands and nitrogen-based ligands.

The electronic configuration of Cr^{3+} makes obtaining information about its environment at biologically relevant concentrations difficult using spectroscopic and magnetic techniques. No charge transfer bands normally occur in the visible region of the electronic spectrum, while two spin-forbidden d-d transitions in the visible region have extinction coefficients well under 100 M^{-1} cm^{-1}. A third d-d transition occurs in the ultraviolet region, but it often obscured by ligand-based transitions. Chromic

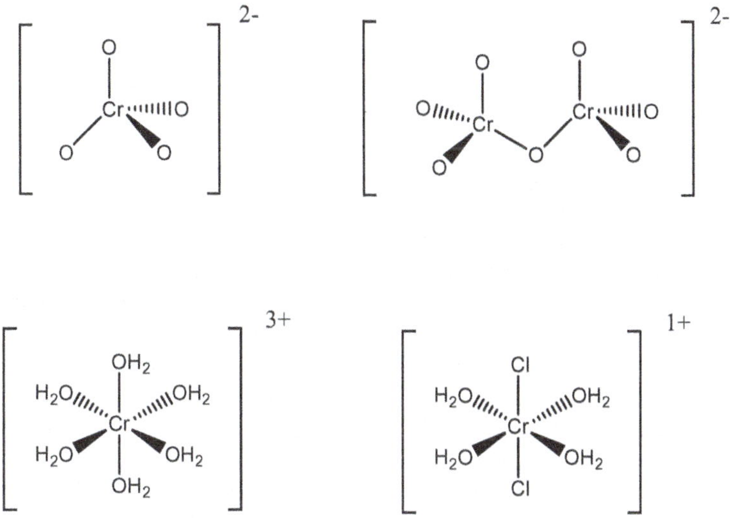

Fig. 3.1 Structures of common chromium compounds: (Top left) chromate anion, (Top right) dichromate anion, (Bottom left) hexaaquochromium(III) cation, and (Bottom right), *trans*-tetraaquadichlorochromium(III) cation

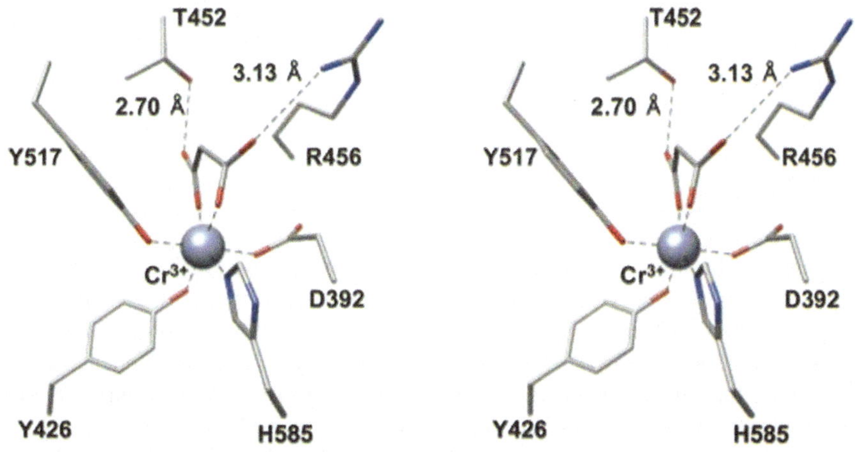

Fig. 3.2 Stereo view of the Cr^{3+}-binding site at the C-lobe of Cr^{3+}-transferrin (PDB entry 6UJ6). Adopted from Petersen et al. (2020). The ligand not provided by the protein is malonate from the crystalloization buffer

complexes, particularly with biological ligands, are generally stable against oxidation or reduction under biological conditions. The $S = 3/2$ center results in greatly broadened (if even observable) and shifted resonances in NMR spectra. As a general rule, species that give rise to sharp features in NMR spectra do not give rise to EPR signals or to very broad EPR signal, and vice versa. Chromic complexes can give rise to sharp features in ESR spectra. Unfortunately, for biomolecules containing bound Cr^{3+}, the features in ESR spectra have been found to be rather broad, limiting interpretation. Thus, Cr^{3+} is one of the exceptions to the general rule where features in both NMR and EPR are broad, if observable, and difficult to interpret. Still, ESR is an underutilized technique in probing Cr^{3+}-containing biomolecules and has recently shown significant value in following the transport of Cr^{3+} in vitro (Vincent 2013a). Applying spectroscopic and magnetic techniques to characterize the environment of Cr^{3+} in a biologically relevant system will be a focus of this chapter.

Chromium has four stable isotopes, with chromium-52 being the most abundant comprising about 84%. Chromium-50 and chromium-53 have been used as stable isotope tracers. Synthetic chromium-51 is stylized experimentally as a radiotracer; the isotope decays via electron capture with associated loss of gamma rays, which can readily be detected.

The objective of the chapter is to summarize the fate of chromium in organisms other than plants for comparison with those of plants in the subsequent chapters.

A note of nomenclature is warranted. The free trivalent chromic ion is symbolized by Cr^{3+}. This is also used to refer in a general sense to the chromic ion. When this ion is part of a coordination complex, then it is properly referred to as Cr(III). In aqueous solution, the free chromic ion does not occur as it is bound by coordinate covalent bonds to water molecules. To avoid going back and forth between Cr(III) and Cr^{3+} in the text depending on context, which is often confusing to readers (particularly those who are not inorganic chemists), Cr^{3+} will be used throughout this chapter to refer to the Cr^{3+} ion regardless of context.

3.2 Yeast

The fields of the biochemistry and nutrition of chromium started with studies utilizing yeast, although the focus was orginally not on chromium nor on the yeasts themselves, but rather on the effects of various yeasts on human nutritional health. The effects of chromium on yeast are still poorly understood, while human and rodent studies examining the effects of the chromium content of yeast are a roadmap for how nutritional and biochemical studies should not be performed.

In the late 1950s and early 1960s, a Torula yeast-based diet fed to rats was found to led to a liver disorder (resulting from the low selenium content of the diet) and an apparent inability to efficiently restore blood glucose levels after an intravenous glucose challenge. The existence of a dietary requirement that was absent from the diet leading to the glucose intolerance, Glucose Tolerance Factor (GTF), was postulated. The addition of inorganic compounds of over three dozen different elements to the diet (at 200–500 μg/kg body mass) failed to restore glucose tolerance while several Cr^{3+} complexes (200 μg Cr/kg body mass) apparently did, leading to the postulate that Cr^{3+} was GTF (Schwarz and Mertz 1959).

Unfortuantely, the studies were flawed, which in hindsight limits ntreperation and value of these studies. For example, the chromium content of the diet was not examined. The rats were kept in cages with metal components (potentially including stainless steel with its large chromium content). As a result, the chromium intake of the rats cannot be ascertained. In fact, at least one study has found that Torula yeast diets are not low in chromium compared to other common rat diets (Shepherd et al. 2002) Additional questions about data handling and the significance of the effect observed have been raised (Woolliscroft and Barbosa 1977). More recently studies with rats generally fail to observe effects on glucose levels in glucose tolerance tests (Vincent 2013a). Also, the amount of chromium used to supplement the diet was about 10^3 times the typical content of commercial rat chows, far above nutritionally relevant levels. This raises the possibility that the effect, if real, was pharmacological in origin.

Brewer's yeast, which accumulates Cr^{3+} for an unknown reason, was also shown to apparently restore glucose tolerance in rats when added to the Torula yeast-based diet. Subsequently, efforts were made to attempt to isolate the Cr^{3+}-containing species in Brewer's yeast with the assupmption that this was the species in Brewer's yeast capable of reversing the purported effects of the Torula yeast diet. The product of this endeavor unfortunately was also termed GTF. This led to confusion as to whether the term GTF refers to the inorganic Cr^{3+} ion or to a complex or Cr^{3+} with a biomolecule(s) or to such a complex specifically isolated from yeast. The latter two became the common usage of the term GTF. GTF was reported to be a complex of Cr^{3+} with nicotinate and the amino acids aspartate, glutamate, and cysteine (Toepfer et al. 1977). Later, a three dimensional structure has even proposed for GTF based largely on this data. This resulted in chromium as the trivalent ion being considered to be an essential trace element as part of the biological molecule GTF for several decades. Unfortunately, the research probing the Brewer's yeast GTF complex was extremely flawed. The isolation procedure required refluxing an extract of Brewer's yeast for 18 h in 5 M HCl, conditions that would have destroyed any protein or peptides or many other biomolecules that could have potentially bound Cr^{3+}. The quantities of chromium, nicotinate, and the amino acids in the material were not reported so that the ratios of the components to one other, if they were stoichiometric, cannot be determined. Finally, the characterization was performed on material that was impure, so that none of the components can definitely be said to be associated the component of the mixture that was active in bioassays (for the potentiation of insulin-dependent glucose metabolism by isolated rat adipocytes); in fact the authors showed the spot isolated from paper chromatography that was active and contained chromium only possessed less than 10% of the chromium from the mixture used to establish the apparent presence of nicotinate and specific amino acids.

Subsequent investigations have shown that the bioactive component of Brewer's yeast in these assays could be separated from Cr^{3+}, so that chromium is not even a component of the species in the yeast responsible for the increased glucose metabolism (reviewed in Vincent 2013a). Amazingly, a recent attempt to isolate GTF has been reported but fell into the same pitfall of reporting the composition of an impure material (Liu et al. 2013; Vincent 2013b). The use of the term GTF should be avoided outside of its historical context.

3.3 Mammals

3.3.1 Essential Element

Chromium is ubiquitous in foods and biological tissues and fluid at very low concentrations, generally parts per billion or lower. As a result, chromium concentrations cannot be determined by flame-based atomic absorption spectrometry (AAS) methods, which lack appropriate sensitivity. Thus, chromium concentrations are normally determined using graphite furnace atomic absorption spectrometry (GFAAS) or more recently also inductively coupled plasma mass spectrometry (ICP-MS). Early GFAAS studies measuring chromium concentrations were plagued with an artifact problem that resulted in concentrations being overestimated by orders of magnitude (Veillon and Patterson 1999). Contamination from contact with stainless steel is also a problem in chromium analyses. The more processed a food the greater its chromium content becomes. Thus, it is likely that humans in more developed nations intake more chromium than their ancestors.

In 1980, the Institute of Medicine of the National Research Council of the Food and Nutrition Board of the National Academies of Science (US) established its first estimated safe and adequate daily dietary intake (ESADDI) for chromium at 50–200 μg for an adult. This was done as the panel found insufficient data to set a recommended daily allowance (RDA). The next time this was reviewed in 1989 the ESADDI was left unchanged. However, in 2001, the recommendation was changed, based largely on the improved techniques for measuring chromium concentrations that indicated chromium intakes were considerably lower than previously believed (Institute of Medicine 2001). Again, insufficient data was found for establishing a RDA, so that an estimated intake (EI) was set at 35 μg per day for adult males and 25 μg per day for adult females. This essentially means that at least 98% of American adults should be chromium sufficient at a dietary intake of ~30 μg chromium daily. Hence, almost every American is basically not chromium deficient.

In Europe, chromium is not considered an essential element for animals nor humans (EFSA 2009, 2014). The European Food Safety Authority has found "no evidence of beneficial effects associated with chromium intake in healthy subjects" and the "setting of an adequate intake for chromium is also not appropriate" (EFSA 2014). The American EI set in 2001 is outdated and needs to be reevaluated, which eventually will probably result in chromium no longer being considered an essential element in the U.S.

This leads to the need to examine the disconnect between chromium not being considered an essential element and reported beneficial effects on carbohydrate and lipid metabolism reported in rodent studies. This has recently been shown to arise from beneficial effects of chromium resulting in increased sensitivity and lower serum cholesterol and triglycerides levels in rodents requiring supra-nutritional, pharmacologically relevant doses of chromium (Di Bona et al. 2011). Starting from a purified diet with the lowest chromium content ever used in a rodent diet and using diets ranging up to almost 100-fold higher chromium contents resulting in a chromium-intake dependent increase in insulin sensitivity in healthy rats. The increase in insulin sensitivity required chromium intakes orders of magnitude above normal dietary levels, making the effect pharmacologically relevant rather than nutritionally relevant. No human clinical trial has used doses approaching those in rodent studies, making observation of beneficial effects from chromium supplementation studies unlikely.

Despite this, claims have been made that supplemental chromium can lead to loss of body mass, loss of fat mass, increases in lean muscle mass, and beneficial effects for subjects with type 2 diabetes and related conditions. In 1997, the Federal Trade Commission (US) disallowed companies from making representations that chromium has beneficial effects on body mass, fat or muscle mass, diabetes, appetite or cravings for sugar, or serum cholesterol or glucose levels (Federal Trade Commission 1997).

Similarly, in response to a request from a nutraceutical company, the Food and Drug Administration (US) examined eight proposed qualified health claims for chromium supplementation, specifically with popular supplement chromium picolinate (Food and Drug Administration 2005). None of the claims were allowed; what was allowed was the following qualified health claim: "One small study suggests that chromium picolinate may reduce the risk of type 2 diabetes. FDA concludes that the existence of such a relationship between chromium picolinate and either insulin resistance or type 2 diabetes is highly uncertain". The FDA did, however, establish that the use of chromium picolinate supplements appeared to be safe up to 1 mg of chromium daily.

The position of the American Diabetes Association is "there is insufficient evidence to support the routine use of herbal supplements and micronutrients, such as…chromium, to improve glycemia in people with diabetes (American Diabetes Association 2023). This position has been consistent for a number of years.

The last decade has seen a flurry of meta-analyses examining the effects of chromium supplementation on body mass or serum glucose, insulin, cholesterol, or triglycerides. The interpretation of these meta-analyses is difficult as the quality of the clinical trials utilized is generally rather poor. Most meta-analyses find no effect or statistically significant but clinically insignificant effects (Costello et al. 2019). Meta-analysis restricting utilized clinical trials to higher quality trials generally fail to observe effects. Thus, to determine if chromium supplementation has any beneficial effects in humans, clinical trials using approximately 10 mg of chromium daily (ten times the dose used in any trials to date) are required.

3.3.2 Transport

As Cr^{3+} is a d^3 transition metal ion, the metal ion is considered substitutionally inert, meaning that chemistry at the metal center that involves loss of coordinated ligands is very slow. This has been a concern about how such a substitutionally inert ion that is redox stable could readily be transported about the body efficiently if the element were essential for animals. However, recent studied have revealed how the ion is readily transported from the blood to the urine, preventing toxic accumulation of the element. The "tricks" utilized to transport Cr^{3+} in animals may have lessons to teach on how the ion might be transported in plants.

Cr^{3+} is poorly absorbed (approximately 1%) from the gastrointesinal tract by passive absorption (reviewed in Vincent 2013a). After passive absorption from the gastrointestinal tract into the blood steam, trivalent chromium initially follows the transport of iron. Transport of iron as the ferric ion from the bloodstream to the tissues of mammals is performed by the protein transferrin. Transferrin is an appropximately 80 kDa glycoprotein with two highly homologous lobes, the N-terminal and C-terminal lobe. Both lobes possesses a metal ion binding site comprised of two tyrosine residues, a histidine residue, and an aspartic acid residue. In addition, a synergistic bicarbonate anion binds concomitantly with the metal ion in the metal binding site. Transferrin is selective for Fe^{3+} in a biological environment as its two metal sites are adapted to bind ions with large charge-to-size ratios; thus, that transferrin appears to also be the transport agent for Cr^{3+}, with a similar charge to size ratio to ferric iron, in the bloodstream is not perhaps surprising (Vincent and Love 2012). A recent X-ray crystal structure has confirmed that Cr^{3+} binds to the same protein-provided ligands of transferrin as Fe^{3+} (Peterson et al. 2020), although under crystrallization conditions with synergistic bicarbonte is replaced by malonate.

The binding of Cr^{3+} to transeferrin has been followed spectroscopically using two techniques: ultraviolet-visible spectroscopy and EPR spectroscopy (Figs. 3.3, 3.4 and 3.5), allowing rate constants and formation constants for binding of Cr^{3+} to be determined (Deng et al. 2015; Edwards et al. 2020b). In this case, electronic spectroscopy at the ultraviolet wavelength of 245 nm can be used to follow the increase in the extinct coefficient of the π–π^* transition of the π system of the tyrosine ligands provided by the metal-binding sites of transferrin upon binding of Cr^{3+}. This increase in extinct coefficient arises from the increased rigidity and delocalization of the π systems upon binding Cr^{3+}.

Currently, such assays can be performed with cultured cells, e.g., 3T3-L1 cells or C2C12 myoblasts, to eliminate the requirement for living vertebrate animals. For example, the peptide EEEEGDD, corresponding to the first seven amino acids of LMWCr, when added concurrently with Cr^{3+} to mouse C2C12 myoblasts results in an increase in the ability of insulin to stimulate glucose uptake by the cells (Arakawa et al. 2016); the addition of Cr^{3+} or the peptide separately had no effect. When the cells are rendered insulin resistant by treating them for 24 h with 25 mM glucose, the additon of the combination of Cr^{3+} and the peptide, but neither by itself, was albe to restore most of the ability of insulin to activate glucose uptake. Intravenous

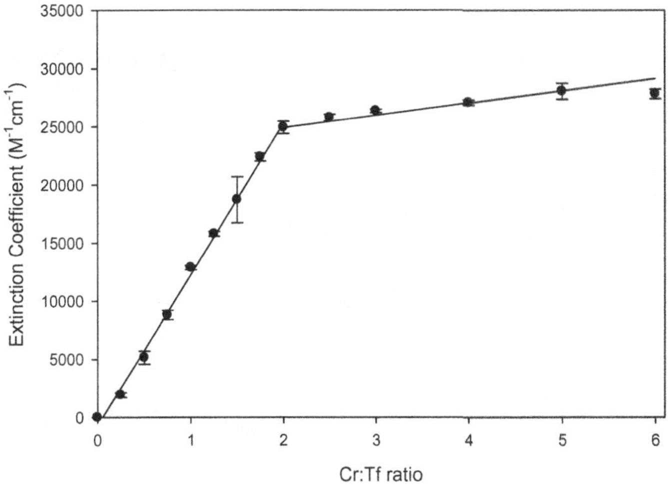

Fig. 3.3 Titration of human serum apo-transferrin with Cr^{3+} in 25 mM HCO_3^-, 0.1 M HEPES, pH 7.4. The time between additions of Cr^{3+} was 3 h. Adopted from Deng et al. (2015)

adminsitraton of Cr^{3+} and the peptide to mice resulted in lowering of the area under the curve (AUC) for glucose in glucose tolerance tests, consistent with the results of the cell studies.

In buffered water with its ambient bicarbonate concentration, Cr^{3+} binds to transferrin slowly, requiring approaching two weeks to be fully loaded with Cr. In contrast ultraviolet spectroscopy studies reveal as a result of the binding being first order in synergistic bicarbonate that Cr^{3+} is fully loaded into transferrin in less than 10 min at 25 mM bicarbonate, the concentration in blood (Fig. 3.4) (Deng et al. 2015). Thus, the binding of Cr^{3+} to transferrin in the bloodstream can take place in a physiologically relevant timeframe.

When Cr^{3+} is allowed to bind for prolonged periods of time, Cr^{3+} in each metal-binding site of the generated $Cr^{3+}{}_2$-transferrin gives rises rise to distinct EPR signals for the human serum protein (Aisen et al. 1969; Edwards et al. 2020b). Cr^{3+} in the N-terminal lobe metal-binding site of human serum transferrin gives rise to EPR signals at $g \sim 5.1$ and 5.6, while Cr^{3+} in the C-terminal lobe metal-binding site gives rise to signal at $g \sim 5.4$ and 2; thus, the amount of Cr^{3+} bound in each lobe can be followed using a window of g between 5 and 6. Combining ultraviolet and EPR studies has revealed that the binding of Cr^{3+} to human serum transferrin is surprisingly coimplicated (Fig. 3.5). Initially, Cr^{3+} binds rapidly to transferrin resulted in a large increases in the molar absorbtivity of the absorbance at 280 nm; this is accompanied by the appearance of an EPR signal centered at $g \sim 2$ from the newly bound trivelent chromic ions, but not the signals between $g = 5$ and 6. With time, the signal at g ~ 2 is replaced by a signal at g ~ 5.2 resulting from chromic ions in *both* metal binding sites. This signal in turn looses intensity with time as new signals at g ~ 5.1 and 5.6 appear. Thus, as a function of time for human serum transferrin, $Cr^{3+}{}_2$-transferrin

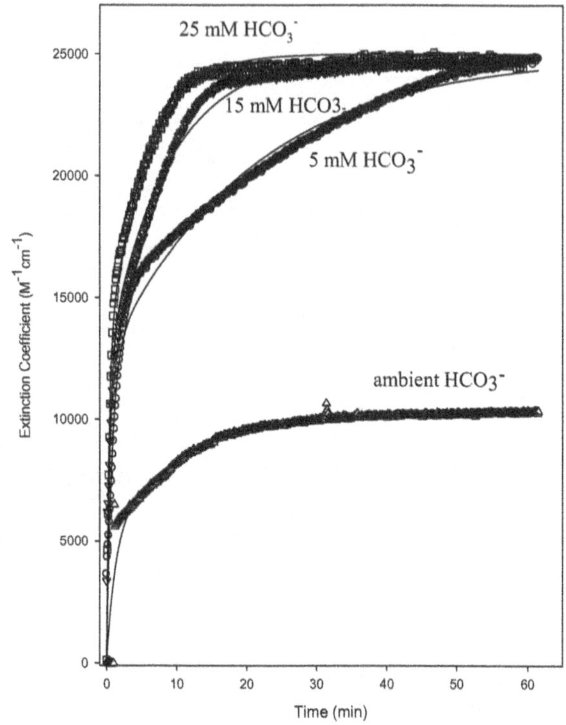

Fig. 3.4 Binding of Cr^{3+} to human serum apo-transferrin as a function of time in 0.1 M HEPES, pH 7.4 with ambient HCO_3^- (squares), 5 mM HCO_3^- (large circles), 15 mM HCO_3^- (small circles), and 25 mM HCO_3^- (triangles). Lines represent fits of the time dependence of the Cr binding to apo-transferrin to the equation $\Delta\varepsilon = a(1 - e^{-bx}) + c(1 - e^{-dx})$ where a and c are the extinction coefficients for each metal binding and b and d are the rate constants. Adopted from Deng et al. (2015)

exists in three different confirmations, giving rise to different EPR signals (and also different extinction coefficients at 280 nm). The middle confirmation is the most biologically relevent one as it is the major conformation from 15 min to few hours after Cr^{3+} addition (Edwards et al. 2020b). In contrast, the addition of chromic ions to bovine transferrin results with time in the formation of the conformation with chromic ions in both metal-binding sites giving rise to the EPR feature at $g \sim 5.4$. The third conformation formed as a function of time for the human protein does not form for the bovine protein.

The exact changes in three-dimensional structure corresponding to the conformation changes are not known. One conformational change can be eliminated. Apotransferrin exists in an open confirmation, where protein-provided ligands in the metal-binding sites are farther apart than in the closed confirmation when metals with large charge to size ratios are bound in the metal-binding sites. Metal ions (or metal-bicarbonate complexes) are believed to first bind to the tyrosine residues, which results in a conformation change, as the protein holds along a hinge in each lobe containing the metal-binding site bringing the aspartate and histidine ligands into close proximity to the metal ion, completing coordination about the metal. This also leavces the metal less exposed than in the open conformation. Three-pulse ESEEM (electron spin echo envelope modulation) spectra of human serum $Cr^{3+}{}_2$-transferrin in each of the three conformations reveals are extremely similar (Edwards et al.

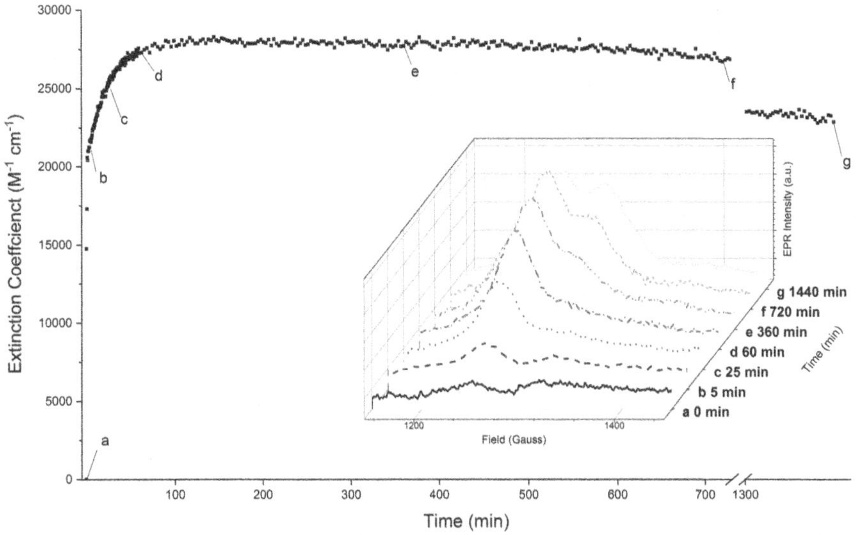

Fig. 3.5 Change in extinction coefficient at 245 nm as a function of time corresponding to the formation of conformations of $Cr^{3+}{}_2$-transferrin following the addition of Cr^{3+} to apo-transferrin in 100 mM HEPES with 25 mM HCO_3^-, pH 7.4, at 37 °C. Insets: EPR spectra of aliquots were taken at prescribed intervals concurrent with the UV measurements. Adopted from Edwards et al. (2020b)

2020b) (Fig. 3.6). All three spectra display coupling of the chromic centers to a nitrogen atom(s), consistent with the chromic ions being bound to histidine; thus, all three conformation ciorrespond to different closed conformations.

When $^{51}CrCl_3$ is administered by gavage to rats, 80% immunoprecipitates with transferrin. This result led to suggestions that the transferrin-bound Cr^{3+} represents a stage in chromium transport or represents a deadend, where transferrin ties up Cr^{3+} to detoxify the ion. For the latter proposal, which would be consistent with the low ligand replacement at trivalent chromium centers, how the chromium-containing transferrin was subsequently to be degraded leading to chromium release and ultimately excretion from the body was not proposed. Intravenous administration of radiolabeled ^{51}Cr–transferrin to rats results in ready incorporation of ^{51}Cr into tissues (Clodfelder and Vincent 2005). Most of the injected ^{51}Cr is transported to the tissues within 30 min so that tissue levels of ^{51}Cr are maximal 30 min after injection. The loss of ^{51}Cr in the tissues with time is mirrored by the appearance of ^{51}Cr in the urine. Approximately 50% of the ^{51}Cr appears in the urine within 6 h of injection.

When ^{51}Cr-containing species found in the blood plasma and urine as a function of time after injection of ^{51}Cr–transferrin have been examined by size-exclusion chromatography, three primary phases are observed (Clodfelder and Vincent 2005). First, the injected ^{51}Cr–transferrin rapidly disappears from the bloodstream. Second, with time, a low-molecular-weight species chromium-contaiing species appears in the bloodstream. Finally, this appearance is rapidly followed by the appearance of a

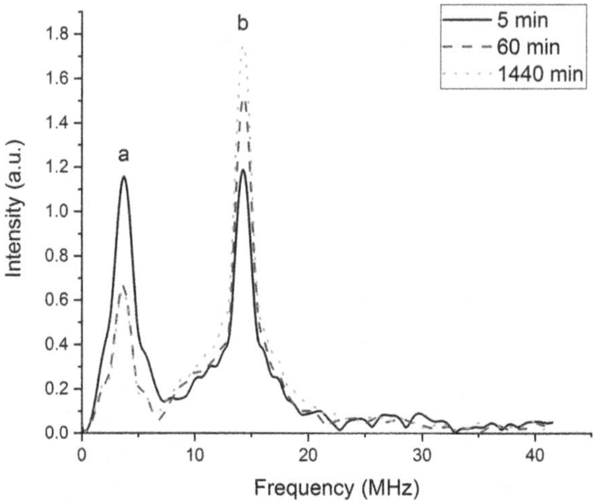

Fig. 3.6 Three-pulse ESEEM spectra of apo-transferrin incubated with 2.0 equivalents of Cr^{3+} at 5, 60, and 1440 min following the addition of Cr^{3+}. **a** Feature at 4 MHz corresponds to coupling between ^{14}N in the vicinity of the Cr^{3+}. **b** Feature at ~15 MHz corresponds to coupling between Cr^{3+} and nearby ^{1}H's. The intensities of these spectral features depend on the tau used during measurement. Adopted from Edwards et al. (2020b)

low-molecular-weight species in the urine. This establishes a clear pathway for the transport of chromium starting from movement of transferrin-bound chromium from the bloodstream into the tissues, followed by subsequent release and processing in the tissues to form a low-molecular-weight chromium-binding species, which is moved into the bloodstream. In the bloodstream, chromium as the low-molecular-weight species is rapidly cleared and ultimately excreted in the urine.

However, at physiological conditions and Cr^{3+} concentrations in the blood stream (~1 ppb), Cr^{3+}, despite its large binding constants for the formation of Cr^{3+}-loaded transferrin, does not load apparently appreciably to apotransferrin (metal-free transferrin) (Levina et al. 2022). However, most transferrin in the bloodstream is not in the form of apo-transferrin, but rather is half loaded with Fe^{3+} in either the C-terminal or N-terminal lobe. These half-loaded forms of Fe^{3+}-containg transferrin bind Cr^{3+} readily under physiologically relevant conditions (Levina et al. 2022). Cr^{3+} loss from mixed Cr^{3+}, Fe^{3+}-transferrins occurs at similar rates to that from Cr^{3+}-loaded transferrin (Vincent et al. 2022). Thus, conditions in the blood stream (e.g., the presence of monferric-transferrin and high bicarbonate concentrations) prime transferrin to be able to bind Cr^{3+}, even at its low concentrations, ultimately for elimination from the body.

Cr^{3+}-containing transferrin is taken up by tissue via endocytosis, similar to the uptake of Fe^{3+}-loaded transferrin, as binding of $Cr^{3+}{}_2$-transferrin to transferrin receptor is inhibited by Fe^{3+}-loaded transferrin (Kornfield 1969); transferrin receptor binds two equivalents of certain metal-loaded transferrin, including transferrin containing Fe^{3+} and/or Cr^{3+} (Bonvin et al. 2017), although Cr^{3+}-containing transeferrins bind weaker to transferrin receptor than fully Fe^{3+}-loaded transferrin. Acidification of the endosome results in loss of metal ions from the transferrin/transferrin receptor complex, although loss of Cr^{3+} is sightly slower than Fe^{3+} (Edwards et al. 2021). This loss can readily be followed by ultraviolet spectroscopy of EPR spectroscopy (Fig. 3.7). The binding of transferrin to its receptor is crucial for the rapid loss of Cr^{3+} upon acidification; Cr^{3+} is lost significantly slower from transferrin, so slow that it could not be released appreciably during the endocytic cycle (Edwards et al. 2020a,b). Thus, the conformation of Cr^{3+}-containing transferrin is crucial for its release in the endosome. Also crucial for the removal of Cr^{3+} from the endosome before it fues with the cell membrane releasing transferrin back to the bloodstream is the oligopeptide low-molecular-weight chromium-binding substance (LMWCr), which binds the Cr^{3+} released from transferrin (Fig. 3.8).

LMWCr was first reported in 1981 by Yamamoto and coworkers (1981). The oligopeptide is about 10 or 11 amino acids in length and comprosed of glutamate, aspartate, glycine, and cysteine. From mammalian, avian, and reptilian liver and human urine, teh oligopeptide N-terminus starts with the sequence EEEEGDD (Chen et al. 2011). The oligopeptide tights binds four equivalents of Cr^{3+} ($K_f \sim 10^{21}$) and cooperatively (Sun et al. 2000) in an anion bridged assembly (Jacquamet et al. 2003).

The anionic LMWCr oligopeptide has recently been shown to be able to dock to the the acidified transferrin/transferrin receptor complex where it fits into a cationic channel beside each metal binding site in transferrin (Edwards et al. 2021). The oligopeptide with its affinity for chromic ions has little propensity to bind Fe^{3+} (Kircheva et al. 2022), making it able to tie up Cr^{3+} released from transferrin in the transferrin/transferrin receptor complex.

After LMWCr is loaded with Cr^{3+} about half exits the endosome and enters the cell, while the remainder fails to exit the cell before the endosome fuses with the cell membrance, releasing its contents to the bloodtsream. Currently, LMWCr is believed to serve as ionophore, exiting the cell while carrying its Cr^{3+} (Edwards et al. 2021).

These EPR and electronic spectroscopy and related studies allow for a detailed model of chromium transport from the bloodtsream to the urine (for ultimate elimination from the body) to be established (Fig. 3.9). After Cr^{3+} enters the bloodstream via passive diffusion from the gastrointestinal tract, the Cr^{3+} binds relatively quickly to monoferric transferrin; the resulting metal-saturated transferrin is incorporated into cells via endosytosis. Acidification of the endosome releases Cr^{3+} from the transferrin/transferrin receptor complex where it is bound by apoLMWCr. About half the Cr^{3+}-loaded LMWCr does migrate out of the endosome rapidly enough to prevent it being carried out of the cell into the bloodstream upon fusion of the endocsoe with the cell membrane. The other half of the Cr^{3+}-loaded LMWCr, serving as an ionophore for Cr^{3+}, crosses the endosomal membrane to enter the cell, where it migrates from the cell into the bloodstream. From the bloodstream, LMWCr is readily eliminated from the body via the urine. LMWCr was low tubular reabsorption rates in the kidneys (Wada et al. 1983). The rate constants determined for the binding and release of Cr^{3+} to transferrin and from the transferrin/transferrin receptor complex can be used in

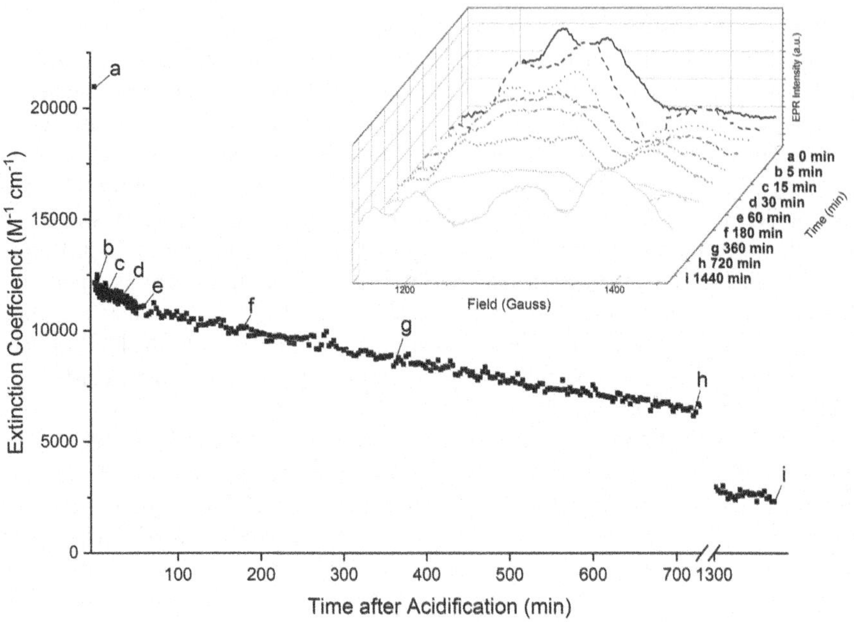

Fig. 3.7 Decrease of the extinction coefficient at 245 nm of Cr_2-transferrin blanked against apo-transferrin following lowering of pH to pH 5.5. Acidification initiated after 24 h of Cr^{3+} incubation with transferrin (2 Cr^{3+}:1 transferrin) at pH 7.4, 37 °C (Fig. 3.5). Insets: EPR spectra of aliquots were taken at prescribed intervals concurrent with the UV measurements. The features in the EPR spectrum at 1440 min after acidification correspond to Cr^{3+} weakly adhering to transferrin. Adopted from Edwards et al. (2020b)

association with this model to reproduce the behavior of Cr^{3+} from $Cr^{3+}{}_2$-transferrin injected via the tail vein into rats (Edwards et al. 2021).

Chromate because of its similarity to phosphate at neutral pH (i.e., HPO_4^{2-}) is transported into cells and subcellular compartments by phosphate and sulfate transporters and, thus, can readily enter cells.

3.3.3 Toxicity

Complexes of Cr^{6+} are well known to be potent carcinogens and mutagens when inhaled, and a current debate exists as to whether these complexes in drinking water pose similar risks. The mechanism(s) by which Cr^{6+} complexes give rise to these problems have not been clearly established but include oxidation by the complexes or more likely the subsequently generated Cr^{4+} and Cr^{5+} intermediates, reactions of reactive oxygen species (ROS) generated as by-products of these reactions, reactions of organic radicals generated in these processes, and the binding of the ultimately generated Cr^{3+} to biomolecules (Levina et al. 2003).

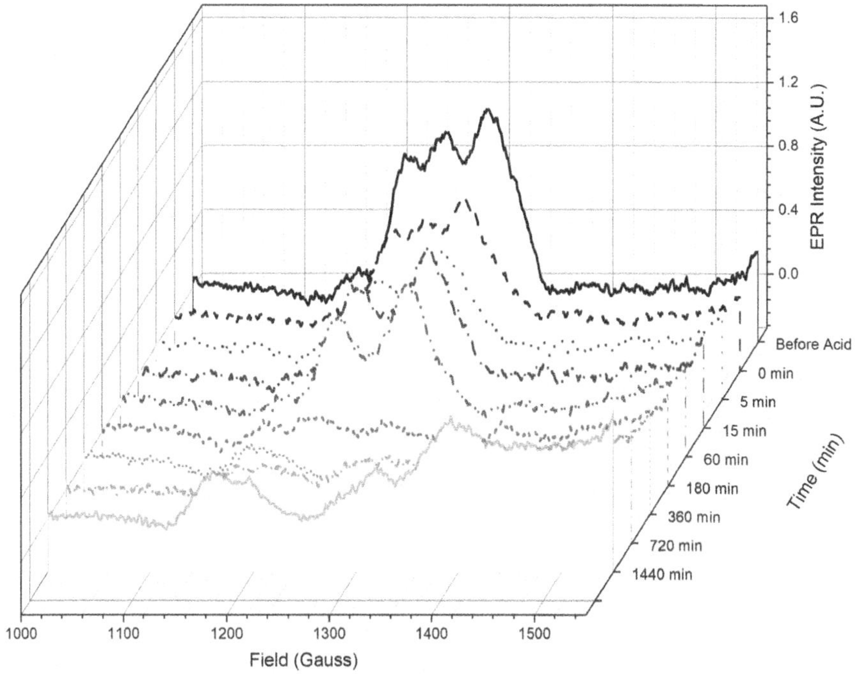

Fig. 3.8 EPR spectra of $Cr^{3+}{}_2$-transferrin in 100 mM HEPES with 25 mM HCO_3^- at 37 °C at various times after the addition of acid to change the pH to 5.5 in the presence of 0.48 mM apoLMWCr. Features at 1210 and 1285 G (g = 5.1 and g = 5.6) correspond to Cr^{3+} in the N-lobe metal-binding site. The feature at 1245 G (g = 5.4) corresponds to Cr^{3+} in the C-lobe metal-binding site. The broad rising feature starting at ~1400 G corresponds to LMWCr. Adopted from Edwards et al. (2021)

The binding of Cr^{3+} to DNA has been proposed to arise to the formation of binary Cr^{3+}-DNA complexes, ternary small molecule-Cr^{3+}-DNA complexes, and Cr^{3+}-based intrastrand crosslinks; the latter two are proposed to be the potentially carcinogenic and mutagenic forms (Zhitkovich 2005). Using a combination of paramagnetic NMR, EPR and other spectroscopic methods, the structure of binary adducts has been determined (Brown et al. 2020). The chromic center in the species $[Cr(H_2O)_5]^{3+}$ binds to N-7 of guanines, while the hydrogens of the coordinated water molecules are hydrogen-bonded to the surrounding DNA. The broadening of the ^1H NMR signals from the hydrogens of the DNA in close proximity to the paramagnetic chromic center in both one-dimensional and two-dimensional NMR experiments (Fig. 3.10) allows for the general position of the Cr^{3+} to be elucidated. The binding site of the Cr^{3+} can be further elucidated using by measuring the T_1 relaxation rates of the ^1H NMR signals. The T_1 relaxation rates change as a function of $1/r^6$ where r is the distance between the chromic center and the hydrogen atom (Rehmann and Barton 1990).

No upper limit (UL) has been established for Cr^{3+} (Institute of Medicine 2001) as excess intake of supplemental chromium could not be associated convincingly with

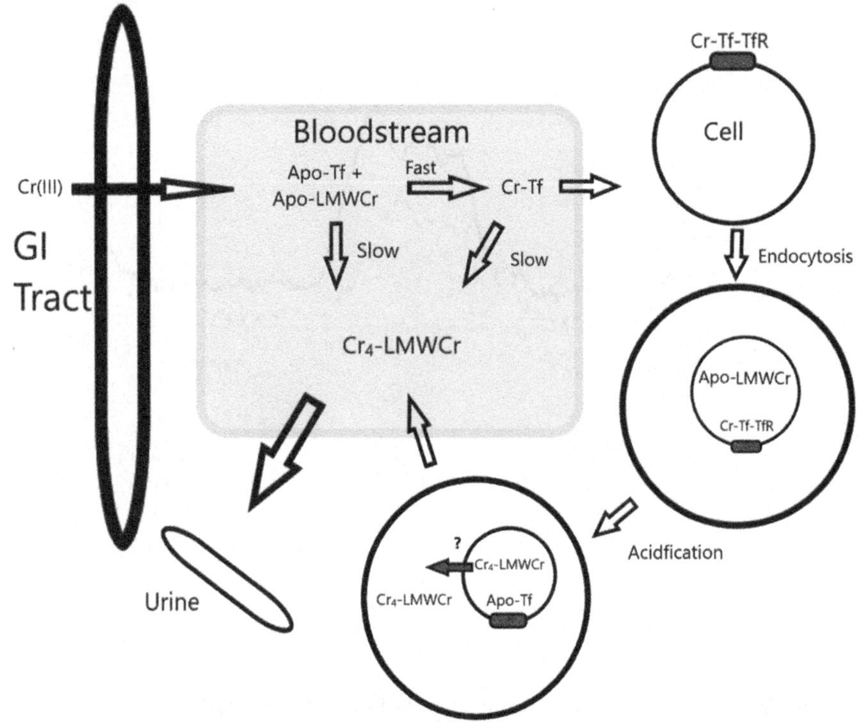

Fig. 3.9 Movement of Cr^{3+} from the gastrointestinal tract to the bloodstream, tissues, and urine. Cr^{3+} is absorbed by passive diffusion into the bloodstream where it binds rapidly at pH 7.4 ~ 25 mM HCO_3^- to transferrin (Tf). Cr^{3+} binding to apoLMWCr and transfer of Cr^{3+} from Cr^{3+}-transferrin to apoLMWCr is slow. Cr^{3+}-transferrin binds to the transferrin receptor (TfR) and undergoes endocytosis. Acidification of the endosome results in the rapid release of Cr^{3+} from the transferrin/transferrin receptor complex with its binding to apoLMWCr. The Cr^{3+}-loaded LMWCr then either passes through the endosome membrane via an unknown mechanism to deliver Cr^{3+}-loaded LMWCr to the cell, where the Cr^{3+}-loaded LMWCr is subsequently is expelled from the cell into the bloodstream or is delivered to the bloodstream by the fusing of the endosome with the cell membrane. Cr^{3+}-loaded LMWCr in the bloodstream is removed by the kidneys and the excreted from the body in urine. Adopted from Edwards et al. (2021)

any adverse health effects. As noted above, the FDA found chromium supplementation up to the highest amounts used in clinical trials (1 mg chromium daily) was safe (Food and Drug Administration 2005).

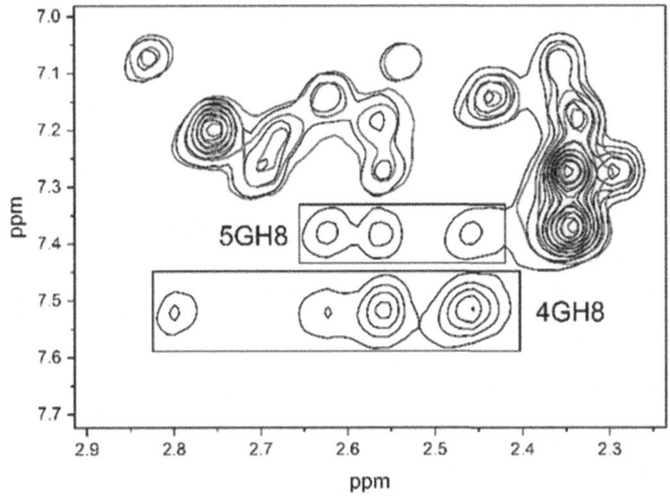

Fig. 3.10 NOESY (Nuclear Overhauser Enhancement Spectroscopy) NMR spectrum of oligonucleotide duplex with no Cr^{3+} (black) or 0.4 Cr^{3+} per oligonucleotide (red). Note the cross peaks remain largely unaffected except for those arising from each H-8 of the guanines whose relaxation times are altered due to binding to Cr^{3+} to the guanines. Adopted from Brown et al. (2020)

3.4 Biological Activity

Another point that arises out of the history of these studies with yeast and rodents is to ask what does it mean for a chromium-containing species to have biological activity or in other words how would one currently want to test a chromium-containing species or extract, such as a new plant extract, for biological activity. Clearly, assays involving Torula yeast should now be avoided. The one method that has seemed to be accepted for the decades is the insulin-dependent stimulation of glucose uptake and/or metabolism by isolated rat adipocytes. This assay was actually proposed initially to replace the use of fat tissue to test for GTF activity or GTF-like activity from Cr^{3+}-containing species other than GTF (Anderson et al. 1978). Curiously, the results of the assays using the rat adipocytes were originally misinterpreted; the assays actually showed that supposed GTF/yeast extracts increased glucose metabolism by the isolated fat cells in the absence of insulin but did not enhance the action of insulin as the GTF was supposed to do (Vincent 1994). This is consistent from hindsight with the subsequent demonstrations that the active species in the yeast extracts did not contain Cr^{3+}. In contrast, LMWCr activates glucose metabolism by rat adipocytes in an insulin-dependent fashion, consistent with activity after the binding of insulin to its receptor (Vincent 1994). A similar effect to adding LMWCr to adipocytes has been observed for healthy and diabetic rats administered pharmacologically relevant doses of Cr^{3+} daily by intraperitoneal injection (20 mg/kg body mass) for four weeks; addition of insulin to isolated adipocytes resulted in greater stimulation of insulin action on glucose metabolism in Cr^{3+}-treated rats compared to controls (Yoshimoto et al. 1992).

References

Aisen P, Aasa R, Redfield AG (1969) The chromium, manganese, and cobalt complexes of transferrin. J Biol Chem 244:4628–4633

American Diabetes Association (2023) Facilitating positive health behaviors and well-being to improve health outcomes: Standards of care in diabetes. Diabetes Care 46:568–596

Anderson RA, Brantner JH, Polansky MM (1978) An improved assay for biologically active chromium. J Agric Food Chem 26:1219–1221

Arakawa H, Kandadi MR, Panzhinskiy E, Belmore K, Deng G, Love E, Robertson PM, Commodore JJ, Cassady CJ, Nair S, Vincent JB (2016) Spectroscopic and biological activity studies of the chromium-binding peptide EEEEGDD. J Biol Inorg Chem 21:369–381

Bonvin G, Bobst CE, Kaltashov IA (2017) Interaction of transferrin with non-cognate metals studied by native electrospray ionization mass spectrometry. Int J Mass Spectrom 420:74–82

Brown S, Lockary MM, Thomas CS, Bowman MK, Woksi SA, Vincent JB (2020) Molecular structure of binary chromium(III)-DNA adducts. Chem Bio Chem 21:628–631

Chen Y, Watson HM, Gao J, Sinha SH, Cassady J, Vincent JB (2011) Characterization of the organic component of low-molecular-weight chromium-binding substance and its binding of chromium. J Nutr 141:1225–1232

Clodfelder BJ, Vincent JB (2005) The time-dependent transport of chromium in adult rats from the bloodstream to the urine. J Biol Inorg Chem 10:383–393

Costello RB, Dwyer JT, Merkel JM (2019) Chromium supplement in health and disease. In: Vincent JB (ed) The nutritional biochemistry of chromium(III), 2nd edn. Elsevier, pp 195–218

Deng G, Wu K, Cruce A, Bowman MK, Vincent JB (2015) Binding of trivalent chromium to serum transferrin is sufficiently rapid to be physiologically relevant. J Inorg Biochem 143:48–55

Di Bona KR, Love S, Rhodes NR, McAdory D, Halder Siinha S, Kern N, Kent J, Strickland J, Wilson A, Beaird J, Ramage J, Rasco J, Vincent JB (2011) Chromium is not an essential trace element for mammals: effects of a "low-chromium" diet. J Biol Inorg Chem 16:381–390

Edwards KC, Gannon MW, Frantom PA, Vincent JB (2021) Low-molecular-weight chromium-binding substance (LMWCr) may bind and carry Cr(III) from the endosome. J Inorg Biochem 223:111555

Edwards KC, Kim H, Vincent JB (2020a) Release of trivalent chromium from serum transferrin is suffciiently rapid to be physiologically relevant. J Inorg Biochem 202:110901

Edwards KC, Kim H, Ferguson R, Lockary MM, Vinecnt JB (2020b) Significance of conformation changes during the binding and release of chromium(III) from human serum transferrin. J Inorg Biochem 206:111040

EFSA Panel on Additives and Products or Substances Used in Animal Feed (FEEDAP) (2009) Safety and efficacy of Availa® (chromium chelate of DL-methionine) as a feed additive for all species. EFSA J 1043:1–69

European Food Safety Authority (2014) Scientific opinion on dietary reference values for chromium. EFSA J 12:3845

Federal Trade Commission (1997) Docket No. C-3758 Decision and Order. https://www.ftc.gov/sites/default/files/documents/cases/1997/07/nutrit2.htm. Accessed 28 Dec 2022

Food and Drug Administration (2005) Qualified health claims: letter of enforcement discretion—Chromium Picolinate and Insulin Resistance Docket No. 2004Q0144. http://wayback.archive-it.org/7993/20171114183739/, https://www.fda.gov/Food/IngredientsPackagingLabeling/LabelingNutrition/ucm073017.htm. Accessed 28 Dec 2022

Institute of Medicine (2001) Dietary reference ıntakes for vitamin A, arsenic, boron, chromium, copper, iodine, iron, manganese, molybdenum, nickel, silicon, vanadium, and zinc. National Academies Press

Jacquamet L, Sun Y, Hatfield J, Gu W, Cramer SP, Crowder MW, Lorigan GA, Vincent JB, Latour GM (2003) Characterization of chromodulin by X-ray absorption and electron paramagnetic resonance spectroscopies and magnetic susceptibilitiy measurements. J Am Chem Soc 125:774–780

Kircheva N, Toshev N, Dudev T (2022) Holo-chromodulin: competition between the native Cr^{3+} and other biogenic cations (Fe^{3+}, Fe^{2+}, Mg^{2+}, and Zn^{2+}) for the binding site. Metallomics 14:mfac082

Kornfield S (1969) The effect of metal attachment to human transferrin on its binding to reticulocytes. Biochim Biophys Acta 194:25–33

Levina A, Codd R, Dillon CT, Lay PA (2003) Chromium in biology: toxicology and nutritional aspects. Prog Inorg Chem 51:145–250

Levina A, Wang B, Lay PA (2022) Urea gel electrophoresis in studies of conformational changes of transferrin on binding and transport of non-ferric metal ions. Gels 8:19

Liu L, Lv J, Uluko H (2013) Purification and characterization of chromium-binding substances from high-chromium yeast. J Agric Food Chem 61:1279–1287

Peterson CM, Edwards KC, Gilber NC, Vincent JB, Thompson MK (2020) X-ray structure of chromium(III)-containing transferrin. J Inorg Biochem 210:111101

Rehmann JP, Barton JK (1990) Proton NMR studies of tris(phenanthroline) metal complexes bound to oligonucleotides: structural characterizations via selective paramagnetic relaxation. Biochemistry 29:1710–1717

Schwarz K, Mertz W (1959) Chromium(III) and the glucose tolerance factor. Arch Biochem Biophys 85:292–295

Shepherd PR, Elwood C, Buckley PD, Blackwell LF (2002) Glucose tolerance factor potentiation of insulin action in adipocytes from rats raised on a Torula yeast diet cannot be attributed to a deficiency of chromium or glucose tolerance activity in the diet. Biol Trace Elem Res 32:109–113

Sun Y, Ramirez J, Woski SA, Vincent JB (2000) The binding of trivalent chromium to low-molecular-weight chromium-binding substance (LMWCr) and the transfer of chromium from transferrin and $Cr(pic)_3$ to LMWCr. J Biol Inorg Chem 5:129–136

Toepfer EW, Mertz W, Polansky MM, Roginski EE, Wolf WR (1977) Preparation of chromium-containing material of glucose tolerance factor activity from brewer's yeast extracts and by synthesis. J Agric Food Chem 25:162–166

Veillon C, Patterson KY (1999) Analytical issues in nutritional chromium research. J Trace Elem Exp Med 12:99–109

Vincent JB (1994) Relationship between glucose tolerance factor and low-molecular-weigh chromium-binding substance. J Nutr 124:117–118

Vincent JB (2013a) The bioinorganic chemistry of chromium. Wiley

Vincent JB (2013b) Comment on purification and characterization of chromium-binding substance from high-chromium yeast. J Agric Food Chem 61:9280–9281

Vincent JB, Love S (2012) The binding and transport of alternative metals by transferrin. Biochim Biophys Acta 1820:362–378

Vincent JB, Keith D, Graham D, Edwards K (2022) Loss of chromium(III) from mixed Cr(III), Fe(III) serum transferrins. Curr Dev Nutr 6:1196

Wada O, Wu GY, Manabe S, Ono T (1983) Purification and chromium-excretory function of low-molecular-weight, chromium-binding substances from dog liver. Environ Res 32:228–239

Wolliscroft J, Barbosa J (1977) Analysis of chromium induced carbohydrate intolerance in the rat. J Nutr 107:1702–1706

Yamamoto A, Wada O, Ono T (1981) A low-molecular weight, chromium-binding substance in mammals. Toxicol Appl Pharmacol 59:515–523

Yoshimoto S, Sakamoto K, Wakabayashi I, Masui H (1992) Effect of chromium administration on glucose tolerance in stroke-prone spontaneously hypertensive rats with streptozotocin-induced diabetes. Metabolism 41:636–642

Zhitkovich A (2005) Importance of chromium-DNA adducts to mutagenicity and toxicity of chromium(VI). Chem Res Toxicol 18:3–11

Chapter 4
A Critical Evaluation of Chromium(III) Ecotoxicity to Aquatic and Terrestrial Plants

Elsa Salles, Vincent Normant, and Davide A. L. Vignati

Abstract Current research on chromium (Cr) ecotoxicity primary focuses on the adverse effects of Cr(VI). Concerns about high levels of Cr(III) in the environment are mostly driven by its possible (re)oxidation to the highly toxic hexavalent form, but trivalent chromium is considered of limited ecotoxicological relevance. However, Cr(III) can also elicit a large range of responses in aquatic and terrestrial plants including inhibition of growth and seed germination, damage to chloroplasts, reduced photosynthesis, oxidative stress, and alteration of nutrient balance, organelles and cellular function. Furthermore, most studies pay little if any attention to the complex chemistry of Cr(III) in the ecotoxicological test media used in controlled laboratory studies. In particular, Cr(III) can rapidly undergo hydrolysis that transforms soluble Cr^{3+} ions into Cr oxy-hydroxides—$Cr(OH)_3$. Given the very low theoretical solubility of $Cr(OH)_3$ (about 5 μg/L), their formation can markedly decrease the Cr(III) levels to which test organisms are actually exposed during the tests. These phenomena make comparison among studies far from straightforward and question the validity of many concentrations *vs.* response relationships reported for Cr(III). Although the high ecotoxicity of Cr(VI) is unquestionable, the critique presented in this chapter suggests that current consensus suffer from a general underestimation of Cr(III) ecotoxicity.

Keywords Plants · Cr(III) toxicity · Ecotoxicology · Cr speciation · Cellular effects

4.1 Introduction

A great deal of research exists on the biological responses induced by Cr in terrestrial and aquatic plants. The vast majority of published studies is concerned with the adverse effects triggered by exposure to Cr(VI) (Cervantes et al. 2001; Shadid et al.

E. Salles · V. Normant · D. A. L. Vignati (✉)
Université de Lorraine, CNRS, LIEC, F-57000 Metz, France
e-mail: david-anselmo.vignati@univ-lorraine.fr

© The Author(s), under exclusive license to Springer Nature Switzerland AG 2023
N. Kumar et al. (eds.), *Chromium in Plants and Environment*, Environmental Science and Engineering, https://doi.org/10.1007/978-3-031-44029-8_4

2017; Shanker et al. 2005) and justifies the large interest on possible techniques for remediation of Cr(VI)-contaminated sites (Beretta et al. 2018; Ao et al. 2022; Murthy et al. 2022). On the other hand, current consensus usually regards Cr(III) as being of little ecotoxicological significance and limits the risks related to the presence of Cr(III) to its potential (re)oxidation to Cr(VI) following abiotic or biologically-mediated reactions (Gorny et al. 2016; Liang et al. 2021).

The high ecotoxicity of Cr(VI) is linked to its structural analogy with phosphate and sulphate anions that facilitates intracellular uptake (Viti et al. 2014). Possible mechanisms of Cr(III) entrance into cells involve Fe(III) transporters, internalization of hydrophobic Cr(III) complexes or endocytosis of Cr-bearing particles (Beyersmann and Hartwig 2008). Inside cells, Cr(VI) is rapidly reduced to Cr(III) (Zhitkovich et al. 2005; Viti et al. 2014). The reduction of Cr(VI) results in the formation of reactive oxygen species that are associated with the severe ecotoxicological effects of Cr(VI), but also in the production of intracellular Cr(III) that reacts with cellular constituents and eventually causes DNA damage (Viti et al. 2014; Medeiros et al. 2003). Indeed, Cr(III) is the predominant or sole oxidation state of Cr inside cells (Zayed et al. 1998; Montes-Holguin et al. 2006). In higher plants, Cr(III) is also the oxidation state that is transported in sap from roots to shoots and leaves, regardless of the Cr form to which the plants are exposed (Marković et al. 2022). A similar situation likely occurs in unicellular algae. Aharchaou et al. (2017) showed that chromium had the same distribution among operationally defined subcellular fractions in cells of *Chlamydomonas reinhardtii* exposed to either Cr(VI) and Cr(III). Overall, a proper understanding of the possible risks linked to Cr contamination requires a solid knowledge of Cr(III) uptake and effects on living organisms.

This chapter provides a critique of the current knowledge on the ecotoxicity of Cr(III) to aquatic and terrestrial plants. It specifically focuses on studies performed under controlled laboratory conditions that allow to establish clear relationships between exposure to Cr and biological responses in the absence of confounding factors that may exist in natural soils and waters. Despite the consensus considering Cr(VI) as much more toxic than Cr(III), studies showing a higher toxicity of Cr(III) are regularly published. In this chapter, we will try to reconcile the results of such studies with the current consensus and to evaluate if studies documenting low Cr(III) ecotoxicity can suffer from unknown bias.

4.2 An Ecotoxicological Perspective of the Chemistry of Cr(III)

Cr has several oxidation states, but only Cr(III) and Cr(VI) are ecotoxicologically relevant for exposures via environmental matrices (Gorny et al. 2016). Cr(VI) occurs in the form of chromate anions that show little reactivity toward environmental particles, usually bearing a negative net charge (Warren and Haak 2001), and exhibit high mobility and long-distance transport (Gorny et al. 2016). At the opposite, Cr(III) predominantly occurs as cationic species (Rai et al. 1989; Giusti and Barakat 2005) that are easily adsorbed onto negatively charged environmental particles. Redox interconversions between the two oxidation states do occur in the environment and are mainly linked to the presence of reduced iron and sulphur for reduction of Cr(VI) to Cr(III) and of Mn oxides for oxidation of Cr(III) to Cr(VI) (Gorny et al. 2016). Bacterial activity can also facilitate both oxidation and reduction reactions. In aquatic ecosystems, particle-bound Cr(III) progressively accumulates into bed sediments via gravitational settling of suspended particulate matter and colloidal pumping (Dominik et al. 2007). In terrestrial (and aquatic) ecosystems, the formation of Cr(III) organic complexes can remobilize Cr (Löv et al. 2017; Liao et al. 2020; Zhu et al. 2022) and oxidation of Cr(III) to Cr(VI) can result in groundwater contamination even in the absence of anthropogenic inputs of chromium (Oze et al. 2007). Although sediments and soils act as large reservoirs of potentially bioavailable Cr, chromium uptake by terrestrial and (rooted) aquatic plants is linked to the presence of soluble Cr pools. It is therefore particularly important to understand the aqueous chemistry of Cr(III), especially when exposure is performed in controlled conditions by the addition of soluble Cr salts, which is common practice in ecotoxicological studies using the aqueous exposure route.

Three aspects of the aqueous chemistry of trivalent chromium are particularly relevant during ecotoxicological experiments: hydrolysis, solubility, and possible oxidation to Cr(VI). Ecotoxicological studies with Cr(III) are usually performed by amending appropriate (aqueous) test media with soluble Cr(III) salts such as $CrCl_3 \cdot 6H_2O$, $Cr(NO_3)_3 \cdot 9H_2O$ and $KCr(SO_4)_2 \cdot 12H_2O$. Following addition to aqueous media, the soluble salts dissociate into Cr(III) cations and the corresponding counter ions. At pH values above 4, free Cr^{3+} ions rapidly undergo hydrolysis according to the following reaction:

$$Cr(H_2O)_6^{3+} + xH_2O = Cr(H_2O)_{6-x}(OH)_x^{(3-x)} + xH_3O^+ \quad (4.1)$$

Because hydrolysis is accompanied by the release of protons (Eq. 4.1), addition of soluble Cr(III) salts can acidify ecotoxicological test media if their buffering capacity is exceeded by e.g., addition of large quantities of Cr(III) for experimental purposes. Following hydrolysis, the predominant Cr(III) species are expected to be $CrOH^{2+}$ in the pH range 3.8–6.3, $Cr(OH)_3$ at pH between 6.3 to 11.5 and $Cr(OH)_4^-$ for pH > 11.5 (Rai et al. 1989).

The species $Cr(OH)_3$ is characterized by a very low solubility product ($K_{sp} = 6.7 \times 10^{-31}$; Gorny et al. 2016), corresponding to a theoretical solubility limit of about 1.5 $\mu g\ L^{-1}$ of chromium. This value agrees well with data reported by Rai et al. (1987) who estimated the solubility of Cr(III) at about 4 $\mu g\ L^{-1}$ in a non-complexing perchlorate medium. Furthermore, Rai et al. (1987) obtained similar results in solutions amended with $Cr(NO_3)_3 \cdot 9H_2O$ or $CrCl_3 \cdot 6H_2O$. Otherwise stated, the formation of $Cr(OH)_3$ precipitates seems independent from the initial composition of the Cr solution; an important observation considering that ecotoxicity testing can be carried out with different Cr(III) salts across different studies. Indeed, Vignati et al. (2008) and Aharchaou et al. (2018) observed a similar decrease in Cr concentrations in algal

ISO medium 8692 (ISO 2012) amended with Cr nitrate, chloride or sulphate. The presence of EDTA in ISO 8692 medium did not prevent Cr(III) precipitation because Cr(III) complexation with multidentate chelators such as EDTA is sluggish compared with the kinetics of hydrolysis (Vignati et al. 2010). The solubility of Cr(III) is further decreased in the presence of iron following the formation of mixed Cr–Fe hydroxides (Sass and Rai 1987). Finally, hydrolysis takes place within microseconds (Giusti and Barakat 2005) and $Cr(OH)_3$ formation can occur within tens of minutes (Pettine et al. 2008; Aharchaou et al. 2018). The kinetics of both processes is therefore very fast compared with the typical duration of ecotoxicity tests (hours to several days). In summary, amending aqueous ecotoxicological test media with soluble Cr(III) salts will result in the rapid formation, and possible precipitation, of insoluble $Cr(OH)_3$ if the added Cr concentrations exceed the corresponding solubility limit of a few $\mu g\ L^{-1}$. These chemical processes have two major ecotoxicological consequences: the decrease of the actual soluble (bioavailable) Cr(III) concentrations in the exposure medium during the test and the formation of a pool of nano-particulate Cr(III) in the exposure medium (Aharchaou et al. 2018). Correct interpretation of Cr(III) ecotoxicity in aqueous media must consider both phenomena and cannot be achieved without an exhaustive knowledge of Cr speciation, including its possible changes over the test duration. In particular, neglecting the decrease in soluble (bioavailable) concentration over time can lead to an underestimation of the actual ecotoxicity of Cr(III).

The examples in Table 4.1 show that few studies provide sufficiently detailed information on Cr(III) chemistry and speciation during ecotoxicity testing. Even basic analytical verification of exposure concentrations is not common practice although the range of added Cr(III) concentrations usually includes concentrations well above the theoretical solubility limit of $Cr(OH)_3$. At the same time, formation of poorly soluble $Cr(OH)_3$ appears very likely in most studies, considering that the pH of most ecotoxicological test media fall in the window favouring $Cr(OH)_3$ formation (6–11 units). In such situations, measurements of total concentrations (e.g., Yu et al. 2008; Yu and Gu 2008b, 2007; Ponce et al. 2019) may include a fraction of Cr-containing nanoparticles whose bioavailability may differ from that of soluble Cr ions. The formation of Cr-containing particles (80–140 nm) has been documented by Aharchaou et al. (2018) in ISO medium 8692 for freshwater algae. Finally, the presence of organic ligands in test medium can affect Cr(III) speciation via the formation of organic-Cr(III) complexes again with possible consequences on chromium bioavailability.

These considerations do not question the quality of the studies listed in Table 4.1 nor the ecotoxicological significance of their results. They simply highlight two major caveats in Cr(III) ecotoxicology. First, Cr(III) speciation markedly changes among test media. Comparisons among studies are therefore far from straightforward when analytical information on (total) exposure concentration is available and close to meaningless when only nominal concentrations are provided. Second, in the absence of analytical verification, relationships between exposure concentrations and biological responses may underestimate Cr(III) ecotoxicity by including both soluble and insoluble Cr species in the pool of chromium bioavailable to the test organisms.

Table 4.1 Examples of studies examining the responses of aquatic and terrestrial plants to Cr(III). The selected items allow to verify possible issues linked with Cr(III) speciation. AAM, Algal Assay medium

Group	Plant species	Added Cr(III) form	Medium	Exposure concentrations (range)	Measured (Y/N)		Other possible issues related to Cr speciation	References
					Cr	pH		
Green alga	*Dictyosphaerium chlorelloides* (wild type)	$CrCl_3 \cdot 6\ H_2O$	BG-11	10–200 μM	N	N	Presence of organic component in the medium	Pereira et al. (2013)
Green alga	*Dictyosphaerium chlorelloides* (Cr-resistant mutant)	$CrCl_3 \cdot 6\ H_2O$	BG-11	300–10000 μM	N	N	Presence of organic components in the medium	Pereira et al. (2013)
Cyanobacteria	*Synechococcus* PCC 7942	$CrCl_3 \cdot 6\ H_2O$	BG-11	50–300 μM	N	**Y**	Presence of organic components in the medium	Thompson et al. (2002)
Cyanobacteria	Nostoc PCC 7120	$CrCl_3 \cdot 6\ H_2O$	BG-11	500 μM–3 mM	N	**Y**	Presence of organic components in the medium	Thompson et al. (2002)
Green alga	*Raphidocelis subcapitata*[$]	Trivalent Cr	AAM	10–1000 $\mu g\ L^{-1}$	N	N	–	Turbak et al. (1986)
Green alga	*Raphidocelis subcapitata*[$]	$CrCl_3 \cdot 6\ H_2O$	AAM	50–400 $\mu g\ L^{-1}$	N	N	–	Greene et al. (1988)
Green alga	*Raphidocelis subcapitata*	$CrCl_3 \cdot 6\ H_2O$ Cr(NO3)3 · 9 H2O $KCr(SO_4)_2 \cdot 12\ H_2O$	ISO 8692	9–2250 $\mu g\ L^{-1}$	Y total and filterable, multiple time points	Y info provided	–	Aharchaou et al. (2018)

(continued)

Table 4.1 (continued)

Group	Plant species	Added Cr(III) form	Medium	Exposure concentrations (range)	Measured (Y/N)		Other possible issues related to Cr speciation	References
					Cr	pH		
Terrestrial plant	*Oryza sativa L. XZX 45*	$Cr(NO_3)_3 \cdot 9\ H_2O$	Modified ISO 8692	12–40 mg L^{-1}	N	N	–	Fan et al. (2020)
Terrestrial plant	*Chenopodium quinoa Willd*	$CrCl_3 \cdot 6\ H_2O$	½ Hoagland + perlite (semi-hydroponic)	0.52–260 mg L^{-1}	N	N	–	Scoccianti et al. (2016)
Terrestrial plant	*Salix babylonica L and Salix matsudana Koidz*	$CrCl_3 \cdot 6\ H_2O$	Modified ISO 8692	1.51 mg L^{-1}	Y total	N	Ascribes temporal decrease in Cr concentration to sequestration to plants	Yu and Gu (2008b)
Terrestrial plant	*Salix matsudana Koidz*	$CrCl_3 \cdot 6\ H_2O$	Modified ISO 8692	About 1–2 mg L^{-1}	Y total start/stop	N	Different nitrogen levels	Yu and Gu (2008a)
Terrestrial plant	*Salix matsudana Koidz x Salix alba L. and Salix babylonica L.)*	$CrCl_3 \cdot 6\ H_2O$	Modified ISO 8692	About 1.5 mg L^{-1}	Y total start/stop	N	Presence of EDTA in some exposure scenarios	Yu and Gu (2008c)
Terrestrial plant	*Salix matsudana Koidz x alba L*	$CrCl_3 \cdot 6\ H_2O$	Modified ISO 8692	2.5–30 mg L^{-1}	Y total Start/stop	N	Different behaviour at <7.5 mg/L versus 15 and 30 mg/L	Yu and Gu (2007)
Terrestrial plant	*Actinidia deliciosa var. deliciosa*	$CrCl_3 \cdot 6\ H_2O$	Agar medium with 0.29 mM sucrose and 0.4 mM boric acid	0–75 μM	N	N	Medium with sucrose	Speranza et al. (2007)
Terrestrial plant	*Convolvulus arvensis L*	$Cr(NO_3)_3 \cdot 9\ H_2O$	Modified Hoagland medium as per Peralta et al. (2001)	1–20 mg L^{-1}	N	N	Agar based medium	Montes-Holguin et al. (2006)

(continued)

Table 4.1 (continued)

Group	Plant species	Added Cr(III) form	Medium	Exposure concentrations (range)	Measured (Y/N)		Other possible issues related to Cr speciation	References
					Cr	pH		
Terrestrial plant	*Salsola kali*	$Cr(NO_3)_3 \cdot 9 H_2O$	Modified Hoagland medium as per Peralta et al. (2001)	0–20 mg L^{-1}	N	N (initially adjusted to 5.8)	Agar based medium. Initial pH value may redude/prevent Cr(III) precipitation	Gardea-Torresday et al. (2005)
Aquatic plant	*Fontinalis antipyretica Hedw*	$Cr(NO_3)_3 \cdot 9H_2O$ $CrCl_3 \cdot 6H_2O$	Volvic mineral water	6.25 × 10^{-5}–50 mM	N	N	$Cr(OH)^{2+}$ predicted as predominant species (MINEQL+) at all concentrations and for both forms	Dazy et al. (2008)
Terrestrial plant	*Apium graveolens (celery)*	$CrCl_3 \cdot 6H_2O$	Heller medium + Fe-EDTA + sucrose + 0.8% agar	0.01–10 mM	N	N	Agar containing medium	Scoccianti et al. (2006)
Terrestrial plant	*Oryza sativa* L. cv. XZX 45	$Cr(NO_3)_3 \cdot 9H_2O$	Modified ISO 8692	0–40 mg L^{-1}	N	N	Modifications not specified. Possibly same as other studies from the same research group	Yu et al. (2018a, b)
Terrestrial plant	*Lycopersicon esculentum* Mill. cv. Juncal (Tomato)	$CrCl_3 \cdot 6H_2O$	Half-stregth Hoagland medium	0–50 µM	N	N		Henriques (2010)

(continued)

Table 4.1 (continued)

Group	Plant species	Added Cr(III) form	Medium	Exposure concentrations (range)	Measured (Y/N)		Other possible issues related to Cr speciation	References
					Cr	pH		
Terrestrial plant	*Vigna unguiculata*	$Cr(NO_3)_3 \cdot 9H_2O$	Hydroponic cultivation setup	0.05–0.5 mM	N	N	Unspecified medium composition	Chow et al. (2018)
Aquatic plant	*Eichhornia crassipes (Mart.) Solms*	Cr_2O_3	Hoagland medium	1 and 10 mM	N	N	Insoluble/sparingly soluble Cr form	Paiva et al. (2009)
Terrestrial plant	*Salvinia rotundifolia*	$CrCl_3 \cdot 6H_2O$	Tap water (±buffer), composition provided	5 and 20 mg L^{-1}	Y (data not shown)	Y (medium buffered at 4.0; 6.0 and 7.6 units, data not shown)	Cr speciation issue considered in experimental set-up	Ponce et al. (2019)

$, Named with its former name of *Selenastrum capricornutum* in the original study

A similar situation is observed with regard to pH, with a very limited number of studies providing information on the temporal stability (or lack thereof) of this parameter over the course of the experiments. Because of hydrolysis (Eq. 4.1), changes in the pH of test medium may occur following the release of protons. One early study warned about possible strong decreases in pH values in a simple test medium (0.05% K_2HPO_4, 0.05% KH_2PO_4, 0.05% $(NH_4)_2SO_4$, 0.05% KNO_3; initial pH just below 7) amended with chromium chloride (Den Dooren de Jong and Roman 1965). Thompson et al. (2002) documented that initial pH differed by about 1.5 units between BG-11 medium amended with 50 μM Cr(III) (pH = 7.27) and 300 μM Cr(III) (pH = 6.14). Most interestingly pH values were comparable at about 9.5 units at the end of the test regardless of the added Cr(III) concentration. However, the differences in pH at the beginning of the test suggest that Cr speciation and its temporal evolution probably depended on the initial concentration of added Cr(III). In practice, exposure conditions may not be fully consistent even within an individual study. As in the case of analytical verification of exposure concentrations over the test duration, monitoring of pH values during the tests should always be performed at least for the lowest and highest tested concentrations.

Information is equally scant as to the possible Cr(III) to Cr(VI) interconversions in the test medium during ecotoxicological experiments. In the absence of biologically mediated reactions, oxidation of Cr(III) to Cr(VI) could be catalyzed only by Mn oxides that are not a standard component of test media. Aharchaou et al. (2018) used ion chromatography ICP-MS to verify the possible occurrence of Cr(III) to Cr(VI) redox interconversion in ISO medium 8692 for freshwater algae (ISO 2012) and did not observe any changes in the oxidation state of chromium. Because Mn enters in the composition of ISO medium 8692 as soluble $MnCl_2$, the general applicability of these results to other aqueous media remains to be verified. The situation is much more complicated in experiments involving the use of synthetic or, especially, natural soils where the presence of some form of organic matter and Fe and Mn oxides is the norm. Similarly, in natural waters, including pore waters and soil solutions, the behavior of Cr(III) can be modified by the presence of natural organic matter (NOM) and other colloidal carrier phases. In particular, adsorption on NOM can increase Cr(III) solubility by avoiding precipitation of Cr hydroxides (Fukushima et al. 1995; Gustafsson et al. 2014). Nonetheless, the presence of colloidal Cr_2O_3 has been documented in soils (Zhu et al. 2022) and polynuclear species have been detected in natural waters (Hu et al. 2016). These considerations confirm that ecotoxicological laboratory studies should start paying much more attention to Cr(III) speciation to correctly assess its actual toxicity and to facilitate extrapolation of laboratory results to real-field conditions.

4.3 Cr(III) Transport and Distribution

4.3.1 Cr(III) Uptake

Cr is not an essential nutrient for plants that, consequently, do not have Cr specific transporters (Panda and Choudhury 2005; Adhikari et al. 2020). However, plants can import both Cr(III) and Cr(VI), the general consensus being that Cr(VI) is more easily taken up than Cr(III) due to its higher transmembrane transport efficiency and solubility (Shanker et al. 2005). However, the accumulation of Cr(VI) and Cr(III) in *Arabidopsis thaliana* was similar (Ding et al. 2019) and Cr(III) is the main form present inside plant tissue (Zayed and Terry 2003; Markovich et al. 2022). Cr(VI) uptake occurs mainly via sulphate or phosphate transporters in some bacteria, fungi, algae and plants because of their structural similarities with chromate anions (Tang et al. 2023; Viti et al. 2014; Xu et al. 2021). Mechanisms involving Cr(III) uptake by plants are not yet completed understood. Cr(III) uptake could mainly be via the same carriers as for essential ion elements (ion channels) such as Fe, Ca, Mg or K or through the simple diffusion of cations exchanges sites in the cell wall (Ding et al. 2019; Singh et al. 2013; Ao et al. 2022). In *Leersia hexandra* Swartz, the antagonistic effect of Fe(III) on Cr(III) uptake by root cells suggests that Cr(III) uptake may be mediated partly by Fe(III)-phytosiderophore complex transporters (Liu et al. 2011). Cr(III) can also be transported by passive mechanism, by cation diffusion facilitators (Skeffington et al., 1976). In this study, *Hordeum vulgare L.* was exposed to Cr(III) and Cr(VI) in presence and absence of metabolic inhibitors and Cr uptake was measured. They demonstrated that the uptake of Cr(VI) was reduced by the inhibitors whereas Cr(III) uptake was not affected, suggesting different uptake mechanisms for the two forms. However, the passive and active uptake mechanisms are not clearly established and evidence of this process is still needed. Precautions needs to be taken as Skeffington et al. (1976) proposed that Cr(VI) was the only form of Cr inside root cells which was corrected later: in this case, Cr(III) was detected in apoplast of root cells (Zayed and Terry 2003). Cr(III) can also be retained by the cation-exchange sites of the cell walls (Marschner 1995). The complexation of Cr(III) with organic acid (e.g. carboxylic acid or amino acid) enhance root uptake of Cr(III), suggesting that organic complexation of Cr(III) would contribute to Cr(III) uptake (Srivastava et al. 1999; Panda and Choudhury 2005). Cr(III) uptake clearly occurs in plant cells and Cr(III) can cross biological membranes, although the exact mechanisms are not yet fully understood.

4.3.2 Cr(III) Translocation and Accumulation

In root cells, Cr(III) ions are highly stabilized by complex formation with organic molecules, such as proteins (glutathione), carbohydrates (especially pentoses),

NAD(P)H, $FADH_2$, and probably also with organic acids, and stored and immobilized in root cell vacuoles in precipitated form (Caldelas et al. 2012; Zeng et al. 2011) or in apoplast in cell walls, which is the reason for restricted mobility of chromium in plants (Shanker et al. 2004; Mangabeira et al. 2011; Babula et al. 2008). Cr has a lower migration rate from root to shoot, than other heavy metals such as Hg, Cd and As (Shanker et al. 2005). For most terrestrial and aquatic plants, Cr distribution in different parts is in the order of roots > stem > leaves > fruits. Many studies showed that Cr(III) is accumulated mostly in roots and only a small part of Cr(III) is translocated to shoots (Paiva et al. 2009). Little translocation of Cr(III) to aerial part was reported in *G. americana*, with a concentration of 45 and 50 mg kg^{-1} in leaves and stems respectively, and most of the Cr(III) immobilized and stored in the roots, with accumulation concentration in the roots of 3841 mg kg^{-1} (Barbosa et al. 2007). In this case, Cr(III) is poorly translocated due to formation of Cr(III) insoluble complexes. Organic compounds, like citrate or EDTA are involved in Cr(III) translocation in xylem vessels and plant distribution. Cr(III)-citrate or Cr(III)-EDTA complexes are therefore more soluble and easily transported by the plants or immobilized and stored after translocation to leaves or fruits (Yu et al. 2008c; Juneja and Prakash 2005). Moreover, a study on *Taraxacum officinale* roots suggested that Cr(III) transport only occurs as Cr(III)-organic complexes with organic acids no matter if the plant is exposed to Cr(VI) or Cr(III), suggesting that Cr(VI) is reduced to Cr(III) inside the plants to be translocated and that Cr(III) is more mobile after complexation with organic compounds as suggested before (Markovich et al. 2022). In parenchyma cells, Cr(III) is accumulated in vacuoles and in the cell wall of xylem cells (Mangabeira et al. 2011; Vazquez et al. 1987). The leafy vegetables that tend to accumulate Fe (i.e., spinach, turnip leaves) appear to be the most effective in translocating Cr to the shoot. The leafy vegetables that do not accumulate relatively high concentrations of Fe in their leaves (i.e. lettuce and cabbage) are substantially less effective in translocation of Cr to the leaves (Cary et al. 1977a, b). Onion, spinach, chive and celery have a higher shoot/root concentration ratio than cabbage, peas, kale, cauliflower and lettuce after Cr(III) exposure (Zayed et al. 1998). Cr(III) is mainly retained in the roots, in epidermal cells. Depending on the chosen biological model, Cr(III) can be transported to the stem (xylem cells), leaf and fruits and could be transported as Cr(III)-organic complex (Fig. 4.1).

4.4 Biological Effects

4.4.1 Effects on Plant Morphology

Studies reported a decrease of total biomass and plant growth (Davies et al. 2001; Arduini et al. 2006; Lopez-Luna et al. 2009). Cr(III) also caused inhibition of growth in *Brassica oleracea* after exposure to 0.5 mM (Chatterjee and Chatterjee 2000).

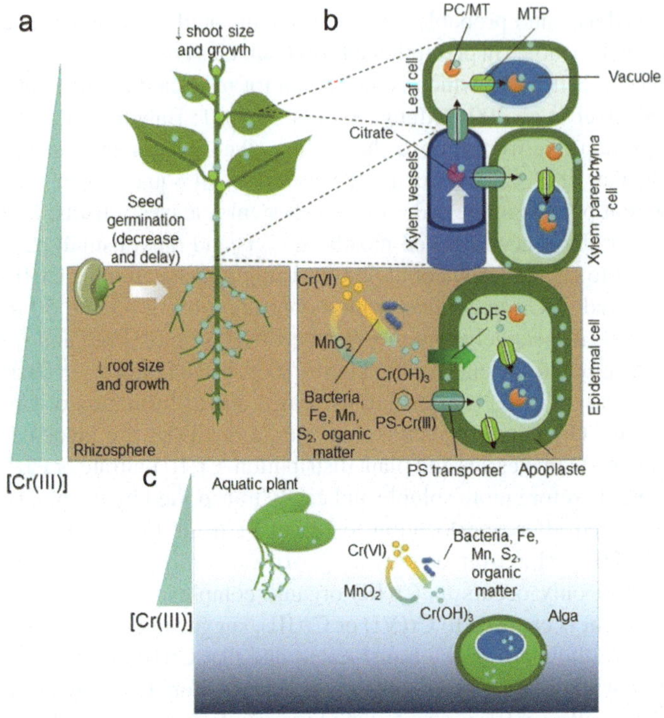

Fig. 4.1 Plant uptake, translocation and accumulation of Cr(III) from soil and water. **a** Cr(III) induce effects on seed germination, shoot and root parts and Cr(III) accumulation is more important in root than shoot. **b** Uptake and translocation of Cr(III) at cellular level in epidermal, xylem parenchyma, and leaf cells and translocation in xylem vessels. **c** Accumulation of Cr(III) in algae and aquatic plants. PS: phytosiderophores, CDFs: cation diffusion facilitators, PC: phytochelatins, MT: metallothioneins MTP: metal tolerance proteins

Roots are the first organ in contact with Cr(III) and Cr(III) preferentially accumulates in plant roots. Several studies showed an inhibition of roots growth, reduction of roots lengths, volume and roots dry weight (Davies et al. 2001; Arduini et al. 2006; Barcelo et al. 1993; Lopez-Luna et al. 2009; Liu et al. 1992; Vajpayee et al. 2011). The reduction of root length is correlated to an increase of the Cr(III) concentration (Table 4.2) (Lopez-Luna et al. 2009; Liu et al. 1992; Barbosa et al. 2007). Opposite effects on root dry weigh, at low concentration of Cr(III) (0.05 mg L^{-1}) were observed on roots of *Phaseolus vulgaris*. Moreover, the root dry weight increased more in presence of Cr(III) when *P. vulgaris* was grown in Fe-deficient conditions (Barcelo et al. 1993). Arduini et al. (2006) observed an increase of root dry weight of *Miscanthus sinensis* after Cr(III) treatment (50 and 100 mg L^{-1}). Changes in root morphology can indicate Cr(III) stress, with a stimulation of root elongation below 150 mg L^{-1} of Cr(III) and a severe inhibition of root length observed at concentrations equal or higher than 150 mg L^{-1} which demonstrates that Cr(III) affects root morphology at all level of concentration (Arduini et al. 2006). In addition, Cr(III) can also affect aerial parts of plants and cause a decrease in leaf size and number (total and green leaves), growth rate, biomass and dry weight and affect the morphology of leaves (Wallace et al.1976; Chatterjee and Chatterjee 2000; Davies et al. 2001; Barbosa et al. 2007; Arduini et al. 2006; Chatterjee and Chatterjee 2000). Increase of leaf dry weight in *P. vulgaris* was observed after exposure to low concentration of Cr(III) (1 μM) (Barcelo et al. 1993).

Cr(III) have negative effects on roots and aerial parts of the plants. Studies reported that root growth was a more sensitive indicator than shoots for Cr(III) toxicity because Cr(III) is uptake via roots and accumulates more on roots than leaves (Chatterjee and Chatterjee 2000; Fargasova et al. 2012). Some studies also reported an opposite

Table 4.2 Effects of Cr(III) on seed germination and plant development

Plant species	Effects	Cr(III) concentration	Compound	References
Allium cepa	Reduction in root growth	0.01–10400 mg L^{-1}	$Cr(NO_3)_3$	Liu et al. (1992)
Avena sativa	Inhibition of root growth	1000–4000 mg kg^{-1}	$CrCl_3$	López-Luna et al. (2009)
Brassica oleracea	Decrease in leaf size, chlorosis and wilting	25 mg L^{-1}	$Cr_2(SO_4)_3$	Chatterjee and Chatterjee (2000)
Genipa americana	Reduction in root, leaf, stem and total biomass	5–30 mg L^{-1}	$CrCl_3$	Barbosa et al. (2007)
Helianthus annuus	Decrease root dry weight	520 mg L^{-1}	$CrCl_3$	Davies et al. (2001)
Miscanthus sinensis	Decrease leaf and flower dry weight	104–520 mg L^{-1}	$CrCl_3$	Davies et al. (2001)
	Increase root length	50–100 mg L^{-1}	$Cr(NO_3)_3$	Arduini et al. (2006)
	Decrease shoot growth and aerial part	50–200 mg L^{-1}	$Cr(NO_3)_3$	Arduini et al. (2006)
	Decrease length and roots biomass	>150 mg L^{-1}	$Cr(NO_3)_3$	Arduini et al. (2006)
Phaseolus vulgaris	Reduction of leaf size and leaf biomass	0.5–5 mg L^{-1}	$Cr_2(SO_4)_3$	Wallace et al. (1976)
	Increase in root and leaf dry weight	0.05 mg L^{-1}	Not specified	Barcelo et al. (1993)
	Decrease in dry weight and chlorophyll content	1–4 mg L^{-1}	Not specified	Barcelo et al. (1993)
Raphanus sativus	Inhibition of roots and shoots growth	50–250 mg L^{-1}	$Cr(NO_3)_3$	Fargašoavà et al. (2012)
Salix alba	Inhibition of roots and shoots growth	50–250 mg L^{-1}	$Cr(NO_3)_3$	Fargašoavà et al. (2012)
Sorghum bicolor	Inhibition of root growth	100–4000 mg kg^{-1}	$CrCl_3$	Lopez-Luna et al. (2009)
Triticum aestivum	Inhibition of root growth	500–4000 mg kg^{-1}	$CrCl_3$	Lopez-Luna et al. (2009)
	Inhibition of germination	25–100 mg L^{-1}	Cr_2O_3	Vajpayee et al. (2011)
Vicia Sativa	Inhibition of roots and shoots growth	50–250 mg L^{-1}	$Cr(NO_3)_3$	Fargašoavà et al. (2012)
Zea mays	Inhibition of roots and shoots growth	50–250 mg L^{-1}	$Cr(NO_3)_3$	Fargašoavà et al. (2012)

effect on roots and leaves of Cr(III) in specific conditions such as low concentration of Cr(III) and imbalanced nutrient supply of Fe (Barcelo et al. 1993; Arduini et al. 2006). However, it is important to consider all parameters for roots, because an increase in length could also indicate a stress reaction when the morphology of the roots is changed (Arduini et al. 2006).

4.4.2 Reproduction and Seed Germination

Cr(III) has a negative effect in the seed germination. Cr(III) exposure and accumulation in seeds delay, decrease and inhibit germination process. For *Triticum aestivum*, a treatment under 10 mg L^{-1} of Cr(III) showed no impact on the germination but treatment of 25, 50 and 100 mg L^{-1} led to 5–19% reduction of germination (Vajpayee et al. 2011). Cr(III) affected germination and growth of wheat (*T. aestivum*) and sorghum (*Sorghum bicolor*) after treatment of 500–1000 mg kg^{-1} of soil, but no effect on germination was observed for oat (*Avena sativa*), more resistant to Cr(III) than the other two species. This is confirmed by the EC_{50} of oat of 2216.84 mg kg^{-1} in roots, two times higher than EC_{50} of wheat and sorghum, 1631.14 and 1089.01 mg kg^{-1} respectively (Lopèz-Luna et al. 2009). Cr(III) has also been reported to interfere with structure and function of male gametophyte in kiwifruit (*Actinidia deliciosa var. deliciosa*) and can inhibit pollen germination and tube growth and induce alterations in pollen tube shape. Modification of callose deposition pattern and arabinogalactan protein distribution in kiwifruit pollen wall was also observed after Cr(III) exposure (Speranza et al. 2007, 2009). The reduction of α-amylase and β-amylase activities observed after Cr(III) treatment and causing a reduction of sugar supply required for the embryo development may be linked to germination reduction rate (Dua and Sawhaney 1991; Zeid 2001; Singh et al. 2013).

4.4.3 Effect of Cr(III) on Photosynthesis and Chloroplast Structure

As other trace elements, Cr(III) can affect plant photosynthesis and cause ultrastructural changes in the chloroplasts leading to inhibition of photosynthesis (Panda and Choudhury 2005; Panda and Patra 2000). Do Nascimiento et al. (2018) observed chloroplast damages after they exposed cocoa plants (*Theobroma cacao*) to a high concentration of Cr(III) (600 mg kg^{-1}). Alteration in shape of leaf chloroplasts resulting in the structural disarrangement of thylakoids and stroma was observed in *Alternanthera philoxeroides* and *Borreria scabiosoides* under Cr(III) stress (Mangabeira et al. 2011). Cr(III) treatment reduced chlorophyll contents in celery seedlings at 1 mM (Scoccianti et al. 2006), in genipayer (*Genipa americana*) seedlings at 30 mg L^{-1} (Barbosa et al, 2007), and in cauliflower (*Brassica oleracea*) at 0.5 mM (Chatterjee and Chatterjee 2000). At the same concentration of Cr(III) and Cr(VI), Cr(III) was much less toxic than Cr(VI) on photosynthesis parameters of water hyacinth (*Eichhornia crassipes*) and might eventually increase photosynthesis and chlorophyll content (Paiva et al. 2009). One mM of Cr(III) stimulated growth and photosynthetic parameters such as photosynthetic rate and stomatal conductance on aquatic hyacinths after a 2 day treatment, but a decrease of photosynthetic rate and signs of toxicity (chlorosis) were observed for plants treated with 10 mM of Cr(III) for 4 days (Paiva et al, 2009). Similar results were shown for *P. vulgaris;* low (1 μM) or moderate (100 μM) concentrations of Cr(III) in irrigation solution increased chlorophyll a and b, and carotenoids content in leaves, but high Cr(III) concentration (10 mM) reduced the contents of chlorophylls and carotenoids (Zeid 2001). In mosses (*Fontinalis antipyretica*), Cr(III) modified chlorophyll a/b ratio. Cr(III) as $Cr(NO_3)_3$ decreased total chlorophyll content whereas Cr(III) as $CrCl_3$ lead to chlorophyll accumulation at low concentration of Cr(III). The effect on chlorophyll seem to depend on Cr(III) form and Cr(III) as a nitrate salt seems to be more toxic (Dazy et al. 2008). Like Pb, Cd or Hg, Cr may reduce δ-aminolevulinic acid dehydratase (ALAD) activity or degrade ALAD, an important enzyme involved in chlorophyll biosynthesis, thereby affecting the δ-aminolevulinic acid (ALA) utilization resulting in the increase of ALA and reducing chlorophyll production (Stobort et al. 1985; Prasad and Prasad 1987; Vajpayee et al. 2011). In cells, Cr(III) may compete with Mg and Fe for assimilation and transport to leaves, affecting therefore pigment biosynthesis (Vernay et al. 2007). Cr(III) exposure can also increase the production of ROS (Shanker and Pathmanabhan 2004). The ROS induce damages in pigment-protein complexes located in thylakoid membranes followed by pheophytinization (two H^+ ions replace the Mg^{2+} ion found in the center of the porphyrin ring of chlorophylls) and destruction of thylakoid membranes (Juarez et al. 2008). Cr(III) decrease the photosystem II (PSII) activity in *Datura innoxia* (Vernay et al. 2008). Barton et al. (2000) observed that Cr(III) at 10 μM increased the ferric chelate reductase activity in alfalfa (*Medicago sativa L.*) roots in iron-limited media. Cr(III) also induced chlorosis on plants (Barton et al. 2000; Schmidt et al. 1996). Chlorosis is generally correlated with Fe-deficiency in plant (Kaya and Ashraf 2019; Jin et al. 2007; Briat et al. 2015). It is possible that chlorosis is due to an inhibitory effect of Cr(III) on iron reductase involved in Fe(III) uptake (Alcántara et al. 1994). Cr(III) could also compete with iron for entry in root cells or interfere with iron uptake (Skeffington et al. 1976).

4.4.4 Gas Exchanges

Leaf gas exchange monitored by photosynthetic rate, stomatal conductance and transpiration was severely affected by Cr(III) in the first 24 h of treatment of *T. cacao* (Do Nascimiento et al. 2018). Severe changes in leaf gas exchange have also been reported for the macrophytes *Alternanthera philoxeroides*, *Borreria scabiosoides*, *Polygonum ferrugineum*, *Eichhornia crassipes* (Mangabeira et al. 2011), *Genipa americana*

(Santana et al. 2012) and *Eichirnia crassipes* (Paiva et al. 2009) subjected to Cr(III) stress. The leaf gas exchanged and stomates opening can be linked to photosynthesis rate, as a decrease in CO_2 will reduce optimal rates of photosynthesis.

4.4.5 Alteration of Organelles and Cellular Functions

Under Cr(III) stress, the shapes of chloroplast and nuclei were altered in two aquatic macrophytes *Alternanthera philoxeroides* (alligator grass) and *Borreria scabiosoides*. At 50 mg L^{-1} of Cr(III), disintegration of the nucleus and deformation of chloroplasts were observed leading to structural disarrangement of thykaloids and stroma (Mangabeira et al. 2011). Damage to chloroplast can affect photosynthesis and plant growth. Alteration of mitochondrial cristae and dense electron material in mitochondria was also observed for both *Allium cepa* and *Borreria scabiosoides* treated with Cr(III) (Mangabeira et al. 2011; Liu and Kottke 2003). In kiwi pollen, similar findings have been reported with an alteration of the shape of mitochondria (swelling and loss of mitochondrial cristae) and the shape of endoplasmic reticulum (Speranza et al. 2007). Cytoplasmic vacuolization was also observed in kiwi pollen after Cr(III) treatment, usually a sign of cell death (Speranza et al. 2007). The impact of Cr(III) on organelles can affect cellular function of the plant.

The presence of Cr(III) produce mitotic irregularities (i.e. anaphase bridges or mitosis lagging), chromosomal aberrations (i.e. chromosome stickiness, chromosome fragmentation) (Liu et al. 1992; Qian 2004; Kumar et al. 2015), chromatin condensation (Speranza et al. 2007) and nuclear abnormalities (nuclear bud, micro nucleus, nuclear notch) (Kumar et al. 2015). These chromosomal irregularities and DNA damage could be linked to the production of ROS (Kumar et al. 2015) or to the formation of DNA adducts with Cr(III) (Viti et al. 2014). The chromosomal aberration observed can be linked with the production of ROS, as Cr(III) induced the formation of ROS and antioxidant enzyme induced to counter oxidative stress can cause chromosomal aberration (Kumar et al. 2015). Cr(III) induces the expression of genes encoding for proteins involved in cellular stress responses. These proteins are also induced in pathogen defence, senescence process and heavy metal stress, suggesting the existence of a common ROS-mediated mechanism of gene regulation (Quaggiotti et al. 2007).

Exposure to Cr(III) induced proteasome misfunction in kiwi (*A. deliciosa var. deliciosa*) pollen that generated accumulation of misfolding and damaged proteins. Similar results were observed after Cr(VI) exposure, but molecular targets at proteasome level may be different (Vannini et al. 2011). The 20S proteasome α-subunit expression was decreased in presence of Cr(III) and the 26S regulatory subunit Rpn11 level was decreased after Cr(VI) exposure (Vannini et al. 2011).

4.4.6 *Effects of Cr(III) on Mineral Nutrition*

Like other trace elements, Cr(III) is structurally similar to other essential elements and may affect plant mineral nutrition. In rhizosphere soil, excessive Cr reduces the accumulation of essential nutriments (Fe, Cu, Mg, Zn Ca, S and P) by masking adsorption sites and forming insoluble or low-bioavailability complexes (Woke et al. 2019; de Oliveira et al. 2015, 2016; Sharma et al. 2020). There is also evidence of increased Fe availability and uptake for plants in presence of Cr(III) in soil (Cary et al. 1977a, b). Yu et al. (2018a, b) found that Cr(III) exposure decreased Mn and Zn concentration in root cells and Zn concentration in shoot cells in rice seedlings. Mn and Zn concentrations were also decreased in tomato root cells after Cr(III) exposure. A decrease of Fe and Cu concentrations was also observed in tomato roots (Moral et al. 1996). Gardea-Torresdey et al. (2005) showed in *Salsola kali* roots a decrease of K, P, Mg and Cu after Cr(III) treatment. In *Phaseolus vulgaris* L., very small quantities of Cr(III) are transported to leaves, but Cr(III) exposure induces a decrease of Fe, Zn and Mo and to a lesser extent a decrease of K, Ca and Mg in leaves (Wallace et al. 1976). Davies et al. (2001) reported that Cr(III) treatment decrease N, P and K levels in *Helianthus annuus* leaves, but enhance Al, Fe and Zn concentration. These effects were enhanced by the presence of mycorrhiza (Davies et al. 2001). These decreases in nutrient uptake are probably due to deterioration of root nutriment penetration under Cr(III) stress and the decline in root growth (Ao et al. 2022; Sharma et al. 2020). The decrease in nutrient uptake could indicate that Cr(III) displaces ions from physiologically important binding sites in plant cells, thus affecting signal transduction, photosynthesis or plant nutrient metabolism (Cipriani et al. 2012; Sharma et al. 2020). Also, the Cr(III) accumulation in the plant cell wall may damage the plasmodesmata, which are important for mineral nutrients transport channels, thus leading to an imbalance in mineral nutrient metabolism (Ao et al. 2022; Fujita 2015; Kitagawa et al. 2015). In presence of 1 μM of Cr(III), nitrate reductase (NR) activity was enhanced suggesting a request in ammonium (NH_4^+) or nitric oxide (NO) during the cellular response to Cr(III), whereas in presence of Cr(VI) ($\geq$2 μM) NR activity decreased in *T. aestivum* (Panda and Patra 2000). Nitrogen is an essential macro-element and plays a role in growth and in plant development and is available as nitrate NO_3^-. An enhanced nitrate reductase could indicate an increased demand of energy, due to a dysfunction of photosynthesis or mitochondrial respiration.

4.4.7 *ROS Production, Lipid Peroxidation and Antioxidative Mechanisms*

Exposure to heavy metals induces the overproduction of ROS (reactive oxygen species) including superoxide radicals ($O_2^{\cdot-}$), hydroxyl radicals ($OH^{\cdot}$), oxygen singlets ($^1O_2$) and hydrogen peroxide (H_2O_2). Hyperaccumulation of ROS affects the growth and development of plants (Maiti et al. 2012; Xie et al. 2019). Redox active metals such as Fe, Cu, Co or Cr have the capacity to produce ROS via Haber-Weiss and Fenton reactions (Sharma et al. 2020; Bokare and Choi 2014). Plants can develop antioxidant enzyme systems for scavenging excessive accumulation of ROS under metal stress. The enzymatic antioxidants include the key enzymes such as superoxide dismutase (SOD), catalase (CAT), peroxidases (POD), guaiacol peroxidase (GPX), ascorbate peroxidase (APX), glutathione reductase (GR), glutathione S-transferases (GST), single dehydroascorbate reductase (MDHAR), dehydroascorbate reductase (DHAR) (Ahmad et al. 2010). Under normal condition, ROS are generated in little quantities in cellular organelles of plants (Maiti et al. 2012) and play important roles in regulating and controlling essential metabolisms, such as signal transduction for programmed cell death, seed dormancy, senescence, and growth (Pourrut et al. 2011). In many studies, Cr(III) exposure induces an increase of antioxidant enzyme activities including SOD, CAT, POD, GPX, APX, GR, MDHAR and DHAR (see references in Table 4.3). Some studies showed a downregulation of antioxidant enzyme activities like CAT and POD in *Brassica oleracea* (Pandey and Sharma 2003; Chatterjee and Chaterjee 2000) and GR in *T. cacao* (Do Nascimiento et al. 2018). CAT uses heme (iron-porphyrin) as a cofactor. Reduction in CAT activity indicates that Cr has the potential to interact with iron in metabolic pool or it may influence the presence of active form of iron (Sharma et al. 2003, 2020). The non-enzymatic antioxidant responses (i.e. ascorbic acid, glutathione (GSH), phenolic acid) are also observed in presence of Cr(III). In *S. bicolor*, after Cr(III) treatment, the GSH/GSSH ratio decreases only in roots but not in leaves, suggesting an increase of oxidative species in root cells (Shanker and Pathmanabhan 2004). Dehydroascorbate (root and leaf) and total ascorbate (root) levels exhibited a high degree of significant increase irrespective of speciation or concentration of Cr(III) in the medium (Shanker and Pathmanabhan 2004). Cr(III) affects the membrane potential by inducing lipid peroxidation. Malondialdehyde (MDA), a biomarker of lipid peroxidation is excessively produced due to lipid peroxidation increase after Cr(III) treatment in root and leaf (Shanker and Pathmanabhan 2004). Oxidative damages resulting from ROS towards biomolecules such as lipids, proteins and nucleic acids is well documented for plant species (Kanazawa et al. 2000; Singh et al. 2006).

4.4.8 Regulation of Phytochelatins, Metallothioneins and Metal Tolerance Proteins

To cope with Cr(III) induced stress, plants have developed different strategies involving morphological, anatomical and molecular defence mechanisms. In order to regulate the uptake and accumulation of trace elements, plants can sequester and chelate metals with metal binding ligands such as metallothioneins (MT), phytochelatins (PC) and metal tolerance proteins (MTP), produced within the plant cells to aid in heavy metal transport and sequestration. These metal chelators protect plants against high heavy metal concentrations through different mechanisms, such

Table 4.3 Effects of Cr(III) on antioxidant enzyme activities

Plant species	Cr(III) concentration	Antioxidant enzyme activities	References
Oryza sativa L. XZX 45	12–40 mg L^{-1}	↑ DHAR ↑ MDHAR ↑ GPX ↑ GR ↑ APX ↑ POD = CAT = SOD	Fan et al. (2020)
Brassica oleracea L var. capitata cv. Snowball	500 μM	↓ CAT ↓ POD	Pandey and Sharma (2003)
Theobroma cacao L	0–600 mg kg^{-1}	↑ CAT, ↑ GPX ↑ SOD = GR	Do Nascimento et al. (2018)
Vigna unguiculata	0.05–0.5 mM	↑ POD ↑ CAT ↑ APX	Chow et al. (2018)
Micrasterias denticulata	10 nM–1 mM	= SOD = CAT	Volland et al. (2012)
Sorghum bicolor (L.) Moench cv CO 27	50–100 μM	↑ SOD, ↑ CAT, ↑ APX, ↑ DHAR, ↑ GR = MDHAR	Shanker and Pathmanabhan (2004)
Brassica oleracea	500 μM	↓ CAT	Chatterjee and Chaterjee (2000)
Parthenium hysterophorus L	1 mM	↑ SOD	UdDin et al. (2015)
Solanium nigrum L	1 mM	↑ SOD	UdDin et al. (2015)
Zea mays	30–150 μmol L^{-1}	↑ SOD	Anjum et al. (2017)
Theobroma cacao	>400 mg kg^{-1}	↓ GR	Do Nascimiento et al. (2018)
Allium cepa	1–100 μg mL^{-1}	↑ SOD	Kumar et al. (2015)
Fontinalis antipyretica	0.625 μM–50 mM	↑ SOD ↑ CAT	Dazy et al. (2008)

↑ Increase enzyme activity ↓ Decrease enzyme activity = activity was not modified

as chelation, sequestration (MT and PC) or efflux (MTP). MT are cysteine-rich proteins that play a crucial role in heavy metals detoxification, metal homeostasis and metabolism via binding through the thiol group (SH) in cysteine residues. MT are transcribed constitutively or induced in response to several types of stress including heavy-metal exposures (Ziller and Fraissinet-Tachet 2018). The increased expression of MT-like protein in sorghum exposed to Cr(III) can indicate a potential role of metal binding ligands in Cr(III) detoxification (Shanker et al. 2004). After chelation, Cr can be compartmentalized in the cell wall and vacuoles. In plants, the cell wall is mainly composed of cellulose, hemicelluloses and pectins (Carpita and McCann 2000; Wolf and Greiner 2012). In the cell wall of root cells, Cr(III) can bind cellulose and pectin (Wang and Lee 2011; Yamada and Shiiba 2015). In *Oryza sativa* tissues, expression of MT genes was increased after Cr(III) exposure suggesting a role of MT in Cr(III) chelation (Yu et al. 2019). PC are synthesized in the cytoplasm under heavy metals toxic stress (Sharma et al. 2016). Biosynthesis of PC is catalysed by phytochelatin synthase (PCS) that is constitutively expressed. However, PCS activity is increased in the presence of heavy metals (Sharma et al. 2016). PT are low-molecular-weight, cysteine-rich small polypeptides with a general structure $(\gamma\text{-Glu-Cys})_n\text{Gly}$ ($n = 2$–11) (Mirza et al. 2014). PC are one of the most important classes of metal chelators. PC-metal complexes are very stable and are formed and sequestered in vacuoles (Sharma et al. 2016). Several studies on metal detoxification via PC have suggested the important role of PC in the detoxification of heavy metals including Cr (Ao et al. 2022). MTP are described as metal efflux transporters such as Fe, Zn, Mn, Cd, Co and Ni from the cytoplasm generally to vacuoles or extracellular spaces to prevent cytoplasmic damages (Ricachenevsky et al. 2013). In *O. sativa*, expression of several mRNA encoding for MTP was induced after Cr(III) exposure in root and shoot (Yu et al. 2018a, b). However, few studies have investigated the detoxification response of MTP to Cr(III) exposure and the transport mechanisms of Cr(III) by MTP in plant remain unclear (Ao et al. 2022).

4.5 Conclusions

Cr(III) clearly has a variety of impacts on terrestrial and aquatic plants and therefore deserves full consideration by ecotoxicologists, stakeholders and regulators. Current consensus regards Cr(VI) as much more toxic than Cr(III) and underpins extensive research efforts to find economically viable processes based on the reduction of Cr(VI) to Cr(III) for remediation purposes. However, Cr(III) chemistry in ecotoxicological studies requires much better consideration to correctly understand the biological effects of this form of chromium. In particular, too few studies have checked the actual speciation of Cr(III) in the exposure media along with the measured biological responses. The lack of information on actual Cr(III) speciation in ecotoxicological studies can lead to an underestimation of Cr(III) toxicity and complicates both comparisons across studies and extrapolation of laboratory findings to real field situations.

The effects of Cr(III) on plants include inhibition of plant growth, seed germination process, damage to chloroplast, reduced photosynthesis, oxidative stress associated with generation of ROS, and alteration of nutrient balance, organelles and cellular function (Fig. 4.2). More knowledge is needed on Cr(III) speciation in ecotoxicological test media to establish reliable concentrations *vs.* responses relationships for all these effects and improve risk assessment for this important oxidation state of chromium in natural environments.

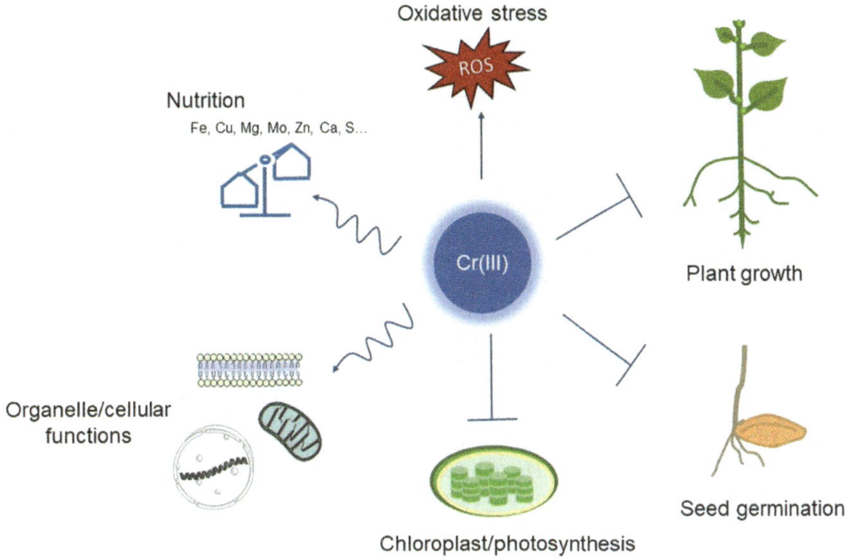

Fig. 4.2 Possible mechanisms of Cr(III) toxicity in terrestrial and aquatic plants

References

Adhikari A, Adhikari S, Ghosh S, Azahar I, Shaw AK, Roy D, Roy S, Saha S, Hossain Z (2020) Imbalance of redox homeostasis and antioxidant defense status in maize under chromium(VI) stress. Environ Exp Bot 169:103873

Aharchaou I, Py JS, Cambier S, Loizeau JL, Cornelis G, Rousselle P, Battaglia E, Vignati DAL (2018) Chromium hazard and risk assessment: new insights from a detailed speciation study in a standard test medium. Environ Toxicol Chem 37:983–992

Aharchaou I, Rosabal M, Liu F, Battaglia E, Vignati DAL, Fortin C (2017) Bioaccumulation and subcellular partitioning of Cr(III) and Cr(VI) in the freshwater green alga Chlamydomonas reinhardtii. Aquat Toxicol 182:49–57

Ahmad P, Jaleel CA, Salem MA, Nabi G, Sharma S (2010) Roles of enzymatic and nonenzymatic antioxidants in plants during abiotic stress. Crit Rev Biotechnol 30:161–175

Alcántara E, Romera FJ, Cañete M, De la Guardia MD (1994) Effects of heavy metals on both induction and function of root Fe(lll) reductase in Fe-deficient cucumber (*Cucumis sativus* L.) plants. J Exp Bot 45:1893–1898

Anjum SA, Ashraf U, Khan I, Tanveer M, Shahid M, Shakoor A, Wang L (2017) Phyto-toxicity of chromium in maize: oxidative damage, osmolyte accumulation, anti-oxidative defense and chromium uptake. Pedosphere 27:262–273

Ao M, Chen X, Deng T, Sun S, Tang Y, Morel JL, Qiu R, Wang S (2022) Chromium biogeochemical behaviour in soil-plant systems and remediation strategies: a critical review. J Hazard Mater 424:127233

Arduini I, Masoni A, Ercoli L (2006) Effects of high chromium applications on miscanthus during the period of maximum growth. Environ Exp Bot 58:234–243

Babula P, Adam V, Opatrilova R, Zehnalek J, Havel L, Kizek R (2008) Uncommon heavy metals, metalloids and their plant toxicity: a review. Environ Chem Lett 6:189–213

Barbosa RMT, de Almeida AAF, Mielke MS, Loguercio LL, Mangabeira PAO, Gomes FP (2007) A physiological analysis of *Genipa americana* L.: a potential phytoremediator tree for chromium polluted watersheds. Environ Exp Bot 61:264–271

Barceló J, Poschenrieder C, Dolores Vázquez M, Gunsé B (1993) Beneficial and toxic effects of chromium in plants: solution culture, pot and field studies. In: Vernet J-P (ed) Studies in environmental science. Elsevier, pp 147–171

Barton LL, Johnson GV, O'Nan AG, Wagener BM (2000) Inhibition of ferric chelate reductase in alfalfa roots by cobalt, nickel, chromium, and copper. J Plant Nutr 23:1833–1845

Beretta G, Mastorgio AF, Pedrali L, Saponaro S, Sezenna E (2018) Support tool for identifying in situ remediation technology for sites contaminated by hexavalent chromium. Water 10:1344

Beyersmann D, Hartwig A (2008) Carcinogenic metal compounds: recent insight into molecular and cellular mechanisms. Arch Toxicol 82:493–512

Bokare AD, Choi W (2014) Review of iron-free Fenton-like systems for activating H_2O_2 in advanced oxidation processes. J Hazard Mater 275:121–135

Briat J-F, Dubos C, Gaymard F (2015) Iron nutrition, biomass production, and plant product quality. Trends Plant Sci 20:33–40

Caldelas C, Araus JL, Febrero A, Bort J (2012) Accumulation and toxic effects of chromium and zinc in Iris pseudacorus L. Acta Physiol Plant 34:217–1228

Carpita NC, McCann MC (2000) The cell wall. In: Buchanan BB, Gruissem W, Jones RL (eds) Biochemistry and molecular biology of plants. American Society of Plant Physiologists, Rockville, MD, pp 52–108

Cary EE, Allaway WH, Olson OE (1977a) Control of chromium concentrations in food plants. 2. Chemistry of chromium in soils and its availability to plants. J Agric Food Chem 25:305–309

Cary EE, Allaway WH, Olson OE (1977b) Control of chromium concentrations in food plants. 1. Absorption and translocation of chromium by plants. J Agric Food Chem 25:300–304

Cervantes C, Campos-García J, Devars S, Gutiérrez-Corona F, Loza-Tavera H, Torres-Guzmán JC, Moreno-Sánchez R (2001) Interactions of chromium with microorganisms and plants. FEMS Microbiol Rev 25:335–347

Chatterjee J, Chatterjee C (2000) Phytotoxicity of cobalt, chromium and copper in cauliflower. Environ Pollut 109:69–74

Chow YN, Lee LK, Zakaria NA, Foo KY (2018) Phytotoxic effects of trivalent chromium-enriched water irrigation in *Vigna unguiculata* seedling. J Clean Prod 202:101–108

Cipriani HN, Bastos ARR, de Carvalho JG, da Costa AL, Oliveira NP (2012) Chromium toxicity in hybrid eucalyptus (*Eucalyptus urophylla* S. T. Blake X Grandis W. Hill Ex. Maiden) cuttings. J Plant Nutr 35:1618–1638

Davies FT, Puryear JD, Newton RJ, Egilla JN, Saraiva Grossi JA (2001) Mycorrhizal fungi enhance accumulation and tolerance of chromium in sunflower (*Helianthus annuus*). J Plant Physiol 158:777–786

Dazy M, Béraud E, Cotelle S, Meux E, Masfaraud JF, Férard JF (2008) Antioxidant enzyme activities as affected by trivalent and hexavalent chromium species in *Fontinalis antipyretica* Hedw. Chemosphere 73:281–290

de Oliveira LM, Lessl JT, Gress J, Tisarum R, Guilherme LRG, Ma LQ (2015) Chromate and phosphate inhibited each other's uptake and translocation in arsenic hyperaccumulator *Pteris vittata* L. Environ Pollut 197:240–246

de Oliveira LM, Gress J, De J, Rathinasabapathi B, Marchi G, Chen Y, Ma LQ (2016) Sulfate and chromate increased each other's uptake and translocation in As-hyperaccumulator *Pteris vittata*. Chemosphere 147:36–43

den Dooren de Jong LE, Roman WB (1965) Tolerance of *Chlorella vulgaris* for metallic and non-metallic ions. Antonie van Leeuwenhoek 31:301–313. https://doi.org/10.1007/BF02045910

Ding G, Jin Z, Han Y, Sun P, Li G, Li W (2019) Mitigation of chromium toxicity in Arabidopsis thaliana by sulfur supplementation. Ecotoxicol Environ Saf 182:109379 do Nascimento JL, de Almeida A-AF, Barroso JP, Mangabeira PAO, Ahnert D, Sousa AGR, Silva JVS, Baligar VC (2018) Physiological, ultrastructural, biochemical and molecular responses of young cocoa plants to the toxicity of Cr (III) in soil. Ecotoxicol Environ Safety 159:272–283

Dominik J, Vignati DL, Koukal B, Pereira de Abreu MH, Kottelat R, Szalinska E, Baś B, Bobrowski A (2007) Speciation and environmental fate of chromium in rivers contaminated with tannery effluents. Eng Life Sci 7:155–169

Dua A, Sawhney SK (1991) Effect of chromium on activities of hydrolytic enzymes in germinating pea seeds. Environ Exp Bot 31:133–139

Fan WJ, Feng YX, Li YH, Lin YJ, Yu XZ (2020) Unraveling genes promoting ROS metabolism in subcellular organelles of *Oryza sativa* in response to trivalent and hexavalent chromium. Sci Total Environ 744:140951

Fargašová A (2012) Plants as models for chromium and nickel risk assessment. Ecotoxicology 21:1476–1483

Fujita T (2015) Plasmodesmata: function and diversity in plant intercellular communication. J Plant Res 128:3–5

Fukushima M, Nakayasu K, Tanaka SHN (1995) Chromium(III) binding abilities of humic acids. Anal Chim Acta 317:195–206

Gardea-Torresdey JL, de la Rosa G, Peralta-Videa JR, Montes M, Cruz-Jimenez G, Cano-Aguilera I (2005) Differential uptake and transport of trivalent and hexavalent chromium by tumbleweed (*Salsola kali*). Arch Environ Contam Toxicol 48:225–232

Giusti L, Barakat S (2005) The monitoring of Cr(III) and Cr(VI) in natural water and synthetic solutions: an assessment of the performance of the DGT and DPC methods. Wat Air Soil Pollut 161:313–334

Gorny J, Billon G, Noiriel C, Dumoulin D, Lesven L, Madé B (2016) Chromium behavior in aquatic environments : a review. Environ Rev 24:503–516

Greene JC, Miller WE, Debacon M, Long MA, Bartels CL (1988) Use of *Selenastrum capricornutum* to assess the toxicity potential of surface and ground water contamination caused by chromium waste. Environ Toxicol Chem 7:35–39

Gustafsson JP, Persson I, Oromieh AG, van Schaik JWJ, Sjöstedt C, Kleja DB (2014) Chromium(III) complexation to natural organic matter: mechanisms and modeling. Environ Sci Technol 48:1753–1761

Henriques FS (2010) Changes in biomass and photosynthetic parameters of tomato plants exposed to trivalent and hexavalent chromium. Biol Plant 54:583–586

Hu L, Cai Y, Jiang G (2016) Occurrence and speciation of polymeric chromium(III), monomeric chromium(III) and chromium(VI) in environmental samples. Chemosphere 156:14–20

ISO (2012) Water quality—Fresh water algal growth inhibition test with unicellular green algae (Norm 8692)

Jin CW, You GY, He YF, Tang C, Wu P, Zheng SJ (2007) Iron deficiency-induced secretion of phenolics facilitates the reutilization of root apoplastic iron in red clover. Plant Physiol 144:278–285

Juarez AB, Barsanti L, Passarelli V, Evangelista V, Vesentini N, Conforti V, Gualtieri P (2008) In vivo microspectroscopy monitoring of chromium effects on the photosynthetic and photoreceptive apparatus of Eudorina unicocca and Chlorella kessleri. J Environ Monit 10:1313–1318

Juneja S, Prakash S (2005) The chemical form of trivalent chromium in xylem sap of maize (Zea mays L.). Chem Speciat Bioavailab 17:161–169

Kanazawa S, Sano S, Koshiba T, Ushimaru T (2000) Changes in antioxidative enzymes in cucumber cotyledons during natural senescence: comparison with those during dark-induced senescence. Physiol Plant 109:211–216

Kaya C, Ashraf M (2019) The mechanism of hydrogen sulfide mitigation of iron deficiency-induced chlorosis in strawberry (Fragaria × ananassa) plants. Protoplasma 256:371–382

Kitagawa M, Paultre D, Rademaker H (2015) Intercellular communication via plasmodesmata. New Phytol 205:970–972

Kumar D, Rajeshwari A, Jadon PS, Chaudhuri G, Mukherjee A, Chandrasekaran N, Mukherjee A (2015) Cytogenetic studies of chromium (III) oxide nanoparticles on Allium cepa root tip cells. J Environ Sci 38:150–157

Liang J, Huang X, Yan J, Li Y, Zhao Z, Liu Y, Ye J, Wei Y (2021) A review of the formation of Cr(VI) via Cr(III) oxidation in soils and groundwater. Sci Total Environ 774:145762

Liao P, Pan C, Ding W, Li W, Yuan S, Fortner JD et al (2020) Formation and transport of Cr(III)-NOM-Fe colloids upon reaction of Cr(VI) with NOM-Fe(II) colloids at Anoxic-Oxic interfaces. Environ Sci Technol 54:4256–4266

Liu D, Jiang W, Li M (1992) Effects of trivalent and hexavalent chromium on root growth and cell division of *Allium cepa*. Hereditas 117:23–29

Liu D, Kottke I (2003) Subcellular localization of chromium and nickel in root cells of *Allium cepa* by EELS and ESI. Cell Biol Toxicol 19:299–311

Liu J, Duan C-Q, Zhang X-H, Zhu Y-N, Hu C (2011) Characteristics of chromium(III) uptake in hyperaccumulator *Leersia hexandra* Swartz. Environ Exp Bot 74:122–126

López-Luna J, González-Chávez MC, Esparza-García FJ, Rodríguez-Vázquez R (2009) Toxicity assessment of soil amended with tannery sludge, trivalent chromium and hexavalent chromium, using wheat, oat and sorghum plants. J Hazard Mater 163:829–834

Löv Å, Sjöstedt C, Larsbo M, Persson I, Gustafsson JP, Cornelis G et al (2017) Solubility and transport of Cr(III) in a historically contaminated soil—Evidence of a rapidly reacting dimeric Cr(III) organic matter complex. Chemosphere 189:709–716

Maiti S, Ghosh N, Mandal C, Das K, Dey N, Adak MK (2012) Responses of the maize plant to chromium stress with reference to antioxidation activity. Braz J Plant Physiol 24:203–212

Mangabeira PA, Ferreira AS, de Almeida AAF, Fernandes VF, Lucena E, Souza VL, dos Santos Júnior AJ, Oliveira AH, Grenier-Loustalot MF, Barbier F, Silva DC (2011) Compartmentalization and ultrastructural alterations induced by chromium in aquatic macrophytes. Biometals 24:1017–1026

Marković S, Levstek L, Žigon D, Ščančar J, Milačič R (2022) Speciation and bio-imaging of chromium in *Taraxacum officinale* using HPLC post-column ID-ICP-MS, high resolution MS and laser ablation ICP-MS techniques. Front Chem 10:863387

Marschner H (1995) Mineral nutrition of higher plants, 2nd edn. Academic Press, 889 p. ISBN: 9780124735439

Medeiros MG, Rodrigues AS, Batoréu MC, Laires A, Rueff J, Zhitkovich A (2003) Elevated levels of DNA–protein crosslinks and micronuclei in peripheral lymphocytes of tannery workers exposed to trivalent chromium. Mutagenesis 18:19–24

Mirza N, Mahmood Q, Maroof Shah M, Pervez A, Sultan S (2014) Plants as useful vectors to reduce environmental toxic arsenic content. Sci World J 921581

Montes-Holguin MO, Peralta-Videa JR, Meitzner G, Martinez-Martinez A, de la Rosa G, Castillo-Michel HA, Gardea-Torresdey JL (2006) Biochemical and spectroscopic studies of the response of *Convolvulus arvensis* L. to chromium(III) and chromium(VI) stress. Environ Toxicol Chem 25:220–226

Moral R, Gómez I, Pedreno JN, Mataix J (1996) Absorption of Cr and effects on micronutrient content in tomato plant (*Lycopersicum esculentum* M.). Agrochimica

Murthy MK, Khandayataray P, Samal D (2022) Chromium toxicity and its remediation by using endophytic bacteria and nanomaterials: a review. J Environ Manag 318:115620

Oze C, Bird DK, Fendorf S (2007) Genesis of hexavalent chromium from natural sources in soil and groundwater. Proc Natl Acad Sci 104:6544–6549

Paiva LB, de Oliveira JG, Azevedo RA, Ribeiro DR, da Silva MG, Vitória AP (2009) Ecophysiological responses of water hyacinth exposed to Cr^{3+} and Cr^{6+}. Environ Exp Bot 65:403–409

Panda SK, Choudhury S (2005) Chromium stress in plants. Braz J Plant Physiol 17:95–102

Panda SK, Patra HK (2000) Nitrate and ammonium ions effect on the chromium toxicity in developing wheat seedlings. Proc Natl Acad Sci India. Section B, Biol Sci 70:75–80

Pandey N, Sharma CP (2003) Chromium interference in iron nutrition and water relations of cabbage. Environ Exp Bot 49(3):195–200. https://doi.org/10.1016/S0098-8472(02)00088-6

Peralta JR, Gardea-Torresdey JL, Tiemann KJ, Gomez E, Arteaga S, Rascon E, Parsons JG (2001) Uptake and effects of five heavy metals on seed germination and plant growth in alfalfa (*Medicago sativa* L.). Bull Environ Contam Toxicol 66:727–734. https://doi.org/10.1007/s001280069

Pereira M, Bartolomé MC, Sánchez-Fortún S (2013) Bioadsorption and bioaccumulation of chromium trivalent in Cr(III)-tolerant microalgae: A mechanisms for chromium resistance. Chemosphere 93:1057–1063

Pettine M, Genneri F, Campanella L, Millero FJ (2008) The effect of organic compounds in the oxidation kinetics of Cr(III) by H_2O_2. Geochim Cosmochim Acta 72:5692–5707

Pourrut B, Shahid M, Dumat C, Winterton P, Pinelli E (2011) Lead uptake, toxicity, and detoxification in plants. In: Whitacre DM (ed) Reviews of environmental contamination and toxicology, vol 213. Springer, New York, NY, pp 113–136

Prasad DDK, Prasad ARK (1987) Effect of lead and mercury on chlorophyll synthesis in mung bean seedlings. Phytochemistry 26:881–883

Qian XW (2004) Mutagenic effects of chromium trioxide on root tip cells of *Vicia faba*. J Zhejiang Univ Sci 5:1570–1576

Quaggiotti S, Barcaccia G, Schiavon M, Nicolé S, Galla G, Rossignolo V, Soattin M, Malagoli M (2007) Phytoremediation of chromium using Salix species: cloning ESTs and candidate genes involved in the Cr response. Gene 402:68–80

Rai D, Sass BM, Moore DA (1987) Chromium(III) hydrolysis constants and solubility of chromium(III) hydroxide. In: Inorganic chemistry, pp 345–349

Rai D, Eary LE, Zachara JM (1989) Environmental chemistry of chromium. Sci Total Env Chromium Paradox Modern Life 86:15–23

Ricachenevsky F, Menguer P, Sperotto R, Williams L, Fett J (2013) Roles of plant metal tolerance proteins (MTP) in metal storage and potential use in biofortification strategies. Front Plant Sci 4

Santana KB, de Almeida A-AF, Souza VL, Mangabeira PAO, Da Silva D, Gomes FP, Dutruch L, Loguercio LL (2012) Physiological analyses of *Genipa americana* L. reveals a tree with ability as phytostabilizer and rhizofilterer of chromium ions for phytoremediation of polluted watersheds. Environ Exp Botany 80:35–42

Sass BM, Rai D (1987) Solubility of amorphous chromium(III)-iron(III) hydroxide solid solutions. Inorg Chem 26:2228–2232

Schmidt W (1996) Influence of chromium(lll) on root-associated Fe(lll) reductase in *Plantago lanceolata* L. J Exp Bot 47:805–810

Scoccianti V, Crinelli R, Tirillini B, Mancinelli V, Speranza A (2006) Uptake and toxicity of Cr(III) in celery seedlings. Chemosphere 64:1695–1703

Scoccianti V, Bucchini AE, Iacobucci M, Ruiz KB, Biondi S (2016) Oxidative stress and antioxidant responses to increasing concentrations of trivalent chromium in the Andean crop species *Chenopodium quinoa* Willd. Ecotoxicol Environ Saf 133:25–35

Shahid M, Shamshad S, Rafiq M, Khalid S, Bibi I, Niazi NK, Dumat C, Rashid MI (2017) Chromium speciation, bioavailability, uptake, toxicity and detoxification in soil-plant system: a review. Chemosphere 178:513–533

Shanker AK, Pathmanabhan G (2004) Speciation dependant antioxidative response in roots and leaves of sorghum (*Sorghum bicolor* (L.) Moench cv CO 27) under Cr(III) and Cr(VI) stress. Plant Soil 265(1–2):141–151. https://doi.org/10.1007/s11104-005-0332-x

Shanker AK, Djanaguiraman M, Sudhagar R, Chandrashekar CN, Pathmanabhan G (2004) Differential antioxidative response of ascorbate glutathione pathway enzymes and metabolites to chromium speciation stress in green gram (*Vigna radiata* (L.) R.Wilczek. cv CO4) roots. Plant Sci 166:1035–1043

Shanker AK, Cervantes C, Loza-Tavera H, Avudainayagam S (2005) Chromium toxicity in plants. Environ Int 31:739–753

Sharma A, Kapoor D, Wang J, Shahzad B, Kumar V, Bali AS, Jasrotia S, Zheng B, Yuan H, Yan D (2020) Chromium bioaccumulation and its impacts on plants: an overview. Plants 9:100

Sharma DC, Sharma CP, Tripathi RD (2003) Phytotoxic lesions of chromium in maize. Chemosphere 51:63–68

Sharma R, Bhardwaj R, Handa N, Gautam V, Kohli SK, Bali S, Kaur P, Thukral AK, Arora S, Ohri P, Vig AP (2016) Responses of phytochelatins and metallothioneins in alleviation of heavy metal stress in plants: an overview. In: Ahmad P (ed) Plant metal interaction. Elsevier, pp 263–283

Singh HP, Batish DR, Kaur S, Arora K, Kohli RK (2006) α-Pinene inhibits growth and induces oxidative stress in roots. Ann Bot 98:1261–1269

Singh HP, Mahajan P, Kaur S, Batish DR, Kohli RK (2013) Chromium toxicity and tolerance in plants. Environ Chem Lett 11:229–254

Skeffington RA, Shewry PR, Peterson PJ (1976) Chromium uptake and transport in barley seedlings (*Hordeum vulgare* L.). Planta 132:209–214

Speranza A, Ferri P, Battistelli M, Falcieri E, Crinelli R, Scoccianti V (2007) Both trivalent and hexavalent chromium strongly alter in vitro germination and ultrastructure of kiwifruit pollen. Chemosphere 66:1165–1174

Speranza A, Taddei AR, Gambellini G, Ovidi E, Scoccianti V (2009) The cell wall of kiwifruit pollen tubes is a target for chromium toxicity: alterations to morphology, callose pattern and arabinogalactan protein distribution. Plant Biol 11:179–193

Srivastava S, Nigam R, Prakash S, Srivastava MM (1999) Mobilization of trivalent chromium in presence of organic acids: a hydroponic study of wheat plant (*Triticum vulgare*). Bull Environ Contam Toxicol 63:524–530

Stobart AK, Griffiths WT, Ameen-Bukhari I, Sherwood RP (1985) The effect of Cd2+ on the biosynthesis of chlorophyll in leaves of barley. Physiol Plant 63:293–298

Tang J, Zhang X, Zhang X, Li X, Chen J, Dong Y (2023) Transformation of chromium and its removal by Fe2O3 during the thermal disposal of municipal solid waste: a study based on density functional theory. Fuel 331:125734

Terrestrial and aquatic ecotoxicity assessment of Cr(VI) by the ReCiPe method calculation (LCIA): application on an old industrial contaminated site | SpringerLink, 2021. https://link.springer.com/article/https://doi.org/10.1007/s11356-012-1254-9. Accessed 10 Dec 21

Thompson SL, Manning FCR, McColl SM (2002) Comparison of the toxicity of chromium III and chromium VI to cyanobacteria. Bull Environ Contam Toxicol 69:286–293

Turbak SC, Olson SB, McFeters GA (1986) Comparison of algal assay systems for detecting waterborne herbicides and metals. Water Res 20:91–96

UdDin I, Bano A, Masood S (2015) Chromium toxicity tolerance of *Solanum nigrum* L. and *Parthenium hysterophorus* L. plants with reference to ion pattern, antioxidation activity and root exudation. Ecotoxicol Environ Saf 113:271–278

Vajpayee P, Khatoon I, Patel CB, Singh G, Gupta KC, Shanker R (2011) Adverse effects of chromium oxide nano-particles on seed germination and growth in *Triticum aestivum* L. J Biomed Nanotechnol 7:205–206

Vannini C, Domingo G, Marsoni M, Bracale M, Sestili S, Ficcadenti N, Speranza A, Crinelli R, Carloni E, Scoccianti V (2011) Proteomic changes and molecular effects associated with Cr(III) and Cr(VI) treatments on germinating kiwifruit pollen. Phytochemistry 72:1786–1795

Vazquez MD, Poschenrieder CH, Barcelo J (1987) Chromium VI induced structural and ultrastructural changes in bush bean plants (*Phaseolus vulgaris* L.). Ann Bot 59:427–438

Vernay P, Gauthier-Moussard C, Hitmi A (2007) Interaction of bioaccumulation of heavy metal chromium with water relation, mineral nutrition and photosynthesis in developed leaves of Lolium perenne L. Chemosphere 68:1563–1575

Vernay P, Gauthier-Moussard C, Jean L, Bordas F, Faure O, Ledoigt G, Hitmi A (2008) Effect of chromium species on phytochemical and physiological parameters in *Datura innoxia*. Chemosphere 72:763–771

Vignati DAL, Beye ML, Dominik J, Klingemann AO, Filella M, Bobrowski A, Ferrari BJ (2008) Temporal decrease of trivalent chromium concentration in a standardized algal culture medium:

experimental results and implications for toxicity evaluation. Bull Environ Contam Toxicol 80:305–310
Vignati DAL, Dominik J, Beye ML, Pettine M, Ferrari BJD (2010) Chromium(VI) is more toxic than chromium(III) to freshwater algae: a paradigm to revise? Ecotoxicol Environ Saf 73:743–749
Viti C, Marchi E, Decorosi F, Giovannetti L (2014) Molecular mechanisms of Cr(VI) resistance in bacteria and fungi. FEMS Microbiol Rev 38:633–659
Volland S, Lütz C, Michalke B, Lütz-Meindl U (2012) Intracellular chromium localization and cell physiological response in the unicellular alga Micrasterias. Aquat Toxicol 109:59–69. https://doi.org/10.1016/j.aquatox.2011.11.013
Wallace A, Soufi SM, Cha JW, Romney EM (1976) Some effects of chromium toxicity on bush bean plants grown in soil. Plant Soil 44:471–473
Wang SL, Lee JF (2011) Reaction mechanism of hexavalent chromium with cellulose. Chem Eng J 174:289–295
Warren LA, Haack EA (2001) Biogeochemical control on metal behaviour in freshwater environments. Earth Sci Rev 54:261–320
Woke J, Osu C, Chukwu U (2019) Dynamic impact of chromium on nutrient uptake from soil by fluted pumpkin (*Telfairia occidentalis*) 7:1–9
Wolf S, Greiner S (2012) Growth control by cell wall pectins. Protoplasma 249:169–175
Xu ZR, Cai ML, Chen SH, Huang XY, Zhao FJ, Wang P (2021) High-affinity sulfate transporter Sultr1; 2 is a major transporter for Cr(VI) uptake in plants. Environ Sci Technol 55:1576–1584
Yamada, M., Shiiba, S., (2015) Preparation of pectin-inorganic composite material as accumulative material of metal ions. Journal of Applied Polymer Science 132.
Yu X-Z, Gu J-D (2007) Accumulation and distribution of trivalent chromium and effects on hybrid willow (*Salix matsudana* Koidz x alba L.) metabolism. Arch Environ Contam Toxicol 52:503–511
Yu XZ, Gu J-D (2008a) Effect of available nitrogen on phytoavailability and bioaccumulation of hexavalent and trivalent chromium in hankow willows (Salix matsudana Koidz). Ecotoxicol Environ Saf 70:216–222
Yu XZ, Gu JD (2008b) The role of EDTA in phytoextraction of hexavalent and trivalent chromium by two willow trees. Ecotoxicology 17:143–152
Yu XZ, Gu JD, Xing LQ (2008c) Differences in uptake and translocation of hexavalent and trivalent chromium by two species of willows. Ecotoxicology 17:747–755
Yu XZ, Fan WJ, Lin YJ (2018a) Analysis of gene expression profiles for metal tolerance protein in rice seedlings exposed to both the toxic hexavalent chromium and trivalent chromium. Int Biodeterior Biodegrad 129:102–108
Yu XZ, Lu CJ, Li YH (2018b) Role of cytochrome c in modulating chromium-induced oxidative stress in *Oryza sativa*. Environ Sci Pollut Res 25:27639–27649
Yu XZ, Lin YJ, Zhang Q (2019) Metallothioneins enhance chromium detoxification through scavenging ROS and stimulating metal chelation in *Oryza sativa*. Chemosphere 220:300–313
Zayed A, Lytle CM, Qian JH, Terry N (1998) Chromium accumulation, translocation and chemical speciation in vegetable crops. Planta 206:293–299
Zayed AM, Terry N (2003) Chromium in the environment: factors affecting biological remediation. Plant Soil 249:139–156
Zeid IM (2001) Responses of *Phaseolus vulgaris* chromium and cobalt treatments. Biol Plant 44:111–115
Zeng F, Zhou W, Qiu B, Ali S, Wu F, Zhang G (2011) Subcellular distribution and chemical forms of chromium in rice plants suffering from different levels of chromium toxicity. J Plant Nutr Soil Sci 174:249–256
Zhitkovich A (2005) Importance of chromium−DNA Adducts in mutagenicity and toxicity of chromium(VI). Chem Res Toxicol 18:3–11

Ziller A, Fraissinet-Tachet L (2018) Metallothionein diversity and distribution in the tree of life: a multifunctional protein. Metallomics 10:1549–1559

Zhu L, Hong C, Zhang J, Qiu Y (2022) Long-distance mobilization of chromium(III) in soil associated with submicron Cr_2O_3. J Hazard Mater 130519

Chapter 5
Antioxidant Defence: A Key Mechanism of Chromium Tolerance

Monika Sharma, Shipa Rani Dey, and Prasann Kumar

Abstract Abiotic stresses, which include high salt accumulation, drought, high temperature, heavy metal stress, light, lack of nutrients, radiation, and many others, pose a constant threat to plants living in an environment that is in a state of constant change. The productivity, as well as the quality of the crops, may be significantly reduced as a result of such stresses. It has been established that Cr is a human carcinogen that can enter the body of a person either through inhalation or the consumption of food products that are contaminated with Cr. Due to the hazardous consequences of the deposition of chromium in the environment, as well as the hazards that the metal may produce, both the Agency for Toxic Substances and Disease Registry and the United States Environmental Protection Agency categorize chromium as a major contaminant. As Cr is found in nature in several valence states, such as Cr^{3+} and Cr^{6+}, it is possible to find it in several different valence states. Chromium (Cr) is a heavy metal that is known to produce reactive oxygen species (ROS), which are especially harmful to vegetation and need to be controlled to safeguard species against osmotic damage caused by high concentrations of Cr. One of the most dangerous and enduring types of Cr in the soil is Cr^{6+}. Reactive oxygen species (ROS), which are produced as a result of Chromium, as well as some cellular and metabolic processes can be disrupted. Researchers who study plant genetics and transcriptional control have discovered that when plants are under Cr stress, various genes involved in detoxification are up-regulated, which confers tolerance on the plants. The higher production of reactive oxygen species (ROS) is an important indicator of the presence of such stresses at the molecular level. ROS are highly reactive in their natural state because they can interact with many different molecules and metabolites found within cells, which can ultimately result in irreversible metabolic dysfunction and death. As ROS were produced and scavenged in various structures of plant cells, the ROS-scavenging routes that arise from the various components of plant cells can also be integrated with the ROS-producing routes that are found in plant cells. New research on plants has demonstrated that extremely small concentrations of ROS may serve as chemical messengers and raise a plant's sensitivity to

M. Sharma · S. R. Dey · P. Kumar (✉)
Department of Agronomy, School of Agriculture, Lovely Professional University, Phagwara, Punjab 144411, India
e-mail: prasann0659@gmail.com

© The Author(s), under exclusive license to Springer Nature Switzerland AG 2023
N. Kumar et al. (eds.), *Chromium in Plants and Environment*, Environmental Science and Engineering, https://doi.org/10.1007/978-3-031-44029-8_5

abiotic and biotic stresses by regulating the activities of protective genes. Several studies have also demonstrated that plants with higher antioxidant levels, whether these antioxidants are inherent or induced, are more resilient to a range of environmental challenges. This phenomenon has been observed in both wild and cultivated plants. We aim to synthesise current findings on the role of ROS in abiotic stress tolerance in this chapter as well as the possible regulatory roles that ROS may play. In addition, We go over the improvements that have been made in the last several decades in terms of enhancing plants' ability to withstand oxidative stress through the application of genetic engineering by various ROS detoxifying systems in plants.

Keywords Abiotic stress · Antioxidants · Chromium · Tolerance · ROS · Heavy metal · Toxicity · Plant · Zero hunger

5.1 Introduction

Abiotic stressors continue to be one of the most significant issues that limit agricultural yields in the world today. According to Rodriguez et al. (2005), Acquaah (2007), researchers have estimated that abiotic stresses directly account for more than fifty per cent of the decrease in yield. Several morphological, physiological, biochemical, and molecular changes in plants have been linked to abiotic stress, according to Wang et al. (2001), all of which negatively affect the growth and productivity of plants. However, it is worth noting that the punctuality and effectiveness of these responses may prove to be deciding factors in determining whether a given species is likely to be able to survive or not. Chromium (Cr) is a heavy metal that has oxidation numbers ranging from Chromium 2^+ to Chromium 6^+. In the modern periodic table, it is in the list of transition elements called VI-B (Abbas et al. 2018). Chromium is a hard metal and silver in colour. The atomic weight of this metal is 51.10 g/M and it has an atomic number of 24. On the list of the highest occurring metals on earth, this metal is ranked 21 on the list of its atomic number, density, and molecular weight (Acquaah 2007). It is estimated that the molecular weight of chrome is 51.10 g per million. There are two most stable forms of chromium found in nature, and they are Cr^{3+} and Cr^{6+} (Adejumo 2019). As the most toxic form of chromium, hexavalent chromium is considered to be the most toxic form because it is more water-soluble, mobile, as well as bioavailable than the other forms of chromium (Afonso et al. 2019). It is also a potentially powerful oxidizing agent. The oxygenated environment is capable of converting Cr^{3+} into Cr^{6+}, and The elements involved in preserving the ideal ratio of various chromium forms include oxygen content, pH, complexing agents, and reducing agents (Agrawal et al. 2009). The mining of chromium has dramatically increased in recent years as a direct result of the material's growing demand across a variety of industrial sectors (Ahmed et al. 2010). The countries of Kazakhstan, South Africa, China, and India are the top four users of chromium anywhere in the world (Al Mahmud et al. 2017; Ali and Alqurainy 2006). The enterprises of tanning leather, metalworking, metal plating, copper alloys, ceramic glazes, timber protection, moisture corrosive environment suppression, heat-resistant masonry, pneumatically wood products, textile materials, as well as dyestuff, powders and acrylics, and paper and pulp making, are to blame for the excessive amount of chromium in the environment. Also, the high levels of Chromium in the environment are caused by things that people do, like dumping liquid and solid wastes that are contaminated with Chromium (Apel and Hirt 2004; Asada 1994; Asada and Takahashi 1987; Ashraf et al. 2017; Augustynowicz et al. 2020). There is a belief that the emissions of Cr from cooling towers of industries are an important source of Chromium (Augustynowicz et al. 2014; Balasaraswathi et al. 2017; Barbosa et al. 2007). Debris or Impurities rising from road banks also seems to be a significant source of Chromium. In agricultural land, the accumulation of more Chromium can adversely affect the growth and development of a plant on multiple levels, including at the organ, cellular, and even genetic levels, depending on the amount accumulated in the soil (Bhargava and Mishra 2018). A higher concentration of chromium in plants can cause a lot of damage to plants because of the induced reactive oxygen species (ROS), which are responsible for both cellular and extracellular damage caused by higher levels of chromium in plants (Blokhnia et al 2003). The ROS in plants have a distinct function: they serve as chemical messengers that trigger the activity of detection systems in response to stressors, which they do if they occur in large amounts and increase cellular injury, in which case they trigger the activity of the detection systems. This in turn can lead to the production of protective molecules that help the plant defend itself against damage and continue healthy growth. There is a direct relationship between the concentration of ROS in the plant's environment and these two functions. This, in turn, may enable the production of specific chemicals that are capable of providing the plant with extra protection against any potential harm. This will enable it to remain healthy and thrive. The greater the amount of ROS present in the air, the stronger the beneficial effects. The formation of reactive oxygen species (ROS) is a by-product of abiotic stress that can endanger tissues if they are allowed to persist for a long period due to the increased formation of abiotic stressors. Thus, by increasing the amount of ROS present in the air, the plant can better protect itself against the harm caused by abiotic stress, and ultimately, lead a healthier and more successful life. As a result, lipids and proteins may peroxide, nucleic acids will be damaged, enzymes will be inhibited, the programmed cell death pathway (PCD) will be activated, and then the cells will eventually die (Mittler 2002; Sharma and Dubey 2005, 2007). Oxidative stress is mostly a controlled process, and the outcome for the plant depends on the equilibrium between reactive oxygen species and antioxidative capacity. Even before abiotic stresses are allowed to continue for just a longer duration, the increased production of reactive oxygen species puts cells at risk (ROS) (Apel and Hirt 2004). Based on their mode of action, these defences can be categorised into either enzymatic antioxidants or non-enzymatic antioxidants to simplify their classification. According to Asada and Takahashi (1987), this antioxidant defence system offers proper protection against active oxygen and free radicals when it is operating under controlled conditions. On the other hand, when an organism is subjected to a stressful situation, the balance between the synthesis and scavenging of relative oxygen species may be disrupted, resulting in a response that is either optimum or less (Gill and Tuteja 2010). The enhanced antioxidant defence has been shown in multiple studies to be effective in the fight against oxidative stress brought on by different abiotic stressors. As the impact of increasingly harsh environmental conditions on crop production, breeders and researchers are facing a pressing issue of creating genes capable of enduring biological variations with the least amount of harm. Therefore, to understand the processes of plant regulation and defence, the first step is to acquire knowledge that will allow you to better understand them. Developing plants that have a higher potential for antioxidation presents an opportunity to develop plants that have a higher tolerance for abiotic stresses. The purpose of this section is to present our current understanding of how plants react to abiotic stressors, both from a physiological and molecular genetic perspective. There has been a particular focus on the physicochemical and non-enzymatic modulation of antioxidant defences under abiotic stress as well as the link between these mechanisms and abiotic stress tolerance.

5.2 Plants Produce ROS as a Result of Their Metabolism

Apel and Hirt (2004) found that reactive oxygen species (ROS) are continuously produced as a result of metabolic activity throughout all the compartments within the plant cells, notably chlorophyll, ATP, and mitochondria, as a result of metabolic activities throughout these compartments. Chloroplasts are the primary organelles in plants that produce reactive oxygen species (ROS). A chlorophyll triplet state may develop when there is an insufficient amount of energy absorbed during photosynthesis. By shifting its energy from this state to oxygen, it can produce a molecule of oxygen by transferring its excitation energy to it (Logan 2005). The electron transport chain (ETC) in photosynthesis makes $O_2^{\cdot -}$ by reducing oxygen (Apel and Hirt 2004). Superoxide dismutase (SOD) then changes O_2^{cdot-}–H_2O_2 (Foyer and Noctor 2000). Reactive oxygen species in visible light are influenced by many physiological and environmental factors, such as a lack of water and access to bright light, which are important in the formation of reactive oxygen species. It appears that ribulose-1,5-bisphosphate carboxylase or oxygenase (RuBisCO) shows an increase in oxygenase activity when conditions prevent chloroplast CO_2 fixation and that the generated glycolate is transported from chloroplasts to peroxisomes under these conditions (Takahashi and Murata 2008). According to Halliwell (2006), the production of hydrogen peroxide in peroxisomes requires glycolate oxidation, which is broken down by glycolate oxidase (GO) and the breakdown of lipids. In contrast, a tiny electron transport chain (ETC) at the level of the peroxisomal membrane and a reaction of xanthine oxidase (XO) within the organelle matrix are both necessary for the generation of O_2 at the level of the peroxisomal membrane. It is important to note that both of these processes take place inside the organelle. It is using the ETC located in the cytoplasm of the plants that reactive oxygen species (ROS) are produced in plant tissues. A multi-complex dehydrogenase complex for the reduction of ubiquinone (Q) within the cell is composed of many small dehydrogenase units that work together. Complex I (NADH dehydrogenase) and the Q zone are likely to be responsible for the majority of ROS production in cells (Mller 2001; Blokhina et al. 2003). Although mitochondrial ROS production is significantly lower than that of chloroplasts, mitochondrial ROS regulate a variety of cellular processes, such as stress adaptation and programmed cell death (PCD), despite the fact that mitochondrial ROS are important regulators of the processes described above (Robson Vanlerberghe 2002). The primary enzyme in charge of producing hydrogen peroxide in glyoxysomes is aryl-CoA oxidase. Plasma membrane-bound NADPH oxidases (NADPHox) and peroxidases connected to cell walls serve as the primary suppliers of reactive oxygen and H_2O_2 generated by apoplastic enzymes (POX). It has been shown that these enzymes become active when they are exposed to different types of stress (Mittler 2002). The catalysis of some reactions which are used in detoxification is done by cytochromes in both the cytoplasm as well as in ER of plant cells are additional sources of reactive oxygen species (ROS) in plant cells (Urban et al. 1989).

5.3 The Effects of Chromium on Oxidative Stress in Plants

There is sufficient evidence to suggest that oxidative injury can be caused to plants when they are subjected to redox-active HM toxicity. Following HM absorption by carriers and movement to plant components, relative oxygen species production occurs. This is triggered either by the heavy metal redox process or by how an HM affects metabolism at a particular subcellular location. After HM is taken up by transporters and distributed to organelles, ROS is produced. Activation of plasma in a manner dependent on HM-membrane-localized NADPH oxidase is another enzyme that plays a role in the production of ROSHMs that is redox-active and facilitate redox processes in the cell include iron, copper, chromium, vanadium, and cobalt, as distinct from HMs that are physiologically non-redox-active, such as Zn^{2+} and Cd^{2+}.They contribute to the generation of hydroxyl radicals from superoxide anion (H_2O_2) via the Haber-Weiss and Fenton reactions, which initiate the process of non-specific lipid oxidation. There is also one element that contributes to the specific increase of lipid peroxidation, and that is the activation of lipoxygenases (LOX) that are dependent on HM (Montillet et al. 2004). The formation of reactive oxygen species (ROS) by plants is a form of self-defence when they are exposed to hostile conditions (Byrne et al. 2017; Chakraborty and Pradhan 2011). There are several kinds of endogenous stress, but the most common is the accumulation of reactive oxygen species (ROS), which can lead to a reduction in plant growth and development (Chalapathi Rao and Reddy 2008). The plants produce a variety of reactive oxygen species (ROS), including hydrogen peroxide (H_2O_2), superoxide anion (O_2), singlet oxygen ($^1O_2$), and hydroxyl ion (HO−), peroxyl (RO−), alkoxyl (RO−), as well as organic hydroperoxide (ROOH) (Chandra 2004; Chen et al. 2003, 2017). Hydrogen peroxide is one of the most prevalent types of ROS in the environment.

According to Conklin (1996), Cui (2017), De Tullio (2004), reactive oxygen species (ROS) are formed as a consequence of a range of metabolic processes occurring in mitochondria, peroxisomes, and chloroplasts. As mentioned above, ROS levels in plants are regulated by several mechanisms, including ROS production, enzymatic scavenging of ROS, and/or non-enzymatic scavenging of ROS (del Ro et al. 2006). As a result of exposure to the following metals, lead (Pb), aluminium (Al), nickel (Ni), cadmium (Cd), and chromium (Cr), there is a significant correlation between the generation and accumulation of reactive oxygen species (ROS) in the body (Dixon 2010; Elsawy et al. 2017; Eltayeb et al. 2006). The induced ROS were accumulated by a wide range of species of plants after they were exposed to either a harmful quantity of Cr or industrial wastes containing a toxic level of Cr. There is considerable evidence that the exposure of plants to chromium results in the formation of reactive oxygen species (ROS), which have a variety of physiological, metabolic, molecular, and morphological effects (Eltayeb et al. 2007). There is a possibility that Cr may interact directly with proteins, lipids, enzymes, and genetic material (DNA and/or RNA) to change physiological and biochemical processes, or it may trigger the accumulation of reactive oxygen species (ROS) within the organism (Fargasova 2012; Florea 2017; Foyer and Noctor 2000). In addition to damaging the membrane, Cr also destroys and inactivates genetic material, proteins, and enzymes, resulting in growth suppression by inhibiting cell division or triggering programmed cell death as a result of interactions with Cr (Fryer 1992; Gapper and Dolan 2006; Ghosh et al. 2017; Gielen et al. 2017). As a result of chromium-induced ROS, morphological changes are induced in a variety of plant tissues in an amount and tissue-specific manner. These morphological changes are irreversible and damage biomolecules of the plant except for DNA, cysteine, and methionine, which can be reconstituted (Hasanuzzaman et al. 2017). As a consequence, chromium-induced ROS are responsible for the destruction of biomolecules. Cr(VI) is reduced to Cr(III) by reactive oxygen species that are formed during the reduction process. The same happens with the Fenton reaction. The Fenton reaction is a catalytic reaction where Cr(III) has a higher catalytic efficiency than iron (Fe), copper (Cu), cobalt (Co), manganese (Mn), and zinc (Zn) (Kalve et al. 2011). There has been little investigation into the role of Cr in such reactions, and it has also been suggested that various other intermediates and variables may also be involved in the production of ROS as a result of Cr (Mittova et al. 2003; Mobin and Khan 2007). Several physiological, biochemical, molecular, and ultrastructural changes were caused by ROS, which acted as a mediator.

5.4 Non-enzymatic Antioxidants

5.4.1 Ascorbate

Ascorbate (AsA) is a vital antioxidant that is found in plant tissues. In higher plants, it is produced in the cytosol, primarily by the conversion of d-glucose to ascorbate, which is a major source of ascorbic acid. As a result of its ability to react with a wide range of reactive oxygen species (ROS), which include H_2O_2, $O^{2\cdot}$, and ${}_1O^2$, it can exert its antioxidant activity. AsA, an electron donor with terminal properties, plays an essential role here by scavenging free radicals from the hydrophilic environment in which plants live. AsA is also an essential molecule in the antioxidant defence mechanism of plants, as it plays a vital role in reacting with ROS to preserve cellular integrity and prevent oxidative stress. In addition to this, it is capable of scavenging OH· at rates controlled by diffusion (Smirnoff 2000). APX uses the AsA-GSH cycle to produce MDHA by converting two molecules of AsA into the water through the reduction of H_2O_2, which is also the result of APX using two molecules of AsA to make H_2O_2. MDHA is a radical with a short half-life that degrades disproportionately into DHA and AsA as a result of its short half-life. As stated by Gapper and Dolan (2006), MDHAR or ferredoxin is responsible for catalyzing the reactions that occur within a chloroplast water-water cycle through the action of ferredoxin. It is common practice to use NADPH as an electron donor for a variety of purposes. In plant cells, AsA is the major reducing substrate that is used for the removal of H_2O_2 (Wu et al. 2007). In addition to the reduction of a-tocopherol, an antioxidant found in chloroplasts, AsA is also assumed to be necessary. According to Conklin et al. (1996) research on AsA in plants, this enzyme might play a role in the formation of the pigment zeaxanthin in plants, which protects against oxidative damage through the elimination of excess light energy from the thylakoid membranes. In addition, according to De Tullio (2004), ASA is also responsible for maintaining the reduced state of prosthetic metal ions, which, in turn, is responsible for the functioning of several antioxidant enzymes. Based on research conducted by Hasanuzzaman et al. (2011a, b), AsA plays a critical role in plant tolerance to abiotic stress. Exogenous administration of AsA reduces the damage caused by oxidative processes by affecting the activity of a large number of enzymes. In addition, it works synergistically with other antioxidants (Shalata and Neumann 2001). It has been shown that glutathione (GSH) is an antioxidant that acts as an antioxidant and is directly involved in the process of lowering the majority of ROS, according to Noctor and Foyer. Moreover, GSH plays a critical role in Foyer and Halliwell's antioxidative defence system, as it plays a key role in the regeneration of other potential water-soluble antioxidants, such as AsA, through the AsA-GSH cycle, which promotes the regeneration of other water-soluble antioxidants. In doing so, it preserves a-tocopherol and zeaxanthin in a diminished state, which is how it indirectly protects membranes by preserving those two compounds in a diminished state. During times of stress, GSH acts to protect proteins from denaturation. This would otherwise result from the oxidation of the thiol groups in these proteins as a result of oxidative stress. GST and GPX both use GSH as a substrate for their enzymes, with both of them contributing to the elimination of ROS in the body (Noctor et al. 2002). Phytochelatins (PCs) are also produced by GSH, which has an affinity for HM and is transported as complexes into the vacuole, enabling plants to be somewhat resistant to HM if they accumulate enough phytochelatins. As a reduced sulfate, GSH is also a catalyst for the breakdown of xenobiotics and acts as a medium for the storage and transportation of this element (Srivalli and Khanna-Chopra 2008). A stress marker can be derived from the fact that glutathione plays a very significant role in the antioxidant defences of the body. The ratio of H_2O_2 reduced (GSH) to oxidized (GSSG) forms changes over the course of the breakdown of the gas. Several redox signalling pathways rely on this modification for their function (Li and Jin 2007). As a result of increased GSH levels, plants are protected from the damaging effects of oxidative stress. GSH functions as a redox sensor for environmental cues. Research has shown that by increasing the levels of GSH in the body, we are better able to cope with a range of abiotic stresses (Hasanuzzaman et al. 2011a).

5.4.2 Tocopherol

It has long been known that tocopherols are abundant in thylakoid membranes, which are also rich in polyunsaturated fatty acids (PUFAs) and are located close to ROS that is produced during photosynthesis (Fryer 1992). Consequently, these compounds are thought to play an important role in protecting thylakoid membranes from oxidative damage. Munne-Bosch and Alegre have found that tocopherol appears to have an important antioxidant role, based on circumstantial and correlative evidence. Additionally, the scientist suggests that tocopherol could be a critical component of the photoprotective system, underscoring its potential antioxidant potential. As a result of the conversion of lipid peroxyl radicals (LOO) to their corresponding hydroperoxides, tocopherols prevent lipid peroxidation from occurring. This action of tocopherols prevents the oxidation of lipids, thus protecting cellular membranes from damage. Tocopherols play an important role in reducing reactive oxygen species (mostly $_1O^2$ and $OH^{\cdot}$) in the membranes of photosynthetic organisms in this way. By preventing lipid peroxidation, tocopherols protect the delicate cellular membranes from oxidative damage caused by reactive oxygen species, thus preserving photosynthetic organisms. Tocopherols are capable of physically quenching oxygen in chloroplasts. Tocopherols can donate electrons to the reactive oxygen species, neutralizing them and preventing them from damaging the cell membrane. This process is known as "quenching" and it is thought to be the primary mechanism by which tocopherols mitigate oxidative damage. According to Munné-Bosch, one molecule of a-tocopherol can deactivate as many as 120 molecules of oxygen in a single reaction. The fact that tocopherols are part of a complex signalling network that is regulated by reactive oxygen species (ROS), antioxidants, and plant hormones also make them an excellent pick for affecting cellular signalling in plants in a positive manner. This means that the presence of tocopherols can help reduce the damaging effects of ROS, which can interfere with the proper functioning of cells. Additionally, tocopherols can act as messengers that help regulate the expression of genes and other plant hormones, which can be beneficial for plant growth and development.

5.4.3 *Components of an Enzyme*

It should be noted that the enzymes involved in removing ROS are located in a variety of places within plant cells, and they work in concert together. There are several enzymes that are involved in the AsA-GSH cycle, as well as SOD, CAT, GPX, and GST, which are considered to be the most important antioxidant enzymes. These enzymes are responsible for removing ROS from the cell and ultimately preventing oxidative damage to the plant. They do this by using the molecules of ascorbate and glutathione, which are found in abundance in the cells, as well as other molecules like superoxide dismutase, catalase, glutathione peroxidase, and glutathione-S-transferase. Together, these enzymes work to neutralize the ROS and prevent it from causing damage. Aside from AsA, GSH, and NADPH, four other enzymes play a significant role in the AsA-GSH cycle, which are known as APX, MDHAR, DHAR, and GR. The enzymes in this cycle, together with the other components of the cycle, play an important role in deactivating H_2O_2 and regenerating AsA and GSH through a series of cyclic processes. APX, MDHAR, DHAR and GR catalyze the redox reactions of the AsA-GSH cycle, and also reduce the amount of oxidizing agents, such as H_2O_2, in the cell. These enzymes also help to maintain the AsA and GSH levels in thc ccll, so they can be used to neutralize the harmful effects of free radicals and other toxic compounds.

5.4.4 *Superoxide Dismutases (SOD)*

In terms of protecting plant cells from reactive oxygen species (ROS), SODs are considered to be the first line of defence. To accomplish this, it catalyzes the dismutation of $O^{2\cdot}$. This results in the conversion of one molecule of $O^{2\cdot}$ into H_2O_2 and the oxidation of another molecule of $O^{2\cdot}$ into O_2. In this way, the presence of $O^{2\cdot}$ is removed from the system. It has been found that metal ions such as manganese (MnSOD), copper and zinc (Cu/ZnSOD), and iron are incorporated into SOD active sites and they are used as a classification of SODs (FeSOD). Although MnSOD can be found in the matrix of mitochondria and peroxisomes, Cu/ZnSOD can be observed in the cytosol and chloroplasts of higher plants, and FeSOD can be found in the chloroplasts of some higher plants, despite its location in mitochondria and peroxisomes (Scandalios 1993). Singh et al. (2008) have found that an increase in the activity of SODs can contribute to the mitigation of the effects of abiotic oxidative stress. MnSOD, Cu/ZnSOD, and FeSOD are all types of superoxide dismutase (SOD) enzymes that have different locations in the cell, but all have the same purpose of mitigating the effects of oxidative stress caused by abiotic stressors. Therefore, an increase in SOD activity can help protect the cell from these kinds of stressors.

5.4.5 *Catalases (CAT)*

Catalases are heme-containing tetrameric enzymes that are involved in the conversion of hydrogen peroxide to water and oxygen. Catalases are essential for many metabolic reactions, as they are extremely efficient in breaking down hydrogen peroxide, which can be harmful to cells. The catalase enzymes are responsible for protecting cells from oxidative damage by using hydrogen peroxide as a substrate and converting it into water and oxygen as a result, Sanchez-Casas and Klesseg (1994). Furthermore, catalase enzymes are widely distributed in organisms and are important in biochemical reactions such as photosynthesis and respiration, where they act as protective agents against oxidative damage. Catalases are found in peroxisomes, glyoxysomes, and other organelles that are connected to peroxisomes, as well as enzymes that produce H_2O_2 (Agarwal et al. 2009). According to Gill and Tuteja (2010), CAT plays a critical role in the removal of H_2O_2, which is generated in the peroxisome by oxidases that are involved in photorespiration, b-oxidation of fatty acids, and purine catabolism within the peroxisome. The reaction between CAT and various hydroperoxides has been demonstrated in addition to the reaction between CAT and molecular oxygen by Ali and Alqurainy (2006). It has been found that different patterns of response to different abiotic stressors and the level of CAT activity have been observed by Fujita.

5.4.6 *AsA-GSH Cycle Enzymes*

In mitochondria, chloroplasts, apoplasts, cytosols, and peroxisomes, the AsA-GSH cycle serves as the first line of defence against reactive oxygen species (ROS) present in mitochondria, chloroplasts, apoplasts, and cytosols. There are four enzymes involved in the AsA-GSH cycle, namely APX, MDHAR, DHAR, and GR, that are also implicated in the AsA-GSH cycle, along with AsA, GSH, and NADPH.

In the process of deactivating Molecular oxygen, these enzymes and the other components of the cycle work together to regenerate AsA and GSH and to regenerate Molecular oxygen through a series of cyclic processes. In this cycle, APX is responsible for catalyzing the reduction of H_2O_2–H_2O, as well as generating monodehydroascorbate (MDHA), which is a precursor to vitamin C. It is then converted into ascorbic acid (AsA) through the action of NADPH-dependent MDHAR, or it can be converted nonenzymatically to ascorbic acid and dehydroascorbic acid (DHA) by nonenzymatic means. As such, the conversion of DHAA to AsA is of significant importance in the biosynthesis of ascorbic acid. Either DHA undergoes a hydrolysis process that cannot be reversed or it undergoes an oxidation process. With the help of DHAR, which utilizes GSH as a reductant, the 2,3-diketogulonic acid can either be converted back to AsA or 2,3-diketogulonic acid. By converting GSSG to GSH, Chen et al. (2003) demonstrate that GSSG will be produced, and GR will then take care of converting it back into GSH. Ascorbate peroxidase (APX) plays a major role in the removal of hydrogen peroxide from the ascorbate-glutathione (GSH) cycle in the first phase. Asada (1994) states that in higher plants this process may be the most critical stage in the process of removing reactive oxygen species and protecting cells from oxidative stress (Asada 1994). Heme-containing enzymes such as APXs play a very important role in the elimination of oxygen molecule H_2 during the hydrological cycle as well as during the AsA-GSH cycle in the body. Several enzymes use AsA as a substrate and help in the transfer of electrons from AsA to H_2O_2 through these enzymes. There are five isoforms of the protein in the APX family, which results in different amounts of DHA and water. The chloroplast stroma soluble form (sAPX), mitochondrial form (mAPX), glyoxisome membrane form (gmAPX), and thylakoid (TapX).form are the four types. In addition to the cytosolic form of APX (CapX), there is also a cytosolic form of APX. In response to a wide range of diverse abiotic stress situations, plants show enhanced APX activity in response to a wide range of abiotic stress responses Hasanuzzaman and Fujita (2011).

5.4.7 Monodehydroascorbate Reductase (MDHAR) and Dehydroascorbate Reductase (DHAR)

AsA is engaged in a univalent oxidation process that leads to the formation of MDHA when it oxidizes univalent. This oxidation of AsA into MDHA is an essential step in the process of cellular energy production. It is imperative to keep in mind that if MDHA is not converted back to AsA by MDHAR, then MDHA will degrade on its own. Additionally, without MDHAR, the concentration of MDHA would increase, leading to inefficient use of energy and potential toxic effects. This will result in AsA and DHA if the process is not interrupted. AsA is then transformed into DHA by DHAR in a subsequent process, which also calls for GSH, in which DHA is regenerated into AsA by DHAR. Therefore, it is essential to ensure that MDHA is converted back to AsA through MDHAR and that the GSH-dependent regeneration of DHA from AsA is also maintained. To maintain the antioxidative capacity of AsA, rapid regeneration is essential. The regeneration of AsA during this cycle is mainly controlled by the activity of MDHAR that is NADPH-dependent. To regenerate AsA and maintain a low level of AsA in the body, this is crucial. An extensive range of crops was tested in this trial, including a variety of different varieties. A study published by Hossain et al. (2011) showed that MDAHR plays a key role in the regulation of oxidative stress tolerance as well as acclimation to environmental conditions. Despite this, there are few reports of Monodehydroascorbate reductase activity in other oxidative stress-related physiological processes. During oxidative stress, MDHAR and DHAR both play a crucial role in the regulation of the quantity of ascorbic acid (AsA) and the redox state of this substance (Eltayeb et al. 2007).

This suggests that MDHAR, along with other enzymes in the antioxidant network, may be involved in the process of acclimation to environmental conditions, and it could be an important factor in the regulation of the ascorbic acid content of cells. It is also likely that MDHAR could play a role in the control of other physiological processes that are associated with oxidative stress. This indicates that MDHAR and DHAR are important in maintaining the AsA concentration, redox state and stress tolerance of an organism. Furthermore, this suggests that MDHAR influences the overall stress tolerance of organisms, which is important for the survival of organisms in their environment. Besides the dehydroascorbate reductase, which maintains the cellular redox state of AsA, the recycling system of AsA is also highly dependent on dehydroascorbate reductase, which regenerates AsA from its oxidized state (DHA) Martinez and Araya (2010). Thus, the ability to be able to withstand various abiotic stressors resulting in the generation of ROS is critical for the ability to survive there. In a study carried out by Hasanuzzaman et al. (2011a, b), it was found that different ROS-inducing stimuli had a positive effect on the level of DHAR activity.

5.4.8 Glutathione Reductase (GR)

There is a possibility that glutathione reductase (GR) may play an important role in the ascorbic acid-glutathione (GSH) cycle. Furthermore, it is also an important part of the body's defence mechanism against the damage caused by oxygen radicals and other reactive chemicals (ROS). In addition to increasing the cell's tolerance to stress, an increase in GR activity can have a significant effect on the oxidation and reduction states of the essential electron transport chain components. It is necessary for the preservation of the GSH pool because GR catalyzes the reduction of the disulfide links in GSSG in an NADPH-dependent manner (Chalapathi Rao and Reddy 2008). The enzyme acts as a reductant for GSH, a compound that plays an important role in a wide range of metabolic functions as well as antioxidative properties in plants. Thus, GR is the gene that controls a high ratio of GSH to GSSG in plant cells, which is not only required but also necessary for the pathway that removes hydrogen peroxide to accelerate, especially under a stressful situation (Pang and Wang 2010). So, GR ensures that plant cells always have a high ratio of GSH to GSSG. GR is an incredibly crucial aspect of plant development that determines how effectively plants will be able to cope with a variety of stresses because it ensures that the cell's antioxidant machinery is functioning properly and, as a result, provides resistance to stress (Hasanuzzaman et al. 2011a).

5.4.9 Glutathione Peroxidases

In this study, glutathione peroxidases (GPXs) will be investigated as they are enzymes that are known to protect plant cells from the detrimental effects of oxidative stress by decreasing the levels of hydrogen peroxide (H_2O_2) as well as organic and lipid peroxides (LOOHs). Additionally, GPXs have been found to reduce the potential for DNA damage, which can cause plant cells to be more susceptible to stress. The GSH family of enzymes is composed of a large number of isozymes. According to Kühn and Borchert, GPX is not only an integral part of the cellular metabolism that may be involved in the re-oxidation of membrane lipids, but it is also a repurposing defence against oxidative damage to the membrane. GPXs are found to reduce the potential for DNA damage by scavenging reactive oxygen species (ROS) and preventing the oxidation of membrane lipids. This helps protect the cell's membrane from oxidative damage and increases the cell's ability to withstand stress. It was reported a few years ago that several GPX genes had been extracted from various plant species and that these genes were associated with H_2O_2 detoxification. Furthermore, GPX also functions as an oxidative signal transducer (Miao et al. 2006). Therefore, GPX plays a crucial role in both protecting the cell from oxidative damage and in promoting cellular stress tolerance.

5.4.10 Glutathione S-Transferases (GST)

As an enzyme that belongs to a domain that has been identified as being involved in catalyzing the conversion of electrophilic xenobiotic substrates into GSH, plant GSTs are versatile enzymes Dixon et al. (2010). Plant GSTs can recognize and modify a wide range of electrophilic xenobiotics, such as herbicides, pesticides and industrial pollutants, by catalyzing the transfer of the electrophilic group from the xenobiotic to the glutathione (GSH) molecule, which is a key component in the plant's defence mechanisms. As reported by Marrs (1996), GST isoenzymes account for about 1% of the total soluble protein of a plant. In the GSH metabolic pathway, these enzymes play an important role. GSTs help protect the plant from environmental damage caused by these xenobiotics by binding to them and detoxifying them. They also act as scavengers of reactive oxygen species, which are generated as a result of oxidative stress caused by these pollutants. This helps the plant to maintain its health by removing these potentially damaging molecules from its environment. According to Edwards et al. GSTs are proteins that catalyze the binding of numerous xenobiotics and their electrophilic metabolites to GSH. Among these xenobiotics are a wide range of pesticides, resulting in the formation of conjugates that are less hazardous and more readily soluble in water due to the reduction in hazardous properties. As a result, GSTs are a crucial defence mechanism for plants against pollutants, as they help reduce the toxicity of xenobiotics and facilitate their excretion from the plant's cells. Furthermore, GSTs also help protect the plant's cells from oxidative damage caused by reactive oxygen species. In a study by Gullner and Komives it was found that GST isoenzymes contain POX activity as well as their ability to catalyze the conjugation of electrophilic molecules to GSH. Abiotic stressors act in a variety of ways as effective inducers of GST activity in plants as a result of diverse abiotic factors. This is because GSTs can act as a scavenger for reactive oxygen species (ROS) like superoxide radicals and hydrogen peroxide, which are generated in large amounts in response to abiotic stress. GSTs can also conjugate these ROS to glutathione, a small peptide, thereby helping to reduce the ROS levels and protect the plant against oxidative damage. Hossain et al. (2011) found that plant GSTs play an important role in how plants adapt to different types of abiotic stress and provide plants with the ability to survive under stress (Tables 5.1 and 5.2).

5.5 Heavy Metal Stress Exposes Plants to a Range of Antioxidant Defence Mechanisms

To detoxify ROS, plants use both non-enzymatic antioxidants (AsA, GSH, a-tocopherol, and carotenoids) and enzymatic antioxidants (APX, SOD, CAT, GR, DHAR, MDHAR, GPX, and GST). It is this ROS-detoxifying antioxidant defence machinery in the plant that protects it from reactive oxygen species damage as well as repairing any damage it may sustain. There is a strong response from plant antioxidative mechanisms to HM exposure, but the direction in which the response takes place depends on the plant organ, plant species, HM utilized, and the intensity of HM stress. In an investigation conducted by Anjum et al. (2011), it was observed that both tolerant and sensitive cultivars of mung bean were found to have a significant decrease in AsA, the AsA/DHA ratio, GSH, and the GSH/GSSG ratio when treated with Cd (100 mg/kg soil). It should be noted that both the vulnerable variety as well as the resistant variety saw declines that were substantially less severe when compared to the vulnerable variety. Cd's stress tolerance is enhanced by the presence of AsA and GSH pools, which lends credence to the idea that these pools serve as protective mechanisms. Under the influence of high levels of HM stress, glutathione participates in bioreductive processes, in which it serves as a critical line of defence against reactive oxygen species (ROS). This aberrance is responsible for the loss of protection of cells from the negative effects of oxidative stress. Inhibiting metal transport into and out of cells, as well as chelating metal ions within cells, are two further methods for reducing metal toxicity. Since GSH is capable of directly scavenging metals, GSH may play a crucial role in HM tolerance and sequestration in a variety of ways (Wójcik and Tukiendorf 2011). It has been shown that the presence of additional GSH in rice plants under controlled conditions had a significant effect on the plants' sensitivity to the effects of cadmium stress. It was discovered that the growth inhibition induced by Cd could be significantly alleviated by applying GSH exogenously to both genotypes and that the uptake of Cd could be significantly reduced by the application of GSH exogenously. On the other hand, Wójcik and Tukiendorf (2011) have discovered that the amount of endogenous GSH present naturally in wild-type Arabidopsis plants is sufficient for them to be resistant to Cd stress. As a result, this contradicts the conclusions that were reached by previous researchers. Plants that have a smaller amount of GSH are less susceptible to Cd, whereas plants that have a higher amount of GSH are less susceptible to Cd, and

Table 5.1 List of chromium-tolerant plant species, their habitats and tolerance mechanisms

Family	Plant	Habitat	Tolerance mechanism	References
Apocynaceae	Calotropis procera (Aiton) W. T. Aiton	Large shrub or small tree	Increased activities of superoxide dismutase (SOD), catalase (CAT), and glutathione reductase (GR)	Usman et al.
Brassicaceae	*Brassica napus* L	Annual or biennialherb	Gentle remediation options (GROs)	Tauqeer et al.
Commeliniaceae	*Tradescantia pallida* (Rose) D. R. Hunt	Succulent perennial herb	Increased anti-oxidant activity	Sinha et al. (2014)
Plantaginaceae	*Callitriche cophocarpa* Sendtn	Water-submerged, macrophyte	Hyper-accumulation	Augustynowicz et al. (2020)
Solanaceae	*Solanum viarum* Dunal	Perennial shrub	Hyper-accumulation	Afonso et al. (2019)
Aizoaceae	*Mesembryanthemum crystallinum* L	Large, mat-forming annual with sprawling stems	Phyto-extraction	Sliwa-Cebula et al. (2020)
Callitraceae	*Callitriche cophocarpa* Sendtn	Aquatic macrophyte	Cr VI reduction	Augustynowicz et al. (2014)
Convolvulaceae	*Ipomoea aquatica* Forssk	Semi-aquatic, tropical plant	Hyper-accumulation	Haokip and Gupta (2020)
Pteridaceae	*Pteris vittata* L	Fern species	Hyper-accumulation	Kalve et al. (2011)
Amaranthaceae	*Gomphrena celosoides* Mart	Perennial herb	Increased proline and antioxidant enzyme activities	Adejumo et al.
Cannabaceae	*Cannabis sativa* L	Annual, herbaceous, flowering	Hyper-accumulation	Ullah et al. (2019), Sajad et al. (2020)
Euphorbiaceae	*Euphorbia helioscopia* L Rumex dentatus L	Desert, herbaceous spurge	Hyper-accumulation	Ullah et al. (2019)
Rubiaceae	*Genipa americana* L	Wood plant	Hyper-accumulation	Barbosa et al. (2007)
Amaryllidaceae	*Allium griffithianum* Boiss	Perennial herb	Hyper-accumulation	Sajad et al. (2020)
Fabaceae	*Medicago sativa* L	Perennial flowering plant	High proline and GST accumulation	Wu et al. (2018)
Salviniaceae	*Salvinia minima* Baker 1886	Aquatic macrophyte	Increased anti-oxidant activity	Prado et al. (2012)
Lamiaceae	*Origanum vulgare* L	Mediterranean, perennial herb	Hyper-accumulation	Levizou et al. (2018)

(continued)

Table 5.1 (continued)

Family	Plant	Habitat	Tolerance mechanism	References
Araliaceae	*Hydrocotyle umbellata* L	Creeping, aquatic herb	Hyper-accumulation	Taufikurahman et al. (2019)
Poaceae	*Leersia hexandra* Sw	Aquatic perennial grass	Iron-biochar nano-complex & hyperaccumulator	Wang et al. (2020)
Cannaceae	*Canna indica* L	Long-lived, perennial herb	Hyper-accumulation	Taufikurahman et al. (2019)
Asteraceae	*Helianthus annuus* L	Annual forb	Hyper-accumulation	Ranieri et al. (2013)
Pontederiaceae	*Eichhornia crassipes* Mart	Aquatic plant	Anti-oxidant activity and accumulation	Mondal and Nayek (2020)

Table 5.2 Chromium-induced genotoxicity in various plant specie

Plant species	Common name	Genotoxicity	Cr-type	References
Hordeum vulgare	Barley	Chromosomal aberrations	Chromosomal aberrations	Truta et al. (2014)
Glycine max	Soybean	DNA damage	Cr(VI)/(III)	Balasaraswathi et al. (2017)
Zea mays	Maize	Chromosomal aberration	Cr(VI)/(III)	Fargasova (2012)
Vicia faba	Faba Bean	Micronucleus, chromosomal fragmentation & bridging, increase in % tail DNA, tail moment and Tail length	Tannery solid waste & Cr(VI)	El Fels et al. (2015), Fargasova (2012), Chandra et al. (2004), Rodriguez (2011)
Vicia sativa	Vetch	Chromosomal aberration, chromosomal fragmentation & bridging	Wastes, Cr(VI)/(III)	Fargasova (2012), Rodriguez (2011), Miadokova et al. (1999)
Arabidopsis thaliana	Arabidopsis	DNA mutation	Cr(VI)	Rodriguez (2011), Labra et al. (2003)
Allium cepa	Onion	Aberrations, micronuclei, chromosomal fragmentation & bridging	Tannery solid waste, tannery effluent & Cr(VI)	Patnaik et al. (2013), Fargasova (2012), Rodriguez (2011), Qian (2004)
Raphanus sativus	Radish	Chromosomal aberration	Cr(VI)/(III)	Fargasova (2012)

even become more hazardous as a result. To protect plants from the oxidative stress caused by toxic HMs, they need to increase the production of antioxidant enzymes, such as SOD, CAT, the enzymes of the AsA-GSH cycle (APX, MDHAR, and GR), as well as GST, and GPX. The combination of these biochemical properties serves as an indicator of how sensitive or resistant different plant species are to HMs on a species-by-species basis (Anjum et al. 2011). According to Gill et al. the increased tolerance to Cd can be attributed to the increased coordination between the antioxidative enzymes in response to Cd exposure. By working together, we can protect the machinery involved in the photosynthesis process. In the presence of an HM stress, the elimination of ROS is carried out by several enzymes that are sequentially and simultaneously activated, forming the enzymatic antioxidant system Gill et al. El-Beltagi et al. found that plants exposed to Cd stress exhibited significantly greater levels of antioxidant enzymes such as CAT, GST, and POX when compared to plants subjected to a control condition. CAT activity in the plant's leaves was measured at a specific activity of 25 ppm of Cd when the CAT activity in the plant's leaves was measured. This value was obtained when the concentration of Cd was raised to a certain level. The activity of CAT in both the leaf and root tissues was reduced when the Cd content was increased to 50 ppm, however, compared to 25 ppm Cd, the activity of CAT was not affected. There was a significant increase in the level of GST-specific activity in both the leaves and roots of plants when cd was used as a fertilizer discovered that the highest concentration of Cadmium, 50 ppm, caused GST activity to reach 459% in the leaves and 756% in the roots of the plants when compared to the control plants. As Dominguez et al. point out, plants have an excellent antioxidant system that allows them to develop normally despite being exposed to cadmium concentrations at the highest possible level. Despite adverse conditions, it has been demonstrated that it is possible to successfully establish resistance to the harmful effects of cadmium even under the most adverse conditions. This study found that activating enzymes that participate in the GPX, CAT, APX, and SOD pathways of the AsA-GSH cycle (APX, MDHAR, DHAR, and GR) was sufficient to inhibit ROS generation and oxidative damage caused by lower Cd concentrations (10 and 100 μM), but not the highest Cd concentration. The effect of this treatment was not enough to reduce the formation of ROS caused by Cd as well as the damage caused by high levels of Cd. Despite the activation of GR, the amount of ROS and oxidative stress that was produced by Cd (1 mM) was not sufficient to diminish the accumulation of ROS and oxidative stress. Anjum et al. (2011) demonstrated that the AsA-GSH cycle metabolism was protective in two different mung bean cultivars. Cd stress was applied to Pusa 9531, a Cd-resistant strain, and PS 16, a Cd-sensitive strain. Different redox states of AsA-GSH in plants treated with Cd, an increased level of asA-GSH-regenerating enzymes such as APX, MDHAR, DHAR, and GR, as well as other antioxidant enzymes such as SOD all strongly suggested overutilization of AsA-GSH in plants treated with Cd. There was a significant increase in lipid peroxidation and H_2O_2 content that was accompanied by a subsequent decrease in reduced ascorbic acid and glutathione pools, suggesting that, as a result of Cd toxicity, oxidative stress could be partially mitigated by a detoxification mechanism based on ascorbic acid and glutathione. Under stress, APX is a crucial component in the removal of H_2O_2 from the atmosphere, according to Gill et al. The amount of Cd used has a direct influence on the level of activity that APX produces. The same was true for both of the genotypes studied in this study. Hossain et al. found that Cd stress resulted in a significant increase in the amount of GSH and GSSG in the blood as well as a notable decline in the amount of AsA, as well as a sharp increase in the amount of H_2O_2 and MDA in the bloodstream. The activities of CAT, MDHAR, DHAR, and GR all significantly decreased in response to Cd stress at a concentration of 1 mM $CdCl_2$ for 24 h. GPX GR, APX, DHAR, GPX, GR, GST, and CAT all increased in response to Cd stress at a concentration of 1 mM $CdCl_2$. Exogenous application of betaine or proline increased GSH and AsA levels as well as the maintenance of a high GSH/GSSG ratio. According to Kachout et al. (2009), there is evidence that antioxidant machinery protects against oxidative stress caused by Cd. The antioxidant enzyme activities of Atriplex plants grown in soils contaminated with heavy metals (Cu, Ni, Pb, and Zn) were changed in plants grown in soils contaminated with heavy metals (Cu, Ni, Pb, and Zn). To get rid of the oxidative stress caused by too much copper, it may be very important for safflower plants to improve the activity of CAT, POX, and SOD enzymes. Several antioxidative mechanisms could be in place in rice seedlings to protect them from the oxidative damage caused by Pb, SOD, POX, and GR, which can play an important role in the protection of seedlings from Pb-induced oxidative damage, according to Verma and Dubey (2003). According to the researchers, the presence of antioxidative activity appears to play a very significant role in how Atriplex plants react to the HM stress that they are exposed to when they are subject to it. In the roots of rice seedlings that had been exposed to 1 mM Pb for 15 days, APX, GPX, and SOD levels were all increased in the roots of the plants. GR activities of the seedlings increased by approximately 128–196% when compared to those of the control seedlings. As a result of the treatment, however, CAT activities decreased as a result of a decrease in CAT activity. The amount of AsA that was present decreased in a dose-dependent manner under Pb stress, while the amount of DHA that was presently increased in a dose-dependent manner under Pb stress. Plants that had been treated with lead had remarkably higher levels of APX, CAT, SOD, GSSG as well as total glutathione than plants that had not been treated with lead (Qureshi et al. 2007). As a result of the addition of 500 mM Pb-acetate to the solution, SOD and APX activities increased dose-dependently, but CAT activities decreased (500 mM Pb-acetate solution). Singh et al. found that when exposed to As, the levels of ASA and GSH detected in the leaves of *P. vittata* were much higher than those found in *P. ensiformis*, and the ratios of AsA to DHA and GSH to GSSG were also much higher than those in *P. ensiformis* when exposed to As. Compared to the fronds of *P. ensiformis*, the leaves of *P. vittata* contained significantly more AsA and GSH. It is important to realize that the higher the level of reactive oxygen species (ROS) that P. ensiform is exposed to, the poorer its ability to scavenge them will be. The activity of APX and SOD decreased when As levels were low. CAT activity, on the other side, demonstrated a rising tendency when the levels of As were less than 1 mg/kg. The levels of antioxidant compounds in P. ensiformis are lower than average (AsA, GSH, and carotenoids) than in P. vittata. Gupta et al. (2009) discovered that the activity of

SOD, GPX, and CAT all rose dramatically in two separate strains of Borrelia burgdorferi termed Varuna and Pusa Bold. Pusa Bold's higher antioxidant enzyme activity is probably responsible for the strain's increased tolerance. juncea in the presence of lower doses of A stress (50 mM). Ascorbic acid (40 mg/kg) can reduce the activity of SOD, APX, POX, and GR. This leads to a higher accumulation of reactive oxygen species (ROS), which in turn causes lipid peroxidation. Shri et al. (2009) found that the levels of various antioxidant enzymes and isozymes were increased with As exposure. Even though this interpretation contrasts with the findings of previous studies, which found that when As was present in rice leaves, there was a dramatic increase in the activity of both SOD and POX. Superoxide dismutase (SOD) is an enzyme family that includes such isoforms as GPX and APX. There was no evident catalase (CAT) enzyme induction. The treatment with Nickel resulted in a considerable increase in the activity of enzymes involved in the AsA-GSH cycle, including MDHAR, DHAR, and GR. Wang et al. discovered that the cotyledons, stems, and roots of Luffa cylindrica all showed significant increases in SOD, CAT, and GPX activities in other studies. They hypothesized that nickel treatment at varied doses could enhance the activity of these antioxidants, resulting in less oxidative damage caused by nickel-induced metal exposure and an increase in the plant's tolerance to nickel. A study conducted by Shanker et al. (2004) found that ROS-scavenging enzymes play an important role in numerous sections of plants when subjected to chromium (Cr) stress. In our study, we observed that a lower concentration of Cr did not result in any scavenging enzymes being induced because there was a regulated quantity of ROS formation. Furthermore, under Cr stress, the synergistic action of SOD and CAT was found to pivotal role in reducing the deleterious consequences of oxidative stress. This was attributed to both enzymes' ability to scavenge H_2O_2 and $O^{2\cdot}$. As a result of its antioxidant defense system, Indian mustard (Brassica juncea) displayed an efficient metabolic defense and adaptive system when it was exposed to mercury (Hg)-induced oxidative stress. A higher concentration of Hg in the plant resulted in a more efficient development of an antioxidant defence system such as CAT (in particular) which was able to scavenge H_2O_2 more efficiently than the plant with a lower concentration of Hg. Therefore, Shiyab et al. (2009) found that there was a decreased level of H_2O_2 in those shoots as a result of this treatment.

5.6 Antioxidant Response

Cr toxicity in plants results in the generation of reactive oxygen species (ROS) via the Fenton and Haber-Weiss reactions (Montillet et al. 2004), which is then followed by altered antioxidant enzyme activity. The enhanced activity of antioxidant enzymes such as POD, catalase (CAT), APX, and SOD protects plants from reactive oxygen species (ROS) formed in response to chromium stress. When *Bacillus thuringiensis* oleracea, *Zea mays*, and *Solanum lycopersicum* were treated with Cr(VI), the roots and leaves had higher glutathione (GSH). These antioxidant enzymes disrupt the chain reaction of free radicals, either entirely stopping or considerably slowing the oxidation process (Munné-Bosch et al. 2009). Cr treatment has also been demonstrated to boost GSH production in *Oryza sativa, Actinidia deliciosa* (A. Chev.) C. F. Liang & A. R. Ferguson, *Brassica napus, Salvinia natans, P. stratiotes, Salvinia rotundifolia,* and *Salvinia minima* (Noctor et al. 2002). When subjected to Cr toxicity, however, Jatropha curcas showed a decrease in GSH activity (Panda 2007). Increased levels of glutathione reductase (GR), a key enzyme in the pathway that leads from ascorbate to glutathione, are seen as a response to Cr stress (Pandey et al. 2016). GR is a metal chelator and a ROS scavenger in addition to being a substrate for PC production. A recent study on Miscanthus sinensis found that when exposed to 0.50–1 mM Cr, 36 proteins involved in oxidative stress, metabolism, molecular chaperones, and other activities were over-expressed (Pang and Wang 2010).

5.7 Chromium-Mediated Alteration in the Enzymatic Antioxidant System

During the process of superoxide dismutase, superoxide, or O_2, is converted into H_2O_2 as a result of superoxide dismutase. It has been developed by the plant an enzymatic antioxidant system that is highly sophisticated and well-organized to combat reactive oxygen species (ROS). As a result of a wide variety of stimuli, residual oxygen species (ROS) are generated in response to a variety of conditions, among them toxic concentrations of Cr Qureshi (2007). Ascorbate peroxidase (APX) and catalase (CAT) are enzymes that are capable of converting H_2O_2 into H_2O_2. To minimize the oxidative stress generated by Chromium, plants use an enzymatic antioxidant system composed of POD, SOD, APX, CAT, dehydroascorbate reductase (DHAR), and GR. It is also well known that plants use an enzymatic antioxidant system to combat oxidative stress. This system includes the enzymes POD, SOD, APX, CAT, dehydroascorbate reductase (DHAR), glutathione reductase (GR), monodehydroascorbate reductase (MDHAR), glutathione peroxidase (GPX), and glutathione S-transferase (GST). This system is in charge of regulating and scavenging reactive oxygen species caused by Cr.

5.8 Conclusion

It is undeniable that abiotic stress in international crop production has an undeniable significance because abiotic factors, collectively, are responsible for the majority of limitations that are placed on crop production around the world due to abiotic stress. Abiotic factors are responsible for the majority of the constraints placed on crop production as a result of abiotic factors. Thus, in the field of agriculture, it is essential to take additional steps to understand the molecular and physiological mechanisms that allow plants to tolerate abiotic stress and to discover how to plant stress tolerance can be increased to improve yield under drought conditions. Abiotic stress is one of the biggest threats to the agricultural production of a country, and as a result, it is possible to minimize this loss of agricultural production by utilizing knowledge of crop physiology and crop husbandry practices strategically. During the development, adaptation, and continuation of the existence of plants, reactive oxygen species (ROS), as well as their metabolism and detoxification, are crucial processes for their survival, development, and adaptation. It is possible to increase a plant's resistance to environmental stresses by artificially inducing the overexpression of novel isoforms of genes that code for ROS-detoxifying enzymes. There are some significant steps involved in the development of a plant's defence mechanism and regulatory mechanisms, such as the production of reactive oxygen species (ROS) and their subsequent removal through scavenging.

Acknowledgements We would like to acknowledge the department of agronomy at the Lovely Professional University, Phagwara, Punjab, India, for their consistent moral support and encouragement throughout the writing process.

References

Abbas G, Murtaza B, Bibi I, Shahid M, Niazi NK, Khan MI, Amjad M, Hussain M, Natasha (2018) Arsenic uptake, toxicity, detoxification, and speciation in plants: physiological, biochemical, and molecular aspects. Int J Environ Res Public Health 15:59

Acquaah G (2007) Principles of plant genetics and breeding. Blackwell, Oxford, p 385

Adejumo SA, Tiwari S, Thul S, Sarangi BK (2019) Evaluation of lead and chromium tolerance and accumulation level in Gomphrenacelosoides: a novel metal accumulator from lead acid battery waste contaminated site in Nigeria. Int J Phytoremediat 21:1341–1355

Afonso TF, Demarco CF, Pieniz S, Camargo FAO, Quadro MS, Andreazza R (2019) Potential of SolanumviarumDunal in use for phytoremediation of heavy metals to mining areas, southern Brazil. Environ Sci Pollut Res 26:24132–24142

Agrawal SB, Singh S, Agrawal M (2009) Ultraviolet-B induced changes in gene expression and antioxidants in plants. Adv Bot Res 52:48–86

Ahmed A, Hasnain A, Akhtar S, Hussain A, Abaid-Ullah YG, Wahid A, Mahmood S (2010) Antioxidant enzymes as bio-markers for copper tolerance in safflower (*Carthamus tinctorius* L.). Afr J Biotechnol 9:5441–5444

Al Mahmud J, Hasanuzzaman M, Nahar K, Rahman A, Hossain MS, Fujita M (2017) Maleic acid assisted improvement of metal chelation and antioxidant metabolism confers chromium tolerance in *Brassica juncea* L. Ecotoxicol Environ Saf 144:216–226

Ali AA, Alqurainy F (2006) Activities of antioxidants in plants under environmental stress. In: Motohashi N (ed) The lutein-prevention and treatment for diseases. Transworld Research Network, Trivandrum, pp 187–256

Anjum NA, Umar S, Iqbal M, Khan NA (2011) Cadmium causes oxidative stress in mung bean by affecting the antioxidant enzyme system and Ascorbate-Glutathione cycle metabolism. Russ J Plant Physiol 58:92–99

Apel K, Hirt H (2004) Reactive oxygen species: metabolism, oxidative stress and signal transduction. Annu Rev Plant Biol 55:373–399

Asada K (1994) Production and action of active oxygen species in photosynthetic tissues. In: Foyer CH, Mullineaux PM (eds) Causes of photooxidative stress and amelioration of defence systems in plants. CRC Press, Boca Raton, pp 77–104

Asada K, Takahashi M (1987) Production and scavenging of active oxygen in chloroplasts. In: Kyle DJ, Osmond CB, Arntzen CJ (eds) Photoinhibition: topics in photosynthesis, vol 9. Elsevier, Amsterdam, pp 227–287

Ashraf A, Bibi I, Niazi NK, Ok YS, Murtaza G, Shahid M, Kunhikrishnan A, Li D, Mahmood T (2017) Chromium(VI) sorption efficiency of acid-activated banana peel over organo-montmorillonite in aqueous solutions. Int J Phytoremediation 19:605–613

Augustynowicz J, Wrobel P, Płachno BJ, Tylko G, Gajewski Z, Wegrzynek D (2014) Chromium distribution in shoots of macrophyte Callitrichecophocarpa Sendtn. Planta 239:1233–1242

Augustynowicz J, Sitek E, Bryniarski T, Baran A, Ostachowicz B, Urbańska-Stopa M, Szklarczyk M (2020) The use of Callitrichecophocarpa Sendtn. For the reclamation of Cr-contaminated freshwater habitat: benefits and limitations. Environ Sci Pollut Res 27:25510–25522

Balasaraswathi K, Jayaveni S, Sridevi J, Sujatha D, Aaron KP, Rose C (2017) Cr-induced cellular injury and necrosis in *Glycine max* L.: biochemical mechanism of oxidative damage in the chloroplast. Plant Physiol Biochem 118:653–666

Barbosa R, de Almeida A, Mielke M, Loguercio L, Mangabeira P, Gomes F (2007) A physiological analysis of *Genipa americana* L.: a potential phytoremediator tree for chromium polluted watersheds. Environ Exp Bot 61:264–271

Bhargava RN, Mishra S (2018) Hexavalent chromium reduction potential of Cellulosimicrobium sp isolated from common effluent treatment plant of tannery industries. Ecotoxicol Environ Saf 147:102–109

Blokhnia O, Virolainen E, Fagerstedt V (2003) Antioxidants, oxidative damage and oxygen deprivation stress: a review. Ann Bot 91:179–194

Byrne P, Taylor KG, Hudson-Edwards KA, Barrett JES (2017) Speciation and potential long-term behaviour of chromium in urban sediment particulates. J Soils Sediments 17:2666–2676

Chakraborty U, Pradhan D (2011) High temperature-induced oxidative stress in Lens culinaris, the role of antioxidants and amelioration of stress by chemical pre-treatments. J Plant Interact 6:43–52

Chalapathi Rao ASV, Reddy AR (2008) Glutathione reductase: a putative redox regulatory system in plant cells. In: Khan NA, Singh S, Umar S (eds) Sulfur assimilation and abiotic stresses in plants. Springer, Berlin/Heidelberg, pp 111–147

Chandra S, Chauhan LKS, Pande PN, Gupta SK (2004) Cytogenetic effects of leachates from tannery solid waste on the somatic cells of *Vicia faba*. Environ Toxicol 19:129–133

Chen Z, Young TE, Ling J, Chang SC, Gallie DR (2003) Increasing vitamin C content of plants through enhanced ascorbate recycling. Proc Natl Acad Sci USA 100:3525–3530

Chen Q, Wu K, Tang Z, Guo Q, Guo X, Wang H (2017) Exogenous ethylene enhanced the cadmium resistance and changed the alkaloid biosynthesis in *Catharanthus roseus* seedlings. Acta Physiol Plant 39:267

Conklin PL, Williams EH, Last RL (1996) Environmental stress sensitivity of an ascorbic aciddefi client Arabidopsis mutant. Proc Natl Acad Sci USA 93:9970–9974

Cui W, Wang H, Song J, Cao X, Rogers HJ, Francis D, Jia C, Sun L, Hou M, Yang Y (2017) Cell cycle arrest mediated by Cd-induced DNA damage in Arabidopsis root tips. Ecotoxicol Environ Saf 145:569–574

De Tullio MC (2004) How does ascorbic acid prevent scurvy? A survey of the nonantioxidant functions of vitamin C. In: Asard H (ed) Vitamin C: its function and biochemistry in animals and plants. Garland Science/BIOS Scientific Publishers, London, New York, pp 176–190

del Río LA, Sandalio LM, Corpas FJ, Palma JM, Barroso JB (2006) Reactive oxygen species and reactive nitrogen species in peroxisomes. Production, scavenging, and role in cell signalling. Plant Physiol 141:330–335

Dixon DP, Skipsey M, Edwards R (2010) Roles for glutathione transferases in plant secondary metabolism. Phytochemistry 71:338–350

El Fels L, Hafidi M, Silvestre J, Kallerhoff J, Merlina G, Pinelli E (2015) Efficiency of the co-composting process to remove genotoxicity from sewage sludge contaminated with hexavalent chromium. Ecol Eng 82:355–360

Elsawy EET, El-Hebeary MR, El Mahallawi ISE (2017) Effect of manganese, silicon and chromium additions on microstructure and wear characteristics of grey cast iron for sugar industries applications. Wear 390–391:113–124

Eltayeb AE, Kawano N, Badawi GH, Kaminaka H, Sanekata T, Morishima I (2006) Enhanced tolerance to ozone and drought in transgenic tobacco overexpressing dehydroascorbate reductase in the cytosol. Physiol Plant 127:57–65

Eltayeb AE, Kawano N, Badawi GH, Kaminaka H, Sanekata T, Shibahara T, Inanaga S, Tanaka K (2007) Overexpression of monodehydroascorbate reductase in transgenic tobacco confers enhanced tolerance to ozone, salt and polyethene glycol stresses. Planta 225:1255–1264

Fargasova A (2012) Plants as models for chromium and nickel risk assessment. Ecotoxicology 21:1476–1483

Florea CD, Carcea I, Cimpoesu R, Toma SL, Sandu IG, Bejinariu C (2017) Experimental analysis of resistance to electrocorosion of a high chromium cast iron with applications in the vehicle industry. Rev Chim 68:2397–2401

Foyer C, Noctor G (2000) Oxygen processing in photosynthesis: regulation and signalling. New Phytol 146:359–388

Fryer MJ (1992) The antioxidant effects of thylakoid vitamin E (a-tocopherol). Plant Cell Environ 15:381–392

Gapper C, Dolan L (2006) Control of plant development by reactive oxygen species. Plant Physiol 141:341–345

Ghosh I, Ghosh M, Mukherjee A (2017) Remediation of mine tailings and fly ash dumpsites: role of poaceae family members and aromatic grasses. In: Enhancing cleanup of environmental pollutants. Springer: Cham, Germany, pp 117–167

Gielen H, Vangronsveld J, Cuypers A (2017) Cd-induced Cu deficiency responses in Arabidopsis thaliana: are phytochelatins involved? Plant Cell Environ 40:390–400

Gill SS, Tuteja N (2010) Reactive oxygen species and antioxidant machinery in abiotic stress tolerance in crop plants. Plant Physiol Biochem 48:909–930

Gupta M, Sharma P, Sarin NB, Sinha AK (2009) Differential response of arsenic stress in two varieties of *Brassica juncea* L. Chemosphere 74:1201–1208

Halliwell B (2006) Reactive species and antioxidants, redox biology is the fundamental theme of aerobic life. Plant Physiol 141:312–322

Haokip N, Gupta A (2020) Phytoremediation of chromium and manganese by Ipomoea aquatica Forssk. from aqueous medium containing chromium-manganese mixtures in microcosms and mesocosms. Water Environ J

Hasanuzzaman M, Fujita M (2011) Selenium pretreatment upregulates the antioxidant defence and methylglyoxal detoxification system and confers enhanced tolerance to drought stress in rapeseed seedlings. Biol Trace Elem Res. https://doi.org/10.1007/s12011-011-8998-9

Hasanuzzaman M, Hossain MA, Fujita M (2011a) Selenium-induced up-regulation of the antioxidant defence and methylglyoxal detoxification system reduces salinity-induced damage in rapeseed seedlings. Biol Trace Elem Res. https://doi.org/10.1007/s12011-011-8958-4

Hasanuzzaman M, Hossain MA, Fujita M (2011b) Nitric oxide modulates antioxidant defence and the methylglyoxal detoxification system and reduces salinity-induced damage of wheat seedlings. Plant Biotechnol Rep. https://doi.org/10.1007/s11816-011-0189-9

Hasanuzzaman M, Nahar K, Gill SS, Alharby HF, Razafindrabe BHN, Fujita M (2017) Hydrogen peroxide pretreatment mitigates cadmium-induced oxidative stress in Brassica napus L.: an intrinsic study on antioxidant defense and glyoxalase systems. Front Plant Sci 8:115

Hossain MA, Hasanuzzaman M, Fujita M (2011) Coordinate induction of antioxidant defence and glyoxalase system by exogenous proline and glycine betaine is correlated with salt tolerance in mung bean. Front Agric China 5:1–14

Kachout SS, Mansoura AB, Leclerc JC, Mechergui R, Rejeb MN, Ouerghi Z (2009) Effects of heavy metals on antioxidant activities of *Atriplex hortensis* and *A. rosea*. J Food Agric Environ 7:938–945

Kalve S, Sarangi BK, Pandey RA, Chakrabarti T (2011) Arsenic and chromium hyperaccumulation by an ecotype of *Pteris vittata*-Prospective for phytoextraction from contaminated water and soil. Curr Sci 100:888–894

Labra M, Di Fabio T, Grassi F, Regondi SMG, Bracale M, Vannini C, Agradi E (2003) AFLP analysis as a biomarker of exposure to organic and inorganic genotoxic substances in plants. Chemosphere 52:1183–1188

Levizou E, Zanni AA, Antoniadis V (2018) Varying concentrations of soil chromium (VI) for the exploration of tolerance thresholds and phytoremediation potential of the oregano (*Origanum vulgare*). Environ Sci Pollut Res 26:14–23

Li JM, Jin H (2007) Regulation of brassinosteroid signaling. Trends Plant Sci 12:37–41

Logan B (2005) Reactive oxygen species and photosynthesis. In: Smirnoff N (ed) Antioxidants and reactive oxygen species in plants. Blackwell, Oxford, pp 250–267

Marrs KA (1996) The functions and regulation of glutathione S-transferases in plants. Annu Rev Plant Physiol Plant Mol Biol 47:127–158

Miadokova E, Duhova V, Vlckova V, Sladkova L, Sucha V, Vlcek D (1999) Genetic risk assessment of acid wastewater containing heavy metals. Gen Physiol Biophys 18:92–98

Miao Y, Lv D, Wang P, Wang XC, Chen J, Miao C, Song CP (2006) An Arabidopsis glutathione peroxidase functions as both a redox transducer and a scavenger in abscisic acid and drought stress responses. Plant Cell 18:2749–2766

Mittler R (2002) Oxidative stress, antioxidants and stress tolerance. Trends Plant Sci 7:405–410

Mittova V, Tal M, Volokita M, Guy M (2003) Up-regulation of the leaf mitochondrial and peroxisomal antioxidative systems in response to salt-induced oxidative stress in the wild salt-tolerant tomato species Lycopersicon pennellii. Plant Cell Environ 26:845–856

Mobin M, Khan NA (2007) Photosynthetic activity, pigment composition and antioxidative response of two mustard (*Brassica juncea*) cultivars differing in photosynthetic capacity subjected to cadmium stress. J Plant Physiol 164:601–610

Møller IM (2001) Plant mitochondria and oxidative stress: electron transport, NADPH turnover, and metabolism of reactive oxygen species. Annu Rev Plant Physiol Plant Mol Biol 52:561–591

Mondal NK, Nayek P (2020) Hexavalent chromium accumulation kinetics and physiological responses exhibited by *Eichhornia* sp. and Pistia sp. Int J Environ Sci Technol 17:1397–1410

Montillet JL, Cacas JL, Garnier L, Montané MH, Douki T, Bessoule JL, Polkowska-Kowalczyk L, Maciejewska U, Agnel JP, Vial A, Triantaphylidès C (2004) The upstream oxylipin profile of Arabidopsis thaliana: a tool to scan for oxidative stress. Plant J 40:439–451

Munné-Bosch S, Falanga V, Pateraki I, Lopez-Carbonell M, Cela J, Kanellis AK (2009) Physiological and molecular responses of the isoprenoid biosynthetic pathway in a drought-resistant Mediterranean shrub, Cistus creticus exposed to water deficit. J Plant Physiol 166:136–145

Noctor G, Gomez L, Vanacker H, Foyer CH (2002) Interactions between biosynthesis, compartmentation and transport in the control of glutathione homeostasis and signalling. J Exp Bot 53:1283–1304

Panda SK (2007) Chromium-mediated oxidative stress and ultrastructural changes in root cells of developing rice seedlings. J Plant Physiol 164:1419–1428

Pandey B, Agrawal M, Singh S (2016) Ecological risk assessment of soil contamination by trace elements around the coal mining area. J Soils Sediments 16:159–168

Pang CH, Wang BS (2010) Role of ascorbate peroxidase and glutathione reductase in Ascorbate—Glutathione cycle and stress tolerance in plants. In: Anjum NA, Chan MT, Umar S (eds) Ascorbate-glutathione pathway and stress tolerance in plants. Springer, Dordrecht, pp 91–112

Patnaik AR, Achary VMM, Panda BB (2013) Chromium(VI)-induced hormesis and genotoxicity are mediated through oxidative stress in root cells of *Allium cepa* L. Plant Growth Regul 71:157–170

Prado C, Pagano E, Prado F, Rosa M (2012) Detoxification of Cr(VI) in *Salvinia minima* is related to seasonal-induced changes of thiols, phenolics and antioxidative enzymes. J Hazard Mater 239:355–361

Qian XW (2004) Mutagenic effects of chromium trioxide on root tip cells of *Vicia faba*. J Zhejiang Univ Sci 5:1570–1576

Qureshi MI, Abdin MZ, Qadir S, Iqbal M (2007) Lead-induced oxidative stress and metabolic alterations in *Cassia angustifolia* Vahl. Biol Plant 51:121–128

Ranieri E, Fratino U, Petruzzelli D, Borges AC (2013) A comparison between *Phragmites australis* and *Helianthus annulus* chromium phytoextraction. Water Air Soil Pollut 224:1–9

Robson CA, Vanlerberghe GC (2002) Transgenic plant cells lacking mitochondrial alternative oxidase have increased susceptibility to mitochondria-dependent and independent pathways of programmed cell death. Plant Physiol 129:1908–1920

Rodríguez M, Canales E, Borrás-Hidalgo O (2005) Molecular aspects of abiotic stress in plants. Biotechnol Appl 22:1–10

Rodriguez E, Azevedo R, Fernandes P, Santos C (2011) Cr(VI) induces DNA damage, cell cycle arrest and polyploidization: a how cytometric and comet assay study in *Pisum sativum*. Chem Res Toxicol 24:1040–1047

Sajad MA, Khan MS, Bahadur S, Naeem A, Ali H, Batool F, Shuaib M, Batool S (2020) Evaluation of chromium phytoremediation potential of some plant species of Dir Lower, Khyber Pakhtunkhwa, Pakistan. Acta Ecol Sin 40:158–165

Sánchez-Casas P, Klesseg DF (1994) A salicylic acid-binding activity and a salicylic acid-inhibitable catalase activity are present in a variety of plant species. Plant Physiol 106:1675–1679

Scandalios JG (1993) Oxygen stress and superoxide dismutases. Plant Physiol 101:7–12

Shahid M, Shamshad S, Rafiq M, Khalid S, Bibi I, Niazi NK, Dumat C, Rashid MI (2017) Chromium speciation, bioavailability, uptake, toxicity and detoxification in soil-plant system: a review. Chemosphere 178:513–533

Shalata A, Neumann PM (2001) Exogenous ascorbic acid (Vitamin C) increases resistance to salt stress and reduces lipid peroxidation. J Exp Bot 52:2207–2211

Shanker AK, Djanaguiraman M, Sudhagar R, Chandrashekar CN, Padmanabhan G (2004) Differential antioxidative response of ascorbate glutathione pathway enzymes and metabolites to chromium speciation stress in green gram (*Vigna radiata* (L.) R. Wilczek, cv CO4) roots. Plant Sci 166:1035–1043

Sharma P, Dubey RS (2005) Drought induces oxidative stress and enhances the activities of antioxidant enzymes in growing rice seedlings. Plant Growth Regul 46:209–221

Sharma P, Dubey RS (2007) Involvement of oxidative stress and role of the antioxidative defence system in growing rice seedlings exposed to toxic levels of aluminium. Plant Cell Rep 26:2027–2038

Shiyab S, Chen J, Han FX, Monts DL, Matta FB, Gu M, Su Y, Masad MA (2009) Mercury-induced oxidative stress in Indian mustard (*Brassica juncea* L.). Environ Toxicol 24:462–471

Shri M, Kumar S, Chakrabarty D, Trivedi PK, Mallick S, Misra P, Shukla D, Mishra S, Srivastava S, Tripathi RD, Tuli R (2009) Effect of arsenic on growth, oxidative stress, and antioxidant system in rice seedlings. Ecotoxicol Environ Saf 72:1102–1110

Singh S, Khan NA, Nazar R, Anjum NA (2008) Photosynthetic traits and activities of antioxidant enzymes in black gram (*Vigna mungo* L. Hepper) under cadmium stress. Am J Plant Physiol 3:25–32

Sinha V, Pakshirajan K, Chaturvedi R (2014) Chromium(VI) accumulation and tolerance by tradescantia pallida: biochemical and antioxidant study. Appl Biochem Biotechnol 173:2297–2306

Sliwa-Cebula M, Kaszycki P, Kaczmarczyk A, Nosek M, Lis-Krzyścin A, Miszalski Z (2020) The common ice plant (*Mesembryanthemum crystallinum* L.)–Phytoremediation potential for cadmium and chromate-contaminated soils. Plants 9:1230

Smirnoff N (2000) Ascorbic acid: metabolism and functions of a multi-faceted molecule. Curr Opin Plant Biol 3:229–235

Srivalli S, Khanna-Chopra R (2008) Role of glutathione in abiotic stress tolerance. In: Khan NA, Singh S, Umar S (eds) Sulfur assimilation and abiotic stress in plants. Springer, Berlin, Heidelberg, pp 207–225

Takahashi S, Murata N (2008) How do environmental stresses accelerate photoinhibition? Trends Plant Sci 13:178–182

Taufikurahman T, Pradisa MAS, Amalia SG, Hutahaean GEM (2019) Phytoremediation of chromium (Cr) using *Typha angustifolia* L., *Canna indica* L. and *Hydrocotyl eumbellata* L. in surface flow system of constructed wetland. In: Proceedings of the IOP conference series: earth and environmental science, Ivano-Frankivsk, Ukraine, vol 308. IOP Publishing: Bristol, UK, p 012020

Truta E, Mihai C, Gherghel D, Vochita G (2014) Assessment of the cytogenetic damage induced by chromium short-term exposure in root tip meristems of barley seedlings. Water Air Soil Pollut 225:1933

Ullah R, Hadi F, Ahmad S, Jan AU, Rongliang Q (2019) Phytoremediation of lead and chromium contaminated soil improves with the endogenous phenolics and proline production in parthenium, Cannabis, Euphorbia, and Rumex Species. Water Air Soil Pollut 230:40

Urban P, Mignotte C, Kazmaier M, Delorme F, Pompon D (1989) Cloning, yeast expression and characterization of the coupling of two distantly related *Arabidopsis thaliana* NADPH-cytochrome P450 reductases with P450 CYP73A5. J Biol Chem 272:19176–19186

Verma S, Dubey RS (2003) Lead toxicity induces lipid peroxidation and alters the activities of antioxidant enzymes in growing rice plants. Plant Sci 164:645–655

Wang WX, Vinocur B, Shoseyov O, Altman A (2001) Biotechnology of plant osmotic stress tolerance: physiological and molecular considerations. Acta Hort 560:285–292

Wang C, Tan H, Li H, Xie Y, Liu H, Xu F, Xu H (2020) Mechanism study of Chromium influenced soil remediated by an uptake-detoxification system using hyperaccumulator, resistant microbe consortium, and nano iron complex. Environ Pollut 257:113558

Wójcik M, Tukiendorf A (2011) Glutathione in adaptation of *Arabidopsis thaliana* to cadmium stress. Biol Plant 55:125–132

Wu G, Wei ZK, Shao HB (2007) The mutual responses of higher plants to the environment: physiological and microbiological aspects. Biointerfaces 59:113–119

Wu S, Hu Y, Zhang X, Sun Y, Wu Z, Li T, Lv J, Li J, Zhang J, Zheng L (2018) Chromium detoxification in arbuscular mycorrhizal symbiosis mediated by sulfur uptake and metabolism. Environ Exp Bot 147:43–52

Chapter 6
Employing Microbes for Cr Alleviation: A Reliant Harmless Approach

S. Venkatesa Prabhu, Mani Jayakumar, C. Gomadurai, Gizachew Assefa Kerga, and Nurelegne Tefera Shibeshi

Abstract Hexavalent chromium [Cr(VI)] is one of the toxic pollutants that creates a serious environmental issue. If Cr(VI) has been persisted to long period, healthy nature of the environment badly effects and makes a deadly impact to the living organisms. Due to extensive use of chromium related compounds by different industries are most responsible for such an environmental contamination. Since Cr(VI) is not easily biodegradable it poses to crucial health risks to wildlife and humans. Studies proven that Cr(VI) is mutagenic, genotoxic, and even carcinogenic. Hence, the concerns should be undertaken for an appropriate remediation for the Cr(VI) remediation/removal. However, currently, different physico-chemical methods are being carried out for Cr(VI) removal, nevertheless, are not environmentally friendly. Furthermore, traditional physico-chemical methods are needed large amount of chemicals that generates significant secondary pollution. To overcome this issue, the techniques with the use of microbes, such as bioaccumulation, biosorption, bio-reduction, bio-precipitation, subsequent bio-efflux have utilized different natural mechanisms to combat chromium toxicity. In this view, the chapter focuses one employing different microbes to respond for effective removal of chromium toxicity. In addition, the research issues and future prospects are also discussed to fill the gaps with respect to the problems associated with recent microbial remediation focusing to real-time applicability.

S. V. Prabhu (✉) · G. A. Kerga
Centre for Food Nanotechnology (CFN), Department of Food Technology, Faculty of Engineering, Karpagam Academy of Higher Education, Coimbatore 641 021, India
e-mail: haiitsvp@gmail.com

M. Jayakumar
Department of Chemical Engineering, Haramaya Institute of Technology, Haramaya University, Haramaya, Dire Dawa, Ethiopia

C. Gomadurai
Department of Chemical Engineering, Kongu Engineering College, Perundurai, Erode, Tamilnadu 638060, India

N. T. Shibeshi
School of Chemical and Bio-Engineering, Addis Ababa Institute of Technology, Addis Ababa University, Addis Ababa, Ethiopia

© The Author(s), under exclusive license to Springer Nature Switzerland AG 2023

N. Kumar et al. (eds.), *Chromium in Plants and Environment*, Environmental Science and Engineering, https://doi.org/10.1007/978-3-031-44029-8_6

Keywords Chromium · Pollutants · Toxicity · Microbes · Bio-remediation · Environment

6.1 Introduction

The element chromium(Cr) can be found in ash, boulders, and dirt from active volcanoes. Due to its strong redox potential, Cr can be found in a wide variety of oxidation states, from (II to IV). The two most stable forms of chromium are the trivalent [Cr(III)] and hexavalent [Chromium(VI)] forms (Sun, Brocato and Costa 2015). A great deal of difference exists between chromium (III) and (IV) in terms of their physical, chemical, and toxicological characteristics. The naturally occurring Cr(III) can be found in ores such ferrochromite, while the more poisonous Cr(VI) is typically produced by human activity. The Cr(III) cation dissolves in water to create the insoluble hydroxide cation, while the Chromium(VI) oxyanion species can exist as either the dichromate, divalent chromate, or monovalent chromate, depending on the pH (DesMarias and Costa 2019). The most mobile form of Chromium in water is Chromium(VI), which is more soluble in water. Since its discovery in 1797, chromium has been put to use in numerous industries due to its adaptability. Chrome plating, dye production, the textile industry, the aerospace sector, wood preservation, leather tanning, and mud drilling all make use of chromium compounds. Examples of chromium compounds with industrial relevance include dichromates, chromates, chromic sulphate, chromic acid, and chromic oxides. Most chemical compounds containing chromium are produced by smelting chromite ore. However, a significant amount of chromium-rich solid and liquid waste and air pollutants are generated throughout the mining and manufacturing processes (Vengosh et al. 2016). In addition to human activities like mining and manufacturing, natural rock formations like ultramafic and mafic rocks can leach Cr(VI) into groundwater. In Mexico, Brazil, Argentina, Italy, California, and Greece, ultramafic aquifers are linked to water storage reservoirs with high Cr(VI) concentrations. Some volcanic and meta-volcanic groundwater and aquifers linked with mixture or more felsic igneous and metamorphic forms in North Carolina display values of up to 25 g/L of Chromium(VI), making this state home to some of the highest Chromium(VI) concentrations in the world. Worldwide, Chromium(VI) pollution has become an urgent problem. In spite of the dangers to human health and environment, many industries throughout the world have illegally dumped hazardous waste or disposed of it in ways that benefit their bottom lines (Georgaki and Charalambous 2022). Chromium pollution and the long-term harm to groundwater are primarily attributable to dumping at these sites. Figure 6.1 illustrates the possible toxicological impacts of chromium on human. A one of the most frequent locales where Cr(VI) are listed below:

- All pigments based on chromates, including dyes, inks, paints, and polymers.
- As the name implies, chrome plating involves applying a thin layer of chromium metal to an object by dipping it in a chromic acid solution.

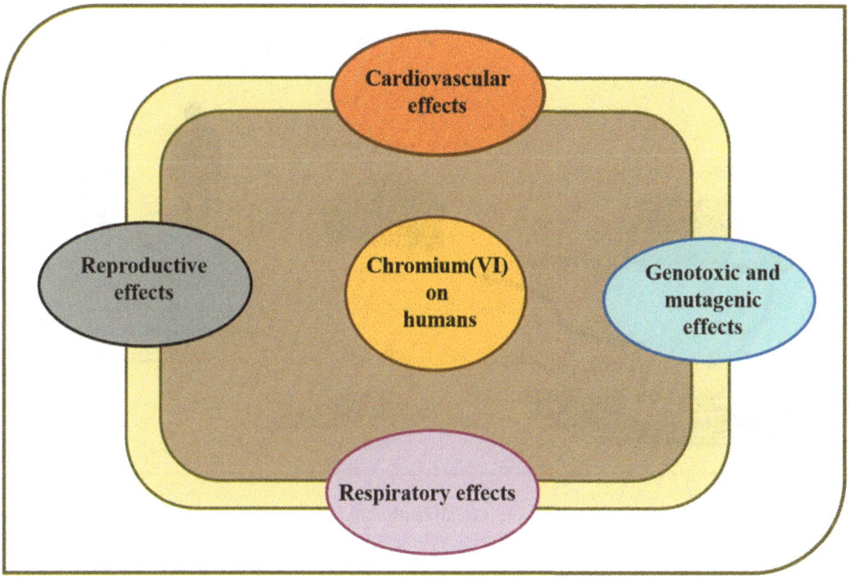

Fig. 6.1 The possible toxicological impacts of chromium(VI) on human

- Particles created during the process of smelting ferrochromium ore.
- Metal fumes produced during welding of nonferrous chromium alloys and stainless steel.

The World Health Organization (WHO) has designated chromium(VI) as a group I carcinogen. According to the guidelines for safe drinking water, chromium levels cannot exceed 50 ug/L. Between 0.2 and 2 g-Chromium(VI)/L is the typical range for Cr(VI) in North American drinking water (Monga et al. 2022). Even though the United States Environmental Protection Agency (EPA) recognizes Chromium(VI) as a toxic substance, it continues to be widely used in many industries. Drinking water is only allowed to have 100 g/L of total chromium [Cr(T)]. Figure 6.2 describes the spreading of chromium(VI) on human health and ecosystem.

To prevent the detrimental effects on human health, strict environmental controls must be imposed immediately on the amount of Chromium(VI) that can be emitted into the environment (Pavesi and Moreira 2020). Adsorption, ion exchange, chemical precipitation, membrane separation, electrocoagulation, and electrodialysis are only few of the methods that can be used to remove chromium(VI) from wastewater. To a large extent, Chromium(VI) is removed via chemical precipitation. Chromium(VI) can be eliminated through the use of chemical precipitators like $Ca(OH)_2$, MgO, NaOH, and calcium magnesium carbonate. The kind of precipitation agent, the volume of sludge, the agitation speed, the pH, the mixing time, and the presence of complexing agents are just few of the variables that might affect the outcome of the precipitation process. So far, it was proven that that the reverse osmosis, membrane filtration, electrocoagulation, ion exchange, and electrodialysis

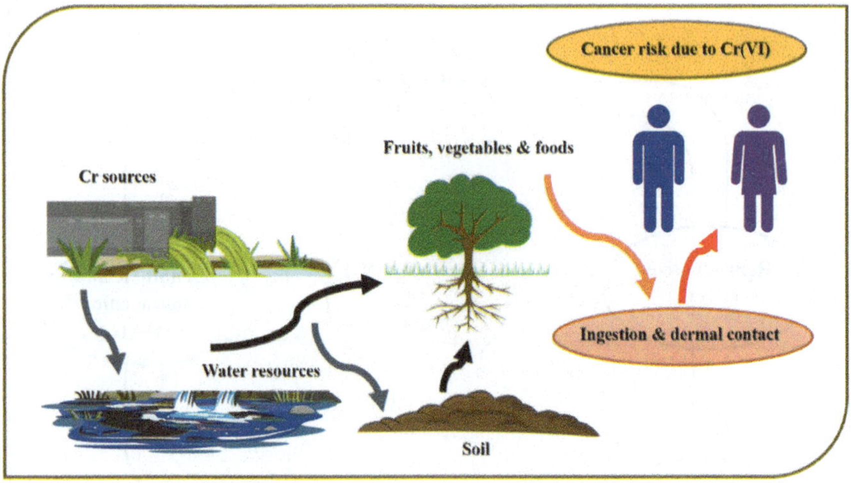

Fig. 6.2 Spreading of chromium(VI) on human health and ecosystem

are all effective methods for removing Chromium(VI), but each process has its own set of problems, such as high prices and concentrated wastes that must be dealt with afterward. Emerging as a potential effective technology for removing Cr(VI) from industrial effluents, bioremediation is on the rise. Bioremediation of chromium contamination by various fungi and bacteria has been demonstrated. Studies looking at the removal of Chromium(VI) from industrial effluent can be more promising while using different microbes, such as *Actinomycetes*, *Streptomyces rimosus*, and *Streptomyces griseus*. There are several examples of affordable agricultural wastes that have the adsorption potential to remove Chromium(VI) from waste-water. These include chitosan, rice husk, pomegranate husk, coconut shell, waste tea leaves, neem leaves, sawdust, watermelon rind, orange peel, and banana rachis. However, remediating chrome-polluted wastewater with microorganisms that are resistant to chromium has not been still thoroughly studied. The potential for hexavalent chromium to induce cancer, teratogenicity, and mutation has made it a widely recognized environmental hazard (Gad 1989). Using microorganisms that can metabolize and break down Chromium(VI) contaminants, this review intends to inform readers about the dangers caused by Chromium(VI) and the methods for eliminating it from polluted places.

6.2 Incidence of Hexavalent Cr(VI) on Human Health

Pollution from heavy metals is increasingly seen as an international environmental emergency. There is growing evidence that hexavalent chromium [Cr(VI)] is neurotoxic and should be treated as a global environmental pollutant. Many different plant and microbial species are essential in the process of decontaminating polluted areas (Oliveira 2012). In humans, Cr(VI) and its metabolites, in particular chromates, enter the body via a unique pathway. Exposure to Cr(VI) typically occurs through inhalation, ingestion, or skin contact. Cr(VI) exposure is broken down into three distinct time periods: short-term (14 days), intermediate (75–364 days), and long-term (more than 364 days). There are a number of ways in which Cr(VI) poisoning can manifest. It is possible to cause modifications to the cellular structure, particularly in the membrane's lipoprotein region. Immune system activity or efficiency can be lowered; key enzymes like oxidative phosphorylation can be suppressed; and competition for cofactor fixation sites can diminish enzyme activity. Chromium(VI) binds to the DNA-polymerase enzyme and damages the molecule, leading to hypersensitivity reactions, nasal irritation, contact dermatitis, ulcers, emphysema, acute bronchitis, liver and kidney sickness, internal bleeding, lung and skin malignancies, and DNA damage. Despite the speed with which Chromium(VI) penetrates cells, it must undergo several modifications in the bloodstream before it can perform as Cr(III) in the tissues of the body (Iqbal, Ashraf and Ashraf 2009). While Chromium(VI) is eliminated from the body, chromate is taken into cells via a transport mechanism that also involves sulphate and phosphate ions. Ions of this type can cause cellular oxidative stress, which has been linked to multiple chronic diseases including cardiovascular disease and neurological disorders. The cellular damage caused by Cr(VI) includes oxidative stress elevation, DNA adduct formation, and chromosomal breakage. Given the substantial body of epidemiological evidence connecting Chromium(VI) to lung cancer, the WHO's International Agency for Research on Cancer (IARC) has classified compounds containing Chromium(VI) as group one human carcinogens with many complex mechanisms of action. Cr(VI) exposure has been associated to many adverse health effects in humans, including eardrum perforation, dermatitis, allergies, respiratory difficulties, ulcers, itchy skin, and even lung cancer. At different times, Cr(VI) radiation can cause oxidative protein alterations, chromosomal damage, and mutations in DNA. It can also cause carcinogenic effects of substances containing Cr(VI) (Sanz-Gallen et al. 2021). Damage to the nasal lining, inflammation, anaemia, stomach ulcers, and other respiratory issues such coughing, nasal blockage, wheezing, and facial erythema can result from inhaling significant amounts of hexavalent chromium. Workplace exposure to hexavalent chromium has the potential to cause the following health problems:

- Inhaling large amounts of hexavalent chromium can irritate or even damage the nasal passages, throat, and lungs (respiratory tract).
- Airborne hexavalent chromium causes lung cancer in workers.
- Hexavalent chromium may cause irritation or even damage to organs if it comes into contact with them in sufficient quantities.

6.3 Microbial Remediation

As can be shown in Table 6.1, many different types of microbes based on their functional groups have evolved in the resistances to Cr(VI). The most well-studied mechanism (Fig. 6.3) of this type of bioremediation is the microbial enzymatic conversion of Chromium(VI) to Chromium(III). Biodegradation of contaminated waste and elimination of Chromium(III) pollution via biological reduction of Chromium(VI) to Chromium hold promise (Song et al. 2016). Chromium-detouring microbes could one day provide a sustainable and green replacement for traditional manufacturing. Bacteria, fungus, and algae are only some of the microorganisms that can be used in these procedures. In place of using biomass (both living and nonliving cells) and biological and agricultural wastes in typical wastewater purification methods, biosorption of Chromium(VI) has been proposed. The bacteria that lead to or aid in the biological decline of Chromium(VI) to less mobile Chromium(VI) may be precipitated for use in cleaning up polluted areas (III). Metal ions can be extracted from the environment by bacteria and used as fuel before being converted to biomass through an enzyme-catalyzed, potentially hazardous chemical breakdown. Microbial remediation involves promoting the breakdown of potentially dangerous compounds in soil, subterranean materials, sludge, water, and leftover bacteria. Multiple bioremediation approaches, such as bioaccumulation, biotransformation, biosorption, and bioleaching, have been shown to be effective at removing Cr and other heavy metals from industrial pollution (Stoltidis et al. 2011). Hexavalent chromium [Cr(VI)] is only absorbed by living organisms through a process called bioaccumulation, which is reliant on the metabolism of these organisms to power the transcellular transit of this toxic metal. There are several stages involved in the bioaccumulation process in bacteria. Toxic heavy metal ions initially bind to a ligand on the cell's outer membrane. Transporter proteins are responsible for bringing the metal-ligand complex from the cell surface inside the cell. Phytochelatins and metallothionein are two types of metal-binding proteins that interact with complexes transported into the cell, triggering reactions like precipitation and methylation (Panda and Choudhury 2005). A larger concentration of metal renders the method ineffective against non-living cells and effectively halts the multiplication of microorganisms. Environmentally beneficial processes such as biosorption, biotransformation, and bioaccumulation degrade and eliminate toxic chromium ions from industrial effluent.

6.4 Biosorption of Chromium(VI)

In contrast to bioaccumulation, which only occurs in actively metabolic cells, biosorption can happen in both actively metabolic cells and decaying microbial biomass. Ion exchange, surface precipitation, or a rigorous manufacturing procedure are used to remove harmful ions like Cr(VI) from the bacterial cell wall. There is a wide variety in composition and organization in the cell walls of bacteria. Algae's cell

Table 6.1 Most studied microbes and responsible functional groups for chromium(VI) remediation

S. No	Microbes	Functional groups	References
1	*Aspergillus Niger*	$-NH_2$, $-OH$, and $-COOH$	Chhikara et al. (2010)
2	*Bacillus marisflavi*	Phosphate groups, OH, free phosphates, $-NH$ acetamido group, and $-CN$	Kim et al. (2022)
3	*Chlorella miniata*	$-CH_3$, O–H, C–H, P = O, COO−, C–O–, and N–H	Congeevaram et al. (2007)
4	*Klebsiella* sp.	O–H, $-NH_2$, $-COOH$, −CONH−, $-CH_2$, and C=C	Han et al. (2008)
5	*Pleurotus ostreatus*	COOH and NH	Pun, Raut and Pant (2013)
6	*Streptomyces werraensis*	N–H, O–H, C–O–, and C–H	Dadrasnia et al. (2015)
7	*Pseudomonas aeruginosa*	Carboxylic group, C–Cl, −NH, −OH, −C–C−, and S−	El-Naggar et al. (2020)
8	*Aspergillus foetidus*	C = O, N = C = S, C–O, PO_{4-3}, amine, and OH	Ahluwalia and Goyal (2010)
9	*Arthrinium malaysianum*	C–O, –OH, C_xOH, C = O, and $-NO_2$	Majumder et al. (no date)
10	*Scenedesmus sp.*	N–H, C–O, C–H, O–H, C–F, −COOH, C–Br, and C–Cl	Han et al. (2008)

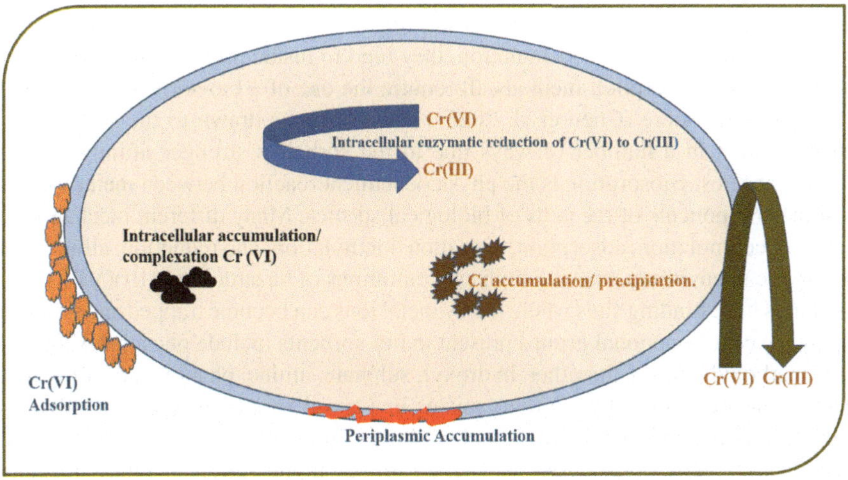

Fig. 6.3 The Microbial resistance mechanisms for Cr(VI)

walls are made up mostly of sulfonated polysaccharides, alginate, and mannans, as opposed to the fungal cell walls' glycoproteins, glucans, melanin, chitin, and peptidoglycan. Both the biomass used in biosorption and the functional groups in the microbial cell wall play significant roles in the biosorption process. For the removal of harmful heavy metals from polluted environments, the biosorption approach is preferred over more traditional kinds of bioremediation. The production of multi-functional groups and the even distribution of binding sites across the cell surface are just two of the many advantages of the biosorption method. In addition to the bio-great sorbent's efficiency and renewability, there is also the potential for metal recovery (De Pauw and Van Vaerenbergh 1983). As a result of these and other advantages, research into the biosorption of heavy metals by diverse microorganisms, especially hexavalent chromium, has expanded. The ability of some organisms to take in heavy metals and then drive their transition into less dangerous forms has captivated environmental protection experts, engineers, and biotechnologists for decades. The Chromium(VI) ion is removed through ion exchange, surface precipitation, or a similar mechanism after it has bound extracellularly to different functional groups of the microbial cell wall. Organisms like microbes have had thousands of years to develop strategies for dealing with environmental pressures. Microorganisms' defense mechanisms against heavy metals are quite diverse. Methods include active transport of metal ions, metal ion reduction, and extracellular and intracellular sequestration are all at play. The metabolic state of the cell will determine which of two biological processes—biosorption or bioaccumulation—will be responsible for the removal of heavy metals. Increased membrane permeability plays a role in the metabolism-dependent process of heavy metal uptake by cells. This happens when metal ions accumulate inside a bio-cell and pollutants are taken up by the cell. Sorbent's Biosorption allows for the rapid, self-sufficient, metabolically passive sequestration of heavy metal ions by dead/inactive biomaterials. During biosorption, heavy metals adhere to the exterior of cells, while during bioaccumulation, they bind to inside proteins such metallothionein. These biosorption methods all require the use of a bio-sorbent that is solid at room temperature (Chen et al. 2023). The sorbate is drawn to the sorbent and attached to it in a number of ways due to the sorbent's stronger affinity for the sorbate species. Biosorption is the physicochemical reaction between metal species and the components of the cells of biological species. Many different mechanisms, such as accumulation, adsorption, oxidation, methylation, and reduction, allow them to survive in environments with high concentrations of hazardous Cr(III)(VI). These creatures have binding sites where heavy metal ions can become trapped and be taken up by the cell. Functional groups present in bio-sorbents include phosphates, imidazole, carboxyl, amino, thioether, hydroxyl, sulphate, amine, phenol, and sulfhydryl. Metal ions are utilized by microorganisms in a broad variety of ways, including as cell wall-associated metals, metal siderophores, intracellular accumulation, extracellular polymeric connections with extracellular mobilization or immobilization of metal ions, and transition and metal volatilization. Physicochemical interactions between ions in solution and the charged surface groups of microorganisms include ion exchange, adsorption, complexation, and microprecipitation. The bioaccumulation process begins with metal uptake and continues through metal binding to metallothionein, Cr localization inside cell component, extracellular precipitation, metal deposition, and complexation. There are three stages in the microbial removal of Chromium(VI): cellular translocation, surface binding, and intracellular reduction (III). Due to the activity of chromate reductase enzymes and the metabolism of

Chromium(VI) metabolites, microbes are able to decrease Chromium(VI) on their surface, in their extracellular and intracellular habitats, and in their food sources (Peng et al. 2023).

6.5 Hexavalent Chromium to Tetravalent Chromium Reduction

Effluents from the textile, galvanizing, leather, tannery, metallurgical, paint, electroplating, and metal processing and refining industries are a global and regional source of harmful metal ions. These companies harm the aquatic environment by discharging metal ions into nearby waterways and open pits. The most likely environmental effect of these metals would be a shift in the total amount of surface and groundwater. As well as posing a threat to human health, these contaminants also pose a risk to animal life (Peng et al. 2023). Discomfort in the body and potentially deadly diseases like kidney failure and cancer stem from this. As opposed to its divalent (Cr^{2+}) form, the trivalent (Cr^{3+}) form is more bioavailable, more stable, and less harmful to humans. EPA and EU regulations limit Cr discharge into surface water bodies to less than 0.05 mg/l and total Chromium output to less than 2 mg/l due to its high toxicity.

6.6 Factor Affecting Bioremediation

Biological therapy refers to the process by which microorganisms including bacteria, fungus, and plants break down, transform, immobilize, and remove several potentially harmful chemicals from the environment. Involving microbes in the process is advantageous because the enzymatic pathways within them work as biocatalysts, speeding up the rate at which biochemical reactions can happen and ultimately destroying the offending contaminant. Microbes can combat pollution because they have access to a wide variety of nutrients, energy sources, and building materials. The success of bioremediation depends on a variety of variables, such as the kind and concentration of contaminants, the state of the surrounding environment, and the existence of appropriate microorganisms (Zhou et al. 2018). Inhibiting interactions between bacteria and pollutants slows down degradation. In addition, bacteria and pollutants do not spread uniformly. Bioremediation processes are difficult to regulate and optimize for a number of reasons. Pollutants being accessible to microbes, and microbes being present that can break down hazardous pollutants.

6.6.1 Availability of Nutrients

The rate and efficiency of biodegradation, as well as microbial growth and reproduction, are all affected by the availability of nutrients. Changing the C:N:P ratio of bacteria, especially by adding essential nutrients like P and N, might enhance their degradation competence. Carbon, nitrogen, and phosphorus are only few of the nutrients that microorganisms need to live and keep reproducing (Xie et al. 2022). The degree to which hydrocarbons break down is also restricted at low concentrations. If the correct nutrients are added, metabolic activity of microorganisms and, thus, the rate of biodegradation, can be increased even in subfreezing conditions. In aquatic environments, biodegradation is hindered due to a lack of accessible nutrients. These nutrients exist in nature, but in minute quantities.

6.6.2 Environmental Factors

Interactions can be predicted during the process by utilizing the metabolic capability of the microorganisms and the physicochemical parameters of the targeted pollutants. However, contextual circumstances at the interaction site alter the interaction's actual success. Many environmental factors, including but not limited to temperature, site characteristics, water solubility, redox potential, pH, nutrients, moisture, oxygen concentration, soil structure, and temperature, affect the growth and activity of microorganisms. The rate of decay is dependent on the aforementioned factors (Xu et al. 2019). Bioremediation can take place in a variety of pH levels, although the optimal range for microbial decomposition in aquatic and terrestrial settings is 6.5–8.5. The rates of degradation of pollutants are affected by a wide variety of factors, including the pH of aquatic and terrestrial ecosystems, the types of soluble elements present, and the quantities of those materials. The survival of microbes and the amount of hydrocarbons present are most strongly influenced by temperature. Most oleophilic bacteria are metabolically quiescent because their cellular transport channels shut down or even freeze due to the extremely cold water in this location. There seems to be a sweet spot for the degradation process when the metabolic cycle of the associated biological enzymes is at its most potent. In addition, a particular temperature is required for the decomposition of a given material. The rate at which bioremediation proceeds is affected by temperature because it affects the physiological characteristics of the microbes involved. At the optimum temperature, microbial activity rates reach their maximum and then gradually decline. The rate of decline picked up speed as the temperature rose or fell, and it levelled off once it reached a certain threshold. Figure 6.4 presents a schematic flow of biosorption process for Cr(VI) by microorganisms.

The acidity, basicity, or alkalinity of a chemical influences the metabolic activity of microorganisms and the pace at which the chemical is removed. Soil pH is a good indicator of the soil's capacity for supporting microbial growth. Because of how pH

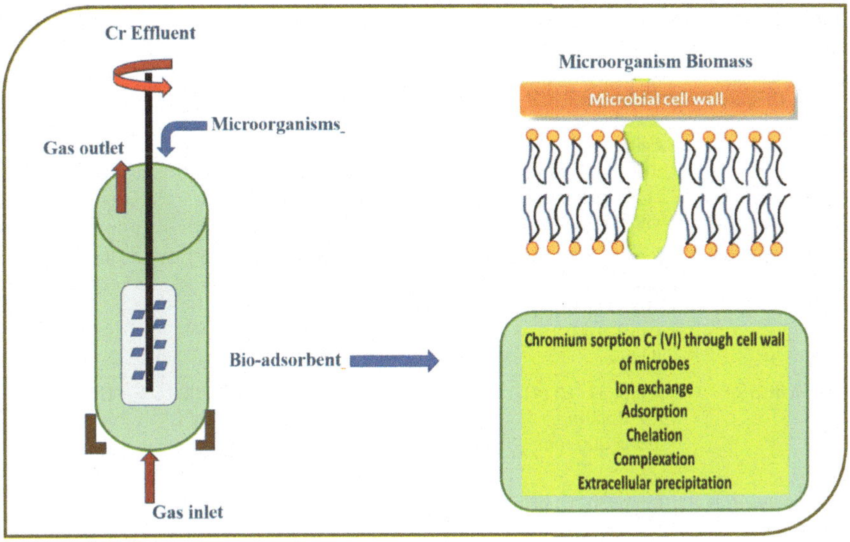

Fig. 6.4 A schematic flow of biosorption process for Cr(VI) by microorganisms

variations affect metabolic processes, the outcomes were unfavorable whether the pH was raised or lowered. Cleanup attempts could be hampered by the toxicity of certain contaminants at high quantities, which has an adverse effect on microorganisms. Toxicants, concentrations, and exposed microorganisms all have a role in determining the extent and mechanisms of toxicity. Specific forms of life are extremely sensitive to a wide variety of organic and inorganic substances (Xu et al. 2022). The different microbes that were reported in terms of their mechanisms, such as transformation, bioaccumulation, and biological removal are presented in Tables 6.2, 6.3 and 6.4.

6.7 Future Prospects

Numerous bioremediation procedures, in particular competent reduction approaches by bacteria, have been developed to tackle the difficult problem of eliminating Cr(VI) contaminants from the environment (Lin et al. 2003). Cleaning, managing, and repairing polluted ecosystems through bacterial metabolism is where microbial degradation shines as a technique. Microorganisms offer electrons to decrease Cr through either endogenous enzymes or externally introduced reducing chemicals (VI). However, the rate at which undesirable waste chemicals are degraded can be slowed by a number of factors, including the presence of competing biological agents, unfavorable external abiotic conditions (pH, moisture, aeration, temperature), a lack of food, and poor pollutant bioavailability. These features make natural biodegradation less effective, leading to unfavorable outcomes. Simply said, bioremediation

Table 6.2 Microbial species that have been reported for the transformation of Cr(VI) and optimal condition

Microorganisms	Optimal condition	Removal efficiency	Isolation source	Initial [Cr(VI)]	References
Bacillus sp.	Agitation: 100 rpm; Temp: 21 °C; pH: 6.9; with 0.5% glucose	100% in72 h	Tannery contaminated soil	10 mg L^{-1}	Zahoor and Rehman (2009)
Bacillus sphaericus	Temp: 25 °C; pH: 6.0; Agitation:120 rpm; with 1.0 g L^{-1} glucose	62% in 48 h	Contaminated soil	20 mg L^{-1}	Ibrahim et al. (2012)
Providencia sp.	Temp: 37 °C; pH: 7.0; Rotating speed: 200 rpm	100% in 96 h (200 mg L^{-1})	Contaminated soil	100, 200, 300 mg L^{-1} $K_2Cr_2O_7$	Thacker et al. (2006)
Acidithiobacillus thiooxidans	Temp: 30 °C; pH: 2.5; Rotating speed: 150 rpm; with Sulphur medium	100% in 1 d	–	2.5 mg L^{-1}	Bhattacharya et al. (2019)
Acinetobacter haemolyticus	Temp: 30 °C; pH: 5–7	100% in 36 h	Effluent from textile	50 mg L^{-1}	Zakaria et al. (2007)
Ochrobactrum sp. CSCr-3	Temp: 35 °C; pH: 10	80% in 30 h	Soil from chromium landfill	200 mg L^{-1}	Wang et al. (2019)
Serratia sp. Cr-10	Temp: 37 °C; pH: 7.0; 1% (w/v), With fructose	100% after 12 h	Soil from chromium contaminated area	10, 20 mg L^{-1}	Zhang et al. (2011)
Cellulosimicrobium sp. MWM81	Temp: 37 °C; pH: 7.0	45% in 48 h	Contaminated soil	10 mM	Zahoor and Rehman (2009)
Acinetobacter guillouiae SFC 500-1A	Temp: 28 ± 2 °C; pH: 8- 10 Agitation:150 rpm (phenol source)	~62% in 72 h	Sludge from tannery	10 mg L^{-1}	Vendruscolo et al. (2017)
Bacillus Subtilis MNU16	Temp: 30 °C; pH: 7.0	75% within 72 h	Soil obtained from coal mining	50 mg L^{-1}	Upadhyay et al. (2017)
Arthrobacter sp. LLW01	Temp: 22 °C; pH: 7–8; Rotation speed: 150 rpm; with 15 mM of glucose	50% in 144 h	Contaminated soil	50 μM	Li et al. (2021)

(continued)

Table 6.2 (continued)

Microorganisms	Optimal condition	Removal efficiency	Isolation source	Initial [Cr(VI)]	References
Penicillium oxalicum SL2	Temp: 30 °C; pH 5–7: Agitation speed: 200 rpm	100% within 144 h	Indoor air	1,000 mg L^{-1}	Yu et al. (2019)
Arthrobacter sp.	Temp: 21 °C; pH: 6.9; Agitation speed: 100 rpm; medium contains 0.5% glucose	100% in 46 h	Tannery contaminated soil	20 mg L^{-1}	Zahoor and Rehman (2009)
Bacillus atrophaeus MM20	Temp: 21 °C; pH: 6–7; Agitation speed:100 rpm	94% after 50 h	Tannery contaminated soil	10 mg L^{-1}	Patra et al. (2010)
Arthrobacter sp. SUK 1201	Temp: 35 °C; pH value: 7; Rotation speed:120 rpm; medium with1.0 g L^{-1} glucose	67% in 7 days	Overburden from chromite mine	2.0 mM	Dey and Paul (2012)
Aspergillus niger (CICC41115)	Temp: 37 °C; pH value: 7.0; roation speed: 150 rpm	100% in 84 h	Soil from commercial	50 mg L^{-1}	Fernández et al. (2018)
Pseudomonas sp. JF122	Temp: 30 °C; pH: 6.5; Agitation speed: 150 rpm	100% in 72 h	Contaminated site	2.0 mg L^{-1}	Islam et al. (2019)
Cellulosimicrobium sp. KX710177	Temp: 35 °C; pH: 6–7; Agitation speed: 120 rpm	62% after 96 h	Wastewater from tannery	300 mg L^{-1}	Bharagava and Mishra (2018)
Bacillus sp. SFC 500-1E	Temp: 28 °C; pH value: 7; Rotation speed:150 rpm	43% after 72 h	Tannery sediments	50 mg L^{-1}	Ahmed (2018)
Bacillus sp.	Temp: 35 °C; pH: 6–7; Agitation speed: 200 rpm;	> 95% for 72 h (40 mg L^{-1})	Soil from chromate pollutant	10, 40 mg L^{-1} $K_2Cr_2O_7$	Elangovan et al. (2006)
Acidithiobacillus ferooxidans	Temp: 30 °C; pH: 1.8; Agitation speed:150 rpm;	100% in 3 d	–	5.0 mg L^{-1}	Bhattacharya et al. (2019)
Burkholderia cepacia MCMB-21	Temp: 35 °C; pH: 9; 2% NaCl; 2% lactose	98% in 36 h	Alkaline crater lake	75 mg L^{-1}	Sanjay et al. (2017)
Pseudomonas sp. G1DM21	Temp: 37 °C; pH value: 7; Rotation speed: 150 rpm	99.7% in 48 h	Landfill from industrial contamination	500 μM	Das et al. (2021)
Bacillus sp. CSB-4	Temp: 35 °C; pH: 7.0; Agitation speed:100 rpm	>90% in 144 h	Soil from chromite mine	100 mg L^{-1}	Das et al. (2021)

Table 6.3 Involvement of microbes for the involvement of bioaccumulation for Cr(VI) and related conditions

Microbe	Optimum conditions	Source of isolation	Initial concentration of [Cr(VI)]	Efficiency of bioaccumulation & time	References
Streptomyces sp. MC1	Temp: 30 °C; agitation speed:220 rpm pH: 6–7	Sediment obtained from contaminated site	50 mg L^{-1}	52% for 72 h	Ahmed et al. (2016)
Acinetobacter sp. PD12S2	Temp: 37 °C; pH: 7.0; medium contains 4.0 g L^{-1} glucose	Tannery waste	8.86 mg L^{-1}	Uptake of 0.19 mg L^{-1} h^{-1}	Panda and Sarkar (2012)
Escherichia coli VITSUKMW3	Temp: 30 °C; pH: 7.5	Water outlet from chromite mining	20 mg L^{-1}	40% for 5 h	Upadhyay et al. (2017)
Arthrobacter sp. Sphe3	Temp: 30 °C; pH: 8.0;	-	45 mg L^{-1}	100% Accumulation	Ramrakhiani et al. (2011)
Baciilus circulans	–	Effluent from tannery	50 mg L^{-1}	Within 24 h	Shukla et al. (2012)
Exiguobacterium sp. ZM2	Temp: 28 °C; pH: 2.5	Contaminated site and tannery effluent	100 mg L^{-1}	29.9 mg g^{-1} for 120 min	Alam et al. (2011)
Acinetobacter sp. AB1	Temp: 30 °C; pH: 10	Tannery	50 mg L^{-1}	100% for 72 h	Essahale et al. (2012)
Enterobacter aerogenes T2	Temp: 37 °C; pH: 7.0; 4.0 g L^{-1} glucose	Effluent from tannery	8.86 mg L^{-1}	0.35 mg L^{-1} h^{-1} (uptake)	Panda and Sarkar (2012)
Bacillus subtilis VITSUKMW1	Temp: 30 °C; pH: 7.5;	Water outlet from chromite mining	20 mg L^{-1}	40% for 8 h	Upadhyay et al. (2017)
Acinetobacter sp. B9	Temp: 30 °C; pH 7.0; agitation speed: 200 rpm	Wastewater of chrome treatment plant	7.0 mg L^{-1}	67% for 24 h	Bhattacharya and Gupta (2013)
Enterobacter sp. DU17	Temp: 37 °C; pH: 7.0; agitation speed: 180 rpm; with 0.2% fructose	Waste dump from tannery	50 mg L^{-1}	Approximately 79%	Chen et al. (2022)

(continued)

Table 6.3 (continued)

Microbe	Optimum conditions	Source of isolation	Initial concentration of [Cr(VI)]	Efficiency of bioaccumulation & time	References
Streptomyces werraensis LD 22	Temp: 41 °C; pH: 7.0; agitation speed: 100 rpm	Residues from animal fecal	250 mg L^{-1}	51.7% for 7 d	Bhattacharya et al. (2019)
B. mycoides 200AsB1	Temp: 30 °C; pH 7.0; agitation speed:180 rpm	Rhizosphere obtained from Pteris vittata	25 mg L^{-1}	100% within 25 h	Bhattacharya et al. (2019)
Aspergillus sydowii	Temp: 28 °C; pH 5.0; agitation speed: 80 rpm	Sediment from Mangrove	50 mg L^{-1}	24.9% for 7 d	Bhattacharya et al. (2019)
Mixed culture	Temp: 20 °C; pH: 9.0; nutrient broth with 4% NaCl	Wastewater from industrial saline effluent	50 mg L^{-1}	89% for 5 d	Koçberber and Dönmez (2007)
Acinetobacter junii ITSUKMW2	Temp: 30 °C; pH: 7.5	Water outlet from chromite mining	20 mg L^{-1}	40% in 8 h	Upadhyay et al. (2017)
Exiguobacterium sp. KSKE41	Temp: 28 °C; pH: 7.0	Polluted soil	10 mM	35% for 48 h	Zahoor and Rehman (2009)
Bacillus subtilis	Temp: 37 °C; pH: 6–9; agitation speed: 180 rpm	Mining samples	0.2 mM	More than 90% within 48 h	Ni et al. (2020)
Saccharomyces cerevisiae	Temp: 25 °C; agitation speed: 100 rpm, pH 5.0	Polluted site	90 mg L^{-1}	99.66% for 3 h	Tang et al. (2021)

can only be effective if conditions are favorable for the growth and development of microorganisms. There have been numerous applications of bioremediation, each with a unique set of circumstances and results. Websites that employ this technique are becoming increasingly common since the benefits generally outweigh the hazards. Many species from many different regions are studied and found to have efficient regulatory processes (De Agostini et al. 2020). However, due to the widespread heavy metal contamination of agricultural land at present, it is imperative that future microbial remediation techniques also focus on the soil and environment,

Table 6.4 Microbe involved for treatment with respect to removal of Cr(VI) and optimal condition

Polluted wastewater	Microbe	Removal efficiency and initial Cr(VI) (mg L^{-1})	Optimal condition	Mode of the treatment	References
Electroplating	*P. aeruginosa A2Chr*	100% for 30 h, 15	Temp: 37 °C; pH: 7.2; agitation speed: 150 rpm	Batch	Chaturvedi (2011)
Electroplating	*P. aeruginosa* A2Chr	93% for 8 h, 10	Temp: 37 °C; pH: 7.2	Rotating bio-contactor using lab-scale	Satarupa and Amal (2010)
Electroplating	*Saccharomyces cerevisiae*	98% for 30 min, 18 ± 1.0	Temp: 25 °C; pH: 2.3; agitation speed: 150 rpm	Batch	Shahida et al. (2017)
Electroplating	(ChromeBac™ system) *Acinetobacter*	99%, 17–81	–	Bioreactor (Pilot scale)	Chen et al. (2022)
Tannery	*P. lilacinus*	100% for 48 h, 50	pH: 8.0	Batch	Wang et al. (2007)
Tannery	*E. aerogenes T2*	84% for 72 h, 1.3	–	Batch	Panda and Sarkar (2012)
Tannery	*B. cereus Cr 1*	73% for 48 h, 2.41	Temp: 35 °C, pH: 8.4; agitation seed: 120 rpm	Batch	Maurya et al. (2022)
Waste Leather Industry	*Arthrinium malaysianum*	30% for 24 h, 2.41	Temp: ambient, pH: 7.3 Shaking condition	Batch	Ramrakhiani et al. (2011)
Electroplating	*P. oxalicumstrain SL2*	100% for 96 h, 96.1	Temp: 30 °C; pH: 7.0	Batch	Fernández et al. (2018)

(continued)

Table 6.4 (continued)

Polluted wastewater	Microbe	Removal efficiency and initial Cr(VI) (mg L^{-1})	Optimal condition	Mode of the treatment	References
Tannery	*P. aeruginosa A2Chr*	60% for 35 h, 40	Temp: 37 °C; pH: 7.0; agitation speed: 150 rpm	Batch	Chaturvedi (2011)
Electroplating	*P. aeruginosa A2Chr*	76% for 4 h, 10	Temp: 37 °C; pH: 7.2	Bioreactor using dialysis	Kumar and Pandey (2006)
Tannery	*Aspergillus sp. FK1*	65% for 7 d, 557	Agitation: 250 rpm; pH: 5.0–55	Lab-scale bioreactor	Yoon et al. (2006)
Electroplating	*Candida lipolytica*	94–100%, 8–30	Temp: 25 °C, pH: 1.92–5.22	Bioreactor (lab-scale)	Konovalova et al. (2003)
Tannery	*Paecilomyces lilacinus*	100% for 18 h, 1.24	pH: 8.0	Batch	Garbisu et al. (1998)
Electroplating	*B. cereus IST105*	76% for 3d, 968	Temp: 30 °C, pH: 7.0	Batch	Ackerley et al. (2004)
Electroplating	*Acinetobacter* sp. B9	93% for 144 h, 30	Temp: 30 °C, pH: 7.0; agitation seed: 200 rpm	Batch	Viti et al. (2003)
Tannery	Fungal Consortia	100% in 36 h, 9.86	Temp: 28 °C, pH: 4.0	Stirred bioreactor	Kotaś and Stasicka (2000)
Electroplating	*Penicillium oxalicumstrain SL2*	100% for 48 h, 40.6	Temp: 30 °C; pH: 7.0; agitation seed: 200 rpm	Batch	Yoon et al. (2006)

as has been reported on bioremediation cutting-edge technologies. Following are some suggestions for addressing the identified gaps in the research:

(1) The complexity of natural environmental variables, especially soil, makes it challenging to achieve the goal of governance using entirely manufactured bacteria. Using a bacterial synergy, mixed cultures of microorganisms improve both environmental adaptability and treatment success.
(2) This is a suitable method for screening microorganisms for their ability to decrease or bind multiple hazardous metals, as polluted areas typically include more than one type of heavy metal.
(3) Bioremediation performs poorly and takes considerably more time than physical and chemical materials for removing heavy metals. The development of a consortia of microorganisms to enhance process efficiency should be the focus of future research.

6.8 Conclusion

This chapter analyzes the impact of metal accumulation pathways on metal removal as it relates to Cr(VI) bioremediation and biosorption by microorganisms. Reducing environmental Chromium(VI) levels with microbial treatment is one of the most effective and long-lasting methods. These bacteria' extraordinary homeostasis and tolerance of toxic metals systems are what have allowed them to thrive in such a harsh environment. Microbe-based technique, or biosorption, is a safe and inexpensive way to remove chromium from water. Furthermore, it shows great promise for future applications. Transport mechanisms such as precipitation, complexation, ion exchange, cell membrane, and physical adsorption are essential for biosorption. Several factors, including contact time, pH, temperature, biomass, and metal content, can drastically affect the efficiency of a bio-biosorption sorbent. Microbe-based technique, or biosorption, is a safe and inexpensive way to remove chromium from water. Furthermore, it shows great promise for future applications. Transport mechanisms such precipitation, cell membrane, ion exchange, complexation, and physical adsorption are essential for biosorption to occur. The effectiveness of a bio-biosorption sorbent is sensitive to a wide range of conditions, such as pH, temperature, biomass, contact time, and metal content. Removing many contaminants at once may be challenging in industrial wastewaters because, unlike laboratory solutions, they can contain dangerous heavy metals. As this review has shown, further study is needed to fully realize the potential of microbial biotechnology for environmental improvement.

References

Ackerley DF et al (2004) Chromate-reducing properties of soluble flavoproteins from *Pseudomonas putida* and *Escherichia coli*. Appl Environ Microbiol 70:873–882

Ahluwalia SS, Goyal D (2010) Removal of Cr(VI) from aqueous solution by fungal biomass. Eng Life Sci 10:480–485

Ahmed A (2018) Micro-remediation of chromium contaminated soils. PeerJ 6:e6076

Ahmed E et al (2016) Remediation and recycling of chromium from tannery wastewater using combined chemical–biological treatment system. Process Saf Environ Prot 104:1–10

Alam MZ, Ahmad S, Malik A (2011) Prevalence of heavy metal resistance in bacteria isolated from tannery effluents and affected soil. Environ Monit Assess 178:281–291

Bharagava RN, Mishra S (2018) Hexavalent chromium reduction potential of Cellulosimicrobium sp. isolated from common effluent treatment plant of tannery industries. Ecotoxicol Environ Saf 147:102–109

Bhattacharya A, Gupta A (2013) Evaluation of Acinetobacter sp. B9 for Cr(VI) resistance and detoxification with potential application in bioremediation of heavy-metals-rich industrial wastewater. Environ Sci Pollut Res 20:6628–6637

Bhattacharya A et al (2019) Alleviation of hexavalent chromium by using microorganisms: insight into the strategies and complications. Water Sci Technol 79:411–424

Chaturvedi MK (2011) Studies on chromate removal by chromium-resistant Bacillus sp. isolated from tannery effluent. J Environ Protect 2:76

Chen F et al (2022) Assessment of chromium toxicity and potential health implications of agriculturally diversely irrigated food crops in the semi-arid regions of South Asia. Agric Water Manag 272:107833

Chen R et al (2023) Functional characterization of an efficient ibuprofen-mineralizing bacterial consortium. J Hazard Mater 130751

Chhikara S et al (2010) Chromium(VI) biosorption by immobilized Aspergillus niger in continuous flow system with special reference to FTIR analysis. J Environ Biol 31:561–566

Congeevaram S et al (2007) Biosorption of chromium and nickel by heavy metal resistant fungal and bacterial isolates. J Hazard Mater 146:270–277

Dadrasnia A et al (2015) Biosorption potential of *Bacillus salmalaya* strain 139SI for removal of Cr (VI) from aqueous solution. Int J Environ Res Public Health 12:15321–15338

Das S et al (2021) Reduction of hexavalent chromium by Exiguobacterium mexicanum isolated from chromite mines soil. Chemosphere 282:131135

De Agostini A et al (2020) Heavy metal tolerance of orchid populations growing on abandoned mine tailings: a case study in Sardinia Island (Italy). Ecotoxicol Environ Saf 189:110018

De Pauw N, Van Vaerenbergh E (1983) Microalgal wastewater treatment systems: potentials and limits. In: Phytodepuration and the Employment of the Biomass Produced. Centro Ric. Produz, Animali, Reggio Emilia, Italy, pp 211–287

DesMarias TL, Costa M (2019) Mechanisms of chromium-induced toxicity. Curr Opin Toxicol 14:1–7

Dey S, Paul AK (2012) Optimization of cultural conditions for growth associated chromate reduction by Arthrobacter sp. SUK 1201 isolated from chromite mine overburden. J Hazard Mater 213:200–206

El-Naggar NEA et al (2020) Innovative low-cost biosorption process of Cr^{6+} by *Pseudomonas alcaliphila* NEWG-2. Sci Rep 10:14043

Elangovan R et al (2006) Reduction of Cr(VI) by a Bacillus sp. Biotech Lett 28:247–252

Essahale A et al (2012) Hexavalent chromium reduction and accumulation by Acinetobacter AB1 isolated from Fez Tanneries in Morocco. Indian J Microbiol 52:48–53

Fernández PM et al (2018) Bioremediation strategies for chromium removal: current research, scale-up approach and future perspectives. Chemosphere 208:139–148

Gad SC (1989) Acute and chronic systemic chromium toxicity. Sci Total Environ 86:149–157

Garbisu C et al (1998) Aerobic chromate reduction by Bacillus subtilis. Biodegradation 9:133

Georgaki MN, Charalambous M (2022) Toxic chromium in water and the effects on the human body: a systematic review. J Water Health

Han X et al (2008) Feasibility of using microalgal biomass cultured in domestic wastewater for the removal of chromium pollutants. Water Environ Res 80:647–653

Ibrahim ASS et al (2012) Hexavalent chromium reduction by novel chromate resistant alkaliphilic Bacillus sp. strain KSUCr9a. Afr J Biotech 11:3832–3841

Iqbal N, Ashraf M, Ashraf MY (2009) Influence of exogenous glycine betaine on gas exchange and biomass production in sunflower (*Helianthus annuus* L.) under water limited conditions. J Agron Crop Sci 195:420–426
Islam MA, Angove MJ, Morton DW (2019) Recent innovative research on chromium(VI) adsorption mechanism. Environ Nanotechnol Monitor Manag 12:100267
Kim B-M et al (2022) Reductive transformation of hexavalent chromium in ice decreases chromium toxicity in aquatic animals. Environ Sci Technol 56:3503–3513
Koçberber N, Dönmez G (2007) Chromium(VI) bioaccumulation capacities of adapted mixed cultures isolated from industrial saline wastewaters. Biores Technol 98:2178–2183
Konovalova VV et al (2003) Chromium(VI) reduction in a membrane bioreactor with immobilized Pseudomonas cells. Enzyme Microb Technol 33:899–907
Kotaś J, Stasicka Z (2000) Chromium occurrence in the environment and methods of its speciation. Environ Pollut 107:263–283
Kumar AAV, Pandey BD (2006) Remediation options for the treatment of electroplating and leather Tanning effluent containig chromium. Bio-processing of solid wastes and secondary resources for metal extraction. Mineral Process Extract Metal Rev 27:99–130
Li D, Li G, Zhang D (2021) Field-scale studies on the change of soil microbial community structure and functions after stabilization at a chromium-contaminated site. J Hazard Mater 415:125727
Lin J, Jiang W, Liu D (2003) Accumulation of copper by roots, hypocotyls, cotyledons and leaves of sunflower (*Helianthus annuus* L.). Biores Technol 86:151–155
Maurya A, Kumar PS, Raj A (2022) Characterization of biofilm formation and reduction of hexavalent chromium by bacteria isolated from tannery sludge. Chemosphere 286:131795
Monga A, Fulke AB, Dasgupta D (2022) Recent developments in essentiality of trivalent chromium and toxicity of hexavalent chromium: Implications on human health and remediation strategies. J Hazard Mater Adv 100113
Ni X et al (2020) The acute toxic effects of hexavalent chromium on the liver of marine medaka (Oryzias melastigma). Comp Biochem Physiol C: Toxicol Pharmacol 231:108734
Oliveira H (2012) Chromium as an environmental pollutant: insights on induced plant toxicity. J Bot 375843
Panda J, Sarkar P (2012) Bioremediation of chromium by novel strains *Enterobacter aerogenes* T2 and *Acinetobacter* sp. PD 12 S2. Environ Sci Pollut Res 19:1809–1817
Panda SK, Choudhury S (2005) Chromium stress in plants. Braz J Plant Physiol 17:95–102
Patra RC et al (2010) Soil Biology & Biochemistry Molecular characterization of chromium(VI) reducing potential in Gram positive bacteria isolated from contaminated sites. Soil Biol Biochem 42:1857–1863
Pavesi T, Moreira JC (2020) Mechanisms and individuality in chromium toxicity in humans. J Appl Toxicol 40:1183–1197
Peng W et al (2023) 'Bioaugmentation removal and microbiome analysis of the synthetic estrogen 17α-ethynylestradiol from hostile conditions and environmental samples by *Pseudomonas citronellolis* SJTE-3. Chemosphere 137893
Pun R, Raut P, Pant BR (2013) Removal of chromium(VI) from leachate using bacterial biomass. Sci World 11:63–65
Ramrakhiani L, Majumder R, Khowala S (2011) Removal of hexavalent chromium by heat inactivated fungal biomass of *Termitomyces clypeatus*: surface characterization and mechanism of biosorption. Chem Eng J 171:1060–1068
Sanjay MS et al (2017) Isolation and identification of chromium reducing bacteria fromtannery effluent
Sanz-Gallen P et al (2021) Increased urine chromium concentrations in a worker exposed to lead chromate due to medicinal herb intake. Int J Occup Environ Safety 5:16–24
Satarupa D, Amal KP (2010) Occurrence and evaluation of chromium reducing bacteria in seepage water from chromite mine quarries of Orissa, India. J Water Resour Protect
Shahida M et al (2017) Chromium speciation, bioavailability, uptake, toxicity and detoxification in soil-plant system: a review. Chemosphere 178:513–533

Shukla D, Vankar PS, Srivastava SK (2012) Bioremediation of hexavalent chromium by a cyanobacterial mat. Appl Water Sci 2:245–251
Song D et al (2016) Adsorptive removal of toxic chromium from waste-water using wheat straw and Eupatorium adenophorum. Plos One 11:e0167037
Stoltidis M et al (2011) Oral ingestion of hexavalent chromium through drinking water and cancer mortality in an industrial area of Greece: an ecological study. Environ Health 10:5057
Sun H, Brocato J, Costa M (2015) Oral chromium exposure and toxicity. Curr Environ Health Rep 2:295–303
Tang X et al (2021) Study on detoxification and removal mechanisms of hexavalent chromium by microorganisms. Ecotoxicol Environ Saf 208:111699
Thacker U et al (2006) Hexavalent chromium reduction by Providencia sp. Process Biochem 41:1332–1337
Upadhyay N et al (2017) Tolerance and reduction of chromium(VI) by *Bacillus* sp. MNU16 isolated from contaminated coal mining soil. Front Plant Sci 8:778
Vendruscolo F, da Rocha Ferreira GL, Antoniosi Filho NR (2017) Biosorption of hexavalent chromium by microorganisms. Int Biodeterior Biodegrad 119:87–95
Vengosh A et al (2016) Origin of hexavalent chromium in drinking water wells from the piedmont aquifers of North Carolina. Environ Sci Technol Lett 3:409–414
Viti C, Pace A, Giovannetti L (2003) Characterization of Cr(VI)-resistant bacteria isolated from chromium-contaminated soil by tannery activity. Curr Microbiol 46:1–5
Wang T et al (2019) Immobilization of hexavalent chromium in soil and groundwater using synthetic pyrite particles. Environ Pollut 255:112992
Wang Y-S et al (2007) Bioleaching of chromium from tannery sludge by indigenous Acidithiobacillus thiooxidans. J Hazard Mater 147(1–2):319–324
Xie Y, Lu L, Chen B (2022) Asymmetrical alternating current electrochemically-mediated washing method for sustainable remediation of Cr(VI)-contaminated soil. J Hazard Mater 437:129088
Xu J et al (2019) High shear-granulated hierarchically porous spheres nanostructure-designed for high-performance supercapacitors. Adv Powder Technol 30:2440–2449
Xu J et al (2022) A nanocubicle-like 3D adsorbent fabricated by in situ growth of 2D heterostructures for removal of aromatic contaminants in water. J Hazard Mater 423:127004
Yoon J et al (2006) Accumulation of Pb, Cu, and Zn in native plants growing on a contaminated Florida site. Sci Total Environ 368(2–3):456–464
Yu C et al (2019) The long-term effects of hexavalent chromium on anaerobic ammonium oxidation process: performance inhibition, hexavalent chromium reduction and unexpected nitrite oxidation. Biores Technol 283:138–147
Zahoor A, Rehman A (2009) Isolation of Cr(VI) reducing bacteria from industrial effluents and their potential use in bioremediation of chromium containing wastewater. J Environ Sci 21:814–820
Zakaria ZA et al (2007) Hexavalent chromium reduction by Acinetobacter haemolyticus isolated from heavy-metal contaminated wastewater. J Hazard Mater 146(1–2):30–38
Zhang XH et al (2011) Chronic occupational exposure to hexavalent chromium causes DNA damage in electroplating workers. BMC Public Health 11:1–8
Zhou M et al (2018) Exchange electrode-electrokinetic remediation of Cr-contaminated soil using solar energy. Sep Purif Technol 190:297–306

Chapter 7
Genetic Engineering for Chromium Removal

Roshnee Bose, Samruddhi Kulkarni, Jaya Lakkakula, and Nilesh S. Wagh

Abstract Chromium(Cr) is a heavy metal and exerts a profusion of toxic effects in a plethora of organisms (bacteria, plants, fungi, archaebacteria, algae, and many more). Cr is increasingly being accumulated in the soil, freshwater, wastewater, etc. because of extensive anthropogenic activities. It is carcinogenic for humans and due to the alarming upsurge of Cr concentration in the environment, it is treated as a priority pollutant. Cr manifests in various valence states of which Cr(VI) (most toxic) and Cr(III) are the most stable. Cr when accumulated inside cells of an organism, leads to the production of reactive oxygen species (ROS) hampering a network of molecular, physiological, and metabolic processes. Several genetic studies have revealed that certain organisms contain specific genes conferring the potential to withstand as well as remediate high concentrations of Cr. These organisms may remove, accumulate or reduce Cr(VI) to the less toxic Cr(III) form. Recent years have shed light on the various genes involved in Cr tolerance. These genes have been exploited using genetic engineering (GE) tools to construct genetically modified organisms (GMOs) with the ability of Cr bioremediation. These GMOs can be allowed to grow in Cr-contaminated regions and ameliorate its toxic effects. This chapter summarizes the Cr toxicity in organisms, the significance of detoxification genes in Cr stress response, mechanisms of Cr tolerance and enzyme activity in Cr reduction and resistance. It also highlights various bacteria, plants, and other organisms such as phages, yeast, etc., utilized for bioremediation in Cr-contaminated regions.

R. Bose · S. Kulkarni · J. Lakkakula (✉) · N. S. Wagh (✉)
Amity Institute of Biotechnology, Amity University Maharashtra, Mumbai-Pune Expressway, Bhatan, Panvel, Mumbai, Maharashtra 410206, India
e-mail: jrlakkakula@mum.amity.edu

N. S. Wagh
e-mail: wnshirish@mum.amity.edu

J. Lakkakula
Centre for Computational Biology and Translational Research, Amity Institute of Biotechnology, Amity University Maharashtra, Mumbai-Pune Expressway, Bhatan, Panvel, Mumbai, Maharashtra 410206, India

N. S. Wagh
Centre for Drug Discovery and Development, Amity Institute of Biotechnology, Amity University Maharashtra, Mumbai-Pune Expressway, Bhatan, Panvel, Mumbai 410206, India

© The Author(s), under exclusive license to Springer Nature Switzerland AG 2023

N. Kumar et al. (eds.), *Chromium in Plants and Environment*, Environmental Science and Engineering, https://doi.org/10.1007/978-3-031-44029-8_7

Graphical Abstract: Schematic representation of genetic engineering techniques used For Cr bioremediation.

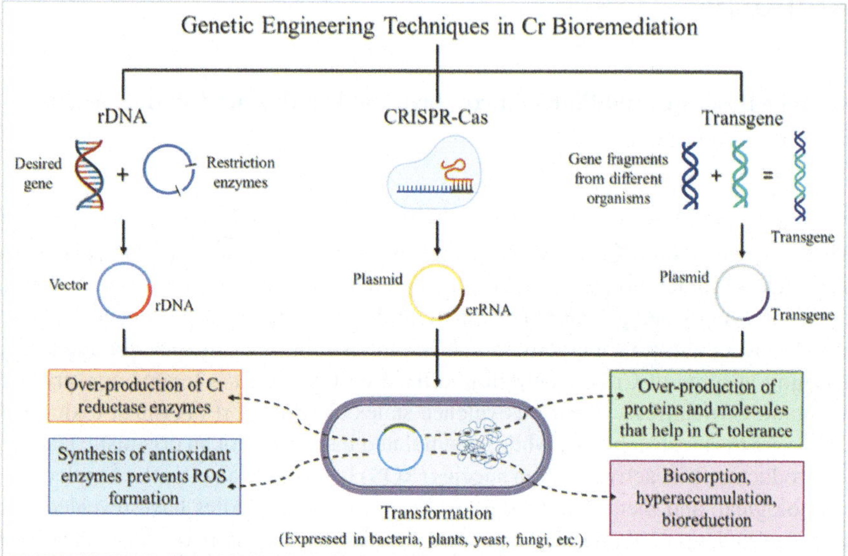

Keywords Chromium (Cr) · Bioremediation · Genetic engineering (GE) · Genetically modified organisms (GMOs)

7.1 Introduction

Chromium(Cr) is the seventeenth most prevalent element found on the earth's crust. For a long duration of time, the scientific community held to the belief that Cr is an essential metal imperative for the proper functioning of the human body. However, a decade ago researchers proved otherwise. Cr is a non-essential metal that beyond trace amounts is found to be detrimental to the health of organisms (di Bona et al. 2011). It hampers the biochemical and metabolic pathways of organisms and if present beyond the threshold level, can cause severe toxic effects in the organism. The subject of chromium contamination in the environment has therefore been a topic of concern worldwide. Naturally, Cr is replenished by the chromium cycle (Johnson et al. 2006). However, Cr can be introduced into the environment via a variety of unregulated anthropogenic activities such as metallurgy, smelting of non-ferrous base metal, electroplating, chemical production, paint, tanneries, dyeing, and paper pulp and steel production, accounting for 70% of the total Cr pollution (Jaishankar et al. 2014).

Being a transition metal, Cr manifests itself in various oxidation states among which Cr(III) and Cr(VI) are found to be the most stable. Cr(VI) is essentially far more harmful than Cr(III) due to its greater solubility and mobility in water. Thus, several detoxification processes employed by organisms revolve around the reduction of Cr(VI)–Cr(III) (Kanagaraj and Elango 2019). One of the most efficient methods of Cr treatment in terms of quality and cost is bioremediation (Dados et al. 2015). Bioremediation is a technique used for removing contaminants from the environment with the help of living organisms such as bacteria, fungi (mycoremediation), plants (phytoremediation), archaea, viruses, etc. All these organisms have developed various mechanisms to detoxify Cr (Verma and Jaiswal 2016).

Most microorganisms are vulnerable to chromate however, some can withstand and are resistant to significant concentrations of Cr(VI). While Cr(VI) reductase genes can be situated on both plasmids as well as the main chromosome, Cr(VI) resistance is primarily plasmid-borne in bacteria. The mechanisms that are better understood include Cr(VI) reduction to Cr(III) and the cellular efflux of chromate ions from the cytoplasm of the cell. Both chromate-sensitive and chromate-resistant bacterial species have the capacity to ameliorate the effects of Cr(VI), which may be attributed to the presence of chromate-reductase activity. The efficiency of Cr resistance is strongly influenced by having a strong efflux mechanism (Thacker et al. 2006). Several Cr resistance mechanisms in bacteria have been accounted for including-(1) Regulation of the sulfate uptake shuttle network that is responsible for initial accumulation in the cell. (2) Reduction of Cr(VI)–Cr(III) extracellularly by reacting with the functional groups present on the cell surface of bacteria. (3) Cr(VI) adsorption to functional groups which are stationed on the surface of the cell and reduction of Cr(VI)–Cr(III) in the cell membrane. (4) Reduction of Cr(VI)–Cr(III) intracellularly and Cr(III) is saltated outside of cells. (5) Ability to activate the enzymes which are responsible for scavenging of the reactive oxygen species (ROS) to counteract the stress induced by chromate (Ngwenya and Chirwa 2011; Tahri Joutey et al. 2014). ROS are known to be inhibited by antioxidant enzymes and non-enzymatic antioxidants (such as flavonoids, thiol antioxidants, etc.). The danger of oxidative stress is known to be reduced by these antioxidants' ability to disperse free radicals. Antioxidants deactivate ROS at the cellular and molecular levels, and at minimal concentrations, they postpone oxidative processes by interfering with the chain reaction of radicals. Additionally, antioxidants chelate the metal ions that produce ROS (Flora 2009).

Cr toxicity in plants causes retardation of seed germination, growth, root and shoot, loss of leaf pigmentation, etc. (Velez et al. 2017). Phytoremediation is usually achieved using hyperaccumulator plants that have a natural tolerance mechanism to remove or reduce Cr stress. Some of these tolerance mechanisms include increased antioxidant activity like catalase, peroxidases, superoxide dismutase, and ascorbate peroxidase to prevent the production of ROS (Gill and Tuteja 2010). Plants can remediate Cr by employing methods like phytoreduction, phytoextraction, rhizofiltration, phytostabilisation, etc. (Marques et al. 2009). Although a plethora of research studies have been conducted to unravel the underlying mechanisms of Cr uptake, regulation, toxicity as well as its detoxification by organisms, applying this knowledge in the development of novel bioremediation techniques remains relatively unexplored.

The most recent developments in the field of bioremediation predominantly include using genetic engineering (GE) techniques to construct genetically modified organisms (GMOs) (Tahri Joutey et al. 2014). GE techniques include recombinant DNA technology, the creation of transgenics, CRISPR (Clustered Regularly Interspaced Short Palindromic Repeats) technology, etc. GE allows the manipulation of the host genome to give it desirable traits that help in enhancing its Cr remediation process. These desired traits may include the upregulation of a gene to amplify the production of enzymes that reduce Cr(VI) (such as chromium reductase), stabilize ROS generation, or produce molecules and proteins (such as trehalose, Metallothioneins, etc.) that remediate Cr (Reisinger et al. 2008). Under most scenarios the desired gene is recognized and isolated from one organism (source) and using vectors, they are transformed into another organism (host) that is directly used for bioremediation. Source and host organisms belonging to the same species, genus, or phyla may be used while organisms originating from different genus or even kingdoms may also be used to form a GMO (such as bacteria and yeast, bacteria, and plants, etc.). GE has broad-spectrum applications that can be exploited to their full potential to harness the best results and optimum output. However, research in the field of GE remains scarce and it is an ongoing research hot topic.

In this chapter, we have focused on the role of GMOs particularly bacteria and plants in the bioremediation of chromium. The mechanisms and techniques developed by researchers to enhance Cr detoxification through the architecture of a multitude of GMOs have been also discussed comprehensively. Insights on research studies conducted on various other organisms (such as fungi, bacteriophages, etc.) have also been included.

7.2 Genetic Engineering Organisms Involved in the Bioremediation of Cr

GE has gained a lot of popularity in the last decade. This is primarily due to its wide range of applications and high precision and efficacy. GE allows scientists to alter the genetic makeup of an organism to enhance its potential or to confer it novel abilities. In the light of bioremediation research, the traditional approaches with wild-type (WT) or mutant organisms have been exploited to great potential. Recent years have seen an upsurge in the use of GMOs for bioremediation purposes. In the following sections, we have discussed the various engineered organisms that have been employed to remediate Cr (Table 7.1).

Table 7.1 Table for GMOs employed for Cr bioremediation

S. No	Source organism	Target organism	Gene transferred	Mechanism	Treatment site	References
1	*Arabidopsis thaliana* (L.) Heynh and *E. coli*	*Brassica juncea* (L.) Czern	APS1 cDNA gshI gene gshII gene	Overexpression of ATP sulfurylase-glutamyl cysteine synthetase Glutathione synthetase	Metal-contaminated mine tailings	Bennett et al. (2000)
2	*Arabidopsis thaliana*	*Brassica juncea* (L.)	APS1	Overexpression of ATP sulfurylase enzyme for Cr accumulations and tolerance	Multi-metal contaminated soil	Wangeline et al. (1999)
3	Not Specified	*Brassica naps* L	BnMP1 gene	Overexpression of the metallothionein gene improved Cr(VI) tolerance and accumulation	Not specified	Terzi and Yıldız (2015)
4	*Agrobacterium tumefaciens* (GV3101)	*Arabidopsis thaliana*	OsGSTL2 (Os03g17470)	Enhanced GST activity leading to detoxification of Cr(VI)	Not specified	Kumar et al. (2013)
5	*Alcaligenes eutrophus* CH34	*Alcaligenes eutrophus* AE104 (pEBZ141)	chr_1 and chr_2 genes	Biosorption	Industrial wastewater	R16
6	*Bacillus cereus* RC607 and *Pseudomonas sp.* K-62	*Escherichia coli* (strain M-BL21)	Gram-positive merP gene (GB) and Gram-negative merP gene (GP)	Bioabsorption: MerP (adsorption protein) was displayed on the host's cell surface and adsorbed Cr(III)	Leather wastewater	Wang et al. (2021)
7	*Bacillus cereus* RC607 and *Pseudomonas sp.* K-62	*E. coli* BL21	merP gene	Biosorption	Wastewater	Kao et al. (2008)

(continued)

Table 7.1 (continued)

S. No	Source organism	Target organism	Gene transferred	Mechanism	Treatment site	References
8	*E. coli*	*E. coli* BL21 (DE3)	yieF gene	Expression of chromate reductase enzyme to function as a four-electron Cr(VI) reducer	Not specified	Ackerley et al. (2004b)
9	*E. coli*	*Shewanella oneidensis* MR-1	dmsE gene	EET enhancement for modulating electron flux using CRISPR-ddAsCpf1 system	Not specified	Li et al. (2020)
10	*E. coli* S17/1	*Alcaligenes eutrophus* AE104	chrBA genes	Biosensor and bioreduction of Cr(VI)	Industrial sewage water	Peitzsch et al. (1998)
11	*E. coli* WM3064	*Shewanella oneidensis* MR-1	luxR and luxI genes	Designing a quorum sensing (QS)-based population-state decision (PSD) system for intelligently enhancing the extracellular electron transfer (EET) regulation system and Cr(VI) bioreduction	Not specified	Li et al. (2020)
12	*E. coli*	MC4100 *E. coli*	ostBA gene	Overproduction of trehalose sugar for Cr(VI) (1 mM) reduction	Not specified	Frederick et al. (2013)

(continued)

Table 7.1 (continued)

S. No	Source organism	Target organism	Gene transferred	Mechanism	Treatment site	References
13	Transgenic *E. coli* and *Bacillus subtilis*	*E. coli@ZIF-8* and *B. subtilis@ZIF-8*		Enhanced biosorption by encapsulating bacteria via microfluidics-based biomimetic mineralization with ZIF-8	Cr-contaminated water	Tang et al. (2021)
14	*Homo sapiens*	*E. coli* Jm109	MT2A and MT3 genes	Biosorption and bioreduction of Cr(VI)	Not specified	R3
15	*Nicotiana tabacum*	*Nicotiana tabacum* and *S. cerevisiae* Y800	NtST1	Accumulation of Cr	Not specified	Kim et al. (2006)
16	Not specified	*Nicotina langsdorffii*	GR gene and *rol*C gene	Higher Cr accumulation-high root metal uptake and capacity to retain it in the root itself,	Not specified	del Bubba et al. (2013)
17	Not specified	Transgenic cotton cultivars (J208 and Z905) and their hybrid (ZD14)	Not specified	Upregulation of antioxidative metabolism	Not specified	Daud et al. (2014)
18	Not specified	*Salix fragilis* L. and *Salix viminalis* L	Not specified	Phytoextraction	Multi-metal contaminated soil	Vandecasteele et al. (2005)
19	*Oryza sativa*	*Arabidopsis thaliana*	OsMYB-R1 (LOC_ Os06g45890)	Enhanced Cr(VI) stress tolerance aided with salicylic acid	Cr-contaminated soil	Tiwari et al. (2020)

(continued)

Table 7.1 (continued)

S. No	Source organism	Target organism	Gene transferred	Mechanism	Treatment site	References
20	*Oryza sativa*	*Schizosaccharomyces pombe*	OsGSTU30 and OsGSTU41	Overexpression of glutathione S-transferase for Cr(VI) detoxification by chelation and sequestration of glutathione-Cr(VI) complexes	Not specified	Huang et al. (2021)
21	*Oryza sativa* L	*Arabidopsis thaliana*	OsGSTU30 (Os10g38600)	Overexpression of OsGSTU30 enzyme shows GST activity leading to detoxification of Cr(VI)	Not specified	Srivastava et al. (2019)
22	Podoviridae, Siphoviridae, Myoviridae, Mimiviridae, Microviridae, phycodnaviridae, inoviridae, ackermannviridae, tectiviridae and genomoviridae families	Pseudomonas, Cronobacter, Salmonella, Enterobacter, Klebsiella, Escherichia and Shigella from the proteobacteria phylum, and actinomyces, micromonospora, and Salinispora from the actinobacteria phylum	AMGs (auxiliary metabolic genes) such as MRGs (HM resistance genes)	Mutualistic symbiosis linkage based on CRISPR conferring the host with high Cr tolerance and detoxification capabilities	Heavily Cr-contaminated soil	Huang et al. (2021)
23	*Pseudomonas ambigua*	*E. coli* BL21 (DE3)	nfsA gene	Cr(VI) reduction via a one-electron transfer mechanism	Not specified	Ackerley et al. (2004a)

(continued)

Table 7.1 (continued)

S. No	Source organism	Target organism	Gene transferred	Mechanism	Treatment site	References
24	*Pseudomonas putida*	*E. coli* BL21 (DE3)	chrR gene	Expression of chromate reductase enzyme for Cr(VI) reduction by the generation of flavin semiquinone		Ackerley et al. (2004b)
25	*Rattus norvegicus*	*Nicotina langsdorffii*	Rat glucocorticoid receptor gene	Interaction between GR receptor and phytosteroids	Not specified	Fuoco et al. (2013)
26	*Saccharomyces cerevisiae* W303α	*Saccharomyces cerevisiae*	CTR1, CTR3, FET4, FTR1, SMF1, SMF2, ZRT1, and ZRT2	Hyperaccumulator	Multi-metal contaminated soil	Sun et al. (2019)
27	*Serratia sp.* CQMUS2	*E. coli* BL21 (DE3)	ChrT gene	Expression of chromate reductase enzyme to reduce Cr(VI)–Cr(III)	Wastewater	Deng et al. (2014)
28	*Serratia sp.* CQMUS2	*E. coli* BL21 (DE3)	ChrT gene	Expression of chromate reductase enzyme to reduce Cr(VI)–Cr(III)	Wastewater and Cr-contaminated soil	Zhou et al. (2017)
29	*Serratia sp. S2*	*E. coli* BL21 (DE3)	ChrA gene	Cr(VI) efflux	Wastewater and Cr-contaminated soil	Simin et al. (2017)

(continued)

Table 7.1 (continued)

S. No	Source organism	Target organism	Gene transferred	Mechanism	Treatment site	References
30	*Serratia sp. S2*	*E. coli* BL21	ChrA and ChrT genes to produce recombinant ChrAT gene	Expression of chromate reductase enzyme to resist and reduce Cr(VI)	Wastewater and Cr-contaminated soil	Gu et al. (2020)
31	*Staphylococcus aureus*	*Caenorhabditis elegans* TJ375 gpls1 (hsp-16.2p::GFP)	hsp-16.2 promoter coupled to the gene encoding green fluorescent protein (GFP)	Expression of Hsp-16.2 leads to the biosynthesis of defensive enzymes in response to oxidative stress resulting in ameliorating the toxic effects of Cr	Cr-contaminated soil and water	Fajardo et al. (2022)

7.2.1 Bacteria-Mediated Cr Removal

A wide variety of bacterial species exhibit the ability to resist, reduce, detoxify, and remediate Cr belonging to the various families like Enterobacteriaceae, Bacillaceae, Pseudomonadaceae, Yersiniaceae, Burkholderiaceae, Shewanellaceae, Alcaligenaceae, Flavobacteriaceae and many more. Bacteria have the ability to carry Cr resistance genes, thus producing Cr reductase enzymes, further synthesizing molecules that can ameliorate the detrimental effects of Cr, and even remove chromate ions via the cellular efflux mechanism. In many cases, the bacterial genome has been genetically modified by either manipulating its own genetic code or by introducing foreign gene sequences for the upregulation of desirable genes and consequently overexpressing proteins, enzymes, molecules, etc. for Cr tolerance. In the following section, genetically modified bacteria with the ability of Cr bioremediation have been discussed.

Fredrick and coworkers (2013) designed a study to test the efficacy of trehalose for bioremediation purposes. Trehalose is a sugar molecule that is known to protect living organisms against reactive oxygen species (ROS), pH, dehydration, thermal shock, osmotic shock, and free radical-induced damage. Therefore, it may be a potential candidate to reduce the toxic effects of Cr. However, its potential in Cr remediation remains unexplored. The biosynthesis of trehalose in *Escherichia coli* is encoded by the ostBA gene. The ostBA operon was therefore cloned into the pBAD18 plasmid and finally transformed into the MC4100 *E. coli* strain. This genetically engineered Trehalose (OstA/OstB) overproducer strain MC4100 *E. coli* produced 262 ± 74 mM of Trehalose while Wild Type (WT) produced only 15 ± 12 mM. Bacterial growth under Cr stress (500 μM potassium dichromate) was compared between Trehalose overproducer and WT for over a week. The OstA/OstB overproducer grew to its maximal level within 2 days whereas WT took almost a week. It was able to reduce 1 mM Cr(VI)–Cr(III) and beyond this concentration, its behaviour was comparable to that of the WT. Additionally, the capacity to reduce the total load of Cr(VI) in the engineered bacteria was more than the WT. The comparative study concluded that Trehalose promotes the growth of *E. coli* under Cr(VI) stress (Frederick et al. 2013).

In another study, Wang, and coworkers (2021) employed engineered *E. coli* strain M-BL21 for the in-situ elimination of Cr(III) from leather wastewater. In this novel approach, the engineered bacterial strain overexpressed a cysteine-rich protein MerP (Mercuric transport protein periplasmic component). MerP is a protein that adsorbs Cr(III). The biosorption process was assisted by immobilized magnetic pellets. A recombinant plasmid pET-23b/M was constructed by synthesizing the MerP gene fragment, amplifying a gene coding for ice nucleation protein (INP), and ligating them to the vector pET-23b. Post expression, the MerP protein was exhibited on the genetically modified *E. coli* cell surface. The recombinant strain M-BL21 effectively adsorbed Cr(III) at 2.38 nmol/g cell. At a cell concentration of 3 g/L, the adsorption efficiency of the engineered strain was maximum at up to 91.2%. This value is significantly higher than using magnetic carrier pellets alone. The desirable dose of the pellets was found to be 40 g/L. In summary, the results demonstrated that the engineered bacteria efficiently removed 88.2% of Cr(III) from tanning wastewater (Wang et al. 2021).

A plethora of microorganisms are known to bind to specific metals and remediate them. However, WT bacteria lack this selectivity. GE tools can be exploited to design bacteria such that they can selectively bind to metals and increase their metal binding capacity. One way to achieve this goal is by introducing metal ion-binding polypeptides such as metallothioneins (MTs) on the recombinant bacterial cell wall. Mammalian MTs comprises 4 distinct isoforms: MT1, MT2A, MT3, and MT4. Akkurt et al. (2022) designed a comparative study in which they cloned MT genes (MT2A and MT3) isolated from *Homo sapiens* into *E. coli* Jm109 (host) for removal of Cr(VI) and reduction in its aqueous solution. The recombinants were then exposed to varying Cr concentrations (0.5, 10, 25, and 50 mg/L). Quantitative analysis revealed that MT2A was the most competent strain in Cr(VI) removal [89% in 25 mg/L Cr(VI)] as well as reduction of Cr(VI) [76% in 25 mg/L Cr(VI)]. On the contrary, MT3 removed 76% of Cr(VI) (25 mg/L). Additionally, the MT2A strain absorbed 22 mg/g of Cr(VI) per dry cell. This experimental data led to the conclusion that both the strains effectively remove/reduce Cr(VI) while MT2A being more efficient may be employed as a biotechnological tool for Cr removal (Akkurt et al. 2022).

In another study, Deng, and coworkers (2014) aimed at cloning full-length ChrT DNA from *Serratia* sp. CQMUS2 as well as researching the derived sequence of amino acid and examining its three-dimensional (3D) structure. *Serratia* sp. CQMUS2 is a WT strain with Cr reduction and resistance capability. In *Serratia* sp. CQMUS2, the two mechanisms for Cr remediation predominantly include: chromate ion efflux from the cell cytoplasm or the NAD(P)H-dependent flavin mononucleotide (FMN) reductase (FMN_red) aided direct Cr(VI) reduction to Cr(III). Upon the analyses of ChrT amino acid sequence, it showed a startling resemblance to the FMN reductase genes originating from *Raoultella ornithinolytica* and *Klebsiella pneumonia.* These species descend from the flavodoxin-2 superfamily. Additionally, the three-dimensional structure of ChrT demonstrated an 85.6% homology to that of *E. coli* ChrR. The 3D structures of the two species share homology with the active sites of four enzymes for Cr(VI) reduction. Thus, it was concluded that the ChrT strain can be a propitious candidate for Cr bioremediation (Deng et al. 2014).

Building further on the foundation laid by the previous experiments, in a de novo study, Zhou et al. (2017) tested the ability of the ChrT gene to reduce Cr(VI) as well as to evaluate its chromate reductase activity in Cr-contaminated soil and water bodies. The ChrT gene codes for a chromium reductase enzyme that catalyzes the reduction of Cr(VI) to the less toxic Cr(III). ChrT gene from *Serratia* sp. CQMUS2 was isolated and cloned in *E. coli* BL21 (DE3) using a pMD19-T plasmid vector (Deng et al. 2014) They were grown in LB liquid media containing Cr(VI) (50 mg/L) at 37 °C. The recombinants were exposed to various culture conditions influencing their reduction efficiency (pH 5.0–11.0; temperatures 20, 25, 30, 37, 40 and 50 °C) for 48 h. Results showed that at pH 7 reduction activity was optimal as there was a residual 20 mg/L Cr(VI) which was substantially lower than the other pH conditions. Similarly, at 37 °C, the engineered strain reduced 50 mg/L Cr(VI) to 19.6 mg/L (up to 40%).

Moreover, the addition of NADPH improved the capability of eliminating Cr(VI) by ChrT as it acts as a coenzyme. The ChrT-engineered bacteria could eliminate Cr(VI), which was a little weaker than WT potentially due to the fact that more than one gene in the WT is responsible for Cr(VI) reductase activity. However, *Serratia* spp. S2 bacteria are associated with biological contamination due to its high pathogenesis. The high level of biosafety of E. coli species helps eliminate this drawback. Additionally, with slight modifications in the chemical and physical environments, the Cr reductase activity of the engineered stain can be optimised. Furthermore, the recombinant can be used to extract large amounts of chromium reductase, which can be then used directly to eliminate Cr(VI) (Zhou et al. 2017).

In another experiment, researchers cloned the ChrA gene from *Serratia* sp. S2 and was linked to the vector pET-28a (+). This was then transformed into *E. coli* BL21 (DE3) strain for ChrA protein expression. *Serratia* sp. S2, the control and the recombinant were cultured to logarithmic phase in $Cr_2O_7^{2-}$ rich (100, 150 and 200 mg/L) LB liquid media at 37 °C. Post 18–24 h, it was observed that the constructed bacteria could tolerate up to 200 mg/L of $Cr_2O_7^{2-}$. Several bacterial species have developed efflux mechanisms allowing them to regulate their internal environment by removal of toxic substances from the cell (X4). In this study, efflux of chromate by *E. coli* BL21 ChrA strains was also examined. The engineered strain was loaded in 50 mg/L $Cr_2O_7^{2-}$ medium for 30 min and resuspended in PBS buffer solution. After 10 min of culture, the efflux of chromium(VI) reached 22%. However, as time increased, the Cr(VI) efflux gradually decreased. The results showed that the efflux ability of the engineered bacteria was greater than that of the control strain, as well as the disparity was statistically crucial, suggesting that the expression of the ChrA gene enhanced the strain's capacity to efflux Cr(VI). ChrA has high specificity as Cr(VI) efflux protein. Resultantly, the recombinant construct can be used to remediate Cr(VI) polluted water or soil (Simin et al. 2017).

Gu and coworkers (2020) cloned Cr(VI) metabolic protein genes encoded by ChrA and ChrT from *Serratia* sp. S2 and linked with plasmid pET-28a (+). These were expressed in *E. coli* BL21 to produce ChrA, ChrT and ChrAT recombinant bacteria. The control, WT and the 3 constructs were exposed to different Cr(VI) levels. On comparing the results, it was found that the Cr-resistance ability of ChrT, ChrA and ChrAT at 5 and 10 mg/L Cr(VI) was greater than the control but lower than the WT. As the Cr concentration was intensified to 15 and 20 mg/L, ChrA and ChrAT showed comparable results and their Cr-resistance capability was greater than ChrT and control strain but lower than the WT (Fig. 7.1). A key observation made during the experiment was the dynamicity in the changes pertaining to Cr(III) content interior and exterior to the cell. The team of researchers hypothesized that ChrT protein in ChrAT strain reduces Cr(VI)–Cr(III) intracellularly by acting as a Cr reductase. During this procedure, Cr(III) can be discharged out of the cell due to the efflux mechanisms channelled by the ChrA protein. Results concluded that the ChrA and ChrAT recombinants demonstrated substantial tolerance to Cr(VI), whereas ChrT and ChrAT strains exhibited a remarkable capacity to reduce Cr(VI) (Gu et al. 2020).

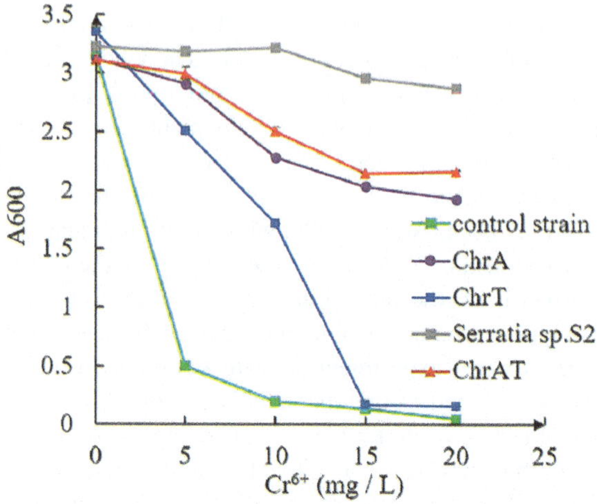

Fig. 7.1 Cr(VI) resistance of strains. Reprinted with permission from Gu et al. (2020)

Bacterial enzymes are essential in catalyzing Cr(VI) reduction to Cr(III). Enzymes that directly catalyze Cr(VI)–Cr(III) are especially of the essence rather than the ones that generate Cr(V) as an intermediate step. The latter leads to the genesis of obnoxious ROS thereby harming the bacteria. In a study led by Ackerley et al. (2004a, b), they studied the enzymatic activity of chromium reductase, ChrR (from *Pseudomonas putida*) and YieF (from *E. coli*). ChrR and YieF are two electrophoretically pure forms of soluble bacterial dimeric flavoproteins. Flavoproteins are primarily dehydrogenases that are involved in biological oxidation processes. The chrR gene was excised from the pCHP4 plasmid and linked to the pET-28a$^+$ vector and thereafter expressed in *E. coli* BL21(DE3). Similarly, the overproduction of the YieF protein was done by cloning the gene in the pET-28a$^+$ and transforming *E. coli* BL21(DE3) strain. Moreover, a *P. putida* KT2440 construct was designed. The chrR gene was disrupted leading to the formation of this mutant. During the course of the study, flavin semiquinone was induced by ChrR dimer during Cr(VI) reduction, transferring >25% of the NADH electrons to ROS. Semiquinone formation was transitory and ROS subsided with time. Hence, these experimental observations suggested that P. ambigua transiently generated Cr(V) before giving Cr(III) as the end product. *E. coli* Fre protein on the other hand reduced Cr(VI) directly to Cr(III). However, both the dimers reduced Cr(VI) with high efficacy to Cr(III) (k_{cat}/K_m = ~2 × 10^4 M^{-1}/s). ChrR is protective against chromate toxicity as established by the studies with mutants; this is presumably as it prevents Cr(VI) reduction by the one-electron reducers in the cell, hence reducing ROS production. The supremacy of the chromate reduction kinetics of this enzyme makes it an excellent candidate for Cr bioremediation. On the other hand, no flavin semiquinone was produced and upto 25% of the NADH electrons were converted to ROS during YieF's chromate reduction process. Therefore, the YieF dimer could be an obligate four-electron Cr(VI) reducer that transfers 3 electrons to chromate and 1 to O_2 in a single step. The direct correlation between YieF function and chromate protection could not be established because a mutant missing this enzyme could not be developed. Nevertheless, the findings imply that YieF might be an even better option for additional research than ChrR (Ackerley et al. 2004b).

To gain further insights into their previous study, Ackerley and coworkers (2004a, b) researched the chromium reductase enzyme NfsA. In their previous study, a direct correlation between the yieF enzyme and its role in protecting E. coli against Cr(VI). Chromate reductases belonging to Class II (such as NfsA) that mainly possess nitro reductase activity are non-homologous to Class I enzymes but show homology to the chromate reductase isolated from *Pseudomonas ambigua.* Ackerley et al. (2004a, b) speculated that NfsA may be helpful in the detoxification of chromate since it reduces nitro compounds by an obligatory two-electron transfer and may also decrease chromate by an analogous mechanism with little ROS production. The experimental analysis led to the understanding that NfsA reduces Cr(VI) via a one-electron transfer mechanism. A comparative study of NfsA ROS-generating activity and chromate transformation with ChrR, LpDH (lipoyl dehydrogenase) and YieF was performed by exposing them to 250 mM K_2CrO_4. ChrR and NfsA transported greater number of electrons to ROS than LpDH, which used 62–74% of the available electrons to produce ROS (34–39% and 49–53%, respectively). ChrR and NfsA demonstrated similar results suggesting that NfsA, like ChrR may play a role in protecting the bacterium against chromate. For this purpose, they examined the Cr(VI) tolerance of WT *E. coli* AB1157 with JVQ1 (its isogenic nfsA knockout mutant). When grown in LB supplemented with 300 mM Cr(VI), JVQ1 did not vary from AB1157 in terms of growth or chromate removal capacity. The mutation had no impact on cell viability, as assessed by plate counts, and dense cell suspensions of the 2 bacterial strains in chromate-modified LB medium converted Cr(VI) at virtually comparable beginning rates. The ineffectiveness of NfsA in this regard can be accounted for by the possible explanation that during chromate reduction, NfsA produced more ROS as compared to ChrR, and its co-existence in the LpDH reaction reduced ROS production to a smaller extent over the presence of ChrR, which may not be enough to have an impact on cell survival (Ackerley et al. 2004a).

As discussed previously, ROS generation while treatment of Cr contaminated water solution leads to bacterial cell damage thereby negatively impacting their biosorption capabilities. In a novel approach based on a microfluidics-based biomimetic mineralization technique, transgenic *E. coli* and *B. subtilis* were treated with zeolitic imidazolate framework-8 (ZIF-8) enabling them to form capsules. A layer of ZIF-8 was allowed to form on the surface of the cell of the two strains in situ. *E. coli*@ZIF-8 and *B. subtilis*@ZIF-8 could then remove Cr(VI) from polluted water and withstand harsh environments (nutrient deficiencies, UV light, etc.). Metal ions or clusters are cross-linked with organic linkers to create ZIF-8. For steady delivery of ligands as well as metal ion fluids via laminar flow, the microfluidic device designed by Tang et. al. (2022) comprises a receiver, connection system and dual-channel microinjection pump, which were coupled by polytetrafluoroethylene tubes with luer joints. After culturing the encapsulated bacteria, they were suspended in Cr(VI) (5, 25, and 50 mg/L) containing aqueous solutions. Within 3 days of the experiment, *B. subtilis*@ZIF-8 could absorb 50% of Cr(VI) ions which is considerably greater than ZIF-8 and WT alone. At 50 mg/L Cr(VI) concentration, Cr(VI) uptake by *E. coli*@ZIF-8 and *B. subtilis*@ZIF-8 was 52% and 68% respectively. Encapsulated *E. coli* adsorbed 55%, 51%, and 33% of Cr(VI) (5, 25, and 50 mg/L respectively) post one day of treatment. The maximum Cr(VI) adsorption by the encapsulated bacteria was observed to be 90%. Therefore, using the microfluidic device for Cr(VI) contaminated water bodies can be considered a promising and elegant approach (Tang et al. 2021).

MerP is a cysteine-containing periplasmic protein. It belongs to a family of Hg^{2+} ion transport proteins. To develop biosorbents, either WT organisms with high selectivity or capacity for heavy metals can be used or genetically modified organisms to produce an abundance of high-affinity metal-binding proteins. In a research study led by Kao et al. (2008), the potential of MerP overexpressed proteins by recombinant *E. coli* strains to remove heavy metals was studied. The merP gene originated from two bacterial strains- Gram-positive (*Bacillus cereus* RC607) (GB) and Gram-negative (*Pseudomonas sp.* K-62) (GP). The adsorption rates in the recombinant bacterial biosorbents (GB and GP) and control (BL21 strain) were examined. The initial Cr(III) metal concentration used was 9.62 mM. Initially, the GE biosorbents (GB and GP) had an adsorption rate (r0) of 33% and 131% higher respectively over BL21. As time progressed, GB recombinants adsorbed Cr(III) at a capacity of 0.98 mmol/g biomass (Fig. 7.2). This was 72% higher than that of the control. Furthermore, the adsorption ability was enhanced by 126% for GP recombinants. The results conclusively support the initial hypothesis that metal-binding affinity improves with MerP protein expression (Kao et al. 2008).

A research study led by Peitzsch et al. (1998), aimed at studying the regulation of cnr and chr with lacZ fusions bringing into light the astonishing dynamicity of the Cr(VI) metabolic pathway of *Alcaligenes eutrophus.* On the plasmid pMOL28 of *A. eutrophus* CH34, the genes encoding resistance to chromate (chr) and to nickel and cobalt (cnr) are close to one another. A cloned portion of the plasmid pMOL28, comprising both determinants, was altered using Tn5-lacZ to create metal-sensing

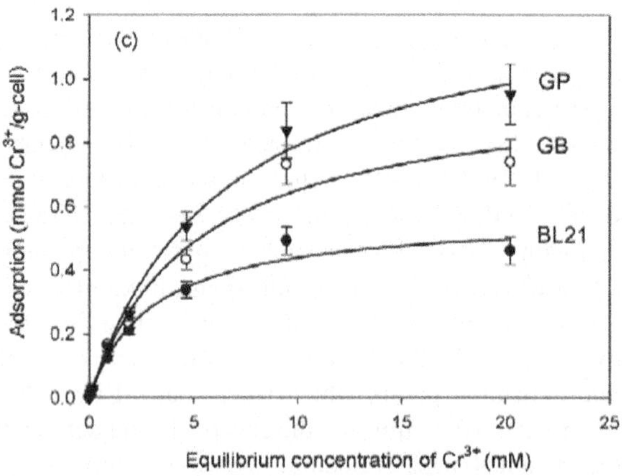

Fig. 7.2 Adsorption isotherms from biosorption of Cr(III) by recombinant E. coli strains expressing MerP proteins (GB and GP) and the MerP-free host strain (BL21). Reprinted with permission from Kao et al. (2008)

bacterial strains. Cr was primarily responsible for inducing the chr::lacZ fusions. The chr::lux fusion-containing broad-host-range IncP1 plasmid pEBZ141 was created. When grown in optimal conditions, *A. eutrophus* AE104 (pEBZ141), which has a chr::lux transcriptional fusion, might be employed as a biosensor for Cr(VI). Fusions in chrA were found to be highly specific and gave the most prominent responses as compared to all other fusions tested in the chr-cnr region. *A. eutrophus* strain AE104 was cultivated for 18 h in 30 μM sulfate and thereafter, 50 μM Potassium chromate was added. Cr(VI) uptake and reduction reached to a maximum of >20 μmol/g DW and 10 μmol/g DW after 2 h and 6 h of incubation respectively. On the other hand, when cultivated in 3 mM sulfate (keeping all other experimental conditions same), Cr(VI) uptake and reduction reached to a maximum of ~10 μmol/g DW and ~3 μmol/g DW after incubating for 2 h and 6 h respectively. Sulfate deprivation led to Cr(VI) uptake and reduction in the bacterial strain. Therefore, the sulfate reduction pathway may act as a catalyst in strain AE104's chromate reduction process. The study led to the conclusion that *A. eutrophus* AE104 (pEBZ141) having a highly specific chromate-sensing process may be readily used as a biosensor even in extreme conditions for Cr(VI) like industrial sewage water (Peitzsch et al. 1998).

A similar study to the earlier experimentation was performed by Srivastava et al. (2010) to study the biosorption capacity of transgenic *A. eutrophus* AE104 (pEBZ141) carrying chr resistance. The recombinant and WT strains were grown in varying concentrations of Cr(VI) (10, 50 and 100 mg/L). The data from this comparative study demonstrated that Cr(VI) (10 mg/L) removal peaked at 93% for the recombinant whereas 60% for the WT post 72 h of incubation. For 100 mg/L Cr(VI) concentration, the mutant was able to remove 48% and WT removed 41%. The results are in agreement that the recombinant *A. eutrophus* AE104 (pEBZ141) can be effectively used to remediate industrial wastewater sites heavily contaminated with chromium ions (Srivastava et al. 2010).

Electroactive bacteria (EAB) are a key class of biological chassis for GE intended for advantageous environmental, energy, bioreduction and earth science applications due to their special extracellular electron transfer (EET) capability. Previous experiments proved inefficient in genetically enhancing EET ability as it hampered the metabolic pathway cellular growth of the microorganism. This impairment was majorly due to the unavailability of cellular resources. In order to tackle this problem a quorum sensing (QS)-based population-state decision (PSD) system was designed for intelligently reprogramming the EET regulation system. This would ensure the rebalanced issuance of the resources needed for bacterial growth and metabolism. Genetically manipulated plasmids constructed from *E. coli* neb10β were conjugated with *E. coli* WM3064 (donor cells) and transformed into *Shewanella oneidensis* cells MR-1 (a model EAB species). The genetic components to construct the PSD system was obtained from the lux QS system. To test for Cr(VI) bioreduction, the PSD-EET strains were suspended in 20 mg/L of $K_2Cr_2O_4$. The reduction rate constant value attained by the PSD-EET strain was 49.8% which was 5.5 times higher as compared to the control. Resultantly, more Cr(III) was found in the supernatant of the PSD-EET group than Cr(VI). On examining the biofilms assembled on the electrodes, the thickness of biofilm formation by the recombinant strain and the control were found to be comparable. Overall, these findings show that *S. oneidensis* MR-1's EET and pollutant bioreduction capacities were improved by the intelligent reprogramming of the EET network using the PSD system. Additionally, this may lay the backbone in the architecture of future new-generation smart bioelectrical devices for manifold applications (Li et al.).

In a novel approach developed by Li et al. (2020), the EET capacity of *Shewanella oneidensis* MR-1 was enhanced using CRISPR-ddAsCpf1 tools. Cpf1 is a family of proteins. CRISPR-Cpf1 possesses many desirable features: (1) Cpf1 protein can be guided by a single crRNA that lacks a trans-activating crRNA (tracrRNA). (2) Cpf1 uses RNase activity to digest its own precursor crRNA. (3) The range of the targeted DNA is increased by Cpf1's recognition of T-rich PAM sequences. These features make it a dynamic tool to be exploited to modulate the electron flux in extracellular respiratory bacteria (ERB). The crRNA plasmids encoding valid various genes for EET enhancement were constructed and transformed into *S. oneidensis* via conjugation. The CRISPR-aided recombinant was treated with 20 mg/L of Cr(VI) pollutant. 6 valid genes were identified of which 5 exhibited an elevation of greater than 25% for the reduction of Cr(VI). The reduction rate of Cr(VI) showed the greatest effect (by 38.14%) in the gene dmsE. CRISPR therefore, is a propitious approach in Cr remediation (Li et al. 2020).

7.2.2 Plants-Mediated Cr Removal

Genetically engineered plants are being put to use to facilitate the reclamation of heavy metal-polluted soils and waters. Phytoremediation refers to the use of naturally occurring plants or genetically engineered plants to deal with the over accumulation of contaminants from soil and water, and have yielded promising results. Overexpression of certain genes and enzymes increases the plant's ability to mobilize, stabilize and accumulate Cr.

The majority of the biological processes including plant maturation and defense responses are regulated by transcription factors. Transcription factors help plants deal with abiotic stress. By directly interacting with the DNA recognition sequence of a number of target genes, they suppress and/or activate their expression. MYB is one such TF gene family that responds to drought stress. The OsMYB-R1 gene in rice imparts biotic and abiotic stress resistance with the help of auxin and salicylic acid (SA). SA plays a pivotal role in fostering the maturation of plants overexpressing OsMYB-R1. Analysis of the promoter region of the OsMYB-R1 gene demonstrated that it confers drought tolerance and Cr(VI) stress. Tiwari et al. (2020) studied the Cr stress tolerance of OsMYB-R1, 15 days old overexpressing WT and OsMYB-R1 plants were provided with Cr(VI) (100 μM media until they reached maturity). The abnormal OsMYB-R1 gene expression contributed to Cr(VI) tolerance, moreover, expanding the potential function of OsMYB-R1 in Arabidopsis as well under Cr stress. The efficiency of the growth of transgenic Arabidopsis was discovered to be more efficient as compared to the WT. On increasing the concentration of

Cr(VI), the overexpressed lines thrive well as compared to the WT plants. RNA seq. Data confirmed that OsMYB-R1 is responsible for gene regulation to improve root structure and in maintaining cell homeostasis. Therefore, OsMYB-R1 gene can be considered a propitious gene source for enhancing Cr(VI) abiotic stress in crops, specifically dicotyledonous plants (Tiwari et al. 2020).

Another experiment was designed to study Cr removal from polluted soil by APS (ATP sulfurylase overexpressing) Brassica juncea L (Indian mustard) transgenic plants with comparison to its WT. To compare metal tolerance between the WT and transgenic plants at seedling level, seeds were grown on agar medium containing Cr. The agar medium was composed of half-strength MS salts and vitamins, 10 g/L sucrose and 4 g/L agargel. These seedlings were enabled to develop for seven days at 25 °C and 16–8 h dark photoperiod. Individual seedlings were washed and root length was measured. The APS seeds did not exhibit improved accumulation/reduced tolerance to Cr. To compare the metal tolerance of the plant at mature stage, seeds were grown in the same agar medium as mentioned above for 4d, then they were planted in the soil, supplied with half-strength Goagland's nutrient solution and grown at 25 °C, 16 h light-18 h dark photoperiod until 35d. Cr metal was supplied at different concentrations, which were replaced after every 3d for a total of 14d. Elemental analysis was performed. The APS showed enhanced accumulation of Cr. Cr was supplied as oxyanion and its uptake was hypothesized to be enhanced because of upregulation of sulfate transported systems. Shoot concentrations of essential elements such as Iron, Molybdenum and Sulfur also showed significant increase after 14d of treatment with Chromium. The APS plants contained 2.5 times higher levels of not only Cr but Cd, Pb, Cu, As, Hg, V, Mo and Win at least one growth phase and are less tolerant to oxyanions. The APS transgenics accumulated more Cr from the soil in comparison to the WT. Enhanced accumulation suggests that overexpressing ATP sulfurase is a promising phytoremediation strategy (Wangeline et al. 1999).

WT Indian mustard and APS, GS and ECS transgenic mustard were gown in metal-polluted soil for 14 weeks. The ECS transgenic had a 170% higher Cr concentration related to WT. No significant difference between shoot metal concentration between WT and APS. No significant difference in metal tolerance and shoot biomass between all the transgenic and the WT. Cr levels were 1.6–2 times higher in ECS plants, and WT and other transgenic plants. Soil samples collected from the pots of the APS, GS and ECS transgenic plants showed lower Cr concentration than in the samples collected and analyzed before the experiment, but the difference was not significant (Bennett et al. 2000).

High environmental Cr levels can lead to the generation of ROS, it can induce the enzymatic antioxidative system and increase the magnitude of low molecular weight non-enzymatic antioxidants. Plants contain certain sulfur-containing compounds such as phytochelatins and metallothioneins to deal with high concentrations of metals. Canola has been as is fast-growing, with high biomass and significant heavy metal tolerance. NK Petrol (Cr(VI)-tolerant) and Sary (Cr(VI)-susceptible) cultivars of *Brassica napus* L. were grown in a Sulfur rich nutrient solution and thereafter exposed to varying sulfur concentrations. These seedlings were nurtured in 100 μM

Cr(VI) for 3d. Sulfur deficiency in combination with stress led to a notable decrement in growth specifications of Sary than of NK Petrol. Cr accumulation in Sary was observed to be substantially greater than in NK Petrol. The level of the BnMP1 (metallothionein (MT)) gene expression in the Sary cultivar was found to be improved by sulfur-deficient Cr treatment, which suggested that MTs are involved in increasing Cr(VI) tolerance when there is a dearth of sulfur (Terzi and Yıldız 2015).

Plants respond to a multitude of biotic and abiotic stresses are modulated by hormones including auxins, brassinosteroids, abscisic acid, 3-indole acetic acid and cytokines along with many metabolites and secondary messengers. In plants, stress can be induced by modifications in their hormonal, genetic as well as physiological pathways. Upregulation of secondary metabolites imparting stress tolerance is activated due to changes in its hormonal networks. The animal glucocorticoid receptor (GR) functions as a steroid-dependent transcription factor responsible for the exhibition of GR genes that regulate cell differentiation, maturation, homeostasis and organ physiology. Genomic integration of GR transgene led to alterations in the physiology and pleiotropic morphology of *Nicotiana glauca* and *Nicotiana langsdorffii*. Auxin levels were found to increase in *N. langsdorffii* and reduced in *N. glauca*. Both the transgenes were exposed to 50 ppm of Cr(VI) and hormonal patterns and alterations in decisive response-related metabolites were studied. Leaf biomass analysis (Fig. 7.3) revealed a decrease in fresh weight was 73% and dry weight was 59% for Cr. The metal uptake by WT was greater than that in the transgene (25%). GR gene transformation resulted in a dramatic rise in abscisic acid (+283%), 3-indoleacetic acid (+200%), Shikimic acid (SHA) (200%) and salicylic acid (+1080%), and showed no significant change in the levels of RSA, CLA and TPH. Phytohormone networks comparative study between WT and transgenic *N. langsdorffii* for the rat GR receptor indicates strong responses on exposure to 50 ppm Cr, in the growth medium. Additionally, on analysis of leaf biomass, it was revealed that the transgene was much more resistant than the WT (Fuoco et al. 2013).

Transformation of *N. langscloffii* with rat GR gene leads to phenotypic changes attributed to the improvement in the ratio of auxin/cytokinin as well as phytohormone concentration. The *rolC* gene in plants leads to a reduction in root growth, leaf area and stem length, and is associated with the enhancement of cytokinin activity. Roots and stems of the transgene and WT were collected, washed and dried. Then the plants were weighed (fresh weight) then the organs of plants were made to freeze in liquid nitrogen, lyophilized and weighed for calculating the DW. Cr concentration was deduced by the digestion of an acidic-oxidant microwave with a Surapur grade HNO_3 and H_2O_2 mixture, followed by examination with inductively coupled plasma optical emission spectroscopy. Substantial Cr accumulation was observed in the treated plants, strongly influenced by *rol*C gene. Differences in chromium concentrations in the WT and the *rol*C transformants were found to be negligible. The shoot-to-root concentration ratio suggested that the *rolC* transformant had a lower translocation ability and accumulated higher Cr concentrations at the root. Effect of exposure to Cr(VI)—reduction in plant dimension and strong biomass decrease in WT. In *rol*C shot, no change in shoot DW was observed. Inserting GR and *rol*C genes granted

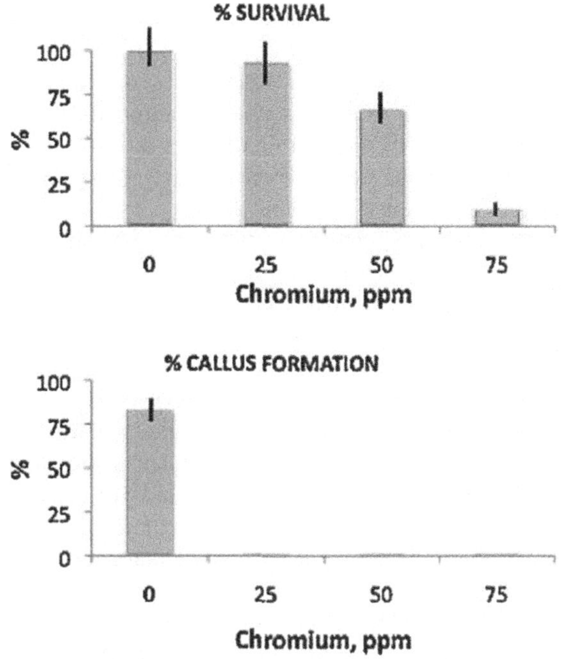

Fig. 7.3 Effect of different concentrations of Cr(VI) on survival and callus formation of *N. langsdorffii* leaf explants determined after one month of culture and on in vitro growth of Nicotiana plants after 15 days of metal treatments. Reprinted with permission from Fuoco et al. (2013)

substantial tolerance of Cr(VI) in the growth media. Cr accumulation was found to be greater in GR transformant than in *rol*C (del Bubba et al. 2013).

A greenhouse pot experiment was carried out on *S. fragilis* and *S. viminali.* Containers were filled with soils at field capacity. *S. fragilis* L. and *S. viminalis* L. were planted in six pots with varying soil types. The moisture content (30%) and irrigation conditions were maintained. Soil solution was nurtured in vacuum tubes and their roots, leaves and stems were sampled post-hundred days of plant maturation. It was observed that Cr was not translocated to the stem and leaves. The shoot length and DW of the root biomass of S. *viminalis* was comparatively lower as compared to S. *fragilis.* It was observed that there was no significant translocation of Cr to stem and leaves and hence the conclusion, that Willows act as root accumulators of Cr, was drawn (Vandecasteele et al. 2005).

The potential of Cr uptake by transgenic cotton (J208 and Z905) and the hybrid (ZD14) was studied. J208 and Z905 are herbicide-resistant and insect-resistant transgenic cultivars of cotton. Cr contents were evaluated in leaves and roots to evaluate accumulation potential. 14 days old seedlings of the transgenic cotton cultivars and their hybrids were exposed to varying Cr(VI) levels (10, 50, and 100 μM). Cr levels in roots and leaves were determined by atomic absorption spectroscopy. At 100 μM Cr(VI) concentration, the root length was found to reduce by 48, 32, and 8% in Z905, J208 and the hybrid respectively. On the other hand, shoot length reduction in Z905, J208 and the hybrid was discovered to be 52, 36 and 34% respectively. Biomass reduction was also observed. FW (59%) reduced more dramatically in comparison to the DW (50%). Experimental analysis revealed the high potential of all the cultivars to accumulate Cr and the translocation factor was low (<1). The low translocation efficacy attributed to the fact that Cr accumulation was found mostly in roots rather than in the leaves. Activities of several antioxidant enzymes also improved along with the increase in the number of vacuoles in root cells. Conclusively, all these results suggested that the 3 cotton cultivars have great potential for Cr uptake, accumulation and detoxification (Daud et al. 2014).

Glutathione S-transferases (GST) are a broad class of enzymes that contribute in ameliorating the detrimental effects of Cr and in combating biotic and abiotic stresses. In a study conducted by P11, Arabidopsis thaliana ecotype Columbia-0 was made to overexpress GST. The gene OsGSTU30 is responsible for the expression of GST. OsGSTU30 cDNA (isolated from *Oryza sativa*) was constructed and transformed in *Agrobacterium tumefaciens* using plant vectors. After a period of maturity of 21 days, the WT and transgene were supplemented with Cr(VI) (400 μM). After 7 days, their leaf samples were analyzed. Results demonstrated that 80–90% of the transgenes survived while 33% of the WT plants recovered. The revival of the transgene can be attributed to the upregulation of various antioxidant enzymes that are involved in ROS scavenging. These enzymes include glutathione S-transferases (GST), glutathione-dependent peroxidase (GPx), superoxide dismutase (SOD) and catalase (CAT). GST, CAT, GPx and SOD activities were found to increase 1.28, 3.15, 2.15 and 1.29 folds in the transgenic line as compared to the WT. These enzymes were responsible for metal chelation and decrease in ROS levels. These activities of the enzymes contributed to maintaining cellular homeostasis under Cr(VI) stress. It was therefore concluded that, the transgenic A. thaliana overexpressing the OsGSTU30 gene were more resistant to Cr(VI) even at high concentrations over WT. These results can be attributed to the functionality of the antioxidant system in the transgene (Srivastava et al. 2019).

In another study GST expressing gene OsGSTL2 was isolated and ligated with a vector procured from *A. tumefaciens* and thereafter transformed into *A. thaliana*. WT and their transgenic products were grown in Cr-rich media (50 and 100 μM). The root length data were analyzed post 10 days of germination. The root length was found to increase in the transgenic *A. thaliana*. Expression of GST in the transgenic lines conferred its ability to tolerate Cr up to 100 μM. In comparison to the WT, the recombinant plant showed enhanced pigmentation and were healthier. Their rate of germination was also greater and the emergence of cotyledon was also observed. Therefore, GST plays a pivotal role in ameliorating the toxic effects of Cr (Kumar et al. 2013).

Jin et al. (2001) isolated a Cr(VI) reductase gene from *Pseudomonas aeruginosa* HP014 to transform *A. tumefaciens* (tobacco plants). *P. aeruginosa* HP014 strain shows NADH-dependent aerobic Cr(VI) reductase activity. The Cr reductase gene in the pBinAR plasmid was manipulated to enhance its reduction ability and cloned in E. coli DH5α cells to achieve higher expression of the reductase activity. Post restriction digestion of the plant vector pBluescript-rtd containing Cr(VI) reductase gene was cloned into the binary plasmid pBinAR leading to the chimeric construct pBinAR-rtd. Plasmid pBinAR-rtd was then introduced into the leaf cells of *A. tumefaciens* strain LBA4404 via electroporation. Thereafter, the transformed plantlets were regenerated from the leaf discs. The Cr(VI) reductase gene was introduced into and expressed in the regenerated plants, according to hybridization experiments. The transgenic plant was grown in a medium containing 0.5 mM chromate and 0.4 mM NADH. In this experiment, to detect the reduction of Cr(VI), a unique calorimetric method was employed. The calorimetric assay measured Cr(VI) concentrations, and the red-violet color of the unknown composition was measured quantitatively using a spectrometer at 540 nm wavelength. The optical density (OD) value for the transgenic was found to be −0.5 whereas for the positive control and non-transformed tobacco plant was −0.2 and 0 respectively. This data suggested that the transgenic plant successfully showed reductase activity while the control and WT did not. Therefore, it can be concluded that the Cr reductase gene isolated from the bacterial strain can be genetically modified and transformed into tobacco plants for the creation of a transgene expressing enhanced Cr(VI) reductase activity. This transgenic plant can be exploited to remediate and treat Cr(VI) contaminated soil (Jin et al. 2001).

7.2.3 Other Organisms Used for Cr Removal

Apart from bacteria and plants, researchers have found other organisms that have exhibited Cr accumulation, tolerance and reduction capability. Many of these organisms are found to have mutualistic synergy while others have been brought together purely for experimental reasons. Some of the GMOs discussed below are formed naturally while others have been created in the laboratory.

Phage-bacterial synergy shows vast diversity in the soil microbiome and exhibits natural CRISPR activity via host-phage linkage. In a heavily Cr-contaminated soil, it was observed that the number of phages (obligate intracellular parasites) associated with HM-tolerant bacterial hosts increased and facilitated transfection, and provided the hosts with additional HM resistance genes. Phages are known to demonstrate parasitic behavior where they integrate their genome into the host in dormant form (prophage), and proliferate at the cost of the host organism and hijack their cellular machinery to produce virions. However, under favorable conditions, a mutualistic relationship between the two is established where the bacteria help the phage to reproduce and the prophage integrated into the host genome gives it microbial environmental resistance. A study led by Huang et al. (2021) aimed at investigating how soil virome variability and interactions between phage and bacteria varied over gradients of environmental stress. Due to the high frequency of occurrence and well-known ecotoxicity of Cr contamination, these sites were selected to represent the contamination of soil by heavy metals. Several soil profiles were chosen for this study: (1) a slightly contaminated group comprising L1, L2, and Z1 (0.11, 0.27, and 0.91 mg/kg of Cr respectively); (2) a moderately contaminated group comprising L3 and Z2 (6.76 and 6.09 mg/kg of Cr respectively); (3) a highly contaminated group comprising Z3 and Z4 (413.84 and 465.42 mg/kg of Cr respectively). It was revealed by metagenomic analysis of species richness that variations in levels of Cr contamination showed a substantial difference in the bacterial populations. 279 bacterial genera were reported that primarily belonged to 8 bacterial phyla of which the most abundant genera were Micropruina, Brevibacterium, and Bacillus. On the other hand, the most abundant phage species in the seven different soil profiles predominantly belonged to the families Siphoviridae, Podoviridae, and Myoviridae. The relative abundance of the Podoviridae, Siphoviridae and Myoviridae families was 0.7–31.9%, 33.3–93.4% and 0.5–12.2%, respectively, among the 7 different soil samples. Other viruses present in minute numbers belonged to Mimiviridae, Microviridae, Phycodnaviridae, Inoviridae, Ackermannviridae, Tectiviridae and Genomoviridae families. By matching the viral CRISPR spacers with the help of the IMG/VR database, the host-viral associations were determined. Pseudomonas, Cronobacter, Salmonella, Enterobacter, Klebsiella, Escherichia and Shigella from the Proteobacteria phylum, and Actinomyces, Micromonospora, and Salinispora from the Actinobacteria phylum were the top ten bacterial genera infected by polyvalent phages in soil contaminated with Cr. Lysogenic phages are characterized by the presence of integrase genes in their genome. These lysogenic phages are encoded for AMGs (auxiliary metabolic genes) that accelerate bacterial metabolism during infection by phage. AMGs such as MRGs (HM resistance genes) help in the improvement of the survival ability of the bacterial species in severe conditions. Additionally, based on viral genomic study, lysogenic phages under more stressful Cr-induced conditions carried more AMGs governing microbial heavy metal detoxification. Due to this reason, bacterial hosts find it beneficial to maintain a mutualistic synergy with their phage counterparts helping them to better adapt to biochemical stress due to Cr and withstand and detoxify high levels of Cr (Huang et al. 2021).

The effect of copper and chromium was evaluated on three strains of *Caenorhabditis elegans*, N2 (WT), RB1072 (mutated allele sod-2) and TJ375 gpls1 (transgenic strain, hsp-16.2 promoter gene fused with GFP reporter). Cr(VI) contamination has a similar effect on this nematode, as observed in plants, i.e., an increase in ROS production. Oxidative stress results in enhanced expression of HSP-16.2, a heat shock protein induced as a defense response. The addition of nZVI (zero-valent iron nanoparticles) showed a significant reduction in oxidative stress, as the nanoparticles have the ability to immobilize the metal for a short time (Fajardo et al. 2022).

Hyperaccumulator plants may contain chelators such as GST, MT, and PCs which limit cellular metal ion toxicity. They bind to metals and remove them from sensitive metal functions. Through overexpression of metal transporters and engineering metal trafficking pathways, transgenic yeast cells were able to absorb/tolerate high concentrations of heavy metals such as Cr. Uptake of chromate was observed to increase by up to 5 times on overexpression of sulfate permeases Sul1 and Sul2 when the transgenic yeast cells were incubated in 100 mM Cr for 4 h (Sun et al. 2019).

Cr(VI) is taken up by *Saccharomyces cerevisiae* through a sulfate transfer system. MSN1 is a transcriptional activator present in the nuclei of *S. cerevisiae*. CrT9 is a Cr(VI) tolerant mutant of yeast. Overexpression of MSN1 in CrT9 showed increased accumulation of Chromium by enhancing the sulfur transport system. Leaves of tobacco plants were grown in a medium containing agar. pJD301, a binary vector, was used to transform MSN1 into LBA4404 (*Agrobacterium tumefaciens* strain) and tobacco leaf discs were transformed with *A. tumefaciens*. MSN1 promotes Cr accumulation via the expression of sulfate transporters and confers Cr tolerance by sequestering it into the plant cell's vacuole. This study showed that transcriptional activators from yeast can be employed to enhance the accumulation and tolerance of Cr in higher plants. The Pj301-MSN1-5 accumulated a two times higher amount of Cr(VI), mainly in the roots, and had a 32–63% higher tolerance to Cr(VI). Tobacco's sulfate transferase 1NtST1 cDNA was cloned and expressed in *S. cerevisiae*. The MSN1 overexpressing yeast accumulated higher levels of Cr(VI) but did not exhibit any significant changes in their tolerance towards Cr (Kim et al. 2006).

OsGSTU30, OsGSTU37 and OsGSTU41 which belong to the 'Tau' class of the gene family of rice showed higher expression in roots under Cr(VI) stress. OsGSTU30 and OsGSTU40 were cloned into *Schizosaccharomyces pombe* and subjected to various Cr(VI) concentrations. Results demonstrated that the transformed strain showed growth in dilutions up to 5 mM, whereas the control was impaired at a small concentration of about 0.1 mM, and at 3 mM growth was completely inhibited. Transformed yeast cells showed enhanced accumulation and resistance to Cr(VI) along with higher GST enzyme activity. Overexpression of GSTs in yeast utilizes GSH and conjugates with Cr(VI) to form a complex which is sequestered into the vacuole (Tripathi et al. 2014).

7.3 Conclusion

Cr is a major pollutant to the environment and can enter into soil and water via various means most of which are due to anthropogenic activities. Although a lot of methods have been developed to deal with overexposure of water and soil to Cr, they are not environmentally friendly and cost-effective. Therefore, the need to make use of factors already available in the biotic world was considered. These methods which make use of organisms such as bacteria, plants, yeast, etc., will help overcome the issues faced earlier. The scientific world today focuses its research on developing novel methods of bioremediation by employing GE techniques that use recombinant DNA technology, constructing transgenes, etc. In a lot of studies, overexpression of enzymes, like ATP Sulfurylase, γ-Glutamylcysteine Synthetase, etc., and genes such as metallothionein gene, rat glucocorticoid receptor gene, GSTs, etc., helped increase plant tolerance and accumulation of Cr. Numerous bacterial forms were also explored. They converted the Cr(VI) form to Cr(III), as Cr(III) is less toxic as compared to Cr(VI). Certain genes/enzymes were cloned into the bacterial cells which helped serve the purpose. Other than reducing Cr(VI)–Cr(III), biosorption was also put to use. Apart from plants and bacteria, studies were carried out with organisms such as yeast, roundworms and even phages.

References

Ackerley DF, Gonzalez CF, Keyhan M et al (2004a) Mechanism of chromate reduction by the Escherichia coli protein, NfsA, and the role of different chromate reductases in minimizing oxidative stress during chromate reduction. Environ Microbiol 6:851–860

Ackerley DF, Gonzalez CF, Park CH et al (2004b) Chromate-reducing properties of Soluble Flavoproteins from Pseudomonas putida and Escherichia coli. Appl Environ Microbiol 70:873–882

Akkurt Ş, Oğuz M, Alkan Uçkun, A (2022) Bioreduction and bioremoval of hexavalent chromium by genetically engineered strains (Escherichia coli MT2A and Escherichia coli MT3). World J Microbiol Biotechnol 38:45

Bennett LE, Burkhead JL, Hale KL et al (2000) Bioremediation and biodegradation analysis of transgenic Indian mustard plants for phytoremediation of metal-contaminated mine tailings. Hirschi

Dados A, Omirou M, Demetriou K et al (2015) Rapid remediation of soil heavily contaminated with hydrocarbons: a comparison of different approaches. Ann Microbiol 65:241–251

Daud MK, Mei L, Variath MT et al (2014) Chromium(VI) uptake and tolerance potential in cotton cultivars: effect on their root physiology, ultramorphology, and oxidative metabolism. Biomed Res Int

del Bubba M, Ancillotti C, Checchini L et al (2013) Chromium accumulation and changes in plant growth, selected phenolics and sugars of wild type and genetically modified *Nicotiana langsdorffii*. J Hazard Mater 262:394–403

Deng P, Tan X, Wu Y et al (2014) Cloning and sequence analysis demonstrate the chromate reduction ability of a novel chromate reductase gene from Serratia sp. Exp Ther Med 9:795–800

di Bona KR, Love S, Rhodes NR et al (2011) Chromium is not an essential trace element for mammals: Effects of a "low-chromium" diet. J Biol Inorg Chem 16:381–390

Fajardo C, Martín C, Garrido E et al (2022) Copper and Chrkomium toxicity is mediated by oxidative stress in *Caenorhabditis elegans*: the use of nanoparticles as an immobilization strategy. Environ Toxicol Pharmacol 92

Flora SJ (2009) Bioscience, structural, chemical and biological aspects of antioxidants for strategies against metal and metalloid exposure. Oxid Med Cell Longev 2:191–206

Frederick TM, Taylor EA, Willis JL et al (2013) Chromate reduction is expedited by bacteria engineered to produce the compatible solute trehalose. Biotechnol Lett 35:1291–1296

Fuoco R, Bogani P, Capodaglio G et al (2013) Response to metal stress of *Nicotiana langsdorffii* plants wild-type and transgenic for the rat glucocorticoid receptor gene. J Plant Physiol 170:668–675

Gill SS, Tuteja N (2010) Reactive oxygen species and antioxidant machinery in abiotic stress tolerance in crop plants. Plant Physiol Biochem 48:909–930

Gu R, Gao J, Dong L et al (2020) Chromium metabolism characteristics of coexpression of ChrA and ChrT gene. Ecotoxicol Environ Saf 204

Huang D, Yu P, Ye M et al (2021) Enhanced mutualistic symbiosis between soil phages and bacteria with elevated chromium-induced environmental stress. Microbiome 9

Jaishankar M, Tseten T, Anbalagan N et al (2014) Toxicity, mechanism and health effects of some heavy metals. Interdiscip Toxicol 7:60–72

Jin TE, Kim IG, Kim WS, Suh SC, Kim BD, Rhim SL (2001) Expression of chromium (VI) reductase gene of heavy metal reducing bacteria in tobacco plants. J Plant Biotechnol 3:13–17

Johnson J, Schewel L, Graedel TE (2006) The contemporary anthropogenic chromium cycle. Environ Sci Technol 40:7060–7069

Kanagaraj G, Elango L (2019) Chromium and fluoride contamination in groundwater around leather tanning industries in southern India: Implications from stable isotopic ratio Δ53Cr/Δ52Cr, geochemical and geostatistical modelling. Chemosphere 220:943–953

Kao WC, Huang CC, Chang JS (2008) Biosorption of nickel, chromium and zinc by MerP-expressing recombinant Escherichia coli. J Hazard Mater 158:100–106

Kim YJ, Kim JH, Lee CE et al (2006) Expression of yeast transcriptional activator MSN1 promotes accumulation of chromium and sulfur by enhancing sulfate transporter level in plants. FEBS Lett 580:206–210

Kumar S, Asif MH, Chakrabarty D et al (2013) Expression of a rice Lambda class of glutathione S-transferase, OsGSTL2, in Arabidopsis provides tolerance to heavy metal and other abiotic stresses. J Hazard Mater 248–249:228–237

Li FH, Tang Q, Fan Y-Y et al (2020) Developing a population-state decision system for intelligently reprogramming extracellular electron transfer in *Shewanella oneidensis*

Li J, Tang Q, Li Y et al (2020) Rediverting electron flux with an engineered CRISPR-ddAsCpf1 system to enhance the pollutant degradation capacity of *Shewanella oneidensis*. Environ Sci Technol 54:3599–3608

Marques APGC, Rangel AOSS, Castro PML (2009) Remediation of heavy metal contaminated soils: phytoremediation as a potentially promising clean-up technology. Crit Rev Environ Sci Technol 39:622–654

Ngwenya N, Chirwa EMN (2011) Biological removal of cationic fission products from nuclear wastewater. Water Sci Technol 63:124–128

Peitzsch N, Eberz N, Nies DH (1998) *Alcaligenes eutrophus* as a bacterial chromate sensor the HEPES-buffered me-dium contained the following (per liter of H_2O): 0.3 mM Na2KPO4, 0.2 mM K_2HPO_4, 50 mM HEPES buffer (pH 7.0), 2 g of NH4Cl, 0.2 g of $MgSO_47H_2O$, 10 mg of $CaCl_22H_2O$, and 5 mg of $FeCl_36H_2O$. Analytical-grade salts of CdCl

Reisinger S, Schiavon M, Terry N, Pilon-Smits EAH (2008) Heavy metal tolerance and accumulation in Indian mustard (*Brassica juncea* L.) expressing bacterial γ-glutamylcysteine synthetase or glutathione synthetase. Int J Phytoremediation 10:440–454

Simin Z, Lanlan D, Yuan HE, Hong X (2017) Characterization of chromate resistance in genetically engineered *Escherichia coli* expressing chromate ion transporter ChrA. J South Med Univ 37:1290–1295

Srivastava NK, Jha MK, Mall ID, Singh D (2010) Application of genetic engineering for chromium removal from industrial wastewater. Int J Environ Ecol Eng 4:633–638

Srivastava D, Verma G, Chauhan AS et al (2019) Rice (*Oryza sativa* L.) tau class glutathione S-transferase (OsGSTU30) overexpression in *Arabidopsis thaliana* modulates a regulatory network leading to heavy metal and drought stress tolerance. Metallomics 11:375–389

Sun GL, Reynolds EE, Belcher AM (2019) Designing yeast as plant-like hyperaccumulators for heavy metals. Nat Commun 10

Tahri Joutey N, Bahafid W, Sayel H et al (2014) Hexavalent chromium removal by a novel *Serratia proteamaculans* isolated from the bank of Sebou River (Morocco). Environ Sci Pollut Res 21:3060–3072

Tang R, Shen L, Yang L et al (2021) Killing two birds with one stone: biomineralized bacteria tolerate adverse environments and absorb hexavalent chromium. ACS Omega

Terzi H, Yıldız M (2015) Interactive effects of sulfur and chromium on antioxidative defense systems and BnMP1 gene expression in canola (*Brassica napus* L.) cultivars differing in Cr(VI) tolerance. Ecotoxicology 24:1171–1182

Thacker U, Parikh R, Shouche Y, Madamwar D (2006) Hexavalent chromium reduction by *Providencia* sp. Process Biochem 41:1332–1337

Tiwari P, Indoliya Y, Chauhan AS et al (2020) Over-expression of rice R1-type MYB transcription factor confers different abiotic stress tolerance in transgenic Arabidopsis. Ecotoxicol Environ Saf 206

Tripathi A, Indoliya Y, Tiwari M et al (2014) Transformed yeast (*Schizosaccharomyces pombe*) overexpressing rice Tau class glutathione S-transferase (OsGSTU30 and OsGSTU41) shows enhanced resistance to hexavalent chromium. Metallomics 6:1549–1557

Vandecasteele B, Meers E, Vervaeke P et al (2005) Growth and trace metal accumulation of two Salix clones on sediment-derived soils with increasing contamination levels. Chemosphere 58:995–1002

Velez PA, Talano MA, Paisio CE et al (2017) Synergistic effect of chickpea plants and Mesorhizobium as a natural system for chromium phytoremediation. Environ Technol (United Kingdom) 38:2164–2172

Verma JP, Jaiswal DK (2016) Book review: advances in biodegradation and bioremediation of industrial waste. Front Microbiol 6

Wang J, Zhao S, Ling Z et al (2021) Enhanced removal of trivalent chromium from leather wastewater using engineered bacteria immobilized on magnetic pellets. Sci Total Environ 775

Wangeline AL, Burkhead JL, Hale KL et al (1999) Overexpression of ATP Sulfurylase in Indian Mustard: effects on tolerance and accumulation of twelve metals

Zhou S, Dong L, Deng P et al (2017) Reducing capacity and enzyme activity of chromate reductase in a ChrT-engineered strain. Exp Ther Med 14:2361–2366

Chapter 8
Chromium Dynamics in the Soil-Plant Continuum

M. L. Dotaniya, C. K. Dotaniya, Dinesh Kumar Yadav, Alaa El Din Mahmoud, R. K. Doutaniya, Dileep Kumar, Ashok Kumar, H. M. Meena, Rajhans Verma, Kuldeep Kumar, and Anita Meena

Abstract Heavy metal use is playing a crucial role in economic development of a country. Another side, generated waste may be affected the quality of natural resources during unscientific disposal. This situation is grimmer in developing countries, where much effort is needed to scientific disposal of waste. Chromium (Cr), one

M. L. Dotaniya (✉)
ICAR-Directorate of Rapeseed-Mustard Research, Bharatpur 321303, India
e-mail: Mohan.Dotaniya@icar.gov.in

C. K. Dotaniya
Department of Soil Science & Agricultural Chemistry, SKRAU, Bikaner 334006, India

D. K. Yadav
ICAR-Indian Institute of Soil Science, Bhopal 462038, India

A. E. D. Mahmoud
Environmental Sciences Department, Faculty of Science, Alexandria University, Alexandria 21511, Egypt

Green Technology Group, Faculty of Science, Alexandria University, Alexandria 21511, Egypt

R. K. Doutaniya
Department of Agronomy, SKN College of Agriculture, Jobner 303329, India

D. Kumar
AICRP on Micronutrients, Anand Agricultural University, Anand 388110, India

A. Kumar
ICAR-National Bureau of Soil Survey and Land Use Planning, Regional Centre, New Delhi 110012, India
e-mail: Ashok.Kumar42@icar.gov.in

H. M. Meena
ICAR-Central Arid Zone Research Institute, Jodhpur 342003, India

R. Verma
Department of Soil Science & Agricultural Chemistry, SKN College of Agriculture, Jobner 303328, India

K. Kumar
ICAR Indian Institute of Soil and Water Conservation, Research Centre, Kota 324002, India
e-mail: Kuldeep.Kumar1@icar.gov.in

© The Author(s), under exclusive license to Springer Nature Switzerland AG 2023
N. Kumar et al. (eds.), *Chromium in Plants and Environment*, Environmental Science and Engineering, https://doi.org/10.1007/978-3-031-44029-8_8

of the heavy metals, is extremely important to the metal, leather, and wooden industries and releases a sizable amount of effluent into water or onto soil surfaces. Crop production potential and soil health indices were both decreased by higher Cr concentrations in the soil. It mediated metabolic activities in plants and organs functionality in human body. An excessive concentration in the human body can have a cancerous effect and shorten life. By the help of scientific tool and techniques, we can manage the Cr pollution prior to discharge in natural bodies. The removal of Cr using physical, chemical, and biological approaches can significantly increase crop production potential. Increase the people's participation and awareness to reduce the Cr toxicity effect through food chain contamination are much needed. Many research and policy organizations are working on Cr toxicity issues to remove/immobilization process without affecting the environmental health. This chapter discusses the significance of Cr, its origins and chemistry in soil, as well as its toxicity to plants and people, effect on soil microbial count and diversity and management options for reducing the Cr toxicity in soil.

Keywords Crop sustainability · Chromium toxicity · Dynamics in soil-plant · Human health · Phytoremediation techniques · Soil contamination

8.1 Introduction

By 2050, India's population will have increased to 1.66 billion, and it would require 333 million tonnes (mt) of grain to feed that population (Minhas et al. 2021). Present scenario of agriculture having improved varieties, tool and techniques of balance fertilizer application, better weather forecasting, friendly production and protection technologies, modern application of extension process are helped to enhance the food grain production 50 mt in 1950 to 309 mt in the year 2022 (Dotaniya et al. 2022c). These achievements are not easy to achieve, but the contentious efforts of researcher and policy maker to strengthen the food production programme in collaboration of national and international institutes. However, growing populations feed on limited natural resources are a big challenge to researchers. Developing countries are increased the rate of industries installation to boost the economic growth of the country. Another side of the coin, these industries are generating huge volume of effluents and discharging into water bodies or on the soils (Solanki et al. 2020). These waste substances are containing huge amount of salt/acid or significant amount of heavy metals (Dotaniya et al. 2022a). While not all types of industries are in the same predicament, the majority of them in developing nations have subpar treatment facilities. There are 269 sewage treatment plant (STP) constructed in India and 38254 million liters per day (MLD) of wastewater is produced, but only 12,000 MLD of processing power has been developed. On an average, 38% of generated

---

A. Meena
ICAR-Central Institute for Arid Horticulture, Bikaner 334006, India

sewage is treated and most of the parts are dumping in different waste channels. It degrades ecosystem quality by raising metal and salt toxicity in soil and water sources (Dotaniya et al. 2018a).

Chromium (Cr), which has an atomic number of 46 and a molecular weight of 52 u, is one of the most poisonous heavy metal. However, it is one of the major metal used in industrial sector and significant amount reached into soils. It is presents in many forms and oxidation states ranged from −4, −2, −1, 0, +1, +2, +3, +4, +5, +6 (Dotaniya et al. 2014a). The Cr hexavalent (Cr^{6+}) is toxic form, whereas, trivalent Cr is nontoxic in nature (Dotaniya et al. 2019). Use of tannery effluent for agricultural crop growing resulted in more than 972 mg/kg total Cr being deposited in the soil (Dotaniya et al. 2016). Applying Cr at a dosage of 100 mg/kg influences the soil's microbial biomass carbon mineralization rate by 66% (Dotaniya et al. 2017a). Similarly, Wyszkowska et al. (2007) mentioned that long term application of Cr, an excessive concentration in the soil solution mediated the number and diversity of soil microorganisms. Dotaniya et al. (2017a) mentioned by an application of Cr upto 100 mg/kg drastically reduced the soil enzymatic activities (71.3% DHA, 40% FDA) and microbial count.

Chromium reaching in food stuffs via food chain contamination. Excessive concentration of Cr intake impaired with metabolic functions of human body. These symptoms are very much detective and extreme cases living organism may die. In crop plants, higher concentration of Cr in soil restricted the mineral and water uptake process by affecting the root connecting tissues (Sharma et al. 2020). It was observed that, root tissues are damaged due to Cr toxicity and poor health of crop was reported in tannery irrigated areas. Inter-conversion of Cr^{3+} to Cr^{6+}, however, by moderating the impact of organic matter and the presence of other metal ions (Mn, Fe, Cu). By immobilizing Cr in the soil, the addition of organic C enhances soil health indicators. Application of organic matter through FYM at 10 t/ha decreased the amount of accessible Cr in the soil solution and directly decreased the amount of Cr that spinach crops absorbed in vertisol (Dotaniya et al. 2022b).

8.2 Role of Metal in Plant Nutrient Dynamics

Metals are the essential part of plant nutrients systems. To complete the life cycles of plant needs 17 essential nutrients based on the criteria of Arnon and Stout (1939). These nutrients include carbon, hydrogen, oxygen, nitrogen, phosphorus, potassium, sulphur, calcium, and magnesium. These nutrients are classed as macronutrients, which means they require a bigger amount than 100 mg/kg dry weight. However, crop plant need smaller amount of nutrient concentration (<100 mg/kg dry weight) classified as micro nutrient (iron, copper, manganese, zinc, molybdenum, boron, nickel, chlorine). Certain essential plant metals are found in greater concentrations in soil, behave like poisonous metals, and slow down the intake of other vital plant nutrients. Reactive oxygen species (ROS), which are produced when metals are accumulated in excess, cause poor plant development and lower yields. In this line many metals are clearly identified as trace metal or heavy metal like cadmium, chromium, arsenic, lead, mercury, arsenic etc. These metals in small amount may retard the adsorption and uptake mechanism of essential nutrients or uptake of more concentration of toxic metal (Xu et al. 2011). These metals initially adsorbed and taken up by root tip/hairs and reach to the cellular levels and mediated the different process of plant metabolism (Fig. 8.1). Some of the sensitive organs are reduced the working capacity or show the toxicity symptoms on different parts of plants. In extreme cases, plant reduces water and mineral nutrient leads dead of organs.

8.3 Heavy Metal Sources and Toxicity

The majority of heavy metals are metals with relative densities greater than water (1 g/cc) (Tchounwou et al. 2012). These are mentioned in periodic table of element with yellow color. Some of the essential plant nutrients are also behaving as heavy metals described in different groups of periodic table (Fig. 8.2). These metals are used for examination of a disease or curing an illness of an organ across the globe. Cobalt −60 used for the detection of different disease in human. Cobalt (Co) alloys have been used in a number of medical devices, such as hip and knee implants, surgical instruments, and vascular stents, for over 70 years because of their excellent biocompatibility, durability, and mechanical qualities (Britannica 2022). When harmful compounds accumulate inside an organism at a pace that is quicker than their rate of breakdown, this is known as bioaccumulation. High blood sugar can result from a Cr deficiency. However, too much exposure to these heavy metals can cause poisoning and other severe health issues. Iron and copper, for instance, might build up in the liver. If this occurs, the liver won't operate normally. Such ill effect is affected the majority of the living organisms. However, some of the microorganism likes bacteria, fungi and algae are less affected by metal toxicity and transformed the metal toxicity form into less toxic form (Figs. 8.3, 8.4 and 8.5).

The majority of heavy metals have a significant negative impact on both human and animal health through modulating plant nutrient absorption dynamics, soil organic carbon (SOC) mineralization dynamics, soil microbial development and diversity. Long-term application of marginal quality agricultural imputes during crop production by knowleging and unknowledgingly; accumulated significant amount of pollutant in soil-water ecosystems. Application of fertilizers and pesticides during the crop production contaminated the soil environment with different heavy metals. These pollution sources may be classified into non point sources of metal pollution. The intensity of metal toxicity is depending on form, eco-system properties, the age of living things and their daily metal consumption. Table 8.1 lists the sources of metal pollution in the food chain and how they affect human health.

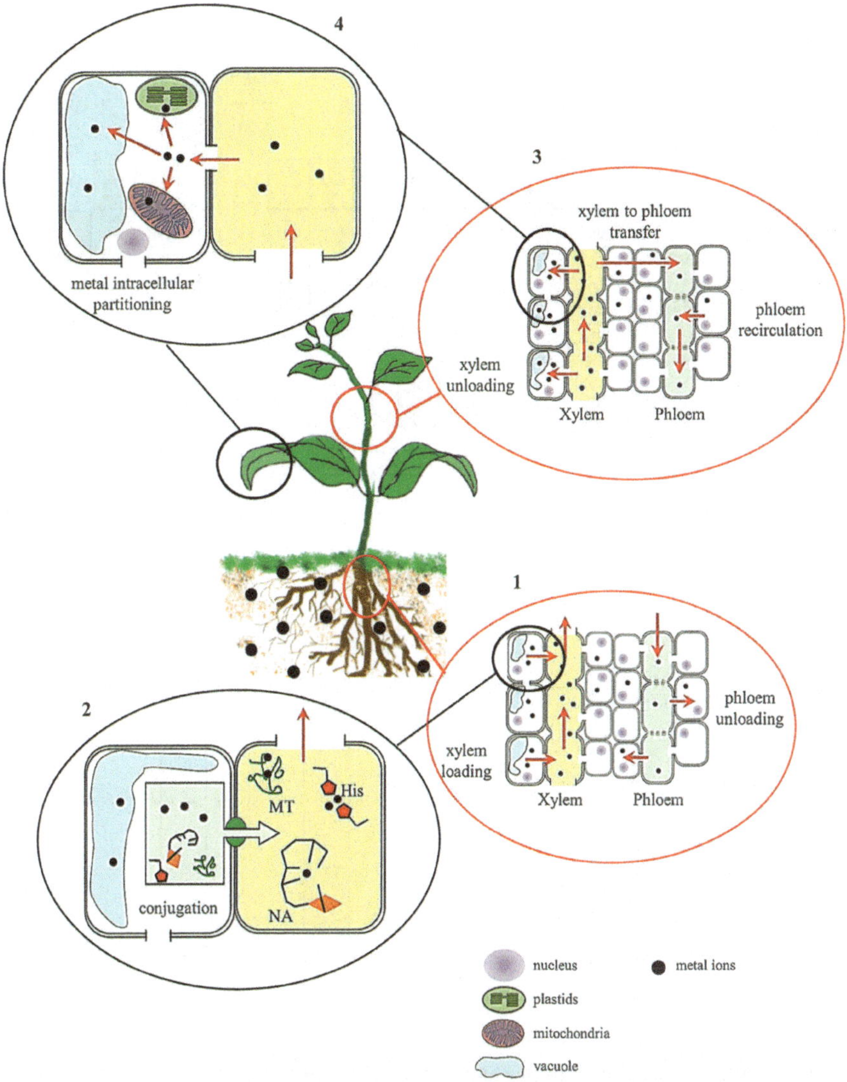

Fig. 8.1 Plant nutrient uptake routs (1) root adsorption metal ions, (2) free ion forms, (3) metal reach to the leaf, (4) metal deliver to the cell/detoxification process. Adopted from DalCorso et al. (2014)

8.4 Chromium Sources and Toxicity

Rapid industrialization and uncontrolled urbanization have caused the entry of heavy metals into water and soil through the inappropriate disposal of industrial wastes directly on land and into water bodies (Mahmoud 2022; Chaukura et al. 2022;

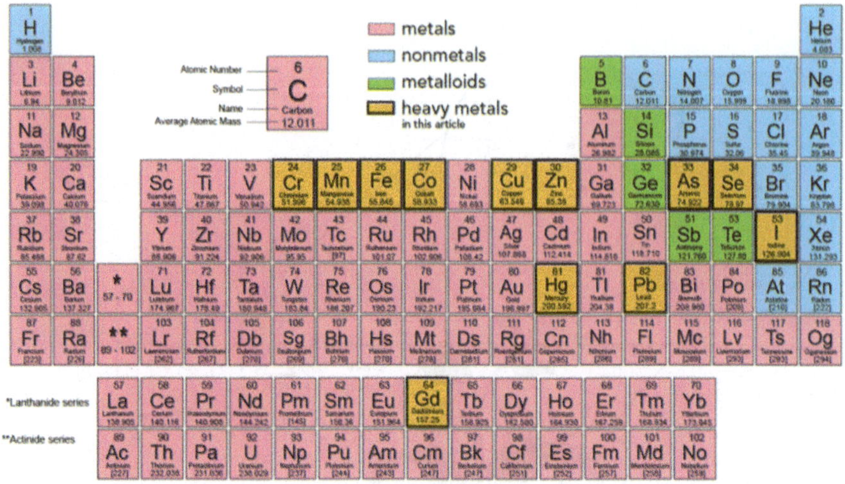

Fig. 8.2 Position of heavy metals in periodic table. Adopted from Letstalk Science (2022)

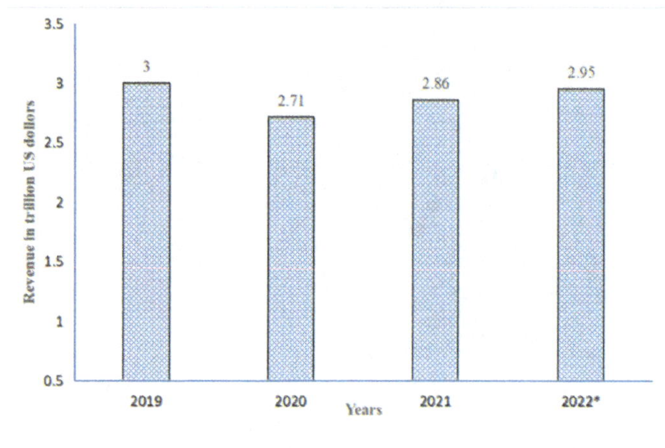

Fig. 8.3 Revenue generation across the globe during 2019–2022 (Statista 2022)

Mahmoud and Kathi 2022; Mahmoud et al. 2021a, b). Pollutants including heavy metals and pesticides cause soil contamination (Sawick et al. 2021; Mahmoud et al. 2016; Mahmoud et al. 2022a). Here, we concentrate on Cr. It is the seventh most prevalent chemical element in the crust of the Earth. Chromium is mostly used in metallurgical processes (67%) as well as refractories (18%) and chemicals (15%) (Saha et al. 2011). It may thus be found in a variety of sectors, including electroplating, tanning, industrial water cooling, paper & pulp manufacture, and petroleum refining. About 35% of the utilised chromium is released as trivalent and hexavalent chromium in the effluent (Sun et al. 2009). The top 20 most dangerous chemicals

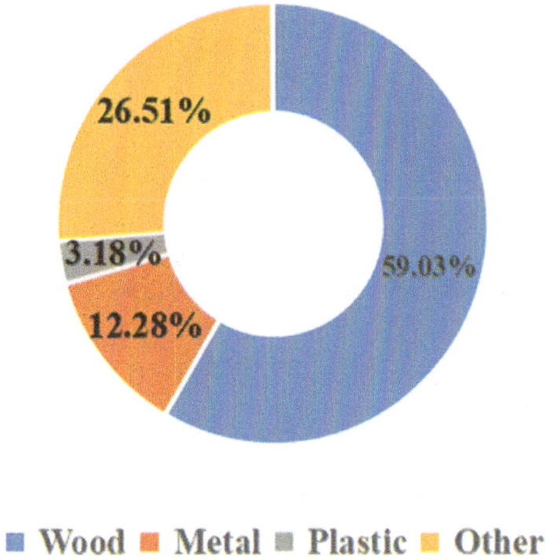

Fig. 8.4 Share of different components in furniture global market in the year 2020

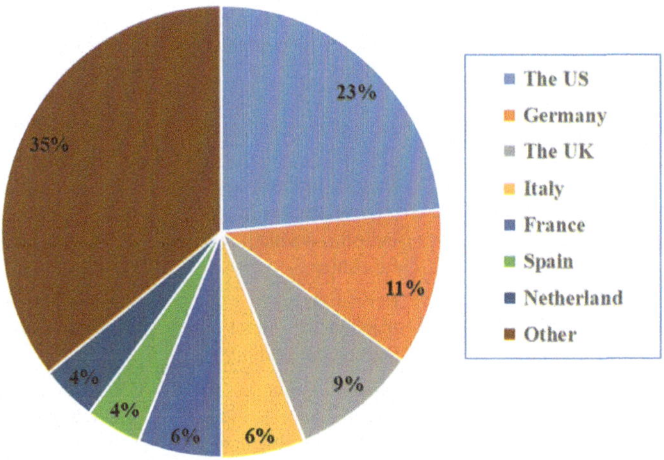

Fig. 8.5 India leather and leather product export during 2021–2022 (DGCI and S 2022)

according to the Agency for Toxic Substances and Disease Registry include the heavy metal chromium (Oh et al. 2007). Due to its nonbiodegradable nature and destructive impacts on living things, chromium is extremely poisonous (Mahmoud et al. 2021a, b, 2022b).

Around 1.29×10^5 tonnes of Cr are discharged into the environment each year, the majority of which has accumulated in soil and resulted in substantial Cr pollution (Ao et al. 2022). The entrance of Cr into cells and its harmful effects are significantly influenced by its chemistry. There are two main oxidation states of Cr in water, ground water, and soil: oxidized hexavalent chromium (Cr^{6+}) and less

Table 8.1 Sources of heavy metals and their impact on human health

Metals	Metal form	Sources	Implication for human health
Arsenic	Trivalent As	Metal smelters, fungicides, and pesticides	Skin, respiration problems
Cadmium	Cd^{2+}	Solding, electroplating, chemical formulations, fertilizers, Cd–Ni dry batteries	Lung functions irregularity and renal dysfunction, bone defects, increase blood pressure, kidney damage, gastrointestinal disorder, cancer
Lead	Organic form of Pb, Pb^{2+}	Paint and varnish, agri-pesticide, smoking, automobile emission, mining, burning of coal	Mental retardation of children, developmental delay, congenital paralysis, sensory neural deafness, malfunctions in nerve system, infection in liver and kidney, mediated function of gastrointestinal
Mercury	Hg^{o}	Insecticide/pesticides, dry batteries, pulp and paper industry	Tremors, gingivitis, minor psychological modifications, premature abortion, malfunctions to nervous system, acute protoplasm poisoning
Chromium	Cr(VI)	Mineral industries, tanning process in leather industry, wooden furnishing	Malfunctions in nervous system, fatigue, irritability
Zinc	Zn^{2+}, Zn^{4+}	Refineries, brass manufacture, metal plating and solding	Skin diseases, nervous system problems
Copper	Cu^{2+}, Cu^{4+}	Mineral & mining sector, insecticide/pesticide production, chemical sector, metal pipe industry	Malfunctions in liver and kidney, stomach itching, problem in intestinal irritation

Adopted from Singh et al. (2011), Dotaniya et al. (2018b)

oxidized trivalent chromium (Cr^{3+}) (Zhang et al. 2022). In comparison to hexavalent chromium compounds, trivalent chromium compounds are very slightly soluble in water. Under acidic circumstances, the resultant hexavalent chromium solutions are potent oxidizing agents; but, under basic conditions, they are less potent (Dotaniya et al. 2014a). In chemical laboratories, for instance, chromic acid (H_2CrO_4) is frequently used to clean glassware by oxidizing organic residues. As a result, groundwater contains substantially more hazardous and mobile hexavalent chromium than it does comparatively stationary trivalent chromium (Fei et al. 2022). Depending on the quantity and acidity, hexavalent chromium can exist as chromate (CrO_4^{2-}) or dichromate (CrO_4^{2-}). The two species of dissolved chromium that are most common are $HCrO_4^-$, CrO_4^{2-}, and CrO_4^{2-} (hexavalent chromium). Which entity will prevail in a particular environment depends on a number of specific factors, including pH, Eh (redox potential), the total concentration of Cr, and the overall aqueous chemistry (Ao et al. 2022). Other metal ions, organic matter, soil moisture, and soil biota are major mediators of the absorption kinetics of Cr in soil (Dotaniya et al. 2017b, 2019). It is well recognized that certain soil characteristics, including pH, cation exchange capacity (CEC), organic matter (OM), and metal oxide concentration, affect both the heterogeneity of natural soils and the bioavailability of metals (Jiang et al. 2020).

Cr toxicity, which typically inhibits plant growth, alters the ultra-structure of the cell membrane and chloroplast, causes chlorosis to affect the root cells, reduces pigment content, disrupts water relations and mineral nutrition, impairs transpiration and nitrogen assimilation, and alters a number of enzymatic activities (Cervantes et al. 2007; Ali et al. 2015; Farooq et al. 2016; Reale et al. 2016; Anjum et al. 2017a, b). An overabundance of reactive oxygen species (ROS), which eventually affects the redox balance in plants, may be the root cause of all of these detrimental effects of Cr (Anjum et al. 2017a, b). The germination, root, and shoot growth of the wheat (Dotaniya et al. 2014b) and pigeon pea (Dotaniya et al. 2014c) crops were inhibited by Cr toxicity. Chromium in crop plants affected different metabolic process (Table 8.2).

When untreated wastewater is used to irrigate these food crops, Cr accumulates in the soil and is transported to the edible sections of the plants. Despite relatively low Cr levels in the effluent, Cr-enhanced soil to plant bioaccumulation did occur. The steady accumulation of Cr in the soil was most definitely caused by the treatment of water sources (Chen et al. 2022). It was confirmed by Zhang et al. (2022) that the toxicity of Cr to microbes varied significantly across soil samples, and it was discovered that Cr toxicity was significantly ($p \leq 0.05$) negatively correlated with soil OM content. This finding was in line with earlier studies that found low bioavailability and toxicity of Cr(VI) to plants in soils with a high OM content. It is generally known that soil organic matter (OM) is important for metal mobility, bioavailability, and sorption/desorption.

OM may move Cr and other heavy metals in soil. Additionally, when employed as an electron donor, soil OM may facilitate the reduction of Cr(VI) (Andrade et al. 2022). Oxyanions may be reduced to trivalent forms by electron donors like OM with ease due to soil's high levels of Cr(VI) oxidation. According to research by Palma et al. (2018), high OM concentrations increased Cr(VI) reduction, which may account for the observed inverse relationship between soil OM and Cr(VI) toxicity and likely explains part of the heterogeneity in Cr toxicity. Depending on the pH and the quantity of hexavalent chromium, oxoforms of various species exist as hexavalent chromium in aqueous solutions. Saha et al. (2011) mentioned three major pH zones for the oxo-species of hexavalent Cr were determined (Table 8.3).

Table 8.2 Chromium toxicity effects in different crops

Plant species	Physiological response	References
Tea	Mediated SOD and CAT action in plants	Sharma et al. (2020); Tang et al. (2014)
Chili pepper	Carotenoid concentration positively improved	Oliveira (2012)
Rice	Lower down glutathione level	Qiu et al. (2013)
Chamomile	Enhanced Malondialdehyde	Kováčik et al. (2013)
Chickpea	Yield and yield attributes	Singh et al. (2020)
Deccan grass	Enhanced catalase and peroxidase biochemical activities	Samantaray et al. (2001)
Kandelia candel (species of mangrove)	Elevated Malondialdehyde level, and level of stress enzymes (Catalase and Superoxide dismutase)	Rahman et al. (2010)
Wheat	Reduced root and shoot growth	Rafique et al. (2022)
Holy basil/tulsi	Stimulated proline concentration	Rai et al. (2004)
Rice	Elevated Peroxidase concentration	Ma et al. (2016)
Maize	Stress created by Increased elevating concentration of lipid peroxidation and hydrogen peroxide	Maiti et al. (2012)
Spinach	Reduced biomass	Dotaniya et al. (2017b)
Rice	Enhanced ethylene production	Trinh et al. (2014)
Rice	Enhanced catalase and superoxide dismutase level	Sharma et al. (2020), Zhang et al. (2010)
Rice	Enhanced peroxidase level	Sharma et al. (2020), Xu et al. (2011)
French bean	Drastically lower down carotenoids level	Aldoobie and Beltagi (2013)
Pea	Lower down ascorbate peroxidase content	Duhan (2012)
Pterogyne	Enhanced spermidine content	Sharma et al. (2020), Paiva et al. (2014)
Radish	Enhanced glycine-betaine level	Sharma et al. (2020), Choudhary et al. (2012)
Wheat	Stimulated Malondialdehyde level	Ali et al. (2015)
Wheat	Enhanced lipid peroxidation activities	Zhang et al. (2010)

(continued)

Table 8.2 (continued)

Plant species	Physiological response	References
Mung bean	Lower down glutathione content	Sharma et al. (2020), Shanker et al. (2004)
Maize	Enhanced Superoxide dismutase and Guaiacol peroxidase level	Maiti et al. (2012)
Pigeon pea	Germination, root elongation and coleoptile growth	Dotaniya et al. (2014c)

Table 8.3 Chromium species affected by soil pH

pH value	Chromium species
pH < 0	H_2CrO_4
2–6	$HCrO_4^-$, $Cr_2O_7^{2-}$
pH > 6	CrO_4^{2-}

8.5 Chromium Pollution is a Necessary Evil?

Chromium is a transitional metal and popular for wooden industries, leather industries, steel industries and medical industries. It is having the hard in nature and mostly used for corrosion resistance. Some of the important industries are having essentially of Cr metal.

8.5.1 Steel Industry

In steel industry, it is mostly preferred for increasing the strength of the alloy. Most of the parts of the automobiles are Cr plated and increasing the hardness protect during the accident. In India, accordingly to the automobile sector expert told that annual transaction of Rs 7.5 lakh crore including 3 lakh crore foreign exchange. Across the globe this sector is growing very fast and contributing significant role in boosting the economic growth of a country. If we were studied the global exchange of automobiles sector depicted approximant 3 trillion US dollars during each year's 2019–2022 (Fig. 8.3). This figure generates huge amount of revenue in associate industries. The prediction of expert from different automobiles firm are expecting more durable, efficient, less corrosive and environmental friendly vehicle lead the world in 2030. In this situation, use of Cr metal would be increased and chunk of it discharged as a waste in the form of solid and liquid. These effluents are partly recycled and major portion of it discharged in different ecosystems. The recovery cost of Cr from automobiles industry effluent needs technological advancement, and lot of cost at initially installation of treatment/recovery plants.

8.5.2 Wooden Industry

It is another priority area of economy of a country. Chromium is preferably used for the preservation of wooden from insect and pest attack, strength, an unmistakable lustre, and a high level of rust resistance. Most of the wooden industries are using Cr as chrome as a trade name. It is manufacturing by adding with different metals like chromium, copper and arsenic as per the need of strength and need of the product (Fig. 8.4). In furniture industries, copper chrome arsenate (CCA) also a popular product used for wooden preservative. Due to significant concentration of different heavy metals, consumers directed not to use for buring and formation of woody items for young children. The global furniture market was US dollors 475.4 billion in the year 2020. It is expected to grow US doller 720.2 billion by the year 2028 with compound annual growth rate (CAGR) 5.5% in the duration 2021–2028-time frame (MRR 2022).

8.5.3 Leather Industries

India's second most revenue generating industry having huge value for economic growth of the country. More than 2000 small and medium sector leather industries are located mostly at Kanpur (UP), Howrah (WB) and Ambur (TN) as household and commercial units. More than 80–90% leather industries are using Cr as a trivalent salt of Cr for tanning purpose. Chromium stabilizes the leather quality by cross-linking the collagen strands during tanning. Approximately, 4–5% of treated Cr strongly bound with leather product component protein. It is adding the strength and smoothness to the leather. Some of the industries are claiming that they are using vegetable dye instead of the Cr metal during the tanning process. India produces roughly 13% of the world's leather hides and skins, and the country's leather industry also produces a substantial amount of leather annually—nearly 3 billion square feet. In India, leather products were exported $402.61 million in May 2022 with a $48.53% increase from May 2021 (IBEF 2022) (Fig. 8.5).

8.5.4 Medical Industries

Many heavy metals are using in different medical industries. Cobalt and selenium are using for mental related medicine. Iron, manganese (Mn), and gadolinium (Gd) are heavy metals that may have been utilized as dyes or contrast agents. These dyes aid in creating a clear image that enables medical professionals to spot tumors or cancerous cells. The treatment of disease is possible with therapeutic radio-pharmaceuticals. For instance, cancer cells can be eliminated from a brain tumour by directing a gamma ray of the isotope cobalt-60 (Co-60) at the tumour.

8.6 Effect of Chromium Toxicity on Soil Health

Worldwide, soil health and its sustainable management are the major areas of concern in the present-day scenario, since soils are a valuable and non-renewable resource (Lal 2015). The dynamic equilibrium of the soil ecosystem is maintained by healthy soil with a good structure, functional state, and buffering capacity. The microbial diversity makes up the majority of arable soil biomass. In order to improve productivity, it is crucial to maintain the health of the soil. Global environmental catastrophes and resource depletion have put achieving food and nutrition security, as well as environmental sustainability, in danger. Therefore, soil health is a focal point for sustainability in food security, plants, animals and human health, and sustainable ecosystems (Babu et al. 2022). The soil's health and crop yield are both seriously threatened by the presence of heavy metals and hazardous chemicals (Li et al. 2020). Due to extended exposure to wastewater and trash disposal, heavy metals that are not biodegradable continue to build and reach harmful levels in the soil. The production of Cr salts, industrial coolants, textile dyeing, leather tanning, chromate plating, and solid waste disposal are the principal anthropogenic activities that cause environmental contamination with Cr (Zhang et al. 2022). Heavy metal pollution has lately emerged as a significant environmental issue worldwide due to its increased concentration over the permitted limits (Srivastava et al. 2021). Some heavy metals, like Cr, for instance, stop nutrients from being absorbed by the soil by forming insoluble compounds. Therefore, there has been a lot of interest in the biological toxicity of heavy metals in soil (Zhang et al. 2022; Louhar et al. 2020). Additionally, the widespread usage of toxic substances like chromium in the environment and a lack of facilities for ethical waste management contribute to significant soil contamination (Paz-Ferreiro et al. 2018). As a result, crops produced in polluted soils accumulate Cr content, posing a major hazard to human health through the food chain (Alengebawy et al. 2021). However, some research has emphasized that compared to other organisms; soil microbes are more vulnerable to heavy metal contamination. Chromium toxicity has been linked to negative impacts on enzyme activity, soil microorganisms, and microbial processes, according to a number of studies (Table 8.4).

8.7 Chromium Chemistry in Soil

The two primary oxidation states of chromium in soils are +3 and +6 in nature. The +3 oxidation state is represented by the Cr^{+3} cation and Cr^{+6} in chromate (CrO_4^{2-}). Both of these states are hazardous, non-biodegradable pollutants (Wang et al. 2022). But in several nations, hexavalent chromium is considered a priority pollutant. Normal soil conditions are good for the Cr^{+3}, which is very immobile and unavailable because of strong complexations and chemi-sorption's with soil oxides, silicates, and organic materials. The persistent form of Cr^{+6} in the soil is extremely dangerous to biota. In the presence of organic matter and acidic conditions, the

Table 8.4 Effect of chromium toxicity on microbial activity, enzymatic activity, and microorganisms in soil

Concentrations (mg/kg)	Effects	Location/type of soil	References
150–300	*Azotobacter* sp. & *Pichia* sp. biomass reduced by Cr (VI) by more than 50%	Romania; ando soil	Diaconu et al. (2020)
0–300	Slowed down the microbial activity	China	Zhang et al. (2022)
0.2–0.6	Reduced Urease activity by 50%	Poland; peat soil	Samborska et al. (2004)
950–2240	Significantly reduces microbial population and dehydrogenase activity	China	Huang et al. (2009)
50–400	Significantly reduces microbial activities	Nigeria, Sandy loam	Chibuzor et al. (2018)
50–2000	Significantly decreased alkaline phosphatase and dehydrogenase activities	China; ferralic cambisol	Peng et al. (2009)
0–150	Decreased the activity of the urease, dehydrogenases, and alkaline phosphatases enzyme	Poland; brown soil	Wyszkowskaw (2002)
0–20	Reduced catalase enzyme activity	Poland; Mollic Gleysol	Stpniewska et al. (2009)
0–100	Fluorescein diacetate, alkaline phosphatase, and DHA activities were all decreased up to 70%	India, Vertisol	Dotaniya et al. (2017a)
0–20	Reduced dehydrogenase activity	Poland; Haplic Luvisol	Stpniewska et al. (2005)
10–100	Decreased the microbial activity	USA; Sandy loam	Ross et al. (1981)
0–800	Reduced dehydrogenase, catalase enzyme and soil respiration activities	China	Quazi et al. (2014)
200–1600	Decreased soil microbial population and enzyme activities	China; paddy soil	Liu et al. (2014)
3–3300	Significantly reduced the microbial activities	USA	Shi et al. (2002)
4700	Significantly slow down the microbial activities	France	Desjardin et al. (2002)

majority or all of the hexavalent chromate ions that enter the soil along with contaminated irrigation are swiftly transformed to Cr^{+3}. On the other hand, the oxidation of Cr(III) to Cr(VI) may increase the availability and toxicity of Cr in soil solution (Saha et al. 2017a). Because, Cr^{+6} is more mobile and soluble than Cr^{+3}, it is thought to be more damaging to living organisms in soil systems. (Vignati et al. 2010). In soil, Cr^{+6} is typically converted to Cr^{+3} in the presence of oxidizing chemicals, which also reduced its carcinogenicity. Additionally, as pH increases, Cr availability falls off quickly. A minor portion of Cr^{+3} may be converted to soluble chromate when the pH is higher (DesMarias et al. 2019).

8.8 Factor Affecting Chromium Availability

The type and concentration of cations and anions in the soil, the availability of other metals, soil moisture, soil microbial activity, and other factors all have a role in the Cr in soil solution and its absorption by crop plants. These factors are mostly converting Cr forms into different meta forms. Major factors are affecting Cr availability and toxicities are as.

8.8.1 Effect of Organic Matter on Chromium Bioavailability

Organic matter (OM), a crucial component of soil, regulates the mobility, bioavailability, and sorption of heavy metals in the ground. The amount of organic matter in the soil is essential for lowering the likelihood of Cr contamination. Soil organic matter has been discovered to have a significant influence on the mobility of chromium in soil because of its tendency to convert mobile Cr(VI) to the more stationary Cr(III). The effect of soil organic matter on the decrease of Cr(VI) has also been the subject of several investigations. It was discovered that a high quantity of organic materials hindered the Cr. A key indicator of toxicity and potential mobility is the degree of chromium oxidation in contaminated soils. Numerous studies have investigated how OM affects movement of Cr in soil-plant dynamics. The ratio of chromate CrO_4^{2-} ion to total Cr was found to be lower in columns with higher OM than in those with lower OM. Consequently, the most significant effect on the mobility of chromium was caused by the presence of organic materials (Banks et al. 2006). Studies have showed that Cr(VI) reduction increased together with soil total organic carbon (TOC) concentration. When exposed to TOC, Cr^{+6} is converted to Cr^{+3}, and in the majority of situations, this causes it to get immobilized in the soil. The chromium oxidation and reduction processes are controlled by organic matter and different acids (Xu et al. 2004) . According to a report, Cr(III) is mostly found in the organic matter-bound fraction in soils in dry zones with saturated conditions (Eckbo et al. 2022; Han et al. 2004).

8.8.2 Effect of pH, Metals, Cations and Anions on Chromium Bioavailability

Soil pH controls the geochemical behaviour of heavy metals in both the solid and solution phases of the soil (Chung and Eum 2001). The pH of the soil affects the sorption and desorption of Cr and other heavy metals in soils. Soil pH has a significant impact on the geochemical behaviour of chromium. Cr(III) is poorly soluble at pH values lower than 5.5. However, beyond this pH, Cr(III) almost totally precipitates. Contrarily, Cr(VI) is incredibly unstable in soil and continues to be mobilized in both acidic and alkaline soil. Additionally, the quantity of Cr(III) that was sorbed rose when soil pH, CEC, clay, and OM increased (Shahid et al. 2017).

Chromium soil toxicity can be reduced by reducing Cr(VI)–Cr(III), which is controlled by the availability of protons and electrons (Choppala et al. 2013). Cr(III) is less toxic to biota and binds firmly to soil minerals because it is relatively insoluble and less easily absorbed by plants. It functions as a Lewis acid and forms complexes with a range of ligands. Cr(III) is a stable cation in soils and is required for human health, in contrast to Cr(VI), which causes cancer and is a mobile anion in soils. Some oxides, such manganese oxides (MnOx), converted soil containing Cr(III) into Cr(VI). On the other hand, Chromium(VI) was reduced by Fe(II), organic matter, and sulphide to Cr(III). Free Cr(VI) ions are easily reduced in the presence of organic materials when they are present in acidic conditions (Han et al. 2004).

Plants may absorb Cr in both its (III) and (VI) valence states. On the other hand, plants passively take up Cr(III) through the cation exchange sites in their cell walls. Through diffusion at the cell wall's cation exchange site, plants passively absorb Cr(III) sulfate transporters help Cr(VI) actively enter plant cells. Whereas, phosphate and sulfate have structural similarities with Cr(VI), its uptake happens via phosphate and sulfate transporters via an active mechanism that requires energy (Srivastava et al. 2021).

8.9 Management of Chromium Toxicity

The amount and type of organic matter, the presence of other metals, the microbial community, root exudates, crop type and stage; all have a significant impact on the chemistry of Cr in soil. Chromium hexavalent converts into trivalent Cr over a period in present of organic matter. This form is less toxic to hexavent chromium. Based on the mode of action and living organism are classified into two groups.

8.9.1 *Immobilization Through Organic and Inorganic Substance*

In this process, different organic and inorganic substances are using to reduce the Cr toxicity in soil. Lime is mostly used for the immobilization of Cd, Ni, Pb; phosphate salt for Pb and Cd; fly-ash for Cr, Cd, Pb; slag for Cr and Pb, Cd; portland cement for Cr and Zn. However, some of the salt containing Ca or sulphate ions reduced the Cr availability in soil solution. Adding organic matter to the soil enhances soil health and crop productivity while lowering the toxicity of Cr. Addition of FYM, poultry manure, crop residues are also reducing the active ions of Cr. Organic substances are acted as bioabsorber to immobilize the Cr in soil (Nagar et al. 2022).

8.9.2 *Bioremediation*

In this living organisms are using for Cr removal/immobilization process to minimize the toxicity. It could consist of green plants, actinomycetes, bacteria, or fungus. Some bacteria may be able to lower metal toxicity through various metabolic processes (Solanki et al. 2019). Many plant species that accumulate larger concentrations of Cr in various sections of their bodies without compromising the development of the plant are known as hyperaccumulator plants, and the process is known as phytoremediation (Dotaniya et al. 2020). Phytoremediation processes are classified in divers group based on the action mechanism and Cr reduction/removal media (Table 8.5).

8.10 Future Suggestions/Research

- Basic and applied research related to uptake kinetics of Cr in presence of multi metal containing effluent.
- Immobilization process of Cr in soil with respect to climate change phenomena.
- Safe disposal of Cr content during phytoremediation process.

8.11 Epilogue

Chromium is a toxic metal; huge adverse impact on soil–plant-human health. It is having many important uses in different industries those contributing significant economic boost up in Gross domestic product (GDP) in India. Chromium contaminated effluents are mediated the plant nutrient dynamic in soil; uptake kinetic in plant, metabolic disorder in living system are the consequences of toxicity. Long-term use of waste water tainted with Cr led to soil pollution, which then reached

Table 8.5 Strategies for phytoremediation techniques

Phytoremediation techniques	Action process	Medium treated
Phytoextraction	Direct pollutant buildup in plant shoots followed by removal of the plant shoots	Soil
Rhizofiltration (Phytofiltration)	Roots of plants that absorb contaminants	Both root-pumped water and surface water
Phytostabilization	Metals precipitate as a result of root exudates, and biomass becomes less bioavailable	Groundwater, soil, mine tailings
Phytovolatilization	Some metal ions and volatile organics are evaporated by plants	Soil, groundwater
Phytodegradation (plant–assisted bioremediation)	Degradation caused by microbes in the rhizosphere	The soil's rhizosphere and groundwater
Phytotransformation	Degradation and organic pollutant absorption by plants	Surface/groundwater
Elimination of airborne pollutants	Different volatile organics are absorbed by leaves	Air

Adopted from Yang et al. (2005), Dotaniya and Lata (2012)

human bodies through the food chain. Researchers are continuously working on different organic and inorganic substances to immobilize the Cr in soil. However, many soil microorganisms are also identified to reduce the Cr toxicity and plant are using for phytoremediation purpose. In all above, we have to use the Cr contaminated waste water for agricultural crop production system with proper treatment. Create awareness among the local peoples with the help of different agencies and periodically quantify the toxicity of metals in polluted environments.

Acknowledgements Writers are highly grateful to Dr. U.S. Meena, Professor, SKN Agriculture University, Jobner, India for motivating to write this manuscript and valuable suggestions.

References

Alengebawy A, Abdelkhalek ST, Qureshi SR, Wang MQ (2021) Heavy metals and pesticides toxicity in agricultural soil and plants: ecological risks and human health implications. Toxics 9:42

Ali S, Bharwana SA, Rizwan M, Farid M, Kanwal S, Ali Q, Ibrahim M, Gill RA, Khan MD (2015) Fulvic acid mediates chromium (Cr) tolerance in wheat (*Triticum aestivum* L.) through lowering of Cr uptake and improved antioxidant defense system. Environ Sci Pollu Res 22:10601–10609

Andrade DC, Đolić MB, Martínez-Huitle CA, dos Santos EV, Silva TFCV, Vilar VJP (2022) Coupling electrokinetic with a cork-based permeable reactive barrier to prevent groundwater pollution: a case study on hexavalent chromium-contaminated soil. Electrochim Acta 429:140936

Anjum SA, Ashraf U, Khan I, Tanveer M, Shahid M, Shakoor A, Wang L (2017a) Phyto-toxicity of chromium in maize: oxidative damage, osmolyte accumulation, anti-oxidative defense and chromium uptake. Pedosphere 27:262–273

Anjum SA, Ashraf U, Tanveer M, Khan I, Hussain S, Shahzad B, Zohaib A, Abbas F, Saleem MF, Ali I (2017b) Drought induced changes in growth, osmolyte accumulation and antioxidant metabolism of three maize hybrids. Front Plant Sci 8:69

Ao M, Chen X, Deng T, Sun S, Tang Y, Morel JL, Qiu R, Wang S (2022) Chromium biogeochemical behaviour in soil-plant systems and remediation strategies: a critical review. J Hazard Mat 424:127233

Arnon DI, Stout PR (1939) The essentiality of certain elements in minute quantity for plants with special reference to copper. Plant Physiol 14:371–375

Babu S, Singh R, Yadav D, Rathore SS, Raj R, Avasthe R, Yadav SK, Das A, Yadav V, Yadav B, Shekhawat K, Upadhyay PK, Yadav DK, Singh VK (2022) Nanofertilizers for agricultural and environmental sustainability. Chemosphere 292:133451

Banks MK, Schwab AP, Henderson C (2006) Leaching and reduction of chromium in soil as affected by soil organic content and plants. Chemosphere 62:255–264

Britannica (2022) The Editors of Encyclopaedia. "cobalt-60". Encyclopedia Britannica. https://www.britannica.com/science/cobalt-60. Accessed 4 December 2022.

Cervantes C, Campos-García J (2007) Reduction and efflux of chromate by bacteria. In: Molecular microbiology of heavy metals. Springer, Berlin/Heidelberg, Germany, pp 407–419

Chaukura N, Muzawazi ES, Katengeza G, Mahmoud AED (2022) Remediation technologies for contaminated soil systems. In: Gwenzi W (ed) Emerging contaminants in the terrestrial-aquatic-atmosphere continuum. Elsevier, pp 353–365

Chen F, Ma J, Akhtar S, Khan ZI, Ahmad K, Ashfaq A, Nawaz H, Nadeem M (2022) Assessment of chromium toxicity and potential health implications of agriculturally diversely irrigated food crops in the semi-arid regions of South Asia. Agric Water Manag 272:107833

Chibuzor NE, Chuks KO, Emmanuel AE, Paul IO, Simeon CE, Uchenna JO (2018) Chromium (III) and its effects on soil microbial activities and phytoremediation potentials of *Arachis hypogea* and *Vigna unguiculata*. Afr J Biotechnol 17:1207–1214

Choppala G, Bolan N, Seshadri B (2013) Chemodynamics of chromium reduction in soils: implications to bioavailability. J Hazard Mater 261:718–724

Choudhary SP, Kanwar M, Bhardwaj R, Yu JQ, Tran LSP (2012) Chromium stress mitigation by polyamine-brassinosteroid application involves phytohormonal and physiological strategies in *Raphanus sativus* L. Plos One 7:e33210

Chung JB, Eum JS (2001) Effects of organic matter and pH on chromium oxidation potential of soil. Korean J Environ Agric 20:346–351

DalCorso G, Manara A, Piasentin S, Furini A (2014) Nutrient metal elements in plants. Metallom 6(10):1770–1788

Desjardin V, Bayard R, Huck N, Manceau A, Gourdon R (2002) Effect of microbial activity on the mobility of chromium in soils. Waste Manag 22:195–200

DesMarias TL, Costa M (2019) Mechanisms of chromium-induced toxicity. Curr Opin Toxicol 14:1–7

DGCI and S (2022) Directorate general of commercial intelligence and statistics. http://www.dgciskol.gov.in/writereaddata/Downloads/20220704130014Council%20for%20Leather%20Exports%20new(1).pdf

Diaconu M, Pavel LV, Hlihor RM, Rosca M, Fertu DI, Lenz M, Gavrilescu M (2020) Characterization of heavy metal toxicity in some plants and microorganisms—A preliminary approach for environmental bioremediation. New Biotechnol 56:130–139

Dotaniya ML, Lata M (2012) Cleaning soils with phytoremediation. GeoGraphy You 12:18–21

Dotaniya ML, Thakur JK, Meena VD, Jajoria DK, Rathor G (2014a) Chromium pollution: a threat to environment. Agric Rev 35:153–157

Dotaniya ML, Das H, Meena VD (2014b) Assessment of chromium efficacy on germination, root elongation, and coleoptile growth of wheat (*Triticum aestivum* L.) at different growth periods. Environ Monit Assess 186:2957–2963

Dotaniya ML, Meena VD, Das H (2014c) Chromium toxicity on seed germination, root elongation and coleoptile growth of pigeon pea (*Cajanus cajan*). Legume Res 37:225–227

Dotaniya ML, Meena VD, Rajendiran S, Coumar MV, Saha JK, Kundu S, Patra AK (2016) Geo-accumulation indices of heavy metals in soil and groundwater of Kanpur, India under long term irrigation of tannery effluent. Bull Environ Conta Toxic. https://doi.org/10.1007/s00128-016-1983-4

Dotaniya ML, Rajendiran S, Meena VD, Saha JK, Coumar MV, Kundu S, Patra AK (2017a) Influence of chromium contamination on carbon mineralization and enzymatic activities in Vertisol. Agric Res 6:91–96

Dotaniya ML, Rajendiran S, Coumar MV, Meena VD, Saha JK, Kundu S, Kumar A, Patra AK (2017b) Interactive effect of cadmium and zinc on chromium uptake in spinach grown on Vertisol of Central India. Intl J Environ Sci Technol 15:441–448

Dotaniya ML, Meena VD, Rajendiran S, Coumar MV, Sahu A, Saha JK, Kundu S, Das H, Patra AK (2018a) Impact of long-term application of Patranala sewage on carbon sequestration and heavy metal accumulation in soils. J Indian Soc Soil Sci 66:310–317

Dotaniya ML, Panwar NR, Meena VD, Regar KL, Lata M, Saha JK (2018b) Bioremediation of metal contaminated soils for sustainable crop production. In: Meena VS (ed) Role of rhizospheric microbes in soil. Springer International, pp 143–173

Dotaniya ML, Saha JK, Rajendiran S, Coumar MV, Meena VD, Das H, Kumar A, Patra AK (2019) Reducing chromium uptake through application of calcium and sodium in spinach. Environ Monit Assess 191:754

Dotaniya ML, Rajendiran S, Meena VD, Coumar MV, Saha JK, Bhogal NS, Patra AK (2020) Comparative evaluation of phytoremediation potential of Indian mustard (*Brassica juncea*) varieties under sewage irrigated sites. J Indian Soc Soil Sci 68:450–457

Dotaniya ML, Dotaniya CK, Doutaniya RK (2022a) Heavy metal dynamics in soil plant system. Lap Lambert Academic Publisher, Germany, p 133

Dotaniya ML, Rajendiran S, Saurabh K, Saha JK, Dotaniya CK, Patra AK (2022b) Immobilization of chromium bioavailability through application of organic waste to Indian mustard (*Brassica juncea*) under chromium contaminated Indian soils. Environ Monit Assess. https://doi.org/10.1007/s10661-022-10625-4

Dotaniya ML, Meena VD, Saha JK, Dotaniya CK, Mahmoud AED, Meena BL, Meena MD, Sanwal RC, Meena RS, Doutaniya RK, Solanki P, Lata M, Rai PK (2022c) Use of poor quality water for sustainable crop production in changing scenario of climate change. Environ Develop Sustain 24(4):31–52

Duhan JS (2012) Chromium stress on peroxidase, ascorbate peroxidase and acid invertase in pea (*Pisum sativum* L.) seedling. Intl J Biotechnol Mol Biol Res 3:15–21

Eckbo C, Okkenhaug G, Hale SE (2022) The effects of soil organic matter on leaching of hexavalent chromium from concrete waste: Batch and column experiments. J Environ Manag 309:114708

Farooq M, Ali S, Hameed A, Bharwana S, Rizwan M, Ishaque W, Farid M, Mahmood K, Iqbal Z (2016) Cadmium stress in cotton seedlings: physiological, photosynthesis and oxidative damages alleviated by glycinebetaine. South Afr J Bot 104:61–68

Fei Y, Chen S, Wang Z, Chen T, Zhang B (2022) Woodchip-sulfur based mixotrophic biotechnology for hexavalent chromium detoxification in the groundwater. J Environ Manag 324:116298

Han FX, Su Y, Sridhar BB, Monts DL (2004) Distribution, transformation and bioavailability of trivalent and hexavalent chromium in contaminated soil. Plant Soil 265:243–252

Huang SH, Bing PENG, Yang ZH, Chai LY, Zhou LC (2009) Chromium accumulation, microorganism population and enzyme activities in soils around chromium-containing slag heap of steel alloy factory. Trans Nonferrous Metals Soc China 19:241–248

IBEF (2022) India brand equity foundation. https://www.ibef.org/exports/leather-industry-India

Jiang B, Ma Y, Zhu G, Li J (2020) Prediction of soil copper phytotoxicity to barley root elongation by an EDTA extraction method. J Hazard Mat 389:121869

Kováčik J, Babula P, Klejdus BI, Hedbavny J (2013) Chromium uptake and consequences for metabolism and oxidative stress in chamomile plants. J Agric Food Chem 61:7864–7873

Lal R (2015) Restoring soil quality to mitigate soil degradation. Sustain 7:5875–5895

ldoobie N, Beltagi M (2013) Physiological, biochemical and molecular responses of common bean (*Phaseolus vulgaris* L.) plants to heavy metals stress. Afr J Biotechno 12:4614–4622

Letstalkscience (2022). https://letstalkscience.ca/educational-resources/stem-in-context/a-taste-heavy-metal

Li H, Watson J, Zhang Y, Lu H, Liu Z (2020) Environment-enhancing process for algal wastewater treatment, heavy metal control and hydrothermal biofuel production: a critical review. Biores Technol 298:122421

Liu J, Zhang YQ, Zhang LM, Zhou XB, Shi XJ (2014) Impact of Cr^{3+} pollution on microbial characteristics in purple paddy soil. Pak J Pharma Sci 27:625–631

Louhar G, Yadav R, Pawar AB, Rekwar RK, Verma AK, Yadav DK (2020) Heavy metals distribution and their correlation with physico-chemical properties of different soil series of northwestern India. Indian J Agric Sci 90:1742–1746

Ma J, Lv C, Xu M, Chen G, Lv C, Gao Z (2016) Photosynthesis performance, antioxidant enzymes, and ultrastructural analyses of rice seedlings under chromium stress. Environ Sci Pollut Res 23:1768–1778

Mahmoud AED, Fawzy M, Radwan A (2016) Optimization of cadmium (Cd^{2+}) removal from aqueous solutions by novel biosorbent. Intl J Phytoremed 18:619–625

Mahmoud AED, Al-Qahtani KM, Alflaij SO, Al-Qahtani SF, Alsamhan FA (2021a) Green copper oxide nanoparticles for lead, nickel, and cadmium removal from contaminated water. Sci Rep 11:12547

Mahmoud AED, Fawzy M, Hosny G, Obaid A (2021b) Equilibrium, kinetic, and diffusion models of chromium(VI) removal using *Phragmites australis and Ziziphus spina-christi* biomass. Intl J Environ Sci Technol 18:2125–2136

Mahmoud AED (2022) Recent advances of TiO_2 nanocomposites for photocatalytic degradation of water contaminants and rechargeable sodium ion batteries. In: Shalan AE, Hamdy Makhlouf AS, Lanceros-Méndez S (eds) Advances in nanocomposite materials for environmental and energy harvesting applications. Springer International Publishing, Cham, pp 757–770

Mahmoud AED, Kathi S (2022) Assessment of biochar application in decontamination of water and wastewater. In: Kathi S, Devipriya S, Thamaraiselvi K (eds) Cost effective technologies for solid waste and wastewater treatment. Elsevier, pp 69–74

Mahmoud AED, Fawzy M, Hosny M (2022a) Feasibility and challenges of biopesticides application. In: Singh HB, Vaishnav A (eds) New and future developments in microbial biotechnology and bioengineering. Elsevier, pp 83–97

Mahmoud AED, Fawzy M, Khairy H, Sorour A (2022b) Environmental bioremediation as an eco-sustainable approach for pesticides: a case study of MENA region. In: Siddiqui S, Meghvansi MK, Chaudhary KK (eds) Pesticides bioremediation. Springer International Publishing, Cham, pp 479–494

Maiti S, Ghosh N, Mandal C, Das K, Dey N, Adak MK (2012) Responses of the maize plant to chromium stress with reference to antioxidation activity. Braz J Plant Physiol 24:203–212

Minhas PS, Saha JK, Dotaniya ML, Saha A, Saha M (2021) Wastewater irrigation in India: current status, impacts and response options. Sci Total Environ 808:1–17

MRR (2022) Market research report. https://www.fortunebusinessinsights.com/furniture-market-106357. Accessed 28 Aug 2022

Nagar MC, Dotaniya ML, Sharma A, Dotaniya CK, Doutaniya RK, Saha JK (2022) Press mud overcome lead toxicity by improving spinach biomass in lead contaminated soils. Environ Monit Assess. https://doi.org/10.1007/s10661-022-10718-0

Oh YJ, Song H, Shin WS, Choi SJ, Kim YH (2007) Effect of amorphous silica and silica sand on removal of chromium (VI) by zero-valent iron. Chemosphere 66:858–865
Oliveira H (2012) Chromium as an environmental pollutant: insights on induced plant toxicity. J Bot 375843:1–8
Paiva LB, Correa SF, Santa Catarina C, Floh EIS, Silva MGD, Vitória AP (2014) Ecophysiological and biochemical parameters for assessing Cr^{+6} stress conditions in *Pterogyne nitens* Tul.: new and usual methods for the management and restoration of degraded areas. Environ Eng Manag J 13:3073–3081
Palma DL, Verdone N, Vilardi G (2018) Kinetic modeling of Cr(VI) reduction by nZVI in soil: the influence of organic matter and manganese oxide. Bull Environ Contam Toxicol 101:692–697
Paz-Ferreiro J, Gasco G, Mendez A, Reichman SM (2018) Soil pollution and remediation. Intl J Environ Res Public Health 15:1657
Peng B, Huang SH, Yang ZH, Chai LY, Xu YZ, Su CQ (2009) Inhibitory effect of Cr(VI) on activities of soil enzymes. J Central South Uni Technol 16:594–598
Qiu B, Zeng F, Cai S, Wu X, Haider SI, Wu F, Zhang G (2013) Alleviation of chromium toxicity in rice seedlings by applying exogenous glutathione. J Plant Physiol 170:772–779
Quazi M, Aery NC (2014) Effect of increasing concentrations of chromium on soil enzymatic activities and soil respiration. J Environ Sci Eng 56:309–314
Rafique MI, Ahmad M, Al-Wabel MI, Ahmad J, Al-Farraj AS (2022) Mitigating the toxic effects of chromium on wheat (*Triticum aestivum* L.) seed germination and seedling growth by using biochar and polymer-modified biochar in contaminated soil. Sustainability 14:16093
Rahman MM, Rahman MM, Islam KS, Chongling Y (2010) Effect of chromium stress on antioxidative enzymes and malondialdehyde content activities in leaves and roots of mangrove seedlings *Kandelia candel* (L.) Druce. J Environ Sci 26:171–179
Rai V, Vajpayee P, Singh SN, Mehrotra S (2004) Effect of chromium accumulation on photosynthetic pigments, oxidative stress defense system, nitrate reduction, proline level and eugenol content of *Ocimum tenuiflorum* L. Plant Sci 167:1159–1169
Reale L, Ferranti F, Mantilacci S, Corboli M, Aversa S, Landucci F, Baldisserotto C, Ferroni L, Pancaldi S, Venanzoni R (2016) Cyto-histological and morpho-physiological responses of common duckweed (*Lemna minor* L.) to chromium. Chemosphere 145:98–105
Ross DS, Sjogren RE, Bartlett RJ (1981) Behavior of chromium in soils: IV. Toxicity to microorganisms. Am Soc Agrono Crop Sci Soc Am Soil Sci Soc Am 10:145–148
Saha R, Nandi R, Saha B (2011) Sources and toxicity of hexavalent chromium. J Coordinat Chem 64:1782–1806
Saha JK, Selladurai R, Coumar MV, Dotaniya ML, Kundu S, Patra AK (2017) Major inorganic pollutants affecting soil and crop quality. Soil pollution-an emerging threat to agriculture. Springer, Singapore, pp 75–104
Samantaray S, Rout GR, Das P (2001) Induction, selection and characterization of Cr and Ni-tolerant cell lines of *Echinochloa colona* (L.) Link in vitro. J Plant Physiol 158:1281–1290
Samborska A, Stepniewska Z, Stepniewski W (2004) Influence of different oxidation states of chromium (VI, III) on soil urease activity. Geoderma 122:317–322
Sawicka B, Umachandran K, Fawzy M, Mahmoud AED (2021) Impacts of inorganic/organic pollutants on agroecosystems and eco-friendly solutions. In: Mtewa AG, Egbuna C (eds) Phytochemistry, the military and health. Elsevier, pp 523–552
Shahid M, Shamshad S, Rafiq M, Khalid S, Bibi I, Niazi NK, Rashid MI (2017) Chromium speciation, bioavailability, uptake, toxicity and detoxification in soil-plant system: a review. Chemosphere 178:513–533
Shanker AK, Djanaguiraman M, Sudhagar R, Chandrashekar C, Pathmanabhan G (2004) Differential antioxidative response of ascorbate glutathione pathway enzymes and metabolites to chromium speciation stress in green gram (*Vigna radiata* (L.) R. Wilczek. cv CO 4) roots. Plant Sci 166:1035–1043
Sharma A, Kapoor D, Wang J, Shahzad B, Kumar V, Bali AS, Jasrotia S, Zheng B, Yuan H, Yan D (2020) Chromium bioaccumulation and its impacts on plants: an overview. Plants 9:1–17

Shi W, Becker J, Bischoff M, Turco RF, Konopka AE (2002) Association of microbial community composition and activity with lead, chromium, and hydrocarbon contamination. Appl Environ Microbiol 68:3859–3866

Singh D, Sharma NL, Singh CK, Sarkar SK, Singh I, Dotaniya ML (2020) Effect of chromium (VI) toxicity on morpho-physiological characteristics, yield, and yield components of two chickpea (*Cicer arietinum* L.) varieties. Plos One 15:e0243032

Singh R, Gautam N, Mishra A, Gupta R (2011) Heavy metals and living systems: an overview. Ind J Pharma 43:246–253

Solanki P, Dotaniya ML, Khanna et al (2019) Phycoremediation of industrial effluents contaminated soils. In: Singh JS (ed) New and future developments in microbial biotechnology and bioengineering. Elsevier, US, pp 245–258

Solanki P, Dotaniya ML, Khanna N, Meena SS, Rabha AK, Rawat S, Dotaniya CK, Srivastava RK (2020) Recent advances in bioremediation for clean-up of inorganic pollutant-contaminated soils. In: Nayak SK, Mishra SK (eds) Frontiers in soil and environmental microbiology. CRC Press, New York, pp 299–310

Srivastava D, Tiwari M, Dutta P, Singh P, Chawda K, Kumari M, Chakrabarty D (2021) Chromium stress in plants: toxicity, tolerance and phytoremediation. Sustain 13:4629

Statista (2022). https://www.statista.com/statistics/574151/global-automotive-industry-revenue/

Stêpniewska Z, Wolinska A (2005) Soil dehydrogenase activity in the presence of chromium [III] and [VI]. Int Agrophysics 19(1)

Stpniewska Z, Wolińska A, Ziomek J (2009) Response of soil catalase activity to chromium contamination. J Environ Sci 21:1142–1147

Sun P, Liu ZT, Liu ZW (2009) Chemically modified chicken feather as sorbent for removing toxic chromium(VI) ions. Ind Eng Chem Res 48:6882–6889

Tang M, Mao D, Xu L, Li D, Song S, Chen C (2014) Integrated analysis of miRNA and mRNA expression profiles in response to Cd exposure in rice seedlings. BMC Genom 15:835

Tchounwou PB, Yedjou CG, Patlolla AK, Sutton DJ (2012) Heavy metal toxicity and the environment. Exp Suppl 101:133–164

Trinh NN, Huang TL, Chi WC, Fu SF, Chen CC, Huang HJ (2014) Chromium stress response effect on signal transduction and expression of signaling genes in rice. Physiol Plant 150:205–224

Vignati DA, Dominik J, Beye ML, Pettine M, Ferrari BJ (2010) Chromium(VI) is more toxic than chromium(III) to freshwater algae: a paradigm to revise? Ecotoxicol Environ Saf 73:743–749

Wang H, Liu J, Gui C, Yan Q, Wang L, Wang S, Li J (2022) Synergistic remediation of Cr(VI) contaminated soil by iron-loaded activated carbon in two-chamber microbial fuel cells. Environ Res 208:112707

Wyszkowska J (2002) Soil contamination by chromium and its enzymatic activity and yielding. Polish J Environ Stud 11:79–84

Wyszkowska J, Boros E, Kucharski J (2007) Effect of interactions between nickel and other heavy metals on the soil microbiological properties. Plant Soil Environ 53:544–552

Xu XR, Li HB, Li XY, Gu JD (2004) Reduction of hexavalent chromium by ascorbic acid in aqueous solutions. Chemosphere 57:609–613

Xu L, Han L, Huang B (2011) Membrane fatty acid composition and saturation levels associated with leaf dehydration tolerance and post-drought rehydration in Kentucky bluegrass. Crop Sci 51:273–281

Yang X, Fenga Y, He Z, Stoffella PJ (2005) Molecular mechanisms of heavy metal hyperaccumulation and phytoremediation. J Trace Ele Med Biol 18:339–353

Zhang X, Zhang S, Xu X, Li T, Gong G, Jia Y, Li Y, Deng L (2010) Tolerance and accumulation characteristics of cadmium in *Amaranthus hybridus* L. J Hazard Mater 180:303–308

Zhang X, Zhang X, Li L, Fu G, Liu X, Xing S, Chen B (2022) The toxicity of hexavalent chromium to soil microbial processes concerning soil properties and aging time. Environ Res 204:111941

Chapter 9
Toxicity of Hexavalent Chromium: Review

Mousumi Layek, Nargis Khatun, Priya Karmakar, Sandip Kundu, Mandira Mitra, Kripasindhu Karmakar, Subhodip mondal, Ajaya Bhattarai, and Bidyut Saha

Abstract Hexavalent chromium is less stable than trivalent chromium. Among trivalent (Cr(III)) and hexavalent (Cr(VI)) chromium has completely different toxicity and essentiality in animal health. Trivalent chromium is mandatory for glucose metabolism in human. In contrary, hexavalent chromium has no recognized biological functions and is a potent carcinogen. Toxicity of hexavalent chromium to humans is mainly linked to occupational exposure. Key paths of exposure are inhalation, ingestion and dermal contact. Severe toxic effects have been recorded among workers of industries which use chromium compounds. Toxic effects to animals are skin allergies, dermatitis, dermal necrosis, dermal corrosion, respiratory sensitization, lung cancer, liver and kidney damage, irritation and ulceration, nasal septum, internal hemorrhage, gastro internal effects, nausea, vomiting and neurodegenerative diseases. Hexavalent chromium is connected to a reduction in nutrient intake and photosynthesis, which contributes to the delayed growth of plants. Several physiological, structural, and biochemical processes in plant cells are severely troubled resulting in the generation of reactive oxygen species. Chlorosis and plant necrosis are two indications of chromium intoxication. In order to protect the environment and assets, the effectiveness of the physiochemical and biological mitigation strategies for Cr(VI) and how they operate have been discussed.

Keywords DNA damages and cancer development · Neurotoxicity · Chlorosis · Plant necrosis · Physiochemical adsorption · Bioremediation

---

M. Layek · N. Khatun · P. Karmakar · S. Kundu · M. Mitra · K. Karmakar · S. mondal · B. Saha (✉)
Environmental Chemistry Laboratory, Department of Chemistry, The University of Burdwan, Burdwan, West Bengal 713104, India
e-mail: bsaha@chem.buruniv.ac.in

A. Bhattarai
Department of Chemistry, M.M.A.M.C, Tribhuvan University, Biratnagar, Nepal

© The Author(s), under exclusive license to Springer Nature Switzerland AG 2023

N. Kumar et al. (eds.), *Chromium in Plants and Environment*, Environmental Science and Engineering, https://doi.org/10.1007/978-3-031-44029-8_9

9.1 Introduction

It has been discovered that Cr can exist in a number of oxidative states, ranging from −2 to +6 (Avudainayagam et al. 2003). The Cr has been reported to exist in two states in soil and groundwater: the highly oxidised hexavalent (+6) state or the trivalent state (+3), which is likewise oxidised but to a smaller extent than the hexavalent state (Saha et al. 2011). Due to the high $E^0_{\mathrm{Cr(III)/Cr(VI)}}$ value of the Cr(III)/Cr(VI) redox pair, the oxidation of Cr(III) is not favoured in nature; instead, only MnO_2 (manganese oxide) appears to be an effective oxidant in the environment. If not, a variety of reducing chemicals, such as Fe(II), phosphate, sulphur, and organic materials, such as humic acid, can quickly reduce Cr(VI)–Cr(III) (Tumolo et al. 2020). In comparison to Cr(III), Cr(VI) is more soluble in water and is therefore more poisonous. Cr(VI) is widely acknowledged as a human carcinogen and an environmental pollution, but Cr(III) is usually thought to be less hazardous and suggested to be potentially medicinal (Haynes et al. 2011). Chromium can be found primarily as Cr(III) or Cr(VI), depending on the pH values of the solution (Unceta et al. 2010). On the Hazards Priority List of the ATSDR (Agency for Toxic Substances and Disease Registry), Cr(VI) is ranked 17th and is classified by the IARC (International Agency for Research on Cancer) as a group 1 human carcinogen (Wise Jr et al. 2022). Despite the fact that Cr(VI) is now controlled as total Cr and there are few reports of Cr(VI) levels in drinking water, it poses the greatest risk of exposure. The maximum amount of Cr allowed in drinking water is 0.01 mg L^{-1}, and the discharge of Cr(VI) to surface water is limited to less than 0.05 mg L^{-1}, according to the USEPA (United States Environmental Protection Agency) (Vaiopoulou and Gikas 2020). Cr, which is toxic to both individuals and animals and poses a serious threat to aquatic life, is one of the main industrial pollutions created by the tannery, electroplating, textile, and metallurgical sectors (Shekhawat et al. 2015). Welders in the stainless-steel manufacturing industries, as well as those working in the chromate and electroplating industries, have been found to be exposed to very high amounts of Cr (Zhitkovich 2005). In addition, chromium is made mobile through natural processes including weathering and biochemical reactions, that infiltrate the soil and disrupt plant development and metabolic processes in all living things (Dhal et al. 2013). Plant roots can uptake both Cr(VI) and Cr(III), Cr(VI) can then be converted by root reductases to Cr(III), which is then retained in the cortex cells of root (Shanker et al. 2005). Due to the consequences of toxins migrating through the food chain from the soil to plants, animals, and ultimately humans, agricultural soil pollution hexavalent compounds are the main source of public concern over chromium (Alemayehu et al. 2011). Inhalation is one of the most frequent ways that Cr is exposed, and it is connected to cardiovascular illness, cardiomyopathies, extensive dermatitis, GI tract damage, human lung cancer, and liver, lung, and kidney damage (Chakraborty et al. 2022). It has been suggested that a number of processes, including the generation of free radicals, the development of stable Cr-DNA adducts, double-strand DNA breaks, single-strand breaks, and cross-links between DNA and proteins, are responsible for the carcinogenic or toxic effects of Cr (Kozlowski et al. 2014). For

Cr-remediation, typically, chemical procedures are employed. However, bioremediation is being considered as a tool to tackle the issue of Cr(VI) contamination thanks to the inference drawn from the variety of Cr-resistance mechanisms exhibited by microorganisms and plants, including biosorption, diminished accumulation, precipitation, and reduction of Cr(VI)–Cr(III) (Dhal et al. 2013). This comprehensive review concentrates on the chromium chemistry in environment, source and the toxicological effect of chromium on the plant, animal and human health. The efficacy of the physiochemical as well as biological mitigation strategies for Cr(VI) and how they work have been discussed in order to safeguard the environment and assets.

9.2 Ubiquitous Forms of Chromium in Environment

Chromium is the most prevalent contaminate in groundwater sources among the potentially harmful impurities (Yan and Chen 2019). Both Cr(III) and Cr(VI), an ionic form of chromium found in soil, are absorbed by plants; Cr(III) is taken more quickly than Cr(VI). The level of organic matter, the availability of red-ox conditions, and pH levels all influence the chromium speciation in environment (Kimbrough et al. 1999). As Cr(III) is the most stable form in soil and water under natural circumstances, comparatively less mobile and permeable. Cr(III) can exist as hydroxyl species in an aquatic environment as $[Cr(OH)_4]$ (Bartlett 1991), $Cr(OH)_3$ (Gorny et al. 2016), $[Cr(OH)]^{2+}$(Rai et al.1989), $[Cr(OH)_2]^{+}$ (Sharma et al. 2008) and Cr(III) also combines with nearby units belonging to the same species to form polynuclear structures (e.g., $[Cr_4(OH)_6]^{6+}$, $[Cr_3(OH)_4]^{5+}$ and $[Cr_2(OH)_2]^{4+}$), which eventually precipitate as $Cr(OH)_3$ (Salem et al. 1989). Only oxo species, like CrO_3 and H_2CrO_4, have Cr(VI). Depending on the medium's pH, H_2CrO_4 and its deprotonated derivatives ($HCrO_4^-$, CrO_4^{2-}, and $Cr_2O_7^{2-}$) are generated when CrO_3 is easily dissolved in water. At alkaline pH, it is present as CrO_4^{2-} (Amy et al. 2004) and in acidic pH, exists as $HCrO_4^-$ and $Cr_2O_7^{2-}$ (James 2007) in solution. Prevalent forms of both Cr(VI) and Cr(III) are listed in tabular form (Table 9.1).

9.3 Environmental Chromium Cycle

The presence of Cr(III) and Cr(VI) in soils, which have diametrically opposed chemical and physical properties, was previously highlighted. The natural cycle of Cr (Fig. 9.1) in the environment, according to Dhal et al. (2013) and Bartlett (1991), The concentration of Cr(VI) in sediments and groundwater increases due to Cr(III) release into the environment and the oxidation of Cr(III)–Cr(VI) by natural oxidants. Additionally, Cr(VI) migration across the natural system increases the risk factor in the environment.

Table 9.1 Speciation of chromium depending on pH

Forms of chromium	pH	References
H_2CrO_4	pH < 1	James (2007)
$HCrO_4^-$	1 < pH < 6.4	Sharma et al. (2008), James (2007)
CrO_4^{2-}	pH ≤ 6.4	Sharma et al. (2008), James (2007)
$Cr_2O_7^{2-}$	pH < 3	James (2007)
$[Cr(OH)]^{2+}$	3.8 < pH < 6.3	Rai et al. (1989)
$[Cr(OH)_2]^+$	6 < pH < 8	Amy et al. (2004)
$[Cr(OH)_4]^-$	pH > 11.5	Rai et al. (1989)
$Cr(OH)_3$	pH ≈ 8 (for solid) and slightly acidic to alkaline conditions (for liquid)	Sharma et al. (2008), Amy et al. (2004), James (2007)

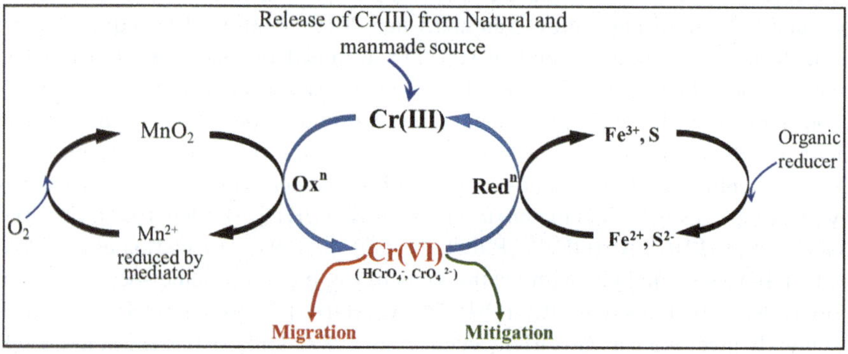

Fig. 9.1 Environmental chromium cycle. Adopted from Dhal et al. (2013)

9.3.1 *Source of Cr(III)*

- **Geogenic source**

Geochemical element chromium is abundantly found in rocks, minerals soil, water, and air (Saha et al. 2011). Between 0.1 and 0.3 mg kg^{-1} of Cr naturally exists in the Earth's crust. This mineral is a compound of various amounts of magnesium, iron, aluminium, and chromium that is found in mafic and ultramafic rocks based on the deposit, in various quantities (Al-Battashi et al. 2016). The main naturally occurring source of Cr (III) in the environment is chromite ore ($FeCr_2O_4$) (Ma et al. 2019). In ultramafic rocks and serpentine soils, chromium is present in large amounts and is typically found as Cr(III). A significant mechanism of Cr(VI) generation in serpentine soils has been identified as chromite ($FeCr_2O_4$)-mediated chromium(III) oxidation (Rajapaksha et al. 2013). The average concentration of Cr in rocks is 100 mg kg^{-1}. For instance, groundwater close to ultramafic rocks and sediments in the Italian province of La Spezia have a Cr(VI) level that ranges from 5 to 73 μg

L^{-1}, exceeding both the Italian drinking water restriction of 5 $\mu g\ L^{-1}$ and the WHO drinking water guideline of 50 $\mu g\ L^{-1}$ (Rajapaksha et al. 2013; Chrysochoou et al. 2016). Mostly wind-blown sand and volcanic activity are responsible for the natural emission of chromium into the atmosphere. Chromium values in air samples from urban regions range from 0.015 to 0.03 mg m^{-3}, while those from distant places range from 59×10^{-6} to 1.29×10^{-3} mg m^{-3} (Xia et al. 2019). The amount of chromium that falls into soil globally is estimated to be 4.6–104 metric tonnes annually, and the annual rates of chromium deposition in remote areas, rural areas, and urban areas are 0.2 kg km^{-2}, 0.5–5.0 kg km^{-2}, and >10 kg km^{-2}, respectively (Nriagu and Pacyna 1988).

- **Anthropogenic source**

There is the way that human-made Cr(III) enters the environment. Chromium is discharged into the water through weathering or contamination from a number of industrial sources, such as chromium-using processes for metal alloys, metal plating, wood treatment, leather tanning, corrosion control, urban storm water runoff, effluent streams from pulp and paper mills, and thermal generating station discharges (Dhal et al. 2013; Moffat et al. 2018; Narayani and Shetty 2013; Chen et al. 2016). The iron and steel sector are the primary human-caused global source of Cr emissions among these industrial processes (McGrath and Smith 1990). After waste residues and wastewater are released into the environment, soil and water sources become seriously contaminated with Cr(VI), which is the second largest cause of a PTE and a serious hazard (Sneddon 2012). Road dust and fugitive chromium emissions from industrial cooling towers are also regarded as two of the most significant sources of Cr (Sneddon 2012). Metallurgical, chemical, and fossil fuel including coal, oil, and gas combustion processes all contribute to chromium emissions into the atmosphere. Additionally, chromium is a naturally existing trace element that is present in most crude oils as well as leftover and distillate oils (Saha et al. 2011). Discharge from chromate mines, improper dumping of mining equipment and materials, and other factors could cause groundwater pollution with Cr (Pandey et al. 2017). Unfortunately, due to pollution from human activity, levels of Cr(VI) in groundwater exceeded 70–90 $\mu g\ L^{-1}$ (Bartlett 1991; Ferronato and Torretta 2019). Cheese, brewer's yeast, wheat germ and calf liver are among the foods that contain chromium, which is primarily found in the Cr(III) form (Saha et al. 2011). Chromium levels in fertilizers, limestone, and animal wastes are of relevance from an agricultural standpoint. Chromium content is higher in phosphate-based fertilizers around 30–3000 mg kg^{-1} (Saha et al. 2011; Steritt et al. 1981).

9.3.2 Oxidation of Cr(III)

However, only a small number of oxidizing agents are present in the environment at sufficient concentrations to start this reaction. Many oxidizing agents have been shown to oxidize Cr(III)–Cr(VI). As good sources of oxygen for oxidizing Cr are water, MnO_2, O_3, H_2O_2, and PbO_2 (III) (Landrot et al. 2009). Numerous variables, such as pH, the shape of MnO_x, the solubility of Cr(III), and organic ligands, have an impact on the efficiency and extent of Cr(III) oxidation by MnO_x. In numerous field soils containing manganese oxides, Bartlett and James (Bartlett 1991) have conducted experimental research on the oxidation of Cr(III)–Cr(VI). According to Apte and colleagues (2005), the oxidation of Cr(III)–Cr(VI) via MnO_2 (external oxidation) under neutral pH circumstances as shown in Eq. 1. According to experimental findings by Dhal et al., soils with MnO_2 are more likely to spontaneously oxidize Cr(III)–Cr(VI) than soils lacking MnO_2 (Dhal et al. 2013). Dissolved oxygen can also convert Cr(III)–Cr(VI), but this process proceeds at a comparatively modest rate. Cr(VI) is also produced via aerobic thermal oxidation of Cr(III), however at room temperature the reaction's kinetics are constrained shown in Eq. 2 (Apte et al. 2005). Cr(III) can also be oxidized by H_2O_2 (hydrogen peroxide) and pH is a factor in Cr(III) oxidation by H_2O_2. Only at pH > 7.5 can H_2O_2 oxidise Cr(III), and the rate at which Cr(III) oxidises rises with rising pH (Pettine et al. 2002; Oze et al. 2007). Photochemical oxidation refers to direct Cr(III) oxidation by ·OH. ·OH is mostly produced through the photolysis of $Fe(OH)^{2+}$ Within the pH range of 2.5–5.5 (Eqs. 3 and 4) (Faust and Hoigné 1990).

$$2Cr^{3+} + 3MnO_2 + 2H_2O \rightarrow 2HCrO_{4^-} + 3Mn^{2+} + 2H^+ \tag{9.1}$$

$$2Cr_2O_3 + 3O_2 \rightarrow H_2CrO_4 \tag{9.2}$$

$$Fe(OH)^{2+} - h\upsilon \quad Fe(II) \tag{9.3}$$

$$Cr(III) + \cdot OH \rightarrow Cr(VI) + OH^- \tag{9.4}$$

9.3.3 *Migration of Cr(VI)*

The primary commercial form of chromium for use in industry is chromite. A rapid migration of soluble pollutants into surface waters and groundwater occurs as a result of improperly disposing of leftovers in open dump sites during the Cr(VI) extraction processes from chromite ore (Ferronato and Torretta 2019; Tumolo et al. 2020) (Fig. 9.2). Birnessite is a Mn(IV) oxide-containing mineral is associated with the Cr (VI) formation from natural Cr(III) in the environment (Rajapaksha et al. 2013). Gattullo et al. investigated at the distribution and speciation of Cr in a contaminated agricultural soil at all scales, from the bulk to the microscopic (Gattullo et al. 2020). Cr(VI) species are As $HCrO_4^-$ and CrO_4^{2-}, Cr(VI) species are easily dissolved and move with groundwater (Richard and Bourg 1991), which is controlled by pH condition. In moderate-to-alkaline soils Cr(VI) migrates easily, cationic colloids in

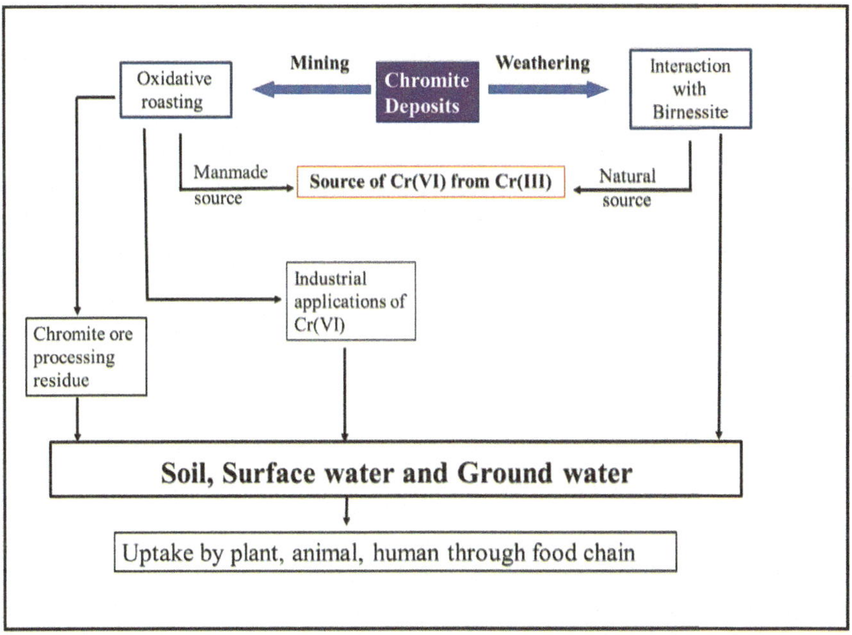

Fig. 9.2 The primary sources and migration of Cr(VI) are shown schematically. Adopted from Tumolo et al. (2020)

soil become negatively charged and prevent Cr(VI) from approaching (Wittbrodt and Palmer 1995) by the precise migration routes like ingestion by plant, horizontal migration to the shallow water, and Vertical migration to the groundwater (McClain et al. 2016).

9.4 Toxicity Effect of Cr(VI)

9.4.1 Human Health Effect

Oxidation state of chromium substantially influences its health impact. An essential micro-nutrition, Cr(III) is harmless and potentially medicinal and is required for normal protein, fat and carbohydrate metabolism, it functions as a glucose tolerance factor in the digestion of carbohydrates (Bagchi et al. 2002). Adults need 0.5–2 μg of absorbable Cr(III) daily, according to estimates. However, if a fractional absorption value of 25% for "biologically integrated" Cr(III) in food is considered, then this is provided by a daily dietary intake of 2–8 μg of Cr(III) (Nordberg and Cherian 2013). People may come into contact with heavy pollutants in a number of ways, including through respiration, skin adsorption, and consumption of foods containing

Cr (Cocârţă et al. 2016). Inhalation and skin contact are the most common methods of occupational exposure and one of the main organs that it primarily affects is the lung through inhalation (Chakraborty et al. 2022). According to several research, chromium was entirely eliminated from all organs (erythrocytes, liver, kidney, and plasma) 140 days after exposure with the exception of the lung and spleen (U.S. Department of health and human Services 2012). Consuming chromium-tainted food and water is the most common method of exposure for the public at large (Mishra and Bharagava 2016). Cr(VI) is converted to Cr(III) by the digestive juices after ingestion (Donaldson and Barreras 1966). Cr(VI) is tremendously risky to all types of life forms, mutagenic in bacteria, and carcinogenic to both animal and human beings (Table 9.2), also responsible for birth abnormalities and declining reproductive health (Dhal et al. 2013).

- **Dermal toxicity**

The body's largest tissue, the skin, is thought to possess some potential towards biotransformation (Alvarez et al. 2021). Cr(VI) compounds can cause skin allergies, dermal corrosion, dermal necrosis, and dermatitis, when they come into contact with the skin, which may lead to "chromium holes" (Saha et al. 2011). A cohort study in Denmark revealed that the fur business was the primary contributor to the chromate skin sensitization in occupational exposure to Cr(VI) (Alvarez et al. 2021). Chromium generates an inflammatory response that can lead to skin ulceration even in nonsensitized people after being exposed to 20 ppm hexavalent chromium (Shelnutt et al. 2007). Additionally, the most well-known response is sensitization and induction of allergic contact dermatitis (ACD), is an immunologic skin reaction that happens after being exposed to environmental irritants. According to the literature, prolonged exposure to Cr(VI) in doses ranging from 4 to 25 ppm can cause sensitization as well as chromium ACD (Shelnutt et al. 2007). Rudolf et al. reported that the shape and motile activity of human dermal fibroblasts (responsible for skin repair and remodelling) were significantly affected by Cr(VI). The suppression of lamellipodia formation and the subsequent motility were among the first alterations brought on by Cr(VI) exposure (Rudolf et al. 2005).

- **Neurotoxicity**

Even though it only makes up 2% of the total mass of the body, the brain is an extremely sophisticated organ that can consume up to 20% of the body's O_2 intake (Wise Jr. et al. 2022). Cr content in the brain as a whole was 0.19 ± 0.03 mg, according to Rajan et al. The hippocampus had the lowest Cr level about 0.10 $\mu g\ g^{-1}$, while the temporal cerebrum (responsible for processing speech and language) had the highest Cr level about 0.18 $\mu g\ g^{-1}$ (Rajan et al. 1997). Cr(VI) likely to have more nuanced adverse effects that may harm a child's ability to learn and pay attention, as well as their olfactory system, social memory, and possibly even cause motor neuron illnesses (Wise Jr. et al. 2022). The cerebrospinal fluid (CSF) may allow Cr(VI) to enter the brain. A modest exchange of Co and Cr from blood plasma to CSF was seen in one study looking at metal-on-metal hip arthroplasty (Harrison-Brown et al.

Table 9.2 Toxicity of Cr in animal and human being

Organ	Model	Toxicity	References
Skin	Human	Skin allergies, dermal corrosion, dermal necrosis, and allergic contact dermatitis (ACD), chromium holes, suppression of lamellipodia formation	Saha et al. (2011), Shelnutt et al. (2007), Rudolf et al. (2005)
	Mouse	Apoptosis or necrosis	Son et al. (2005)
Brain	Human	Olfactory system, social memory, and possibly even cause motor neuron illnesses, Autism spectrum disorder, brain tumours, risk of acquiring glioma	Hara et al. (2010), Yorbik et al. (2010), Parent et al. (2017)
	Freshwater fish	Exhibit excitability and irregular swimming	Velma and Tchounwou (2009), Azmat and Javed (2011)
	Female mice	Increased generation of ROS, lipid peroxidation	Bagchi et al. (2002)
Lung	Human	Bronchial cancer, bronchitis, bronchopneumonia, asthma, lung cancer, increase in miR-21 and a decrease in PDCD4 levels	Alvarez et al. (2021), Abreu et al. (2018), Horie et al. (2013)
Kidney	Human	Increased in caspase-3, Bax, caspase-8, FoxO1, IGF-1, PARP and suppress of Bcl-2, Bcl-xl level, apoptosis shrinking the size of the kidney tissue	Chakraborty et al. (2022), Avila-Rojas et al. (2020)
	Goldfish (*Carassius auratus*)	Disruption of the DNA in kidney	Velma and Tchounwou (2009)
	European eels	Increase in ROS levels	Ahmad et al. (2006)
Liver	Human	ROS generation of HepG2 liver cells, cell death	Chakraborty et al. (2022), Das et al. (2015)
	Mice	Apoptosis	Wang et al. (2010a, b)
	Goldfish (*Carassius auratus*)	Disruption of the DNA in liver	Velma and Tchounwou (2009)
	Broiler	Cell death by increasing the level of Caspase-3 and Bax	Xueting et al. (2018)

2020). In a cohort study of 1193 Japanese men, Hara et al. in 2010 observed that Cr platers had an increased chance of dying from brain tumours, with a SMR (standard mortality ratio) of 9.14 (Hara et al. 2010). Cr contributes to the development of ASD (Autism spectrum disorder) (Yorbik et al. 2010). Parent et al. (2017) evaluated the link between workplace welding fume exposure and the risk of acquiring glioma (Parent et al. 2017).

- **Effect on the lungs**

In 1932, a correlation regarding enhanced lung cancer risk and exposure to Cr(VI) had first been identified (Beukes et al. 2010). Inhalation of Cr(VI) compounds has been linked to bronchial cancer, bronchitis, bronchopneumonia, asthma, and the perforation of the nasal septum (Alvarez et al. 2021). The shape of the lungs is unaltered by chromium inhalation, but macrophages grow bigger, become multi-nucleated or vacuolated, and lumps develop in intra-alveolar areas (Sharma et al. 2022). While humoral immune function and the alveolar macrophage phagocytic function are inhibited by elevated doses of Cr(VI), these processes are encouraged by lower doses of Cr(VI) (Sharma et al. 2022). Rare occurrences of bronchial asthma have been observed among chrome plating workers, and the condition is caused by exposure to hexavalent chromium compounds (Costa, 1997; Menezes et al. 2004). Cr-initiation research on human lung cell cultures (bronchial) revealed an increase in reactive oxygen species (ROS) levels, which resulted in an increase in miR-21 and a decrease in PDCD4 levels, resulting in oncogenesis (Abreu et al. 2018). When Cr_2O_3 employed as nanoparticles also boosted the quantities of intracellular ROS in human lung cancer cells A549 (Horie et al. 2013). CrO_4^{2-} generation has repeatedly and distinctly been linked to increased lung cancer risks in scientific studies on cancer in people.

- **Effect on the liver and kidney**

The liver and kidney may be harmed by prolonged exposure to Cr(VI) (Martin and Griswold 2009). Hexavalent chromium poisoning observations on human (HepG2) liver cells revealed that ROS were produced, and a surge in their contents was seen that was based on the concentration of Cr (Chakraborty et al. 2022; Das et al. 2015). Following exposure to Cr, caspase-3, Bax, caspase-8, FoxO1, IGF-1, and PARP were increased in the kidney, while Bcl-2 and Bcl-xl were reported to suppress. These tissues underwent apoptosis as a result of all these alterations (Chakraborty et al. 2022). Additionally, current studies have demonstrated that exposure to Cr in the form of potassium dichromate damages mitochondria by fragmenting them, changing their structure, and shrinking the size of the kidney tissue (Avila-Rojas et al. 2020).

- **Effect on the genome**

Cr(V) or Cr(III), produced as intracellular reduction intermediates after Cr(VI) has penetrated into the cell through the sulphate transportation network, is the mechanism of Cr(VI) mutagenicity (Fig. 9.3) (Tumolo et al. 2020; Mishra and Bharagava 2016; Sugiyama 1992). Cellular reducing agents such glutathione, ascorbic acid, and flavoenzymes like glutathione reductase, cytochrome P-450, and riboflavin efficiently convert Cr(VI) into the Cr(III) (Mishra and Bharagava 2016; Bianchi et al. 1983). When Cr(VI) reduction takes place inside of a cell, it destroys cell organelles, causes oxidative-mediated toxicity, and results in DNA mutation (Wakeel et al. 2020).

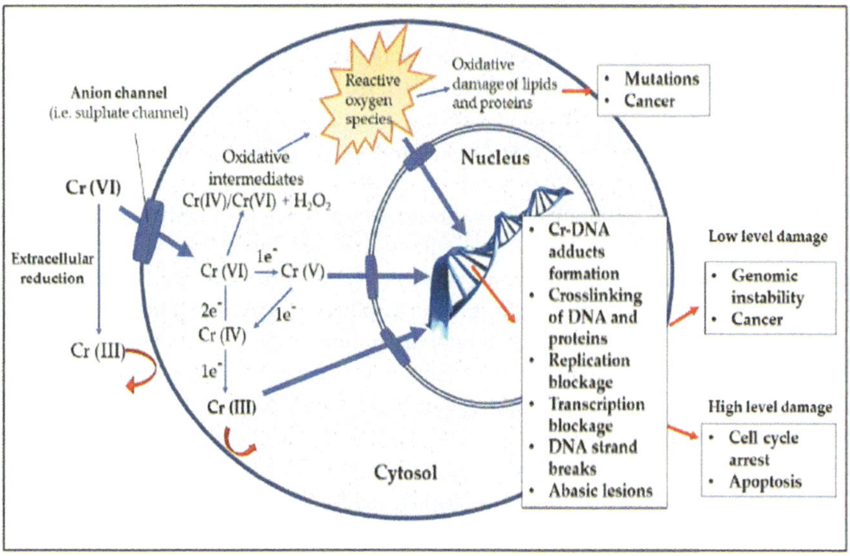

Fig. 9.3 The primary routes and cellular absorption in chromium genotoxicity. Adopted from Tumolo et al. (2020)

Cr(VI) exposure for an extended period of time promotes oxidative stress in the tissue and triggers ROS-mediated cell apoptosis (Xu et al. 2018). It has been suggested that a number of processes, including the generation of free radicals, single-strand breaks, double-strand DNA breaks, the development of stable Cr-DNA adducts, cross-links between DNA and proteins, protein-Cr-DNA adducts and mutagenesis are responsible for the carcinogenic effects of Cr (Kozlowski et al. 2014). The most prevalent type of genetic diseases caused by Cr(VI) in mammalian cells are small Cr-DNA adducts (Zhitkovich 2005). In erythrocytes, Cr(VI) stimulates the activity of superoxide dismutase (SOD) which promoting the production of H_2O_2 (hydrogen peroxide), which can then combine with Cr(V) and Cr(VI) to form ·OH (hydroxyl radicals) (Alvarez et al. 2021). Double-stranded DNA breakage emerges from the combination of ·OH with DNA nucleotides (Nimse and Pal 2015).

9.4.2 *Effect on Animals*

For oral toxicity in rats, the LD50 (dose that kills 50% of a specified animal population) ranges from 1900 to 3000 mg kg^{-1} for Cr(III) and 50–100 mg kg^{-1} for Cr(VI) (De flora et al. 1990). In mice livers, Cr(VI) triggered apoptosis or cell death has been shown with increase in caspase-3, Bax expression, p53, and cytochrome C and also decrease in Bcl-2 levels (Wang et al. 2010a, b). According to Son et al., mouse skin epidermal cells died from Cr(VI)-induced cell death by apoptosis or necrosis

(Son et al. 2010). In the case of fish, the gills are the primary route via which chromium enters their bodies (Velma and Tchounwou 2009). Yilmaz et al. investigated the uptake of hexavalent chromium by a freshwater Tilapia fish (*Oreochromis aureus*) and reported the Cr content in fish muscle tissue was least, ranging from 0.86 to 12.34 $\mu g\ g^{-1}$ w.w., while the amount in the gills was high, ranging from 3.11 to 45.23 $\mu g\ g^{-1}$ w.w. (Yilmaz et al. 2010). When freshwater fish have been introduced to a Cr-contaminated condition, they exhibit excitability and irregular swimming (Velma and Tchounwou 2009; Azmat and Javed 2011). Regardless of its oxidation states Cr is a potent suppressor of Cyt-P450-reductase function in fish (Bozcaarmutlu and Arinc 2007). In another comprehensive research study, goldfish (*Carassius auratus*) displayed substantial disruption of the DNA in kidney and liver tissue upon introduction to Cr, and the damage was the most severe during the initial seven days of exposure (Velma and Tchounwou 2009). According to a study, when broiler liver was treated with Cr ($K_2Cr_2O_7$), it underwent cell death by increasing the level of Caspase-3 and Bax (Xueting et al. 2018). Female C57BL/6Ntac and p53-deficient C57BL/6TSG p53 mice were used to study the dose-dependent effects of Cr (VI). According to Bagchi et al. 2002 in the hepatic and brain tissues of female mice, there is an increased generation of ROS, lipid peroxidation and in mice lacking the p53 gene, Cr(VI) increased the severity of oxidative damage in several target organs. Another investigation using European eels revealed that kidney ROS levels significantly increased in the presence of Cr (Ahmad et al. 2006).

9.4.3 Effect on Plants

Plant tissue cell membranes are harmed by Cr(VI), a strong oxidising substance (Mei et al. 2002). In addition to physiological and biochemical changes, Cr chronic effects in plants include inhibition of root growth, seedling growth and development, germination, and biomass, as well as induction of leaf chlorosis and necrosis (Singh et al. 2013). Because plants tend to accumulate Cr(VI) in their biomass due to its higher solubility than Cr(III), lead in is produced. By oxidising the cell wall, Cr(VI) destroys the shoots (Shahandeh and Hossner 2000). Because Cr(III) and Fe(III) are comparable chemical species, plants that are less able to accumulate Fe(III) in the vegetable leaves also tend to be less able to translocate Cr(III) into the leaves. Transport of Cr(III) from roots into shoots begins after the build-up of Cr(III) in the roots reaches a concentration of 50 mg kg^{-1}. High Cr levels in the soil impair the formation of fibrous roots, which reduces nutrient intake and water content (Zou et al. 2006). At low doses, Cr(III) is not dangerous, however Cr(VI) ions are known to have mutagenic and carcinogenic effects on a variety of bacteria (Chen and Hao 1998). The anionic surface elements of the majority of bacteria cannot trap Cr(VI) ions (Volesky and Holan 1995). When Cr(III) precipitates as hydroxides, the amount of Cr(III) that may pass through a cell wall is decreased (Kamaludeen et al. 2003). The sulphate transport system and Cr(VI), which is anionic in nature, can easily traverse the cell wall (Cervantes et al. 2001). Since seed germination is the first physiological activity that Cr affects, a seed's capacity to germinate in a medium containing Cr would be an indication of its level of metal tolerance (Peralta et al. 2001). Under conditions of Cr stress, seeds may germinate less frequently due to a depressive effect of Cr on the activity of amylases and the subsequent transport of carbohydrates to the embryo axis (Zeid 2001). whereas root length was more influenced by Cr than by other heavy metals investigated, it has been observed that the order of metal toxicity to new root primordia in *Salix viminalis* is Cd greater than Cr and Cr larger than Pb (Prasad et al. 2001). The general response of reduced root growth caused by Cr toxicity may be attributable to the inhibition of root cell division, root elongation, or the extension of the cell cycle in the roots. When both Cr species are present in high concentrations, the root growth may be reduced as a result of the roots of seedlings coming into direct contact with the Cr in the medium, which may cause a collapse and impair their ability to absorb water from the medium (Barcelo et al. 1986). When Cr was applied to *Sinapsis alba* plants at rates of 200 or 400 mg kg^{-1} soil together with N, P, K, and S fertilisers, there was a noticeable decrease in plant height (Hanus and Tomas 1993). A high level of Cr(VI) about 200 ppm had a negative impact on *Albizia lebbek* seedlings leaf area and biomass (Tripathi and Tripathi 1999). Bush bean plants cultivated in 1–10 $\mu g\ cm^{-3}$ Cr revealed both primary and trifoliate leaves that significantly reduced in leaf area; trifoliate leaves were more impacted by Cr than primary leaves (Barcelo et al. 1985). According to a study on the effects of Cr(III) and Cr(VI) on spinach, when Cr was given at levels of 60 mg kg^{-1} and above, the size of the leaves was reduced, the leaf tips or margins were burned, and the pace of leaf growth was slowed (Singh 2001). Subrahmanyam 2008 investigated effects of chromium toxicity on leaf photosynthetic characteristics and oxidative changes in wheat (*Triticum aestivum L.*) and observed that Cr(VI) significantly reduced rates of net photosynthesis and transpiration and of stomatal conductance. In a study involving multiple heavy metals, it was discovered that Cr had a notable impact on leaf development and preferred to influence early leaves in tomato plants (Pedreno et al. 1997). The suppression of RNA transcription and DNA replication demonstrates the negative impacts of Cr(VI) on plants (Ukhurebor et al. 2021).

9.5 Mitigation of Hexavalent Chromium

The risks and threats that Cr poses to plants, animals and human life have already been covered in the section above. Several efforts have been conducted over the past few years to migrate Cr toxicity in polluted environment. Some strategies used for chromium contaminant remediation includes different, physiochemical and biological remediation techniques are as follows (Fig. 9.4) and some of them have been discussed below.

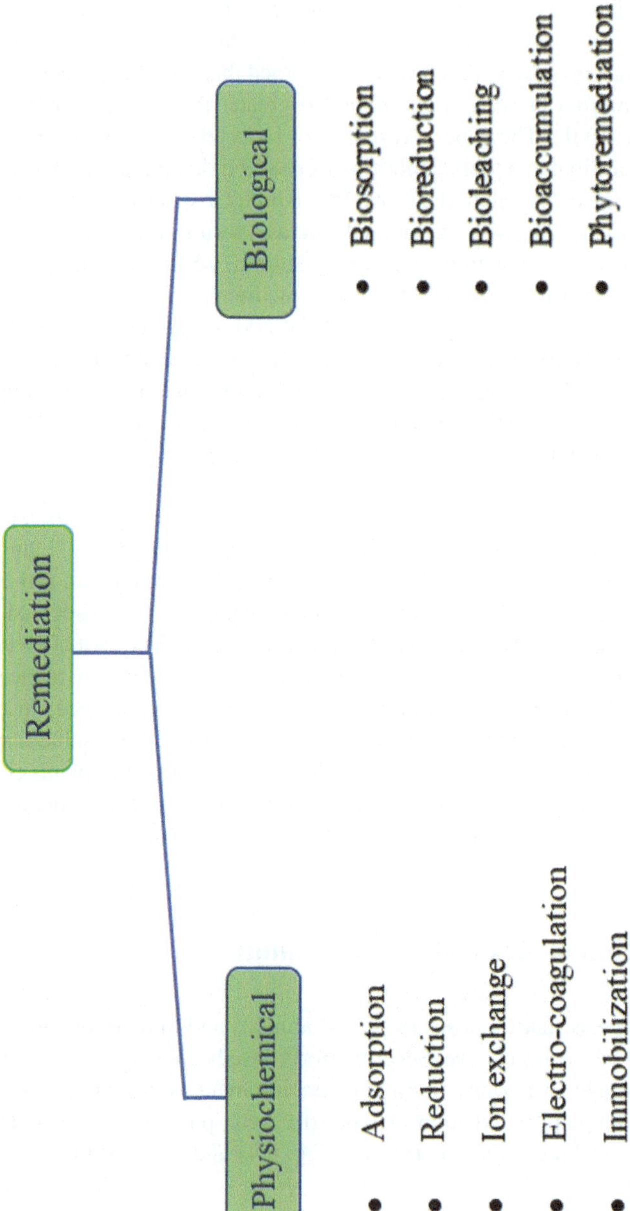

Fig. 9.4 A flowchart illustrating a number of the popular remediation strategies

9.5.1 Physiochemical Remediation

- **Adsorption**

Numerous studies have concentrated on the use of natural and manufactured adsorbent materials, such as, carbon nanotubes, volcanic rocks, modified clay and sand, and activated carbon, for the remediation of chromium (Dhal et al. 2013; Alemayehu et al. 2011). According to reports, modified montmorillonite with cetylpyridinium-bromide removes Cr(VI) from water with a 98% effectiveness rate (Brum et al. 2010). Recently, utilizing a composite adsorbent made of graphene and sand, a high adsorption capacity of 2859.38 mg g^{-1} for chromium was found (Dubey et al. 2015). Adsorption of Cr(VI) into soil is pH dependent. The forms of Cr(VI), $HCrO_4^-$ are captured by positively charged colloids such as hydrated oxides of iron/aluminum (FeO(OH) and Al_2O_3) in acidic soils of pH < 6 due to the H-bonding interaction with iron/aluminum oxides and the most common form of Cr(VI) (Richard and Bourg 1991). Recently, the effectiveness of Amberlite XAD 1180 resin impregnated with trioctylamine was shown in the elimination of chromium. A high adsorption capacity of 171.8 mg g^{-1} was demonstrated by this resin adsorbent (Rajesh et al. 2011). According to Jardine et al. research in 2013, Cr(VI) adsorption rises with falling soil pH and rising Total Organic Carbon content because these two factors induce cationic colloids to accumulate more positive charges (Jardine et al. 2013). A unique metal organic hybrid Cu(I) metallo-gel adsorbent with a high adsorption capacity of 331 mg g^{-1} at pH 2.7 was also examined for the reductive elimination of Cr(VI)–Cr(III), followed by adsorption (Sarkar et al. 2014).

- **Reduction**

By donating electrons, chemical reducing agents like S^{2-}, Fe(0), or Fe(II) transform hexavalent chromium into less toxic trivalent chromium. Malic acid acts as a catalyst to accelerate the reduction of Cr(VI) in Fe-rich soils at low pH (Zhong and Yang 2011). Because Cu(II) has a synergistic impact on the adsorption of maleic acid on ultisol (Fe-rich soil), the addition of Cu(II) enhanced the reduction rate. A possible autocatalytic effect of Cr(VI) concentration on its reduction by ferrous iron was proposed by the experiment by Katsoyiannis et al. (2020). The scientific literature has established ZVI (zero-valent iron) and in particular nZVI (nanoscale zero-valent iron) as a widely accessible and reasonably priced reducing agent for the elimination of Cr(VI) (Shi et al. 2011). Additionally, for solutions containing 10 mg/l of Cr, the efficiency is increased to 100% by adding carboxymethyl cellulose (CMC) as a stabiliser for nZVI (Wang et al. 2010a, b). Long et al. in 2007 studied the electrokinetic mitigation of CrO_4^{2-} in soils is accelerated in the presence of an oxidising agent, such as NaClO (sodium hypochlorite) (Cang et al. 2007). The reduction of Cr(VI) from Cromite Ore Processing Residues has also been accomplished using CaSx (calcium polysulfide) (Graham et al. 2006).

- **Ion-exchange**

Ion exchange techniques are a common physicochemical method for eliminating chromium from waste water (Rengaraj et al. 2003). A typical anionic resin that removes Cr(VI) from acidic medium by possessing a quaternary amine group (Kalidhasana et al. 2016). Strongly basic anion exchangers are more suitable for eliminating Cr(VI). It calls for a polymeric resin that is inert yet has interchangeable surface and interfacial functional groups (Plummer et al. 2018). Fast exchange kinetics and significant metal ion adsorption capacity are displayed by solvent impregnated resins (SIRs). Muraviev provides a thorough discussion of the elements that contribute to the stabilization of SIRs in a recent review (Muraviev 1998). Cation exchange resins are more efficient for the treatment of effluents containing Cr(III) due to the fact that trivalent chromium can present as cationic hydroxides. The removal of Cr(VI) from aqueous solutions was examined by Rafati et al. (2010) using Lewatit FO36, an ion exchanger that has a layer of iron oxide that binds Cr(VI) in a particular surface complex (Rafati et al. 2010). In batch and column systems, the adsorption performance of EDE-D301 anion exchange resin for eliminating Cr(VI) impurities was examined by Han et al. in 2020 (Han et al. 2020). Cr(VI) adsorption capacity onto EDE-D301 was evaluated at a maximum level of 298 mg g^{-1} (Han et al. 2020).

9.5.2 Biological Remediation

- **Biosorption**

Due to their bioaccumulation qualities, living and non-living organisms utilized in biosorption have been extensively exploited in the removal of Cr(VI). Beveridge and Murray have studied metal sequestration from water is significantly influenced by the presence of anionic ligands on the cell walls of bacteria, specifically carboxyl, amine, hydroxyl, phosphate, and sulfhydryl groups (Beveridge and Murray 1980). A native strain of cynanobacteria isolated from soil and the microalgal isolate Chlorellaminiata have the capacity to effectively reduce Cr(VI) in waste water (Han et al. 2007; Anjana et al. 2007). Kumar et al. (2008) investigated that 87.9% of the Cr(VI) may be removed from electroplating effluent by *Aspergillus sydoni* is a dead fungal biomass (Kumar et al. 2008). Recently, a montmorillonite-starch bionanocomposite material with an adsorption capacity of 20.6 mg g^{-1} was explored to scavenge Cr(VI) (Koriche et al. 2014). Glucoproteins, lipids, pigments, inorganic salts, cellulose, chitin, b-glucan, a-glucan, polyuranides, and chitosans are all components of the fungal cell wall. Ali Redha in 2020 have investigated in both living and dead forms, fungus can be used as biosorbent materials because of the remarkable metal-binding qualities of their cell walls (Redha 2020).

- **Bioreduction**

In order to reduce Cr(VI) and subsequently immobilize Cr(III) in soils, microorganisms are essential. Algae, yeasts, and fungus as well as bacteria including *Pseudomonas, Micrococcus, Escherichia, Enterobacter, Bacillus, Aeromonas, Achromobacter*, and *Desulfotomaculum* reduce Cr(VI)–Cr(III) (Cervantes et al. 2001). Numerous publications have explored the microbial reduction of Cr(VI) in the presence of molasses (Jayasingh and Philip 2005) and the utilization of microorganisms found in animal waste (Sethunathan et al. 2005) from tannery-contaminated soils and sediments. Using anaerobic metabolism, sulfate- and iron-reducing microorganisms, for example, can directly reduce hexavalent chromium and generate S(II) and Fe(II), respectively (Sethunathan et al. 2005). *Acidithiobacillus ferrooxidans* and *Acidithiobacillus thiooxidans* bacteria have been observed to reduce Cr(VI) by varying percentages. Cr(VI) can be reduced abiotically by organic matter directly, and electron-rich functional groups (such as phenolic, hydroxyl, carboxyl, and aldehyde groups) play a key part in this process (Aldmour et al. 2019; Hou et al. 2019). In agricultural soil, elemental sulphur is typically utilised to deliver nutrients to promote plant growth. Additionally, it has a strong redox activity and the ability to convert polluted soil containing Cr(VI)–Cr(III) (Choppala et al. 2018). Also, Inside the cell membrane intermediate oxygen radicals may occur during the biotransformation of Cr(VI)–Cr(III) in cells, and intermediate oxidation states of Cr, such as Cr(V) and Cr(IV), may have a more harmful effect than Cr(III) (Costa 2003). To address these issues in cellular-mediated conversion of Cr(VI)–Cr(III), Cr-tolerant bacterial species have different pathways such as chromate-resistant plasmids and iron efflux systems (Cervantes and Silver 1992; Nies and Silver 1995). Bacillus sp. CRB-B1 was a highly efficient Bacillus sp. for Cr(VI) reduction and adsorption, according to research by Tan et al. published in 2020 (Tan et al. 2020).

- **Phytoremediation**

One of the most affordable green technologies is phytoremediation, in which plants and microorganism absorb heavy metals from contaminated soils (Ali et al. 2013). This method, which uses several types of plants that are intended to capture or corrupt atypical contaminants, is one of the best techniques among other conventional methods for treating contaminated sites, according to Saravanan et al. (2019). Experimental evidence has also shown that aquatic plants that have grown in water that contains hexavalent Cr, such as *Eichhornia crassipes*, Pistia stratiotes, and Salvinianatans, accumulate decreased Cr (III) in their leaves and roots (Sundaramoorthy et al. 2010). *Hydrocotyle umbellata L.* and *Typha angustifolia L.* (99.78%) had elimination efficacy of 86.36 and 99.67% for Cr, respectively, according to Taufikurahman et al.'s nine-day investigation of certain wetland plants, including *Canna indica L., Hydrocotyle umbellata L.* and *Typha angustifolia L.* (Taufikurahman et al. 2019). According to the research done by Kassaye et al. in 2017, with the *Polygonum coccineum, Cyprus papyrus*, and *Brachiara mutica* plant showed *Brachiara mutica* and *Cyprus papyrus* exhibited considerably higher elimination efficacies in removing both Cr(III) and Cr(VI), from wastewater (Kassaye et al. 2017). By altering root architecture, emitting metal-eliminatting compounds in rhizomes, and reducing phytotoxicity caused by chromium, plant growth promoting bacteria (PGPB) increase plants' ability to absorb Cr (Ahemad 2015).

9.6 Conclusion and Future Perspective

Environmental scientists around the world are extremely concerned about the problem of health risks emerging from environmental degradation. According to numerous studies, Cr is one of the main potentially inorganic pollutants that are contributed to the environment by a variety of anthropogenic and geogenic sources, especially in shallow water and soil. This evaluation focuses on chromium's prevalent pH-dependent environmental forms, its many natural and artificial sources, its chemistry, toxicity, and a variety of physiochemical and biological mitigation techniques. Hexavalent chromium is the most poisonous type of the metal since it poses a serious risk to human health. Once chromium is injected into human cells, its genotoxicity can emerge promoting genomic instability, the initiation of cancer, cell cycle inhibition and cell death. Several technologies have been investigated over the past few years to see if they are effective at chromium decontamination. Because microbial reduction of Cr(VI)–Cr(III), which is comparatively less mobile and significantly less hazardous, eco-friendly and safe, and maybe affordable as well, holds tremendous potential for developing an acceptable technique, microbial Cr(VI) bioremediation is a promising option. Prior to doing scientific experiments and in situ investigations, for improvement of soil decontamination must be done at the research laboratory. Monitoring the redox potential of soils and supplying them with industrial reductants like Fe(II) and sulphides or natural reductants like humus on a regular basis are crucial to preventing the re-oxidation of Cr(III) in soils that have undergone Cr(VI)-remediation. As paddy fields, which are located in anaerobic circumstances, it is proposed that agricultural soils be employed. The development of graphene oxide-ionic liquid, clay-ionic liquid, and graphene oxide-biopolymer based composite friendlier adsorbent materials should be prioritised given the various ways to reduce Cr(VI) pollution.

Acknowledgements Sincere gratitude is extended by the authors to the authors whose research papers and publications were consulted for this review study. There was no funding provided for this study because it did not get any funding.

References

Abreu PL, Cunha-Oliveira T, Ferreira LM, Urbano AM (2018) Hexavalent chromium, a lung carcinogen, confers resistance to thermal stress and interferes with heat shock protein expression in human bronchial epithelial cells. Biometals 31:477–487

Ahemad M (2015) Enhancing phytoremediation of chromium-stressed soils through plant-growth-promoting bacteria. J Genet Eng Biotechnol 13:51–58

Ahmad I, Maria V, Oliveira M, Pacheco M, Santos M (2006) Oxidative stress and genotoxic effects in gill and kidney of *Anguilla anguilla* L. exposed to chromium with or without pre-exposure to β-naphthoflavone. Mutat Res Genet Toxicol Environ Mutagen 608:16–28

Al-Battashi H, Joshi SJ, Pracejus B, Al-Ansari A (2016) The geomicrobiology of chromium(VI) pollution: microbial diversity and its bioremediation potential. Open Biotechnol J 10:379–389

Aldmour ST, Burke IT, Bray AW, Baker DL, Ross AB, Gill FL et al (2019) Abiotic reduction of Cr(VI) by humic acids derived from peat and lignite: kinetics and removal mechanism. Environ Sci Pollut Res 26:4717–4729

Alemayehu E, Thiele-Bruhn S, Lennartz B (2011) Adsorption behaviour of Cr(VI) onto macro and micro-vesicular volcanic rocks from water. Sep Purif Technol 78:55–61

Ali H, Khan E, Sajad MA (2013) Phytoremediation of heavy metals—Concepts and applications. Chemosphere 91:869–881

Alvarez CC, Gòmez MEB, Zavala AH (2021) Hexavalent chromium: regulation and health effects. J Trace Elem Med Biol 65:126729

Amy G, Yoon J, McNeill L, Banerjee K (2004) Chapter 2 Literature review. In: Brandhuber P, Frey M, McGuire MJ, Chao P, Seidel C, (eds) Low-level hexavalent chromium treatment options: bench-scale evaluation. American Water Works Association Research Foundation, Denver, CO, USA, pp 8–9

Anjana K, Kaushik A, Kiran B, Nisha R (2007) Biosorption of Cr(VI) by immobilized biomass of two indigenous strains of cyanobacteria isolated from metal contaminated soil. J Hazard Mater 148:383

Apte AD, Verma S, Tare V, Bose P (2005) Oxidation of Cr(III) in tannery sludge to Cr(VI): field observations and theoretical assessment. J Hazard Mater 121:215–222

Avila-Rojas SH, Aparicio-Trejo OE, Briones-Herrera A, Medina-Campos ON, Reyes-Fermín LM, Martínez-Klimova E, León-Contreras JC, Hernández-Pando R, Tapia E, Pedraza-Chaverri J (2020) Alterations in mitochondrial homeostasis in a potassium dichromate model of acute kidney injury and their mitigation by curcumin. Food Chem Toxicol 145:111774

Avudainayagam S, Megharaj M, Owens G, Kookana RS, Chittleborough D, Naidu R (2003) Chemistry of chromium in soils with emphasis on tannery waste sites. Rev Environ Contam Toxicol 178:53–91

Azmat H, Javed M (2011) Acute toxicity of chromium to *Catla catla*, *Labeo rohita* and *Cirrhinus mrigala* under laboratory conditions. Int J Agric Biol 13:961–965

Bagchi D, Stohs SJ, Downs BW, Bagchi M, Preuss HG (2002) Cytotoxicity and oxidative mechanisms of different forms of chromium. Toxicology 180:5–22

Barcelo J, Poschenriender C, Ruano A, Gunse B (1985) Leaf water potential in Cr(VI) treated bean plants (Phaseolus vulgaris L). Plant Physiol Suppl 77:163–164

Barcelo J, Poschenrieder C, Gunse B (1986) Water relations of chromium VI treated bush bean plants (*Phaseolus vulgaris* L cv Contender) under both normal and water stress conditions. J Exp Bot 37:178–187

Bartlett RJ (1991) Chromium cycling in soils and water: links, gaps, and methods. Environ Health Perspect 92:17–24

Beukes JP, Dawson NF, Van Zyl PG (2010) Theoretical and practical aspects of Cr(VI) in the South African ferrochrome industry. J S Afr Inst Min Metall 110:743–750

Beveridge TJ, Murray RG (1980) Sites of metal deposition in the cell wall of Bacillus subtilis. J Bacteriol 141:876–887

Bianchi V, Celotti L, Lanfreanchi G, Majone F, Marin G, Di Montal, A, Sponza G, Tamino G, Venier P, Zantideschi A, Levis AG (1983) Genetic effects of chromium compounds. Mutat Res 117:279–300
Bozcaarmutlu A, Arinc E (2007) Effect of mercury, cadmium, nickel, chromium and zinc of kinetic properties of NADPH-cytochrome P450 reductase purified from leaping mullet (Lizasaliens). Toxicol in Vitro 21:408–416
Brum MC, Capitaneo JL, Oliveira JF (2010) Removal of hexavalent chromium from water by adsorption onto surfactant modified montmorillonite. Miner Eng 23:270–272
Cang L, Zhou DM, Alshawabkeh AN, Chen HF (2007) Effects of sodium hypochlorite and high pH buffer solution in electrokinetic soil treatment on soil chromium removal and the functional diversity of soil microbial community. J Hazard Mater 142:111–117
Cervantes C, Silver S (1992) Plasmid chromate resistance and chromate reduction. Plasmid 27:65–71
Cervantes C, Campos-García J, Devars S, Gutiérrez-Corona F, Loza-Tavera H, Torres-Guzmán JC, Moreno-Sánchez R (2001a) Interactions of chromium with microorganisms and plants. FEMS Microbiol Rev 25:335–347
Chakraborty R, Renu K, Eladl MA (2022) Mechanism of chromium-induced toxicity in lungs, liver, and kidney and their ameliorative agents. Biomed Pharmacother 151:113119
Chen JM, Hao OJ (1998) Microbial chromium(VI) reduction. Crit Rev Environ Sci Technol 28:219–251
Chen T, Chang QR, Liu J, Clevers PW, Kooistra L (2016) Identification of soil heavy metal sources and improvement in spatial mapping based on soil spectral information: a case study in northwest China. Sci Total Environ 565:155–164
Choppala G, Kunhikrishnan A, Seshadri B, Park JH, Bush R, Bolan N (2018) Comparative sorption of chromium species as influenced by pH, surface charge and organic matter content in contaminated soils. J Geochem Explor 184:255–260
Chrysochoou M, Theologou E, Bompoti N, Dermatas D, Panagiotakis I (2016) Occurrence, origin and transformation processes of geogenic chromium in soils and sediments. Curr Pollut Rep 2:224–235
Cocârţă DM, Neamţu S, Deac AR (2016) Carcinogenic risk evaluation for human health risk assessment from soils contaminated with heavy metals. Int J Environ Sci Technol 13:2025–2036
Costa M (1997) Toxicity and carcinogenicity of Cr(VI) in animal models and humans. Crit Rev Toxicol 27:431
Costa M (2003) Potential hazards of hexavalent chromate in our drinking water. Toxicol Appl Pharmacol 188:1–5
Das J, Sarkar A, Sil PC (2015) Hexavalent chromium induces apoptosis in human liver (HepG2) cells via redox imbalance. Toxicol Rep 2:600–608
De flora S, Bagnasco M, Serra D, Zanacchi P (1990) Genotoxicity of chromium compounds—A review. Mutat Res 238:99–172
Dhal B, Thatoi HN, Das NN, Pandey BD (2013) Chemical and microbial remediation of hexavalent chromium from contaminated soil and mining/metallurgical solid waste: a review. J Hazard Mater 250–251:272–291
Donaldson RM, Barreras RF (1966) Intestinal absorption of trace quantities of chromium. J Lab Clin Med 68:484–493
Dubey R, Bajpai J, Bajpai AK (2015) Green synthesis of graphene sand composite (GSC) as novel adsorbent for efficient removal of Cr(VI) ions from aqueous solution. J Water Process Eng 5:83
Faust BC, Hoigné J (1990) Photolysis of Fe(III)-hydroxy complexes as sources of OH radicals in clouds, fog and rain. Atmos Environ Part A 24:79–89
Ferronato N, Torretta V (2019) Waste mismanagement in developing countries: a review of global issues. Int J Environ Res Public Health 16:1060
Gattullo CE, Allegretta I, Porfido C, Rascio I, Spagnuolo M, Terzano R (2020) Assessing chromium pollution and natural stabilization processes in agricultural soils by bulk and micro X-ray analyses. Environ Sci Pollut Res 27:22967–22979

Gorny J, Billon G, Noiriel C, Dumoulin D, Lesven L, Madé B (2016) Chromium behavior in aquatic environments: a review. Environ Rev 24:503–516

Graham MC, Farmer JG, Anderson P, Paterson E, Hillier S, Lumsdon DG, Bewley RJF (2006) Calcium polysulfide remediation of hexavalent chromium contamination from chromite ore processing residue. Sci Total Environ 364:32–44

Han X, Wong YS, Wong MH, Tam NF (2007) Biosorption and bioreduction of Cr(VI) by a microalgal isolate, Chlorella miniate. J Hazard Mater146:65

Han S, Zang Y, Gao Y, Yue Q, Zhang P, Kong W, Jin B, Xu X, Gao B (2020) Co-monomer polymer anion exchange resin for removing Cr(VI) contaminants: adsorption kinetics, mechanism and performance. Sci Total Environ 709:136002. https://doi.org/10.1016/j.scitotenv.2019.136002

Hanus J, Tomas J (1993) An investigation of chromium content and its uptake from soil in white mustard. Acta Fytotech. 48:39–47

Hara T, Hoshuyama T, Takahashi K, Delgermaa V, Sorahan T (2010) Cancer risk among Japanese chromium platers, 1976–2003. Scand J Work Environ Health 36(3):216–221

Harrison-Brown M, Scholes C, Field C, McQuilty R, Farah SB, Nizam I, Kerr D, Kohan L (2020) Limited penetration of cobalt and chromium ions into the cerebrospinal fluid following metal on metal arthroplasty: a cross-sectional analysis. Clin Toxicol 58:233–240

Haynes EN, Chen A, Ryan P, Succop P, Wright J, Dietrich KN (2011) Exposure to airborne metals and particulate matter and risk for youth adjudicated for criminal activity. Environ Res 111:1243–1248

Horie M, Nishio K, Endoh S, Kato H, Fujita K, Miyauchi A, Nakamura A, Kinugasa S, Yamamoto K, Niki E (2013) Chromium (III) oxide nanoparticles induced remarkable oxidative stress and apoptosis on culture cells. Environ Toxicol 28:61–75

Hou S, Wu B, Peng D, Wang Z, Wang Y, Xu H (2019) Remediation performance and mechanism of hexavalent chromium in alkaline soil using multi-layer loaded nanozero-valent iron. Environ Pollut 252:553–561

James BR (2007) Chromium. In: Encyclopedia of water science. CRC Press, Boca Raton, FL, USA, pp 105–109

Jardine PM, Stewart MA, Barnett MO, Basta NT, Brooks SC, Fendorf S, Mehlhorn TL (2013) Influence of soil geochemical and physical properties on chromium(VI) sorption and bioaccessibility. Environ Sci Technol 47:11241–11248

Jayasingh J, Philip L (2005) Bioremediation of chromium contaminated soil: optimization of operating parameters under laboratory condition. J Hazard Mater B 118:113–120

Kalidhasana S, Santhana Krishna Kumara A, Rajeshb V, Rajesha N (2016) The journey traversed in the remediation of hexavalent chromium and the road ahead toward greener alternatives—A perspective. Coord Chem Rev 317:157–166

Kamaludeen SPB, Megharaj M, Juhasz AL, Sethunathan N, Naidu R (2003) Chromium-microorganism interactions in soils: remediation implications. Rev Environ Contam Toxicol 178:93–164

Kassaye G, Gabbiye N, Alemu A (2017) Phytoremediation of chromium from tannery wastewater using local plant species. Water Pract Technol 12:894–901

Katsoyiannis IA, Xanthopoulou M, Zouboulis AI (2020) Cr(VI) removal from ground waters by ferrous iron redox-assisted coagulation in a continuous treatment unit comprising a plug flow pipe reactor and downflow sand filtration. Appl Sci 10:802

Kimbrough DE, Cohen Y, Winer AM, Creelman L, Mabuni C (1999) A critical assessment of chromium in the environment. Crit Rev Environ Sci Technol 29:1–46

Koriche Y, Darder M, Aranda P, Semsari S, Hitzky ER (2014) Bionanocomposites based on layered silicates and cationic starch as eco-friendly adsorbents for hexavalent chromium removal. Dalton Trans 43:10512

Kozlowski H, Kolkowska P, Watly J, Krzywoszynska K, Potocki S (2014) General aspects of metal toxicity. Curr Med Chem 21:3721–3740

Kumar R, Bishnoi NR, Bishnoi GK (2008) Biosorption of chromium(VI) from aqueous solution and electroplating wastewater using fungal biomass. Chem Eng J 135:202

Landrot G, Ginder-Vogel M, Sparks DL (2009) Kinetics of chromium (III) oxidation by manganese (IV) oxides using quick scanning X-ray absorption fine structure spectroscopy (Q-XAFS). Environ Sci Technol 44:143–149

Ma L, Xu J, Chen N, Li M, Feng C (2019) Microbial reduction fate of chromium(Cr) in aqueous solution by mixed bacterial consortium. Ecotoxicol Environ Saf 170:763–770

Martin S, Griswold W (2009) Human health effects of heavy metals. Environ Sci Technol Briefs Citizens 15:1–6

McClain CN, Fendorf S, Webb SM, Maher K (2016) Quantifying Cr(VI) production and export from serpentine soil of the California coast range. Environ Sci Technol 51:141–149

McGrath SP, Smith S (1990) Chromium and nickel. In: Alloway BJ (ed) Heavy metals in soils, Glasgow, Balckie, pp 125–150

Mei B, Puryear JD, Newton RJ (2002) Assessment of Cr tolerance and accumulation in selected plant species. Plant Soil 247:223–231

Menezes MÂBC, Sabino CVS, Franco MB, Maia ECP, Albinati CCB (2004) Assessment of workers' contamination caused by air pollution exposure in industry using biomonitors. J Atmos Chem 49:403

Mishra S, Bharagava RN (2016) Toxic and genotoxic effects of hexavalent chromium in environment and its bioremediation strategies. J Environ Sci Health Part C 34(1):1–32

Moffat I, Martinova N, Eidel C, Thompson CM (2018) Hexavalent chromium in drinking water. J Am Water Works Assn 110:E22–E35

Muraviev D (1998) Surface impregnated sulfonate ion exchangers: preparation, properties and application. Solvent Extr Ion Exch 16:381

Narayani M, Shetty KV (2013) Chromium-resistant bacteria and their environmental condition for hexavalent chromium removal: a review. Crit Rev Env Sci Technol 43:955–1009

Nies DH, Silver S (1995) Ion efflux system involved in bacterial metal resistances. J Indus Microbiol 14:186–199

Nimse SB, Pal D (2015) Free radicals, natural antioxidants, and their reaction mechanisms. RSC Adv 5:27986–28006

Nordberg M, Cherian GM (2013) Chapter 9, Biological response of elements. In: Essentials of medical geology, Revised edn. Springer Science & Business Media, Dordrecht, The Netherlands, pp 195–213

Nriagu JO, Pacyna JM (1988) Quantitative assessment of worldwide contamination of air, water and soils by trace metals. Nature 333:134–139

Oze C, Bird DK, Fendorf S (2007) Genesis of hexavalent chromium from natural sources in soil and groundwater. Proc Natl Acad Sci U S A 104:6544–6549

Pandey N, Shukla SK, Singh NB (2017) Water purification by polymer nanocomposites: an overview. Nanocomposites 3:47–66

Parent ME, Turner MC, Lavoúe J, Richard H, Figuerola J, Kincl L, Richardson L, Benke G, Blettner M, Fleming S et al (2017) Lifetime occupational exposure to metals and welding fumes, and risk of glioma: a 7-country population-based case-control study. Environ Health 16(90)

Pedreno NJI, Gomez R, Moral G, Palacios J, Mataix J (1997) Heavy metals and plant nutrition and development. Recent Res Dev Phytochem 11:73–79

Peralta JR, Torresdey G, Tiemann JL, Gomez E, Arteaga S, Rascon E et al (2001) Uptake and effects of five heavy metals on seed germination and plant growth in alfalfa (*Medicago sativa* L.). Bull Environ Contam Toxicol 66:727–734

Pettine M, Campanella L, Millero FJ (2002) Reduction of hexavalent chromium by H_2O_2 in acidic solutions. Environ Sci Technol 36:901–907

Plummer S, Gorman C, Henrie T, Shimabuku K, Thompson R, Seidel C (2018) Optimization of strong-base anion exchange O&M costs for hexavalent chromium treatment. Water Res 139:420–433

Prasad MNV, Greger M, Landberg T (2001) Acacia nilotica L bark removes toxic elements from solution: corroboration from toxicity bioassay using Salix viminalis L in hydroponic system. Int J Phytoremed 3:289–300

Rafati L, Mahvi AH, Asgari AR, Hosseini SS (2010) Removal of chromium(VI) from aqueous solutions using Lewatit FO36 nano ion exchange resin. Int J Environ Sci Tech 7:147–156
Rai D, Eary LE, Zachara JM (1989) Environmental chemistry of chromium. Sci Total Environ 86:15–23
Rajan MT, Rao KSJ, Mamatha BM, Rao RV, Shanmugavelu P, Menon RB, Pavithran MV (1997) Quantification of trace elements in normal human brain by inductively coupled plasma atomic emission spectrometry. J Neurol Sci 146:153–166
Rajapaksha AU, Vithanage M, Ok YS, Oze C (2013) Cr(VI) formation related to Cr(III)-Muscovite and Birnessite interactions in ultramafic environments. Environ Sci Technol 47:9722–9729
Rajesh N, Kumar ASK, Kalidhasan S, Rajesh V (2011) Trialkylamine impregnated macroporous polymeric sorbent for the effective removal of chromium from industrial wastewater. J Chem Eng Data 56:2295
Redha AA (2020) Removal of heavy metals from aqueous media by biosorption. Arab J Basic Appl Sci 27:183–193
Rengaraj S, Joo CK, Kim Y, Yi J (2003) Kinetics of removal of chromium from water and electronic process wastewater by ion exchange resins: 1200H, 1500H and IRN97H. J Hazard Mater 102:257
Richard FC, Bourg ACM (1991) Aqueous geochemistry of chromium: a review. Water Res 25:807–816
Rudolf E, Cervinka M, Cerman J, Schroterova L (2005) Hexavalent chromium disrupts the actin cytoskeleton and induces mitochondria-dependent apoptosis in human dermal fibroblasts. Toxicol in Vitro 19:713–723
Saha R, Nandi R, Saha B (2011) Sources and toxicity of hexavalent chromium. J Coord Chem 64:1782–1806
Salem FY, Parkerton TF, Lewis RV, Huang JH, Dickson KL (1989) Kinetics of chromium transformations in the environment. Sci Total Environ 86:25–41
Saravanan A, Jayasree R, Hemavathy RV, Jeevanantham S, Hamsini S, Kumar PS, Yaashikaa PR, Manivasagan V, Yuvaraj D (2019) Phytoremediation of Cr(VI) ion contaminated soil using Black gram (*Vigna mungo*): assessment of removal capacity. J Environ Chem Eng 7:103052
Sarkar S, Dutta S, Bairi P, Pal T (2014) Redox-responsive copper(I) metallogel: a metal-organic hybrid sorbent for reductive removal of chromium(VI) from aqueous solution. Langmuir 30:7833
Sethunathan N, Megharaj M, Smith L, Kamaludeen SPB, Avudainayagam SR, Naidu R (2005) Microbial role in the failure of natural attenuation of chromium(VI) in long-term tannery waste contaminated soil. Agric Ecosyst Environ 105:657–661
Shahandeh H, Hossner L (2000) Plant screening for chromium phytoremediation. Int J Phytoremediation 2:31–51
Shanker A, Cervantes K, Loza-Taverac CH, Avudainayagam S (2005) Chromium toxicity in plants. Environ Int 31:739–753
Sharma SK, Petrusevski B, Amy G (2008) Chromium removal from water: a review. J Water Supply Res Technol Aqua 57:541–553
Sharma P, Singh SP, Parakh SK, Tong YW (2022) Health hazards of hexavalent chromium (Cr(VI)) and its microbial reduction. Bioengineered 13:4923–4938
Shekhawat K, Chatterjee S, Joshi B (2015) Chromium toxicity and its health hazards. Int J Adv Res 3:167–172
Shelnutt SR, Goad P, Belsito DV (2007) Dermatological Toxicity of Hexavalent Chromium. Crit Rev Toxicol 37:375–387
Shi L, Zhang X, Chen Z (2011) Removal of chromium(VI) from wastewater using bentonite-supported nanoscale zero-valent iron. Water Res 45:886–892
Singh AK (2001) Effect of trivalent and hexavalent chromium on spinach (*Spinacea oleracea L.*). Environ Ecol 19:807–810
Singh HP, Mahajan P, Kaur S, Batish DR, Kohli RK (2013) Chromium toxicity and tolerance in plants. Environ Chem Lett 11:229–254
Sneddon C (2012) Chromium and its negative effects on the environment. Case Study. Department of Earth Sciences. Montana State University

Son YO, Hitron JA, Wang X, Chang Q, Pan J, Zhang Z, Liu J, Wang S, Lee JC, Shi X (2010) Cr(VI) induces mitochondrial mediated and caspase-dependent apoptosis through reactive oxygen species-mediated p53 activation in JB6 Cl41 cells. Toxicol Appl Pharmacol 245:226–235

Steritt RM, Brown MJ, Lester JN (1981) Metal removal by adsorption and precipitation in the activated sludge process. Environ Pollut 24:313

Subrahmanyam D (2008) Effects of chromium toxicity on leaf photosynthetic characteristics and oxidative changes in wheat (*Triticum aestivum L.*). Photosynthetica 46:339–345

Sugiyama M (1992) Role of physiological antioxidants in Cr(VI)-induced cellular injury. Free Radic Biol Med 12:397–407

Sundaramoorthy P, Chidambaram A, Ganesh KS, Unnikannan P, Baskaran L (2010) Chromium stress in paddy: (i) nutrient status of paddy under chromium stress; (ii) phytoremediation of chromium by aquatic and terrestrial weeds. C R Biol 333:597–607

Tan H, Wang C, Zeng G, Luo Y, Li H, Xu H (2020) Bioreduction and biosorption of Cr(VI) by a novel Bacillus sp. CRB-B1 strain. J Hazard Mater 386:121628

Taufikurahman T, Pradisa MAS, Amalia SG, Hutahaean GEM (2019) Phytoremediation of chromium(Cr) using *Typha angustifolia* L., *Canna indica* L. and *Hydrocotyle umbellata* L. in surface flow system of constructed wetland. IOP Conf Ser Earth Environ Sci 308:12020

Tripathi AK, Tripathi S (1999) Changes in some physiological and biochemical characters in *Albizia lebbek* as bio-indicators of heavy metal toxicity. J Environ Biol 20:93–98

Tumolo M, Ancona V, De Paola D, Losacco D et al (2020) Chromium pollution in European water, sources, health risk, and remediation strategies: an overview. Int J Environ Res Public Health 17:5438

U.S. Department of health and human Services, Toxicological Profile for Chromium (2012)

Ukhurebor KE, Aigbe UO, Onyancha RB, Nwankwo W, Osibote OA (2021) Effect of hexavalent chromium on the environment and removal techniques: a review. J Environ Manag 280:111809

Unceta N, Séby F, Malherbe J, Donard OFX (2010) Chromium speciation in solid matrices and regulation: a review. Anal Bioanal Chem 397:97–1111

Vaiopoulou E, Gikas P (2020) Regulations for chromium emissions to the aquatic environment in Europe and elsewhere. Chemosphere 254:126876

Velma V, Tchounwou PB (2009) Hexavalent chromium-induced multiple biomarker responses in liver and kidney of Gold fish *Carassius auratus*. Environ Toxicol 26:649–656

Volesky B, Holan Z (1995) Biosorption of heavy metals. Biotechnol Prog 11:235–250

Wakeel A, Xu M, Gan Y (2020) Chromium-induced reactive oxygen species accumulation by altering the enzymatic antioxidant system and associated cytotoxic, genotoxic, ultrastructural, and photosynthetic changes in plants. Int J Mol Sci 21:728

Wang XF, Lou XM, Shen Y, Xing ML, Xu LH (2010a) Apoptotic-related protein changes induced by hexavalent chromium in mice liver. Environ Toxicol Int J 25(1):77–82

Wang Q, Qian H, Yang Y, Zhang Z, Naman C, Xu X (2010b) Reduction of hexavalent chromium by carboxymethyl cellulose-stabilized zerovalent iron nanoparticles. J Contam Hydrol 114:35–42

Wise JP Jr, Young JL, Cai J, Cai L (2022) Current understanding of hexavalent chromium [Cr(VI)] neurotoxicity and new perspectives. Environ Int 158:106877

Wittbrodt PR, Palmer CD (1995) Reduction of Cr(VI) in the presence of excess soil fulvic acid. Environ Sci Technol 29:255–263

Xia S, Song Z, Jeyakumar P, Shaheen SM, Rinklebe J, Ok YS, Bolan N, Wang H (2019) A critical review on bioremediation technologies for Cr(VI)-contaminated soils and wastewater. Critic Rev Environ Sci Technol 49:1027–1078

Xu J, Zhao M, Pei L, Zhang R, Liu X, Wei L, Yang M, Xu Q (2018) Oxidative stress and DNA damage in a long-term hexavalent chromium-exposed population in North China: a cross-sectional study. BMJ Open 8:e021470

Xueting L, Rehman MU, Mehmood K, Huang S, Tian X, Wu X, Zhou D (2018) Ameliorative effects of nano-elemental selenium against hexavalent chromium-induced apoptosis in broiler liver. Environ Sci Pollut Res 25:15609–15615

Yan B, Chen Z (2019) Influence of pH on Cr(VI) reduction by organic reducing substances from sugarcane molasses. Appl Water Sci 9:61

Yılmaz AB, Turan C, Toker T (2010) Uptake and distribution of hexavalent chromium in tissues (gill, skin and muscle) of a freshwater fish, Tilapia, *Oreochromis aureus*. J Environ Chem Ecotoxicol 2:28–33

Yorbik Ö, Kurt İ, Haşimi A, Öztürk Ö (2010) Chromium, cadmium, and lead levels in urine of children with autism and typically developing controls. Biol Trace Elem Res 135:10–15

Zeid IM (2001) Responses of Phaseolus vulgaris to chromium and cobalt treatments. Biol Plant 44:111–115

Zhitkovich A (2005) Importance of chromium-DNA adducts in mutagenicity and toxicity of chromium(VI). Chem Res Toxicol 1:3–11

Zhong L, Yang J (2011) Reduction of Cr(VI) by malic acid in aqueous Fe-rich soil suspensions. Chemosphere 86:973–978

Zou J, Wang M, Jiang W, Liu D (2006) Chromium accumulation and its effects on other mineral elements in *Amaranthus viridis L.* Acta Biol Crac Ser Bot 48:7–12

Chapter 10
Health Risk Linked to Cr Toxicity in Food and Environment

Abhishek Pathak, Varun Asediya, Pranav Anjaria, and Satya Pal Singh

Abstract The increasing prevalence of exposure to high concentrations of chromium in food and the environment has made chromium toxicity a pressing public health issue. This review provides an overview of chromium toxicology, sources of exposure, and strategies for mitigating the associated health risks and consequences. Hexavalent chromium, a toxic form of the metal, has been causally linked to numerous negative health outcomes, including carcinogenicity, genotoxicity, oxidative stress, and inflammation. Exposure to hexavalent chromium primarily occurs through contaminated food and the environment, especially in areas close to industrial sites. Food can become contaminated through soil, water, air, and the use of chromium in food packaging and processing equipment. The highest chromium concentrations have been detected in certain food items such as grains, spices, and meat products. To minimize exposure to chromium toxicity, it is crucial to be aware of the sources of chromium in one's diet and environment. Opting for chromium-free food packaging and limiting consumption of foods with high chromium concentrations can reduce exposure. The government and industry must also take steps to reduce chromium release into the environment and ensure the safety of food and water supplies. In conclusion, this review highlights the significance of comprehending the health risks posed by chromium toxicity and taking necessary precautions to reduce exposure. Further research is necessary to gain a more comprehensive understanding of the toxic effects of chromium and develop effective prevention strategies.

A. Pathak (✉)
Department of Veterinary Pharmacology and Toxicology, Apollo College of Veterinary Medicine, Jaipur, Rajasthan, India
e-mail: drpathak1704@gmail.com

V. Asediya
Animal Nutrition Research Station, AAU, Anand, Gujarat, India

P. Anjaria
Department of Veterinary Public Health and Epidemiology, C. V. Sc & A.H, Kamdhenu University, Anand, Gujarat, India

S. P. Singh
Department of Veterinary Pharmacology and Toxicology, G. B. Pant University of Agriculture and Technology, Pantnagar, Uttarakhand, India

© The Author(s), under exclusive license to Springer Nature Switzerland AG 2023
N. Kumar et al. (eds.), *Chromium in Plants and Environment*, Environmental Science and Engineering, https://doi.org/10.1007/978-3-031-44029-8_10

Keywords Chromium toxicity · Environmental toxicant · Genotoxicity · Carcinogenicity · Adsorption and bioremediation

10.1 Introduction

The overexploitation of Earth's natural resources has resulted in the depletion of these finite resources, and the presence of environmental pollutants in the air, soil, water, and food chain is a significant global concern. These pollutants impact ecosystem functions and decrease food and environmental safety. With ongoing urbanization, driven by population migration to urban areas, the discharge of untreated industrial waste into the environment is expected to increase, exacerbating the presence of hazardous inorganic and organic contaminants in aquatic ecosystems, including rivers, lakes, estuaries, and coastal regions.

Humans are exposed to toxic metals, including Chromium(VI) compounds, through various pathways, including skin absorption, ingestion of contaminated soil, or inhalation of air particles. These metals can accumulate in both plants and animals, entering the food chain and posing potential health risks for humans. Approximately 90% of human exposure to hazardous metals is through contaminated food, accounting for 30% of all human cancers globally. Chromium, element 24 on the periodic table, was discovered in 1797 by French scientist Louis Vauquelin and its name is derived from the Greek word "chroma," referring to the distinctive hues present in its compounds. Chromium is a hard, glossy, steely-grey metal with an atomic weight of approximately 52 g/mol and a high melting point, as well as reflective qualities.

The potential health hazards of Chromium (VI) compounds in urban areas have been recognized since the first report of cancer in workers exposed to Chromium over 130 years ago. Subsequent epidemiological studies, particularly in occupational settings, have established a correlation between exposure to Chromium (VI) and specific disease phenotypes, such as bronchogenic carcinoma. Recent studies have implicated Chromium in the development of stomach cancer through various modes of exposure and mechanisms, such as genomic instability or epigenetic alterations, and have associated it with a range of health problems, including respiratory, hepatic, renal, reproductive, and neurological issues. It is important to note that the relationship between human susceptibility and health problems caused by exposure to toxic agents is dependent on individual factors, rather than direct.

10.2 Physiochemical Properties

Chromium, which has an atomic number of 24 and a relative atomic mass of 51.996, exhibits multiple oxidation states, with the most prevalent being +2, +3, and +6. The instability of divalent chromium in air presents difficulties in comprehending its biological implications, necessitating the evaluation of the chemical and biological properties of its +3 and +6 oxidation states. Hexavalent chromium, in particular, is characterized by its strong oxidizing potential, requiring 1.33 eV of energy to oxidize trivalent chromium to hexavalent chromium. In biological systems, trivalent chromium remains unoxidized, while hexavalent chromium is reduced to an insoluble form upon entering red blood cell membranes. This reduction results in the incorporation of trivalent chromium into cellular components and its subsequent prevention from release from erythrocytes. The most common forms of chromium compounds and their respective applications are illustrated in Fig. 10.1. For more than a century, chromium has been utilized extensively in various industries, primarily in the form of alloys and soluble salts that contain hexavalent chromium ions. Exposure to excessive levels of chromium, commonly encountered in industries such as chromate manufacturing, metal plating, alloy production, and metal welding and forming processes, has been linked to numerous hazardous effects on human health, including irritation, carcinogenicity, and allergenicity. The hexavalent chromium ion has been identified as the primary culprit of these adverse health effects, while trivalent chromium compounds, widely used in certain industrial processes, have only been acknowledged as irritants, but not carcinogens or allergens. The incidence and prevention of dermatitis, asthma, and cancer resulting from chromium exposure have been widely studied.

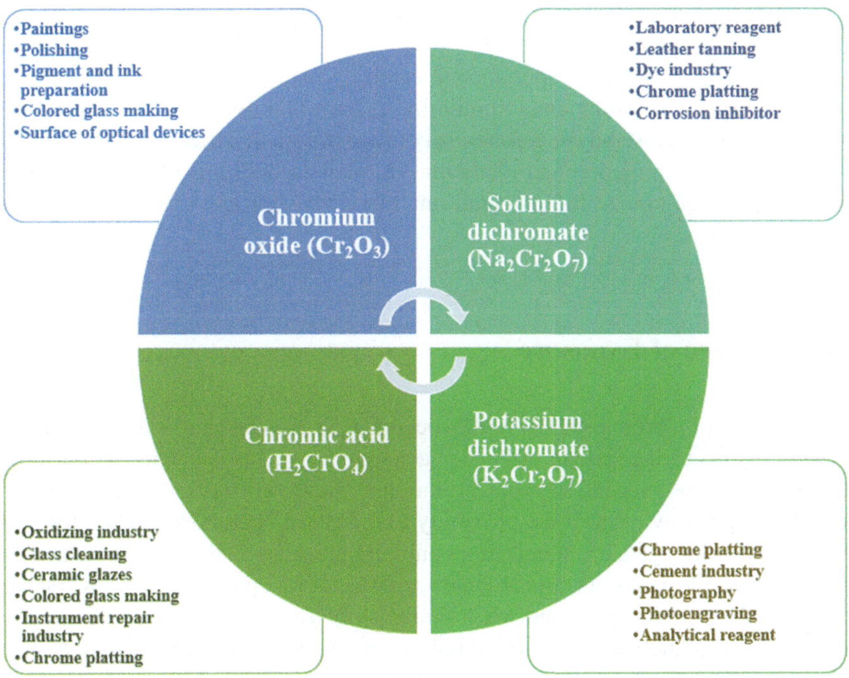

Fig. 10.1 Common forms of chromium and their uses

10.3 Geology of Chromium

Chromium is a transition metal that is widely distributed in the Earth's crust and is found in a variety of mineral forms. Out of the 82 documented mineral species, 23 contain chromium in the hexavalent state (Cr[VI]), which is of great scientific and industrial interest due to its unique properties. The most abundant chromium mineral is chromite ($FeCr_2O_4$), which is a complex oxide mineral composed of iron, magnesium, aluminum, and chromium in both the trivalent (Cr[III]) and tetravalent (Cr[IV]) states. The chromium content of chromite ranges from 45 to 80% and is the primary source of chromium metal, which is widely used in the production of stainless steel and other alloys. Chromite is also used as a refractory material, due to its high melting point and good thermal stability.

Chromite ore is typically formed through the intrusion and solidification of molten magma, and is typically associated with ultramafic rocks such as peridotites, pyroxenites, and dunites. Chromite deposits can be found in the form of stratiform or podiform deposits, and can contain up to 400 parts per million (ppm) of chromium. Although the occurrence of Cr[VI] in nature is limited, it can be produced through the oxidation of Cr[III] in chromite through interaction with birnessite, a manganese oxide mineral. The presence of Cr[VI] in the environment can pose a risk to human health and the environment, as it is a strong oxidizing agent and has been linked to respiratory problems, skin irritation, cancer, and changes in the composition and function of soil microflora.

Therefore, it is of great importance to carefully manage and control the release of Cr[VI] into the environment, to minimize its harmful effects. Further research is needed to fully understand the geochemistry of chromium minerals and the processes involved in their formation and alteration, in order to optimize the production of chromium and minimize the potential environmental impacts associated with its extraction and use.

10.4 Source of Exposure

The presence of Chromium (III) in the diets of both animals and humans is due to its naturally occurring state in soil and vegetation, as well as its inclusion in animal feed. However, the concentration of Chromium(Cr) may vary based on several factors such as the geology of the area, proximity to industrial operations, and its form of availability. Although Chromium compounds are prevalent in the environment, they are recognized as persistent contaminants as a result of their extensive industrial use during the 21st century. Exposure to Chromium(VI) in humans and animals primarily occurs via inhalation of airborne emissions, including industrial and automotive fumes, cigarette smoke, and other sources. Contamination of water is another avenue of exposure and results from industrial releases and the leaching of wastewater, often due to inadequate waste management practices (Welling et al. 2015). This mode of exposure is particularly concerning as it affects the entire population through oral consumption of contaminated water, unlike inhalation, which is typically limited to occupational exposure.

Under normal environmental conditions, Chromium(III) exhibits low solubility in water within a pH range of 6–9. The oxidation of Chromium(III) to the hexavalent state (Cr-VI) is often facilitated by catalytic reactions with Mn-bearing minerals, while reduction of hexavalent chromium in the environment is primarily driven by Fe(II) in solution or Fe(II)-bearing minerals, sulfides, and organic matter (Loyaux-Lawniczak et al. 2001). The presence of hexavalent chromium (Cr-VI) in the environment is primarily a result of human activities. The largest contributors to anthropogenic Cr contamination are cement manufacturing and handling practices during construction, accounting for 70% of steel production. The widespread use of chromium sulfate in the tanning industry, with nearly 90% of all leather produced globally containing the compound, is another significant contributor. Occupational exposure to Cr is substantial among workers in over 80 industries, including, but not limited to, welding, electroplating, metal finishing, wood preservation, and the production of textiles, dyes, and pigments (IARC 1990). Human activities result in the release of approximately 75,000 tons of chromium into the environment, of which 33% is the toxic form of Cr-VI, while natural sources account for the remaining 54,000 tons (Kieber et al. 2002). Research has shown that precipitation can effectively remove Cr from the atmosphere due to its high solubility and mobility, leading to leaching into surface water from soil. The estimated atmospheric residence time of Cr is 10 days, and the transmission of Cr through the food chain is depicted in Fig. 10.2.

Approximately 12 billion tons of shipping-grade chromite, a significant portion of the global reserves, can be found in several key locations. The largest concentration of these resources is located in South Africa, accounting for 84% of the total, followed by Zimbabwe with 6%, Kazakhstan with 5%, and India with 2%. The remaining 3% is contributed by countries including Brazil, the United States, Canada, Russia, Finland, and others. India is the world's second-largest producer of chromite ore, further emphasizing the significance of its chromite reserves (Das and Singh 2011).

10.5 Pharmacokinetics

Chromium (Cr) exists in two stable oxidation states in nature, trivalent chromium (Cr III) and hexavalent chromium (Cr VI) (IARC 1990). The toxicity resulting from exposure to chromium is dependent on various factors, including the chemical form, oxidation state, solubility, and bioavailability of the relevant chromium compounds. Chromium compounds can alter their oxidation state through alkylation and interact with biological molecules upon ingestion. The physiological consequences of exposure to chromium, especially Cr VI, have been shown to impact multiple systems within the human body. This includes the skin, where exposure can lead to allergic reactions, skin rashes, dryness, and swelling (IARC 1990), the respiratory tract,

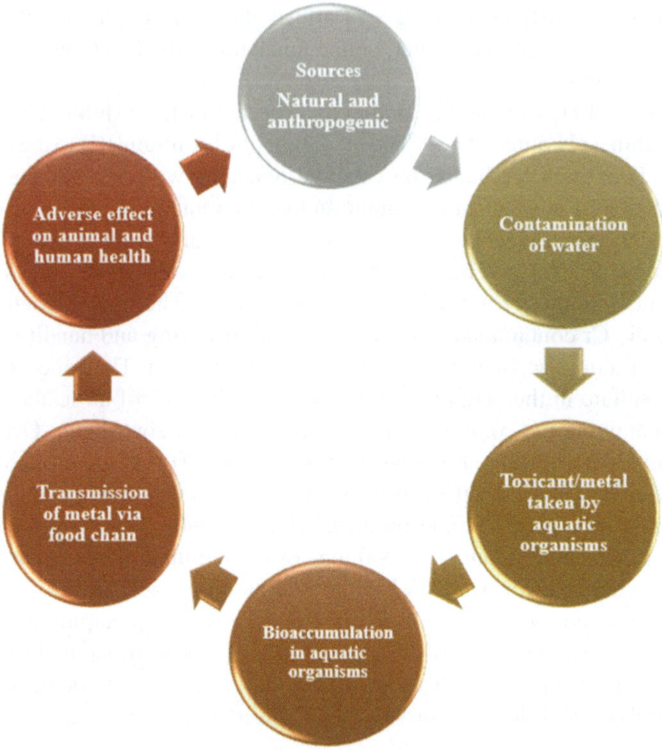

Fig. 10.2 Transmission of chromium via the food chain

which can experience ulceration and perforation of the nasal septum, irritation of the pharynx and larynx, asthmatic bronchitis, nose irritations, nosebleeds, and even lung cancer (IARC 1990), and the gastrointestinal system, causing upset stomachs and ulcers, as well as the renal system (IARC 1990). The mechanisms of Cr III entry into cells have been partially characterized, including slow processes such as passive diffusion or phagocytosis (IARC 1990). Although limited, our understanding of these processes suggests that Cr III exhibits low membrane permeability, is capable of accumulating within cells to levels hundreds of times greater than its extracellular concentration, and therefore presents a high chemical potential for intracellular reactions (Fang et al. 2014). Chromium III transport across the cell membrane may be facilitated by transferrin (Tf), a globulin and main iron III transport protein in the bloodstream, which has been demonstrated to transport various metals, including Cr III, due to its similarities in size and charge to the ferric ion. However, *In-vitro* studies using HepG2 (human hepatoma) cells, which are rich in Tf receptors, suggest that Tf functions as a natural defense against the toxicity and genotoxicity of ingested Cr III by inhibiting its cellular accumulation (Fang et al. 2014). Interferometry studies have indicated that the mechanism of exclusion and efflux of Cr III from cells is due to a lower affinity of cell receptors for Cr III transferrin (Tf) compared to iron III Tf, which is a critical step in the Tf cycle (Edwards et al. 2020). The debate continues regarding the adequacy or slow rate of release of Cr III from serum Tf to have a physiologically relevant effect. Over 38 variants of the highly polymorphic Tf gene have been identified. Most studies have suggested that Cr VI cannot interact directly with DNA and does not exhibit genotoxic effects under conditions that do not favor reduction (Wakeman et al. 2017). The "uptake-reduction model," which describes the intracellular reduction and generation of intermediate species, including Cr V and Cr IV, is considered essential in understanding the pathological effects of this element (DeLoughery et al. 2014).

10.6 Mechanism of Toxicity

Extensive research on hexavalent chromium (Cr[VI]) has established that its toxicity surpasses that of trivalent chromium (Cr[III]). The superior toxicity of Cr[VI] can be attributed to its enhanced permeability, enabling its cellular uptake and eventual reduction to Cr[III]. Following intracellular reduction, the occurrence of short-lived pentavalent and tetravalent chromium species is documented, which exhibit dissimilar binding tendencies to cellular components as compared to Cr[III].

The pentavalent form of chromium is stabilized by glutathione and is considered a detoxification process if its reduction occurs away from the site of toxicity or genotoxicity. However, if the reduction takes place within or near the cell nucleus, it may activate Cr and contribute to the initial binding of cellular macromolecules. The intracellular reduction of Cr[VI] is largely dependent on the presence of glutathione and cysteine, although ascorbic acid, microsomes in the presence of NAD/NADH, microsomal cytochrome P450, mitochondria, hemoglobin, and glutathione reductase may also play a role. The mechanism of toxicity of chromium is depicted in Fig. 10.3.

The chromium requirement for humans and animals remains a subject of debate, with some sources positing that it is an essential trace element for insulin-mediated glucose metabolism, while others question its biological or toxicological effects. Despite its poor permeability across cell membranes, Cr[III] is capable of interacting with biomolecules such as DNA. However, at present, there is a lack of conclusive evidence demonstrating the essentiality of Cr in biomolecular or physiological processes (Fig. 10.4).

The transport of Cr[VI] compounds, such as chromate anions, into cells is facilitated through protein channels, driven by both passive diffusion and saturable transport. The anion exchanger 1 (AE1) is the primary carrier responsible for transporting chromate, sulfate, and phosphate ions. In contrast, other proteins do not bind significantly to Cr[VI] under similar conditions. The entry of Cr[III] into the nucleus is hindered by the formation of kinetically inert octahedral complexes.

A minor portion of Cr[VI] may penetrate the nucleus through an active protein transport mechanism that involves histone sequences, lysine- or arginine-rich peptides, or nuclear proteins. Chromatin formation is favored by the association of histones with DNA polyanions, and the reversible binding through electrostatic

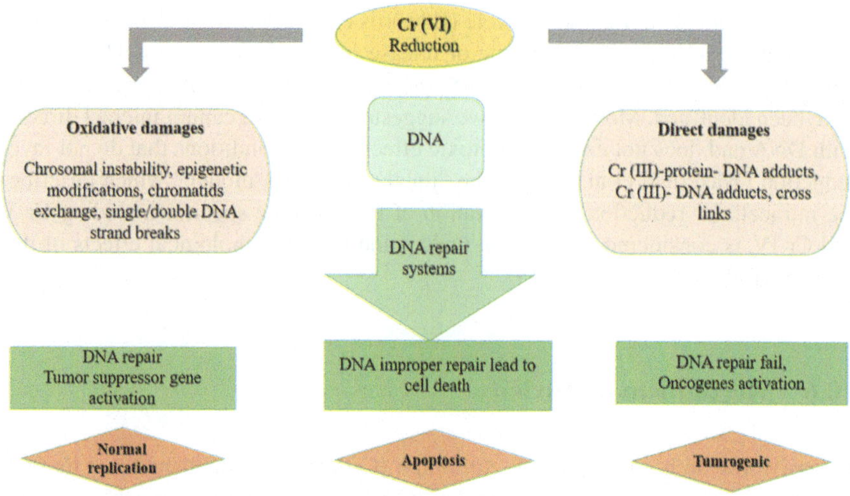

Fig. 10.3 Mechanism of action of chromium toxicity

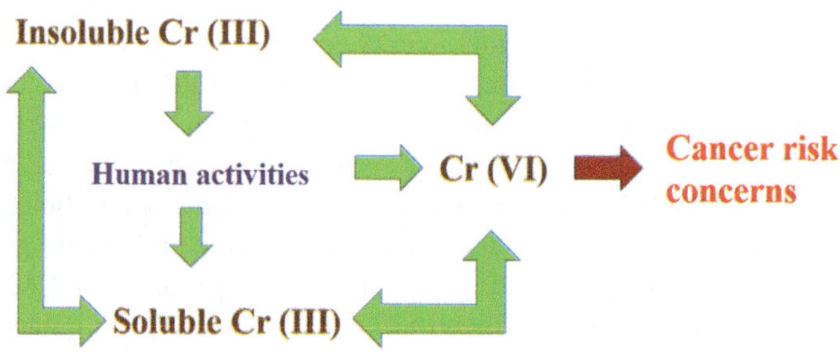

Fig. 10.4 Chromium in drinking water and cancer risk

attraction and hydrogen bonds may contribute to this process. The interaction of Cr[VI] with cytoplasmic elements may also impact the uptake of nuclear protein complexes and subsequent binding, leading to the generation of highly genotoxic Cr[III]-DNA adducts.

10.7 Health Risks Linked with Chromium Exposure in the Food

The bioavailability of chromium in dietary sources has been the focus of extensive research in recent decades. Evidence from studies has revealed that the extent to which chromium is absorbed from food sources is influenced by several factors, including the chemical form in which it exists, the processing and cooking techniques utilized, as well as the presence of other nutrients, such as ascorbic acid, which can augment its absorption. Furthermore, individual factors such as age, gender, and nutritional status may also impact the bioavailability of chromium from dietary sources. Investigations have established that trivalent chromium in the form of chromium III picolinate or chromium yeast demonstrates greater bioavailability compared to other forms of chromium, such as chromium chloride, which is characterized by low bioavailability. Food sources that are abundant in chromium III picolinate include whole grains, fruits, vegetables, and meats, while chromium yeast is commonly present in beer. The bioavailability of chromium from food sources is dependent on several factors, including the chemical form in which it exists, processing and cooking techniques, and the presence of other nutrients that may influence its absorption. As a result, it is crucial to consume a diverse diet that comprises of chromium-rich food sources to meet the daily recommended intake of chromium.

10.7.1 Absorption of Chromium in the Gut

The absorption of chromium is regulated by various proteins, such as the low molecular weight chromium-binding substance (LMWCr) and glucose tolerance factor (GTF), which enhance its uptake and utilization within the body. Renal excretion of chromium is limited, and its half-life in the body has been estimated to be approximately 20–30 days. The daily requirement of chromium is typically considered to be in the range of 20–35 μg, however, the precise requirement may vary based on factors such as age, sex, health status, and other dietary components. Chronic exposure to elevated levels of chromium, particularly hexavalent chromium (Cr(VI)), can result in toxicity and associated health problems such as skin irritation, respiratory issues, and genotoxicity. Hence, it is essential to maintain a balanced and appropriate intake of chromium to promote optimal health and wellness.

10.7.2 Bioavailability of Chromium in Food

Chromium, a trace mineral, plays a critical role in the regulation of glucose metabolism through its interaction with insulin. The extent to which chromium is absorbed and utilized by the body, known as bioavailability, is subject to a multitude of factors, including the chemical form of chromium in food, processing and preparation methods, and the presence of co-occurring minerals and nutrients. Studies have shown that processing methods, such as refining, milling, and cooking, can result in a substantial reduction of chromium content in food, with the degree of reduction differing among food types. For instance, refined grains, such as white flour and white rice, have been found to have lower chromium levels compared to their whole grain counterparts. The interaction of chromium with other minerals and nutrients also affects its bioavailability. The absorption of chromium may be hindered by high levels of iron, zinc, and calcium, while its uptake can be facilitated by the presence of ascorbic acid, glucose, and nicotinic acid. In conclusion, to attain sufficient chromium intake, it is recommended to consume a varied diet comprising unprocessed whole foods, and to limit the consumption of refined and processed foods. Furthermore, adequate ascorbic acid intake may enhance the bioavailability and utilization of chromium.

10.7.3 Dietary Sources of Chromium

Chromium is an essential trace mineral that has been demonstrated to play crucial roles in glucose and insulin metabolism, as well as lipid metabolism. The recommended daily intake of chromium for adults is estimated to be in the range of 20–35 μg, although this may vary based on various factors such as age, gender, and overall health status. Dietary sources of chromium include whole grains, fruits, vegetables, animal-derived products, and dairy products. Some foods that have been reported to contain higher concentrations of chromium include whole wheat, beer yeast, potatoes, dairy cheese, and leguminous plants such as beans, lentils, and peanuts. Of the various forms of chromium present in the diet, trivalent chromium (chromium III) is considered the most biologically active and is best absorbed by the body. This form of chromium is present in foods such as brewer's yeast, whole grain bread, and potatoes.

In contrast, hexavalent chromium (chromium VI) is a less biologically active form that is toxic in high doses, and is primarily found in industrial processes and can contaminate food and water sources. It is well established that the bioavailability of chromium from dietary sources can be influenced by a range of factors, including processing and preparation methods, the concurrent presence of other minerals and nutrients, and the chemical form of the chromium. To ensure adequate chromium intake, it is recommended to consume a varied diet that includes a range of chromium-rich foods.

10.7.4 Chromium Levels in Food and Drinking Water

Chromium is found in drinking water in both trivalent (Cr(III)) and hexavalent (Cr(VI)) forms. Cr(III) is considered an essential nutrient with no adverse effects on human health, while Cr(VI) has been identified as a highly toxic substance associated with an increased risk of cancer and other health problems (NTP, 2011). To protect public health, the US Environmental Protection Agency (EPA) has established a Maximum Contaminant Level (MCL) of 100 μg per liter (μg/L) for total chromium in drinking water (EPA 2019). Chronic exposure to elevated levels of Cr(VI) in food and drinking water can pose a significant threat to human health, with potential consequences including lung cancer, respiratory problems, skin irritation, and allergic reactions (NTP 2011). Long-term exposure to elevated levels of Cr(VI) has also been linked to kidney and liver damage, as well as an increased risk of cardiovascular disease (Schrauzer and Shrestha 2002). Adequate chromium intake, estimated to be between 50 and 200 μg per day, can be obtained through a well-balanced diet. Monitoring the levels of chromium in food and drinking water is crucial in ensuring their safety and mitigating potential health hazards.

10.7.5 Effects of Chromium in Food

The genotoxic potential of chromium represents a significant health concern as research has established that exposure to the element may lead to DNA damage both In vitro and In vivo, thus increasing the risk of neoplastic transformation (Akerstrom et al. 2013). Moreover, chromium has been implicated in inducing oxidative stress and inflammation, thereby contributing to the progression of chronic diseases, including cardiovascular disease and diabetes (Bhatnagar et al. 2007). The cardiovascular system is also susceptible to the impacts of chromium exposure, as evidenced by the findings of Kaur et al. (2010), which demonstrate an association between high chromium levels and elevated risk of cardiovascular disease, including coronary heart disease, myocardial infarction, and cerebrovascular incidents. Chromium has also been shown to disrupt lipid and glucose metabolism, thus increasing susceptibility to type 2 diabetes (Schneider et al. 2013).

The impact of chromium on the endocrine system has garnered significant attention in the scientific community, with research indicating that chromium can alter insulin sensitivity, potentially contributing to the development of type 2 diabetes (Schneider et al. 2013). Additionally, chromium has been shown to possess estrogenic properties, which can interfere with the normal functioning of the endocrine system and result in reproductive and developmental consequences (Bhatnagar et al. 2007).

10.7.5.1 Effect of Chromium on Liver

Chromium is a trace element that is essential for proper glucose and lipid metabolism. However, excessive exposure to chromium, particularly in its toxic form (chromium VI), can lead to serious health effects, including liver toxicity. Studies have shown that chronic exposure to elevated levels of chromium can cause liver injury, as evidenced by elevated levels of liver enzymes and histopathological alterations. The liver is a critical metabolic organ that is involved in the processing and elimination of toxic substances, and as a result, it is particularly susceptible to chromium toxicity. The mechanism by which chromium toxicity causes liver injury involves the induction of oxidative stress and inflammation.

Oxidative stress is a condition in which there is an imbalance between the production of free radicals and the ability of the body to neutralize them. This can result in cellular damage and contribute to liver injury. Inflammation, on the other hand, is a response to injury or damage that can further contribute to liver injury in the presence of excessive chromium. A study by Eskandari et al. (2020) investigated the toxic effects of chromium on the liver and found that chronic exposure to elevated levels of chromium led to elevated levels of liver enzymes, histopathological alterations, and oxidative stress. Another study by Akinola et al. (2019) similarly found that chromium toxicity induced oxidative stress and inflammation in the liver, ultimately leading to liver damage and increasing the risk of liver disease. The literature supports the idea that chronic exposure to elevated levels of chromium can result in liver injury through the induction of oxidative stress and inflammation. It is important to limit exposure to excessive levels of chromium to prevent adverse health effects, particularly to the liver.

10.7.5.2 Effect of Chromium on Kidney

Studies have shown that chronic exposure to elevated levels of chromium can have a damaging effect on the kidney, leading to renal dysfunction and heightened susceptibility to kidney disease. The nephrotoxic effects of chromium have been linked to oxidative stress and inflammation within the renal tissue, which can result in decreased renal function. A study by Abbas et al. (2018) investigated the nephrotoxic effects of chromium on the kidney and found that exposure to elevated levels of chromium induced oxidative stress and inflammation in renal tissue, leading to decreased renal function and increased susceptibility to kidney disease.

Chromium is eliminated from the body primarily through the kidneys, which makes the kidney particularly vulnerable to the toxic effects of this element. Oxidative stress and inflammation can cause damage to renal tissue and impair renal function, ultimately leading to decreased kidney function and an increased risk of kidney disease. Exposure to elevated levels of chromium can result in renal dysfunction and damage through the induction of oxidative stress and inflammation within the renal tissue. It is important to limit exposure to excessive levels of chromium to prevent adverse health effects, particularly to the kidney.

10.7.5.3 Effect of Chromium on the Cardiovascular System

The cardiovascular system is susceptible to the detrimental effects of chromium toxicity, as demonstrated by the correlations between elevated levels of chromium exposure and increased risk of cardiovascular diseases, including but not limited to atherosclerosis and coronary artery disease (Powers 2009). Furthermore, chromium toxicity has been shown to elevate blood pressure, thereby increasing the likelihood of hypertension (Flynn 2014). Given the crucial role of the cardiovascular system in facilitating the transport of oxygen and nutrients throughout the organism, it is a vital target for the toxic effects of chromium.

10.7.5.4 Chromium and Carcinogenicity: Possible Link to Cancers

The study conducted by Wang et al. (2012) aimed to assess the relationship between exposure to chromium (VI) in drinking water and colorectal cancer using a mouse model. The findings revealed a heightened incidence and growth of tumors as a result of exposure to chromium. This phenomenon was attributed to the activation of the Wnt/beta-catenin signaling pathway triggered by reactive oxygen species (ROS). Furthermore, it is believed that the interaction between ROS and reactive nitrogen species (RNS) with cellular macromolecules, such as DNA, lipids, and proteins, plays a vital role in cancer toxicity and progression (Aggarwal et al. 2019).

In another investigation by Patlolla et al. (2009), Sprague-Dawley rats were given potassium dichromate through intraperitoneal injection at doses of 2.5, 5.0, 7.5, and 10 mg/kg body weight per day for 5 consecutive days. The results showed a considerable rise in both ROS and malondialdehyde levels in the liver and kidney, as well as a dose-dependent increase in the activities of superoxide dismutase and catalase. Additionally, the study demonstrated a dose- and time-dependent DNA damage 24-, 48-, 72-, and 96 h post-treatment.

Lung cancer

Various studies demonstrated that the genotoxic effects of hexavalent chromium on human lung epithelial cells were investigated. The results showed that exposure to hexavalent chromium caused significant DNA damage and alterations in the p53 and p16 tumor suppressor genes. Additionally, the study found that hexavalent chromium exposure increased oxidative stress, as evidenced by the upregulation of reactive oxygen species and the depletion of antioxidant enzymes. These findings provide further evidence of the carcinogenic potential of hexavalent chromium in the lung and its ability to induce oxidative stress and DNA damage.

Another study by Wang et al. (2019) investigated the effect of hexavalent chromium on lung tumorigenesis in a mouse model. The results revealed that hexavalent chromium exposure significantly enhanced lung tumor formation and growth, as well as the activation of multiple oncogenic pathways, including the epidermal growth factor receptor (EGFR) and phosphoinositide 3-kinase (PI3K) signaling pathways. These findings suggest that hexavalent chromium plays a critical role in the development of lung cancer and highlights the need for increased protective measures for workers exposed to this carcinogenic form of chromium.

Stomach cancer

A study by Chen et al. (2014) showed that chromium (VI) exposure can lead to an increase in oxidative stress and DNA damage in gastric mucosal cells, thereby increasing the risk of stomach cancer development. The findings were supported by subsequent research which showed that hexavalent chromium can trigger oxidative stress, genomic instability, and inflammation in the stomach, which can result in increased susceptibility to gastric cancer. Moreover, studies have also reported that chromium-induced oxidative stress can affect the normal function of the cell cycle and cause DNA damage, thereby increasing the risk of stomach cancer (Zhou et al. 2017). These findings highlight the importance of limiting exposure to chromium, especially among individuals who are at an elevated risk of stomach cancer.

10.7.6 Interactions with Minerals and Nutrients in Food

Chromium toxicity has been implicated in a wide range of negative health outcomes, including nutritional deficiencies. Elevated levels of chromium have been shown to obstruct the absorption of vital nutrients, such as iron, zinc, and calcium, leading to adverse effects on health and well-being, including anemia, compromised immune function, and impaired bone health. One widely recognized manifestation of chromium toxicity is the association between elevated levels of chromium and decreased absorption of iron, which can result in iron-deficiency anemia. Characterized by a reduced red blood cell count and diminished capacity of the blood to transport oxygen, iron-deficiency anemia is a prevalent issue, particularly among women and children, and has far-reaching impacts on health and well-being.

In addition to its effects on iron absorption, chromium toxicity can also negatively impact the uptake of other crucial minerals, such as zinc. As a vital nutrient, zinc plays a key role in several physiological processes, including immune response, wound healing, and cell division. Zinc deficiency can result in various health problems, including impaired immune function, dermatological issues, and impaired wound healing. Furthermore, chromium toxicity has the potential to affect bone health by interfering with calcium metabolism, a critical component of healthy bones. This can result in decreased bone density and elevated risk of fractures and osteoporosis.

10.8 Health Risks Linked with Chromium Exposure in the Environment

10.8.1 Toxic Effect of Chromium in Human

The International Agency for Research on Cancer (IARC), a subsidiary of the World Health Organization (WHO), has classified hexavalent chromium compounds (Cr(VI)) as Category 1 human carcinogens based on comprehensive epidemiological evidence linking exposure to lung cancer (Seidler et al. 2013). Human exposure to Cr(VI) can result in a plethora of adverse health outcomes, including, but not limited to, skin and nasal irritation, ulcers, eczema, dermatitis, and lung cancer (Focardi et al. 2013). The toxicity of Cr(VI) can arise through multiple mechanisms, including oxidative protein modifications, DNA mutations, genotoxicity, cytotoxicity, carcinogenicity, and chromosomal damage (Al Osman et al. 2019). Furthermore, high levels of Cr(VI) exposure via inhalation can lead to respiratory issues such as asthma and coughing, as well as anemia, ulcers, and gastrointestinal irritation (Yang et al. 2020). Prolonged exposure to elevated levels of Cr(VI) and related compounds can result in significant health problems.

Studies have indicated a positive correlation between Cr(VI) exposure and the incidence and mortality of several cancers, including lung, bladder, larynx, kidney, testicles, thyroid, and bone (Deng et al. 2019). An ecological study conducted in Greece in 2011 found elevated levels of Cr(IV) in drinking water, with concentrations ranging from 41 to 156 g/L, and an increase in the incidence of genitourinary, lung, and liver cancers among women (Linos et al. 2011). A study conducted in India in 2012 found an increased prevalence of gastrointestinal and dermatological issues in individuals exposed to groundwater contaminated with Cr(VI) (Sharma et al. 2012). The mechanisms underlying Cr toxicity and carcinogenicity include DNA damage, genomic instability, and production of reactive oxygen species (ROS) (Pavesi and Moreira 2020).

Both Cr(VI) and Cr(III) have the potential to generate ROS (Pavesi and Moreira 2020). Chromium-induced carcinogenicity results in transcriptional regulation disruption and DNA damage. This disruption of chromatin architecture leads to alterations in chromatin accessibility and nucleosomal positions at both local and genome-wide levels. Chromium-induced DNA damage can take various forms, including DNA-Cr-protein crosslinks, DNA inter- and intrastrand crosslinks, single- and double-strand breaks (SSBs and DSBs, respectively), and p53 point mutations (Thompson et al. 2011). The toxicological effects of chromium on humans and animals are depicted in Fig. 10.5.

Numerous In-vitro studies have established the interaction of chromium(III), Cr(III), with nucleic acids (Fang et al. 2014). The formation of binary and ternary DNA adducts has been observed in Cr(III) exposure, with ternary adducts being more toxicologically significant in mammalian cells (Zhitkovich 2005). The four primary ternary adducts identified are Cr(III) ascorbate, Cr(III) cysteine, Cr(III) histidine, and Cr(III) glutathione (Salnikow and Zhitkovich 2008). The presence of DNA-protein crosslinks has also been detected in human lung adenocarcinoma cells (A549) (Macfie et al. 2010). Rapidly dividing cells, such as cancer cells, can exhibit collapse of the replication fork during DNA replication, resulting in cell cycle arrest and activation of the p53 tumor suppressor, leading to apoptosis (Wilhelm et al. 2020).

Studies have established that exposure to hexavalent chromium (Cr(VI)) can induce DNA damage and carcinogenic potential, as demonstrated by the detection of DNA strand breaks (SSBs and DSBs) and protein-chromium (Cr)-DNA crosslinks in liver cells of F344 rats following administration of potassium chromate via drinking water for three weeks (Coogan et al. 1991). The mechanisms behind the carcinogenic effects of Cr(VI) exposure are multifactorial, encompassing factors such as tissue type, cell specificity, concentration of Cr(VI), duration of exposure, and the reactivity of intermediate forms of Cr(V), Cr(IV), and Cr(III) (Ferreira et al. 2019).

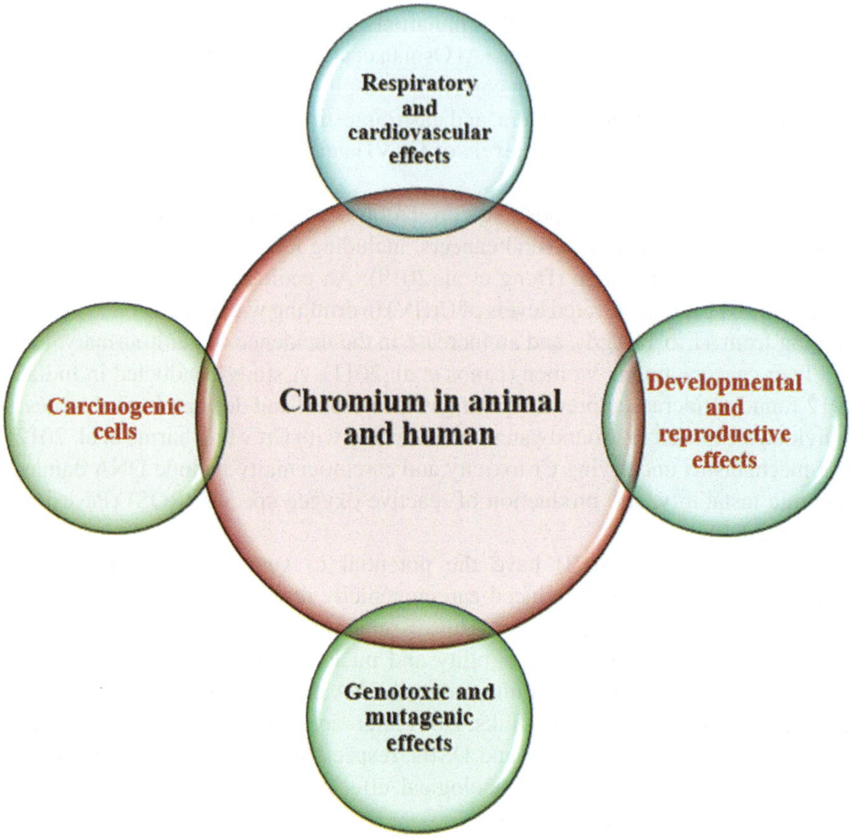

Fig. 10.5 Toxicological effects of chromium on animal and human

Cr(VI) does not bind directly to macromolecules or DNA, but instead forms intermediates with cellular reductants, including Cr(V), Cr(IV), and Cr(III) (Zhitkovich 2005). The intermediates generated during Cr(VI) reduction can trigger Fenton-type reactions in the presence of hydrogen peroxide, resulting in hydroxyl radical formation and oxidative stress (Yao et al. 2008). Additionally, the formation of Cr-Asc (ascorbate), Cr-GSH, and Cr-Cys (cysteine) crosslinks also contributes to oxidative stress (Yao et al. 2008). Cr(VI)-induced oxidative stress targets cells' DNA and lipid contents, causing DNA damage and lipid peroxidation, respectively. This oxidative stress can result in additional cellular harm, cell death through necrosis and apoptosis, and even promote tumor growth and development at low to medium levels of ROS production.

10.8.2 Chromium and Antimicrobial Resistance

The unique characteristics of chromium, particularly its antimicrobial properties, have garnered significant scientific attention. Chromium compounds have been utilized in the medical field for their ability to inhibit the growth of various microorganisms, including bacteria and fungi. However, their use as antimicrobial agent's raises concern about the development of antibiotic resistance, a major public health concern. Antibiotic resistance occurs when microorganisms evolve mechanisms that enable them to withstand the effects of antimicrobial agents, potentially leading to the selection of antibiotic-resistant strains of microorganisms over time. Hexavalent chromium has been demonstrated to possess potent antimicrobial activity against a wide range of microorganisms. However, the challenge of comprehending resistance mechanisms employed by these microorganisms to counteract chromium's antimicrobial properties remains. For instance, studies have shown that bacteria like Escherichia coli have evolved mechanisms such as reducing hexavalent chromium to less toxic trivalent chromium or utilizing efflux pumps to pump out chromium ions, leading to reduced toxicity and decreased efficacy of chromium compounds as antimicrobial agents. Furthermore, there is concern regarding the toxic effects of hexavalent chromium on human health, including respiratory problems, skin irritation, and genetic damage due to long-term exposure. The ongoing research and regulatory evaluations surrounding the safety and efficacy of chromium compounds as antimicrobial agents aim to determine the best approach for utilizing their antimicrobial properties while minimizing potential harm to human health and the environment.

10.8.3 Chromium Induced Genotoxicity

Chromium and its various compounds have been extensively studied with regards to their genotoxic effects. A wealth of evidence from a range of genetic tests demonstrates that hexavalent chromium compounds, with varying water solubility, exhibit significant genotoxicity. Chromates, dichromates, and other forms of chromium (VI) such as chromium trioxide, have been found to induce DNA damage, gene mutations, sister chromatid exchanges, chromosomal aberrations, cell transformations, and dominant lethal mutations in various In vitro and In vivo test systems, including human and animal cells.

In contrast, the genotoxicity of trivalent chromium compounds remains less well understood. While these compounds have been shown to exhibit reactivity with purified DNA and isolated nuclei, few studies using cellular test systems have shown positive results, and these were often observed at high concentrations and under specific conditions. The lack of cellular reducing agents in cell-free systems precludes the reduction of hexavalent chromium to less toxic forms, and the prevalent hypothesis posits that the genotoxicity of hexavalent chromium is due to the highly reactive intermediates generated during cellular reduction, such as Cr(V) and Cr(IV). The role of cellular reducing agents such as ascorbate and sulfhydryl compounds in reducing Cr(VI) remains unclear, as does the contribution of intermediates such as hydroxyl, cysteinyl, and thionyl radicals to chromium-induced carcinogenesis.

Chromium-mediated reactions have also been shown to generate oxygen-free radicals and activate the transcription factor NF-kB, which regulates genes involved in inflammation, immunity, and apoptosis. The intricate and complex nature of these mechanisms underscores the ongoing challenge of fully understanding chromium carcinogenicity. Currently, inhaling hexavalent chromium is classified as a known human carcinogen by the International Agency for Research on Cancer (IARC), while the significance of other valence states, the underlying mechanisms, and the impact of solubility remain subjects of ongoing investigation and debate.

10.8.4 Chromium Induced Hepatotoxicity

The carcinogenic properties of chromium (Cr) and its derivatives have far-reaching implications for the environment, industry, and occupational health. Chromium plays a vital role in the metabolic processes of carbohydrates, lipids, and proteins (Zhao et al. 2019). It can enter the bloodstream through various pathways, including the skin, lungs, and perforations, before being eliminated by the liver (Xiao et al. 2012). The reduction of trivalent Cr to tetravalent Cr can generate reactive oxygen species (ROS), leading to oxidative damage and various forms of toxicity, including hepatotoxicity, cardiotoxicity, and gene toxicity (Balakrishnan et al. 2013). High doses of hexavalent Cr have been shown to cause liver damage (Xiao et al. 2012). Accumulation of Cr in the nuclei and mitochondria of cells can result in DNA modifications and changes in cellular regulatory mechanisms (Venter et al. 2015). Histopathological studies have indicated that Cr exposure is associated with nuclear pyknosis, central phlebectasia, and hepatocyte degeneration (Tian et al. 2018).

10.8.5 Effects on Immune Response

The findings of the studies on the effects of hexavalent chromium and other metal elements on cultured human lymphocytes emphasize the criticality of comprehending the toxicity of chromium, particularly hexavalent chromium, in human health. The observed decreases in blastogenesis and immunoglobulin synthesis following chromium uptake by human lymphocytes have far-reaching implications for the immune system and overall health, as the proper functioning of these cells is crucial for maintaining the body's defense against infections and diseases. Extensive research has revealed that exposure to high levels of hexavalent chromium can lead to DNA damage, which in turn increases the likelihood of cell mutations and the development of cancer. Despite these findings, the exact mechanisms by which hexavalent chromium affects the functioning of human lymphocytes remain unclear and require further investigation.

It is imperative to consider the potential impacts of chromium exposure on vulnerable populations, such as pregnant women, infants, and individuals with pre-existing health conditions, as they may be more susceptible to the toxic effects of chromium. Further research is needed to gain a comprehensive understanding of the potential harm to the health of these populations. The results of the studies on the effects of hexavalent chromium on human lymphocytes highlight the urgency for continued research on the toxicity of chromium and the need for caution in the use and exposure to chromium compounds. These findings have far-reaching implications for public health and inform the development of safe and effective chromium-based products.

10.8.6 Chromium Induced Allergic Reactions

In light of the limited current understanding of chromium-induced allergies, further research is crucial to gain a comprehensive insight into the potential of chromium compounds to elicit allergic responses. A multidisciplinary approach incorporating a thorough understanding of the underlying immune mechanisms, the extent and nature of exposure, and the susceptibility of various populations to chromium-induced allergies is necessary to establish the foundation for the development of effective strategies for the prevention and management of these reactions.

Studies have demonstrated that chromium-based products, such as chromium-plated jewelry, can elicit skin reactions through the process of contact sensitization, whereby the immune system recognizes chromium as a foreign substance and generates an immune response upon subsequent exposures. There is also evidence suggesting that chromium-containing alloys used in dental implants may result in systemic allergic reactions, as a result of the release of chromium ions into the bloodstream. Given the significance of immune responses in the development of allergies, a better understanding of the mechanisms involved in chromium-induced allergies is imperative. This will enable the creation of effective strategies for the prevention and management of these reactions, and ensure the safety and well-being of the public, particularly among susceptible populations such as individuals with pre-existing health conditions or those who are exposed to high levels of chromium.

10.8.7 Skin Allergy

The relationship between exposure to chromium compounds and the incidence of contact dermatitis has been extensively documented in the scientific literature. Notably, a decrease in the prevalence of this condition in Scandinavia can be attributed to reduced exposure to chrome-plated objects. However, the underlying mechanisms by which chromium induces sensitization and elicits a skin reaction, particularly at low levels of exposure, remain poorly understood. The observation of cross-reactivity among sensitized individuals and the results of lymphocyte transformation and migration inhibition tests provide evidence that trivalent chromium compounds serve as the underlying allergens for adverse skin reactions. Furthermore, oral ingestion of hexavalent chromium salts has been shown to induce sensitization, suggesting that hexavalent chromium can enter the body and trigger a systemic immune response.

Clinically, chromium-induced contact dermatitis is characterized by a delayed hypersensitivity response, which progresses over time and is evidenced by histological findings on biopsy. However, the minimum duration and amount of exposure required for sensitization remain unclear and require further investigation. While the evidence linking chromium exposure, particularly hexavalent chromium, to contact dermatitis is substantial, many questions about the underlying mechanisms and risk factors associated with this condition remain unanswered. Further research is needed to advance our understanding of chromium-induced contact dermatitis and to develop effective strategies for its prevention and management.

10.8.8 Effects on Macrophages

Inhaling chromium compounds can have a range of effects on human health, depending on the chemical form and concentration of chromium as well as the duration of exposure. The most common route of chromium exposure is via inhalation, and the lung is particularly susceptible to the effects of hexavalent chromium (Cr[VI]). Studies have shown that low-level exposure to Cr[VI] can result in an increase in the immune response and phagocytic activity of alveolar macrophages, which are the primary defense mechanism against inhaled particulate matter. This enhancement of immune response and phagocytic activity is beneficial in protecting the lung from further damage.

However, exposure to high concentrations of Cr[VI] leads to a reduction in the function of alveolar macrophages, which can result in the formation of nodules in the intra-alveolar spaces. These nodules can cause structural damage to the lung and may contribute to further health problems, such as fibrosis and scarring. Furthermore, Cr[VI] exposure has also been associated with limited morphological alterations in the lung, including changes in the size and structure of alveoli and changes in the epithelial cells that line the airways. These alterations can also contribute to the formation of nodules and the compromise of lung function.

It is worth noting that the toxicity of chromium is complex and dependent on multiple factors, including the chemical form of chromium, concentration, and duration of exposure. Trivalent chromium (Cr[III]), for instance, is relatively non-toxic and is required for human nutrition, whereas Cr[VI] is highly toxic and has been classified as a human carcinogen. Inhaling chromium compounds can have a significant impact on the lung, especially in the case of Cr[VI] exposure. Low-level exposure to Cr[VI] can enhance the function of alveolar macrophages, whereas high-level exposure can reduce their function and contribute to the formation of nodules in the intra-alveolar spaces.

10.9 Toxic Effects of Chromium in Microorganisms

The presence of chromium ions in living systems is considered essential for their role in the electron transport chain. Despite its essentiality, chromium and its compounds have been shown to elicit elevated levels of sensitivity in microorganisms, similar to those observed in humans and plants (Mishra and Bharagava 2016). Chromium(VI) exposure has been demonstrated to significantly impact the diversity, genetic makeup, and behavior of microflora. This exposure can lead to structural modifications of nucleic acids and cellular impairments, such as restriction of enzymatic activity and oxidative phosphorylation, membrane disruption, osmotic imbalances, and lipid peroxidation, which, collectively, have the potential to disrupt the morphology, metabolism, and proliferation of microorganisms (Ayangbenro and Babalola 2017). Chromium exposure has also been demonstrated to inhibit cell growth, as it induces bacterial cell expansion and dilatation and hinders cell division. Therefore, both an insufficient and excessive amount of chromium can have adverse effects on microorganisms, such as inhibiting growth and slowing cellular processes.

10.10 Toxic Effects of Chromium in Plants

The proliferation of chromium complexes in soil, where various plant species thrive, has become a widespread issue, with the most prevalent forms being $HCrO_4$ and CrO_4. These forms of chromium have been observed to readily penetrate plant tissues and seep into deeper soil layers, thereby endangering surface and groundwater (Elahi et al. 2020). Compared to other oxidation states, Cr(III) is considered to pose less of a hazard due to its low solubility that restricts leaching into freshwater bodies and plant uptake. However, Cr(VI) has been demonstrated to inflict significant harm to living cells (Cervantes et al. 2001). Moreover, Cr(VI) exposure has been linked to a reduction in nutrient uptake and photosynthesis, leading to stunted plant growth. The production of reactive oxygen species (ROS) in plant cells arises from disruptions in various physiological, morphological, and biochemical processes, resulting in symptoms such as chlorosis and necrosis (Jobby et al. 2018). The principal detrimental effects of chromium on plants are summarized in Fig. 10.6.

The effect of chromium on plant growth is primarily manifested in the primary photosynthetic organ, the leaf. Elevated chromium concentrations can significantly decrease leaf area, biomass, and photosynthetic activity, resulting in the onset of chlorosis and necrosis in leaves. Chromium exposure triggers a multitude of detrimental impacts on leaves, including inhibition of chlorophyll synthesis, disruption of chloroplast structure and function, suppression of photosynthetic electron transfer,

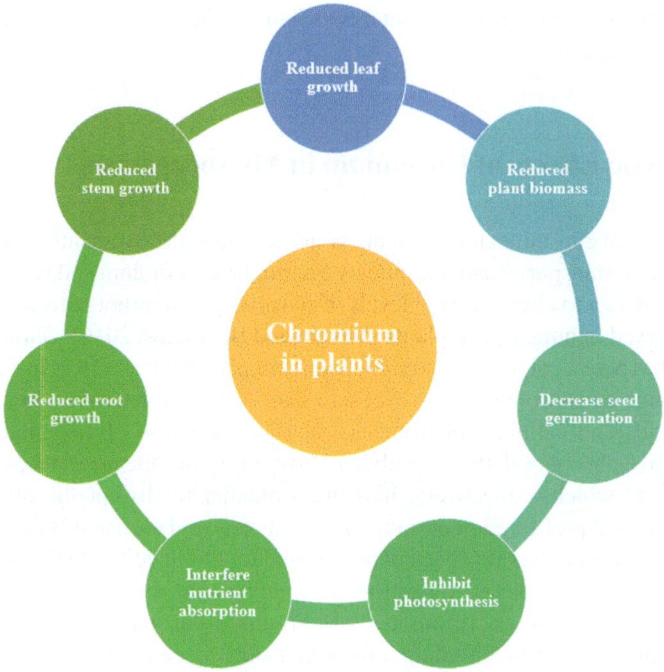

Fig. 10.6 Harmful effects of chromium on plants

and liberation of magnesium ions from the chlorophyll molecule (Stambulska et al. 2018). The detrimental effects of Cr(VI) exposure on plants encompasses stunted growth, necrotic and malformed leaves, reduced enzyme activity, impaired nutrient uptake and transport, hampered photosynthesis, root cell damage, lipid peroxidation, chromosomal abnormalities, and DNA strand breakage (Guo et al. 2021). As such, chromium exposure can impede photosynthesis, nutrient uptake, seed germination, and overall growth and performance of various plant components.

10.11 Prevention and Control of Cr Toxicity in Food and Environment

10.11.1 Exogenous Salicylic Acid Remediation of Chromium Toxicity in Rice

The toxicity of chromium is primarily attributed to its hexavalent form, Cr(VI), which is commonly encountered as chromate or dichromate oxyanions in the presence of oxygen. This form of chromium has been shown to be highly toxic to plants, disrupting crucial physiological processes such as photosynthesis, nutrient uptake, and overall yield (Shanker et al. 2005). Rice (Oryza sativa L.), being a widely cultivated and stable crop globally, is particularly vulnerable to soil contamination with chromium, which may result in elevated levels of this heavy metal in rice-based foods, with potential implications for human health, including asthma, chronic irritation, and cancer (Hayat et al. 2012). In response to heavy metal toxicity, plants have developed several adaptive strategies, including limiting heavy metal uptake by roots, vacuolar sequestration of heavy metals within plant tissues, and activation of antioxidant activities to mitigate reactive oxygen species (ROS) stress. Absorption of chromium by plants occurs through essential transporters such as iron, sulfate, and phosphate (Hayat et al. 2012). To cope with increased chromium levels, plants may also reduce iron uptake by converting Fe(III)–Fe(II) or competing with Fe(II) at the absorption site (Shanker et al. 2005). Additionally, some plants synthesize low-molecular-weight chelators, such as phytochelatins (PCs) and metallothioneins (MTs), to detoxify heavy metals. These chelators form complexes with heavy metals and transport them to the vacuole, minimizing transport to the aerial parts of the plant (Emamverdian et al. 2015). MTs play a crucial role in heavy metal detoxification by regulating sequestration and intracellular metal homeostasis, as well as adjusting metal transport.

The phenolic signaling molecule, Salicylic Acid (SA), has demonstrated the ability to induce tolerance mechanisms in plants under abiotic stress conditions, when applied through various methods such as seed soaking, nutrient supplementation, irrigation, or spraying (Anwar et al. 2013). Evidence shows that SA can modulate the antioxidant system and mitigate membrane lipid peroxidation in rice exposed to copper and lead (Zengi 2014). In the presence of heavy metal stress, SA has been observed to regulate photosynthesis, photosystem II, photosynthetic pigments, Rubisco, and carbonic anhydrase activities (Noriega et al. 2012). The role of SA as a signaling molecule in response to heavy metals and abiotic stress has been well established. Studies have found that SA supplementation during chromium cultivation can restore the growth features of roots and shoots while reducing chromium accumulation in shoots. This reduction in chromium accumulation in shoots is likely due to the trapping of elevated chromium in roots, which is indicated by the lack of negative impact on plants despite increased chromium presence in roots. At the biochemical level, SA supplementation under chromium stress has been shown to restore membrane stability and total soluble protein content in roots. These findings suggest that SA may serve as a remediation agent, reducing chromium toxicity in plants, and ultimately reducing chromium contamination in crops and food materials.

10.11.2 Chemoprotection by N-Acetylcysteine

Reactive oxygen species (ROS) have been implicated in a wide range of biological processes, including cell signaling, oxidative stress, and adverse drug reactions. The increase in ROS levels can result in oxidative stress, which can cause damage to cellular macromolecules, such as DNA, proteins, and lipids. To counter this oxidative stress, various antioxidants have been used, including N-acetylcysteine (NAC). NAC has been widely used to investigate the role of ROS in the toxicity of carcinogenic metals, as it has been shown to have potent antioxidant activity. This antioxidant activity of NAC is due to its ability to scavenge ROS and maintain the redox balance within cells. The inhibition of ROS by NAC has been shown to mitigate the toxicity of carcinogenic metals, such as chromium, cadmium, and lead, by reducing oxidative stress and cellular damage. While the antioxidant activity of NAC has been shown to mitigate the toxicity of carcinogenic metals, the role of ROS in the toxicity of these metals is still a subject of ongoing research. Some studies have shown that oxidative stress plays a significant role in the toxicity of these metals, while others have shown that it is only a minor contributor. In addition, genetic models have shown that ROS may not play a role in the toxicity of certain metals, such as mercury and cadmium, as these metals do not exhibit redox activity. In conclusion, the relationship between ROS and the toxicity of carcinogenic metals is complex and still not fully understood. While NAC has been shown to have potent antioxidant activity and to mitigate the toxicity of some carcinogenic metals, the role of ROS in the toxicity of these metals is still being researched. Further studies are needed to fully understand the intricate relationship between ROS and carcinogenic metals and to develop more effective strategies for mitigating their toxicity.

10.11.3 Soil Metal Detoxification

10.11.3.1 Bioremediation

The bioremediation potential of bacteria in remediating chromium(VI) contaminated environments has been the focus of several studies. The ability of Bacillus cereus strain XMCr-6 to efficiently remediate chromium(VI) has been demonstrated, with a complete reduction of 100 mg/L of chromium(VI) within 48 h being reported by Dong et al. (2013). Additionally, Pseudochrobactrum spp. and Proteus spp. isolated from chromium(VI) contaminated seawater were found to effectively reduce 1000 mg/L of chromium(VI) within 96 or 144 h, respectively, while tolerating concentrations of up to 2000 and 1500 mg/L of chromium(VI). Furthermore, Annamalai et al. (2014) reported the ability of Bacillus subtilis to effectively reduce up to 99% of 100 mg/L of chromium(VI) through the biosorption of chromium(VI) and its conversion into chromium(III), forming nanometer-scale chromium oxide particles. The isolation of chromium(VI) reducing bacteria from diverse genera and environments highlights their potential for bioremediation. These microorganisms undergo gradual adaptation in response to heavy metal exposure, leading to the production of metallothionein, which confers resistance. Chromate reductase within the cytoplasm reduces chromium(VI) that enters microbial cells through the cell membrane or accumulates within the cells (Chen et al. 2014).

The role of extracellular polymeric substances (EPS) in the biosorption of chromium(VI) is critical. Polysaccharides, a type of negatively charged cell secretion, serve as a means of adherence for chromium(VI) to the surface of cells. Conversely, the functional groups of extracellular proteins, glycoproteins, and glycolipids chemically bond with chromium(VI) and prevent its infiltration into the cell, thus reducing its toxicity. Studies have indicated that the number of polysaccharides and proteins in the EPS of aerobic granular sludge increases with increasing chromium(VI) concentrations. Fluorescence spectroscopy has revealed the presence of humic acid-like and fulvic acid-like substances. This increase may result from the degradation of organic macromolecules and dead cell debris (Wang and Na 2014). Both Gram-negative and Gram-positive bacteria exhibit high biosorption capabilities (Philip et al. 1998).

10.11.4 Cleaning Chromium Pollution from Aquatic Environments

Chromium pollution in aquatic environments is a growing concern due to the widespread use of chromium in various industrial processes and its persistent toxic nature. This contamination poses a threat to aquatic life and human health as chromium is known to cause carcinogenic and mutagenic effects. Several conventional methods have been employed for cleaning chromium pollution from aquatic environments, including chemical precipitation, ion exchange, membrane filtration, and adsorption. However, these methods are limited by high costs, technical complexity, the production of secondary pollutants, and inconsistent results.

Recently, the integration of nanotechnology and bioremediation, commonly referred to as nano-bioremediation, has garnered attention as a promising approach to mitigate chromium pollution in aquatic environments. This strategy leverages the functionalities of microorganisms to effectively detoxify chromium from contaminated water. The efficacy of nano-bioremediation has been demonstrated in numerous studies, which have shown varying levels of effectiveness in the removal of chromium from water. The efficacy of nano-bioremediation has been demonstrated in numerous studies, which have shown varying levels of effectiveness in the removal of specific heavy metals, such as palladium and chromium. For instance, metal removal efficiencies of nano bioremediation have been reported to range from 12% with biogenic palladium nanoparticles derived from *Spirulina platensis* to 100% with iron oxide nanoparticles of *Geobacter sulforeducens*. The physical, chemical, and biological methods for chromium remediation are comprehensively presented in Table 10.1.

Nanomaterials

The removal of trace elements from water is a crucial aspect of water purification, as the presence of these elements can have adverse effects on human health and the environment. In order to achieve this elimination, various sorbents have been developed and employed, including the utilization of nanomaterials. Nanomaterials,

Table 10.1 Different physical, chemical and biological methods available for chromium remediation

S. No	Physical methods	Chemical methods	Biological methods
1	Membrane filtration	Sodium metabisulfite ($NaHSO_3$)	Reverse osmosis
2	Ion exchange	Sodium dithionite ($Na_2 S_2 O_4$)	Biosorption
3	Biomineralization	Hydrogen sulfide (H_2S)	Electrodialysis
4	Adsorption	Calcium polysulfide (CaS_5)	Bioaccumulation
5	Extracellular precipitation	Ferrous sulfate ($FeSO_4$)	Bio-reduction

defined as materials and structures with dimensions ranging from 1 to 100 nm, have unique properties that distinguish them from larger structures with similar compositions. These unique properties are shaped by factors such as shape, size, surface characteristics, and internal structure, and have been found to have significant effects on the mechanisms of sorption.

Sorption mechanisms by nanomaterial sorbents are highly dependent on both the properties of the sorbent and the physical–chemical conditions of the solution being treated. The chemical functionalization and specific coatings of nanomaterials provide additional opportunities for tailoring their properties to the specific requirements of a given water purification application. The use of nanomaterials in water purification provides a promising solution to the challenge of trace element removal, as they offer the potential for high sorption efficiency, low cost, and low environmental impact. However, it is important to consider the potential ecological and health risks associated with the release of nanomaterials into the environment and the importance of responsible and sustainable use of these materials in water purification applications.

10.11.5 Removal of Chromium from Wastewater

10.11.5.1 Membrane Filtration

Membrane filtration processes involve separating substances from a liquid by utilizing a selective membrane that allows some substances to pass through while retaining others. This separation process can be achieved through various membrane processes, including liquid, pressure-driven, and hybrid membranes. The selection of the appropriate membrane type, pore size, and material composition can greatly impact the effectiveness and cost-efficiency of the membrane filtration process. Five prominent membrane processes that are commonly used in water treatment include reverse osmosis, ultrafiltration, microfiltration, nanofiltration, and electrodialysis.

Giagnorio et al. (2018) investigated the utilization of nanofiltration membranes for chromium removal and reported that the use of nanofiltration membranes NF270 and NF90 led to chromium removal efficiencies of 98.8% and 76.5%, respectively. The authors addressed challenges such as fouling and system performance in their study. Shi et al. (2015) demonstrated the excellent stability and strength of a magnetite membrane derived from oxidized 316 L stainless steel, which effectively removed chromium at pH 4. The authors reported that this technology offers a promising alternative for chromium removal in water treatment processes.

Other technologies, such as ion exchange, adsorption, and electrochemical technology, can be integrated with membrane technology to enhance the performance of water treatment processes. The integration of multiple technologies provides a comprehensive solution that addresses the challenges and limitations of individual technologies.

The utilization of membrane filtration processes for water treatment has become increasingly popular due to its effectiveness in waste removal, low generation of pollutants, and potential for reduced energy consumption. The selection of the appropriate membrane type, pore size, and material composition can greatly impact the effectiveness and cost-efficiency of the membrane filtration process. Nanofiltration and magnetite membrane technologies have been demonstrated to be effective in chromium removal. Additionally, the integration of membrane technology with other technologies provides a comprehensive solution for water treatment processes.

10.11.5.2 Chemical Precipitation

The oxidation states of chromium in aqueous solutions are primarily Cr(III) and Cr(VI). The reduction of Cr(VI) to the more stable and less toxic Cr(III) form is a commonly used approach in the removal process of chromium from aqueous solutions (Azimi et al. 2017). Chemical precipitation is a widely used method for the treatment of chromium-contaminated water, where heavy metal ions react with suitable chemical precipitant agents to form insoluble solid particles (Fu and Wang 2011). The solid particles formed can be separated from the solution through sedimentation or filtration techniques (Zamboulis et al. 2004).

The selection of appropriate chemical precipitant agents is critical to the success of the precipitation process, as it affects the solubility and stability of the precipitates, as well as the metal removal efficiency (2011). The pH of the solution, the metal ion concentration, and the presence of other contaminants can also have a significant impact on the precipitation process and the formation of solid particles (Azimi et al. 2017). Chemical precipitation is a widely used method for the treatment of chromium-contaminated water, where heavy metal ions are precipitated through the reaction with suitable chemical precipitant agents. The choice of chemical precipitant agents, as well as solution parameters, plays a crucial role in the success of the precipitation process and the metal removal efficiency.

10.11.5.3 Ion Exchange

The Ion Exchange process, a technique first documented over a century ago, has undergone a significant evolution in the intervening years. Its applications have broadened and expanded to encompass a range of uses, including water softening and the integral role it plays in numerous industrial and technological processes. The process involves a stoichiometric, reversible chemical reaction, in which a charged ion from an electrolyte, solution, or molten salt is exchanged for a similarly charged ion attached to an immobile, insoluble solid substance (Dharnaik and Ghosh 2014).

In the Ion Exchange process, an insoluble resin material is utilized to remove Cr(III) from wastewater, releasing other ions with comparable charges while preserving the structural integrity of the resin. This highlights the effectiveness of this process in enabling ion separation and recovery. The ion exchange process, therefore, represents a critical component in the treatment of wastewater, especially in situations where selective removal of specific ions is required.

10.11.5.4 Adsorption

The utilization of activated carbon and its composites as adsorbent materials has garnered considerable attention in recent literature due to their demonstrated efficacy in removing a broad spectrum of pollutants, including heavy metals and dyes (Shakoor et al. 2020). The intraparticle model of adsorption mechanisms outlines four steps that are frequently described by pseudo-second-order models. These steps include (1) bulk transport in the solution phase, (2) film transport through the hydrodynamic boundary layer from the bulk liquid to the external surface of the adsorbent, (3) intraparticle diffusion of heavy metal ions from the exterior into the pores of the adsorbent, and (4) adsorption.

Biosorption has emerged as an eco-friendly alternative to conventional techniques for removing heavy metal ions from aqueous media due to its low cost and high adsorption capacity (Escudero et al. 2006). The interaction between heavy metal ions and biomaterials results in the formation of complex compounds, which is facilitated by the presence of functional groups and ligands on the biosorbent material (Mishra et al. 2020). The pH of the reaction medium plays a significant role in determining the efficiency of Cr(VI) adsorption during biosorption. Protonation of active sites on the biosorbent material at low pH levels results in the formation of electrostatic interactions with anionic species, leading to their binding to the biosorbent. Other studies have found that pH values of 3.0 and 1.5 are optimal for the sorption of hexavalent chromium onto cork and grape stalks, respectively (Machado et al. 2002).

10.11.5.5 Electrochemical Technology

The utilization of electrochemical technology for the remediation of hexavalent chromium (Cr(VI)) has garnered substantial attention in recent years, due to its potential to promote environmental sustainability and possess technical advantages. A key advantage of this process is the utilization of clean electrons as a reagent, contributing to its compatibility with the environment and broad application potential (Duarte et al. 1998). Additionally, electrochemical technology offers benefits such as operational flexibility, potential for automation, and the ability to operate under mild reaction conditions, which can be considered safe (Breslin et al. 2019). A comparative analysis by Golder between electrocoagulation and chemical coagulation for the removal of Cr(VI) found that electrocoagulation was nearly three times more efficient than chemical coagulation utilizing aluminum or aluminum sulfate (Akbal and Camc 2010). This highlights the potential for electrochemical methods to provide efficient, cost-effective, and environmentally responsible solutions for the removal of Cr(VI) while complying with strict environmental regulations.

10.11.5.6 Electrocoagulation

The electrocoagulation process is a method for removing heavy metals and organic pollutants from aqueous solutions through the application of electrical current to sacrificial anodes, which are typically constructed from materials such as aluminum or iron. The efficacy of removal of these contaminants has been shown to be dependent on the type of electrode material used (Kim et al. 2020). Iron, stainless steel, and aluminum are the most commonly employed electrode materials in electrocoagulation due to their availability, low cost, and favorable electrochemical properties.

It has been suggested that the composition, surface properties, and size of the anode material play a crucial role in the performance of the electrocoagulation process. For instance, the presence of certain impurities in the electrode material may affect the generation of destabilizing agents, leading to changes in the efficiency of contaminant removal. Moreover, the surface area and morphological features of the electrode material can also impact the efficiency of electrocoagulation.

To optimize the performance of the electrocoagulation process and enhance the removal of heavy metals and organic pollutants, various parameters such as current density, reaction time, and electrode configuration must be considered. Furthermore, it is important to understand the underlying electrochemical mechanisms and the interactions between the anode material, contaminants, and reaction environment to develop effective strategies for controlling the process.

10.11.5.7 Electrochemical Reduction

The electrochemical reduction of Cr(VI) has been widely recognized as an efficient method for removing this toxic metal ion from aqueous solutions. The effectiveness of this process is largely dependent on the electrochemical surface area of the electrode and the material properties of the chosen cathode. Traditionally, electrodes made of titanium with dimensionally stable anodes or PbO_2 coatings have been widely used in the electrochemical reduction of Cr(VI) (Almaguer-Busso et al. 2009). However, recent advancements in the field have motivated the exploration of novel cathode materials with the aim of enhancing the removal efficiency of Cr(VI).

A number of cathode materials have been investigated, including carbon-based electrodes, mercury-based electrodes, bismuth film electrodes, gold electrodes, conducting polymers, and fuel cell systems. The use of these materials has garnered significant attention in the field due to their potential to offer superior removal efficiency and longer operational stability compared to traditional electrodes. However, it is worth noting that the effective application of these novel cathode materials for the electrochemical reduction of Cr(VI) requires a thorough understanding of the underlying electrochemical mechanisms and the optimization of the reaction conditions. Further research is needed to fully assess the feasibility and practicality of these novel cathode materials for large-scale Cr(VI) removal applications.

10.11.5.8 Photocatalysis

Photocatalysis, a phenomenon that relies on the interaction of a photocatalyst with electromagnetic radiation in the ultraviolet (UV) or visible light spectrum, has emerged as a promising strategy for the treatment of toxic metal ions in water. Among the metals of concern, Hg(II), Cd(II), Cu(II), and Cr(VI) have been identified as the most frequently studied and are known to pose significant environmental and human health risks. The process of photocatalysis is initiated by the absorption of light by the photocatalyst, leading to the generation of reactive oxygen species (ROS) such as hydroxyl radicals and superoxide anions. These ROS species interact with the metal ions, leading to their oxidation or reduction, and eventually to their removal from the aqueous solution. The advantages of photocatalysis over traditional treatment methods include lower levels of secondary pollutants, environmentally benign end-products, and reduced reaction times.

However, the efficacy of photocatalytic removal of Cr(VI) is known to be impacted by several key factors. One of the primary mechanisms influencing the removal efficiency is the interfacial charge transfer between the photocatalyst and Cr(VI) ions, which can impact the rate and extent of Cr(VI) reduction. Additionally, the dependence of photocatalytic efficacy on the bandgap of the photocatalyst has also been reported, with narrow bandgap materials generally exhibiting higher removal efficiency compared to their broad bandgap counterparts. Photocatalysis has been established as a promising technology for the treatment of toxic metal ions in water, with a particular focus on Hg(II), Cd(II), Cu(II), and Cr(VI). Further research is needed to fully understand the underlying mechanisms governing the photocatalytic removal of Cr(VI) and to optimize the photocatalytic process for enhanced removal efficiency and practical applicability.

10.11.5.9 Nanotechnology

The utilization of nanoscale objects as adsorbents in wastewater treatment has gained significant attention in recent years due to their high surface area to volume ratio, making them highly effective for this purpose (Nogueira et al. 2015). These nanoscale adsorbents can be categorized into three groups: adsorptive, reactive, and hybrid magnetic nanoparticles. Nanomagnetic oxides, possessing mesoporous structures, stability, and large surface areas, are widely employed in wastewater treatment processes and have been demonstrated to be effective in removing Cr(VI) from wastewater. The efficacy of these materials is influenced by reaction conditions and the specific adsorbent used (Hua et al. 2012).

In addition to nanomagnetic oxides, nano-zero-valent iron has also been demonstrated as a promising alternative for the removal of Cr(VI) from wastewater. The addition of nano-zero-valent iron reduces the redox potential and pH, which enables the rapid transformation of Cr(VI) into the environmentally benign Cr(III) (Vilardi et al. 2019). Hybrid magnetic nanoparticles, characterized by their magnetic properties, low toxicity, affordability, and high surface area, are also commonly used in Cr(VI) removal due to their effectiveness (Yu et al. 2013). The results of a study by Hu et al. (2007) on the use of different types of nanoparticles for Cr(VI) removal showed that the order of Cr(VI) adsorption capacity was $MnFe_2O_4 > MgFe_2O_4 > ZnFe_2O_4 > CuFe_2O_4 > NiFe_2O_4 > CoFe_2O_4$. In conclusion, the use of nanoscale adsorbents in wastewater treatment presents a promising approach for removing Cr(VI) from wastewater and requires further investigation to fully understand the impact of reaction conditions and the type of adsorbent used.

10.11.5.10 Treatment Media Used for Chromium Removal

The remediation of Cr(VI) in wastewater presents a significant challenge due to factors such as high operational costs, complex treatment processes, and substantial sludge production. To address these challenges, Scientists introduced a cost-efficient solution by constructing a composite biosorbent consisting of acid-treated oil palm shell charcoal coated with chitosan (CCAB). Chitosan has been demonstrated to possess a high sorption capability for metal ions, including chromium, and its efficacy can be influenced by various environmental conditions such as initial pH, agitation, biosorbent dose, and contact duration. A case study has shown that 92% of chromium can be effectively treated with the composite biosorbent at a pH of 5.

Vermiculite has also been investigated for its potential in eliminating chromium from tannery effluent. A study conducted in Ethiopia and found that the combination of horizontal settling tanks and an advanced integrated wastewater pond system (AIWPS) was capable of removing 58–95% of trivalent chromium after a one-day retention period at a pH close to 8, which is the optimal precipitation pH for trivalent chromium. A comparative study of mono and mixed columns of vermiculite of varying grades revealed that the mixed column was more efficient in removing chromium (74.6%) compared to mono vermiculite (63.6%).

In the preliminary stage of wastewater treatment, the removal of particulate and colloidal matter is crucial. The chemically enhanced primary treatment (CEPT) technology utilizes various coagulants to attain improved pollutant removal, with alum being the most effective coagulant for tannery wastewater. When applied in a dose range of 200–240 mg/L as $Al_2(SO_4)_3$, alum effectively removes 98.7–99.8% of chromium, leading to an effluent that complies with the national effluent quality standards for chromium and total suspended solids. However, secondary treatment is still necessary to address other chemical oxygen demand contents. In conclusion, the CEPT method has been demonstrated as an effective approach for the removal of chromium from tannery wastewater.

10.12 Conclusion

The toxicity of the heavy metal chromium has become a matter of significant concern due to its widespread presence in food and the environment. The metal is commonly utilized as an additive in industrial processes, such as leather processing and textile manufacturing, and has been shown to enter the food chain through contaminated water and soil. Chronic exposure to chromium has been linked to a range of deleterious health effects, including respiratory dysfunctions, cutaneous irritation, and an elevated risk of carcinogenesis. Vulnerable populations, including children, pregnant women, and those with weakened immune systems, may be particularly susceptible to the toxic effects of chromium. Despite the well-documented health risks posed by chromium toxicity, our understanding of its impact on human health remains limited. Further investigation is required to fully comprehend the issue and to develop effective strategies for mitigating risk. This involves studying the underlying mechanisms of chromium toxicity and identifying populations that may be at increased risk. In light of these considerations, it is imperative that regulatory bodies take prompt action to monitor and regulate chromium levels in food and the environment. This should include setting acceptable limits for chromium in food additives and monitoring environmental concentrations. In addition, public education campaigns aimed at raising awareness of the risks posed by chromium toxicity and promoting individual efforts to reduce exposure should be a priority. As researchers and public health experts, it is our responsibility to continuously examine the health risks posed by chromium toxicity and to advocate for public health protective policies. By collaborating and pooling our resources, we can mitigate the harm caused by this toxic metal and secure a healthy future for all.

References

Abbas S (2018) Chromium toxicity and its effects on kidney function. J Environ Sci Health 53(12):1033–1042

Aggarwal V, Tuli HS, Varol A, Thakral F, Yerer MB, Sak K et al (2019). Role of reactive oxygen species in cancer progression: molecular mechanisms and recent advancements. Biomolecules 9(11):735

Akbal F, Camcı S (2010) Comparison of electrocoagulation and chemical coagulation for heavy metal removal. Chem Eng Technol 33(10):1655–1664

Akerstrom B, Eriksson P, Vahter M (2013) Chromium and its toxic effects on living organisms. Environ Toxicol Pharmacol 36(2):526–535

Akinola O (2019) Chromium toxicity and liver injury. Toxins 11(5):365

Al Osman M, Yang F, Massey IY (2019) Exposure routes and health effects of heavy metals on children. Biometals 32:563–573

Almaguer-Busso G, Velasco-Martínez G, Carreño-Aguilera G, Gutiérrez-Granados S, Torres-Reyes E, Alatorre-Ordaz A (2009) A comparative study of global hexavalent chromium removal by chemical and electrochemical processes. Electrochem Commun 11(6):1097–1100

Annamalai K, Nair AM, Chinnaraju S, Kuppusamy S (2014) Removal of chromium from contaminated effluent and simultaneously Green nanoparticle synthesis using *Bacillus subtilis*. Malaya J Biosci 1(1):13–18

Ayangbenro AS, Babalola OO (2017) A new strategy for heavy metal polluted environments: a review of microbial biosorbents. Int J Environ Res Public Health 14(1):94

Azimi A, Azari A, Rezakazemi M, Ansarpour M (2017) Removal of heavy metals from industrial wastewaters: a review. ChemBioEng Rev 4(1):37–59

Balakrishnan R, Kumar CSS, Rani MU, Srikanth MK, Boobalan G, Reddy AG (2013) An evaluation of the protective role of α-tocopherol on free radical induced hepatotoxicity and nephrotoxicity due to chromium in rats. Indian J Pharmacol 45(5):490

Bhatnagar A, Ansari GA (2007) Chromium and its adverse effects on environment and human health. J Environ Biol 28(1):411–424

Breslin CB, Branagan D, Garry LM (2019) Electrochemical detection of Cr(VI) with carbon nanotubes decorated with gold nanoparticles. J Appl Electrochem 49:195–205

Cervantes C, Campos-García J, Devars S, Gutiérrez-Corona F, Loza-Tavera H, Torres-Guzmán JC, Moreno-Sánchez R (2001) Interactions of chromium with microorganisms and plants. FEMS Microbiol Rev 25(3):335–347

Chen L, Ma L, Bai Q, Zhu X, Zhang J, Wei Q et al (2014) Heavy metal-induced metallothionein expression is regulated by specific protein phosphatase 2A complexes. J Biol Chem 289(32):22413–22426

Coogan TP, Motz J, Snyder CA, Squibb KS, Costa M (1991) Differential DNA-protein crosslinking in lymphocytes and liver following chronic drinking water exposure of rats to potassium chromate. Toxicol Appl Pharmacol 109(1):60–72

Das AP, Singh S (2011) Occupational health assessment of chromite toxicity among Indian miners. Indian J Occup Environ Med 15(1):6

DeLoughery Z, Luczak MW, Zhitkovich A (2014) Monitoring Cr intermediates and reactive oxygen species with fluorescent probes during chromate reduction. Chem Res Toxicol 27(5):843–851

Deng Y, Wang M, Tian T, Lin S, Xu P, Zhou L et al (2019) The effect of hexavalent chromium on the incidence and mortality of human cancers: a meta-analysis based on published epidemiological cohort studies. Front Oncol 9:24

Dharnaik AS, Ghosh PK (2014) Hexavalent chromium [Cr(VI)] removal by the electrochemical ion-exchange process. Environ Technol 35(18):2272–2279

Dong G, Wang Y, Gong L, Wang M, Wang H, He N et al (2013) Formation of soluble Cr(III) end-products and nanoparticles during Cr(VI) reduction by Bacillus cereus strain XMCr-6. Biochem Eng J 70:166–172

Duarte HA, Jha K, Weidner JW (1998) Electrochemical reduction of nitrates and nitrites in alkaline media in the presence of hexavalent chromium. J Appl Electrochem 28:811–817

Edwards KC, Kim H, Vincent JB (2020) Release of trivalent chromium from serum transferrin is sufficiently rapid to be physiologically relevant. J Inorg Biochem 202:110901

Elahi A, Arooj I, Bukhari DA, Rehman A (2020) Successive use of microorganisms to remove chromium from wastewater. Appl Microbiol Biotechnol 104:3729–3743

Emamverdian A, Ding Y, Mokhberdoran F, Xie Y (2015) Heavy metal stress and some mechanisms of plant defense response. Sci World J

Environmental Protection Agency (2019) National primary drinking water regulations. https://www.epa.gov/ground-water-and-drinking-water/national-primary-drinking-water-regulations

Escudero C, Fiol N, Villaescusa I (2006) Chromium sorption on grape stalks encapsulated in calcium alginate beads. Environ Chem Lett 4:239–242

Eskandari F (2020) Chromium toxicity and liver injury: a review of the literature. J Trace Elem Med Biol 58:126–131

Fang Z, Zhao M, Zhen H, Chen L, Shi P, Huang Z (2014) Genotoxicity of tri-and hexavalent chromium compounds In vivo and their modes of action on DNA damage In vitro. Plos One 9(8):e103194

Ferreira LM, Cunha-Oliveira T, Sobral MC, Abreu PL, Alpoim MC, Urbano AM (2019) Impact of carcinogenic chromium on the cellular response to proteotoxic stress. Int J Mol Sci 20(19):4901

Flynn A (2014) Chromium toxicity and its impact on cardiovascular health. Nutr Metabol Insights 7:49–53

Focardi S, Pepi M, Focardi SE (2013) Microbial reduction of hexavalent chromium as a mechanism of detoxification and possible bioremediation applications. Biodegrad Life Sci 321–347

Fu F, Wang Q (2011) Removal of heavy metal ions from wastewaters: a review. J Environ Manag 92(3):407–418

Giagnorio M, Steffenino S, Meucci L, Zanetti MC, Tiraferri A (2018) Design and performance of a nanofiltration plant for the removal of chromium aimed at the production of safe potable water. J Environ Chem Eng 6(4):4467–4475

Guo S, Xiao C, Zhou N, Chi R (2021) Speciation, toxicity, microbial remediation and phytoremediation of soil chromium contamination. Environ Chem Lett 19:1413–1431

Hayat S, Khalique G, Irfan M, Wani AS, Tripathi BN, Ahmad A (2012) Physiological changes induced by chromium stress in plants: an overview. Protoplasma 249:599–611

Hu J, Lo IM, Chen G (2007) Comparative study of various magnetic nanoparticles for Cr(VI) removal. Sep Purif Technol 56(3):249–256

Hua M, Zhang S, Pan B, Zhang W, Lv L, Zhang Q (2012) Heavy metal removal from water/wastewater by nanosized metal oxides: a review. J Hazard Mater 211:317–331

International Agency for Research on Cancer (IARC) (1990) Chromium, nickel and welding. In: Monographs on the evaluation of carcinogenic risks to humans, pp 491–648

Jobby R, Jha P, Yadav AK, Desai N (2018) Biosorption and biotransformation of hexavalent chromium [Cr(VI)]: a comprehensive review. Chemosphere 207:255–266

Kaur G, Kaur J (2010) Chromium and its adverse effects on human health. Indian J Med Res 132(5):557

Kieber RJ, Willey JD, Zvalaren SD (2002) Chromium speciation in rainwater: temporal variability and atmospheric deposition. Environ Sci Technol 36(24):5321–5327

Kim T, Kim TK, Zoh KD (2020) Removal mechanism of heavy metal (Cu, Ni, Zn, and Cr) in the presence of cyanide during electrocoagulation using Fe and Al electrodes. J Water Process Eng 33:101109

Linos A, Petralias A, Christophi CA, Christoforidou E, Kouroutou P, Stoltidis M et al (2011) Oral ingestion of hexavalent chromium through drinking water and cancer mortality in an industrial area of Greece-an ecological study. Environ Health 10(1):1–8

Loyaux-Lawniczak S, Lecomte P, Ehrhardt JJ (2001) Behavior of hexavalent chromium in a polluted groundwater: redox processes and immobilization in soils. Environ Sci Technol 35(7):1350–1357

Macfie A, Hagan E, Zhitkovich A (2010) Mechanism of DNA-protein cross-linking by chromium. Chem Res Toxicol 23(2):341–347

Machado R, Carvalho JR, Joana Neiva Correia M (2002) Removal of trivalent chromium (III) from solution by biosorption in cork powder. J Chem Technol Biotechnol Int Res Process Environ Clean Technol 77(12):1340–1348
Mishra S, Bharagava RN (2016) Toxic and genotoxic effects of hexavalent chromium in environment and its bioremediation strategies. J Environ Sci Health C 34(1):1–32
Mishra A, Gupta B, Kumar N, Singh R, Varma A, Thakur IS (2020) Synthesis of calcite-based bio-composite biochar for enhanced biosorption and detoxification of chromium Cr(VI) by Zhihengliuella sp. ISTPL4. Bioresour Technol 307:123262
National Toxicology Program (2011) Report on Carcinogens. Research Triangle Park, NC, U.S. Department of Health and Human Services, Public Health Service
Noriega G, Caggiano E, Lecube ML, Cruz DS, Batlle A, Tomaro M, Balestrasse KB (2012) The role of salicylic acid in the prevention of oxidative stress elicited by cadmium in soybean plants. Biometals 25:1155–1165
Nogueira V, Lopes I, Rocha-Santos T, Gonçalves F, Pereira R (2015) Toxicity of solid residues resulting from wastewater treatment with nanomaterials. Aquat Toxicol 165:172–178
Patlolla AK, Barnes C, Yedjou C, Velma VR, Tchounwou PB (2009) Oxidative stress, DNA damage, and antioxidant enzyme activity induced by hexavalent chromium in Sprague-Dawley rats. Environ Toxicol Int J 24(1):66–73
Pavesi T, Moreira JC (2020) Mechanisms and individuality in chromium toxicity in humans. J Appl Toxicol 40(9):1183–1197
Philip L, Iyengar L, Venkobachar C (1998) Cr (VI) reduction by Bacillus coagulans isolated from contaminated soils. J Environ Eng 124(12):1165–1170
Powers J (2009) Chromium toxicity and its effects on glucose metabolism and insulin sensitivity. Diabetes Obes Metab 11(7):613–619
Saba Anwar MI, Raza SH, Iqbal NAEEM (2013) Efficacy of seed preconditioning with salicylic and ascorbic acid in increasing vigor of rice (Oryza sativa L.) seedling. Pak J Bot 45(1):157–162
Salnikow K, Zhitkovich A (2008) Genetic and epigenetic mechanisms in metal carcinogenesis and cocarcinogenesis: nickel, arsenic, and chromium. Chem Res Toxicol 21(1):28–44
Schneider HJ, Waller DP, Svec F (2013) Chromium in human nutrition and health. Nutr Rev 71(8):548–558
Schrauzer GN, Shrestha KP (2002) Chromium in the natural environment. Annu Rev Nutr 22(1):277–300
Seidler A, Jähnichen S, Hegewald J, Fishta A, Krug O, Rüter L et al (2013) Systematic review and quantification of respiratory cancer risk for occupational exposure to hexavalent chromium. Int Arch Occupat Environ Health 86:943–955
Shakoor MB, Ali S, Rizwan M, Abbas F, Bibi I, Riaz M, Rinklebe J (2020) A review of biochar-based sorbents for separation of heavy metals from water. Int J Phytorem 22(2):111–126
Shanker AK, Cervantes C, Loza-Tavera H, Avudainayagam S (2005) Chromium toxicity in plants. Environ Int 31(5):739–753
Sharma P, Bihari V, Agarwal SK, Verma V, Kesavachandran CN, Pangtey BS, Goel SK (2012) Groundwater contaminated with hexavalent chromium [Cr(VI)]: a health survey and clinical examination of community inhabitants (Kanpur, India). Plos One 7(10):e47877
Shi M, Li Z, Yuan Y, Yue T, Wang J, Li R, Chen J (2015) In situ oxidized magnetite membranes from 316L porous stainless steel via a two-stage sintering process for hexavalent chromium [Cr(VI)] removal from aqueous solutions. Chem Eng J 265:84–92
Stambulska UY, Bayliak MM, Lushchak VI (2018) Chromium(VI) toxicity in legume plants: modulation effects of rhizobial symbiosis. BioMed Res Int
Thompson CM, Proctor DM, Haws LC, Hébert CD, Grimes SD, Shertzer HG, Harris MA (2011) Investigation of the mode of action underlying the tumorigenic response induced in B6C3F1 mice exposed orally to hexavalent chromium. Toxicol Sci 123(1):58–70
Tian X, Zhang H, Zhao Y, Mehmood K, Wu X et al (2018) Transcriptome analysis reveals the molecular mechanism of hepatic metabolism disorder caused by chromium poisoning in chickens. Environ Sci Pollut Res 25:15411–15421

Venter C, Oberholzer HM, Taute H, Cummings FR, Bester MJ (2015) An in ovo investigation into the hepatotoxicity of cadmium and chromium evaluated with light-and transmission electron microscopy and electron energy-loss spectroscopy. J Environ Sci Health Part A 50(8):830–838

Vilardi G, Di Palma L, Verdone N (2019) A physical-based interpretation of mechanism and kinetics of Cr(VI) reduction in aqueous solution by zero-valent iron nanoparticles. Chemosphere 220:590–599

Wakeman TP, Yang A, Dalal NS, Boohaker RJ, Zeng Q, Ding Q, Xu B (2017) DNA mismatch repair protein Mlh1 is required for tetravalent chromium intermediate-induced DNA damage. Oncotarget 8(48):83975

Wang X, Mandal AK, Saito H, Pulliam JF, Lee EY, Ke ZJ, Shi X (2012) Arsenic and chromium in drinking water promote tumorigenesis in a mouse colitis-associated colorectal cancer model and the potential mechanism is ROS-mediated Wnt/β-catenin signaling pathway. Toxicol Appl Pharmacol 262(1):11–21

Wang H, Na C (2014) Binder-free carbon nanotube electrode for electrochemical removal of chromium. ACS Appl Mater Interfaces 6(22):20309–20316

Wang Y (2019) Hexavalent chromium and its impact on lung cancer risk. Environ Res 170:138–145

Welling R, Beaumont JJ, Petersen SJ, Alexeeff GV, Steinmaus C (2015) Chromium VI and stomach cancer: a meta-analysis of the current epidemiological evidence. Occup Environ Med 72(2):151–159

Wilhelm T, Said M, Naim V (2020) DNA replication stress and chromosomal instability: dangerous liaisons. Genes 11(6):642

World Health Organisation (1990) Chromium (Environmental Health Criteria 61) international programme on chemical safety. Switzerland, Geneva

Xiao F, Feng X, Zeng M, Guan L, Hu Q, Zhong C (2012) Hexavalent chromium induces energy metabolism disturbance and p53-dependent cell cycle arrest via reactive oxygen species in L-02 hepatocytes. Mol Cell Biochem 371:65–76

Yang W, Song W, Li J, Zhang X (2020) Bioleaching of heavy metals from wastewater sludge with the aim of land application. Chemosphere 249:126134

Yao H, Guo L, Jiang BH, Luo J, Shi X (2008) Oxidative stress and chromium(VI) carcinogenesis. J Environ Pathol Toxicol Oncol 27(2)

Yu X, Tong S, Ge M, Wu L, Zuo J, Cao C, Song W (2013) Adsorption of heavy metal ions from aqueous solution by carboxylated cellulose nanocrystals. J Environ Sci 25(5):933–943

Zamboulis D, Pataroudi SI, Zouboulis AI, Matis KA (2004) The application of sorptive flotation for the removal of metal ions. Desalination 162:159–168

Zengin F (2014) Exogenous treatment with salicylic acid alleviating copper toxicity in bean seedlings. Proc Natl Acad Sci India Sect B: Biol Sci 84:749–755

Zhao Y, Zhang H, Wu X, Zhang T, Shen K, Li L et al (2019) Metabonomic analysis of the hepatic injury suffer from hexavalent chromium poisoning in broilers. Environ Sci Pollut Res 26:18181–18190

Zhitkovich A (2005) Importance of chromium–DNA adducts in mutagenicity and toxicity of chromium(VI). Chem Res Toxicol 1(18):3–11

Zhou Y (2017) Chromium in drinking water and the risk of gastric cancer: a systematic review. Environ Res 155:662–668

Chapter 11
Phytoremediation of Chromium from Soil and Water

Monalisa Mohanty and Hemanta Kumar Patra

Abstract Chromium (Cr) is a common heavy metal pollutant found in mine and industrial areas. Restoration of soil and water quality in these locations is highly crucial for healthy living and sustainable development. Mostly Chromium (Cr) persists in the environment in stable hexavalent (Cr^{+6}) and trivalent (Cr^{+3}) forms. The former is more toxic than the later. Cr^{+6} form of chromium is abundant in mine soil and wastewater and causes serious health problems. Therefore, there is a high need for sustainable eco-friendly low-cost technology for Cr removal from soil and wastewater to substitute costly physicochemical treatments. Phytoremediation is a cost-effective, eco-friendly plant-based approach to remediate toxic Cr from contaminated soil and water. To improve the effectiveness of Cr removal by plants, thorough understanding on Cr in a soil-water-plant relationship and its accumulation in the plants is indispensable. This chapter describes the modes of bio-sorption and remediation of Cr from soil and water using algae, fungi, and higher plants. The roles of aquatic, tolerant, and hyperaccumulator plant species are described. The means of Cr removal by bio-sorption, phytosiderophore and chelate-based remediation are discussed. The use of phytoremediation techniques like rhizofiltration, phytoextraction, phytostabilisation and rhizodegradation/rhizosphere remediation for Cr removal including the application of genetic engineering and rhizospheric microbes are emphasized.

Keywords Chromium · Bio-sorption · Phytoremediation · Soil · Water

---

M. Mohanty (✉)
Assistant Professor, Department of Biotechnology, Ramadevi Women's University, Vidya Vihar, Bhubaneswar 751022, Odisha, India
e-mail: 18.monalisa@gmail.com; monalisamohanty@rdwu.ac.in

H. K. Patra
Visitng Professor, Department of Botany, Utkal University, Bhubaneswar 752004, Odisha, India

© The Author(s), under exclusive license to Springer Nature Switzerland AG 2023

N. Kumar et al. (eds.), *Chromium in Plants and Environment*, Environmental Science and Engineering, https://doi.org/10.1007/978-3-031-44029-8_11

11.1 Introduction

In recent years the toxic nature of heavy metals viz. Chromium (Cr) has created much interest and attention by the researchers in fields of biology and environmental science. The commercial exploitations of Cr from chromite ore (Primarily of Fe^{+2} and Cr^{+3} oxide compounds viz. $FeCr_2O_4$) by mining and industrial operations lead to very serious environmental hazards due to the formation of toxic hexavalent chromium (Cr^{+6}) in soil and water. Cr^{+6} has wide use for the synthesis of chemicals, paints, and metal coating, etc. On the other side, Cr^{+3} has been considered as less toxic as compared to Cr^{+6}. Cr^{+3} can be used as an essential and nutritional supplements for human, however, it is not yet established to consider Cr as a required nutrient for plants (WHO 1988).

Toxicity nature of Cr attracted more attention because of its availability to food chain. Plants in soil and water environments with wide diversity responds to Cr toxicity by tolerance and defence mechanisms along with several cytological, genetical and physiological responses (Shanker et al. 2005, Singh et al. 2013; Jutsz and Gnida 2015; Shahid et al. 2017b; Ao et al. 2022). During last three decades many research papers and reviews are published on Cr chemistry, its impacts on humans, plants and other organisms (WHO 1988; Katz and Salem1994; Greenwood and Eamshaw 1997; Shanker et al. 2005; Hansel et al. 2015; Bhalerao and Sharma 2015; Gomes et al. 2017; Shahid et al. 2017b, Ao et al. 2022). At the same time, attempts are being taken on Cr removal from soil and water using different plants as tool of remediation and the technologies behind the process (Salt et al. 1998; Zayed and Terry 2003; Mohanty and Patra 2011, 2012a; Patra et al. 2018a, b, 2019, 2020a, b, 2021). The availability of Cr in soil-water to plant relationship and its accumulation in plant organs is inevitable. In this chapter, we describe the modes of remediation of Cr from soil and water in a wider range using algae, fungi and higher plant species. The key roles of terrestrial land hyperaccumulators and aquatic species are described along with the modes of tolerance and resistance mechanisms. The means of Cr removal by bio-sorption, phytosiderophore and chelate-based remediation are analyzed. The use of phytoremediation techniques like rhizofiltration, phytoextraction, phytostabilisation and rhizodegradation/rhizosphere remediation for Cr removal including the application of genetic engineering and rhizospheric microbes are discussed.

11.2 Chromium in the Environment and Its Toxicity

Chromium is a transitional metal of group six (atomic number 24, relative atomic mass 51.996) and occurs in different oxidation states (WHO 1988; Greenwood and Eamshaw 1997). Among different charges ranging from −2 to +6, the +3 and +6 oxidation states occur mostly in soil environment (Greenwood and Eamshaw 1997). The chemistry of Cr in +3 and +6 oxidation states are described by Holleman et al. (1985).

$$Cr_2O_7^{2-} + 14H_3O^+ + 6e^- \rightarrow 2Cr^{3+} + 21H_2O$$

$$CrO_4^{2-} + 4H_2O + 3e^- \rightarrow Cr(OH)_3 + 5OH^-$$

Among different 'heavy metals', chromium (Cr) is the 7th most available metal (Katz and Salem 1994; Singh et al. 2013) with density 7.15 g cm^{-3}. Chromium exists in all environmental segments (soil, water and air) including biota at different concentrations which has been described vividly (Shanker et al. 2005; Ao et al. 2022). Insoluble trivalent form of chromium in chromite is oxidized to hexavalent one by weathering process (Hansel et al. 2015), which is then cycled to the marine water through rivers. Two main oxidation forms of Chromium (Cr^{+3}/Cr^{+6}) produced by natural weathering of Cr-containing rocks and anthropogenic activities (mining and industrial) enters the biological system via waters and soils. The origin and cycling of Cr in environment and biota are given in Fig. 11.1.

Cr in aquatic environment undergoes several geo-chemical conversions like oxidation, reduction, sorption, dissolution, and suspension (Bhalerao and Sharma 2015). Like water, Cr in soil also undergoes a variety of conversions like oxidation, reduction etc. The aqueous solubility of Cr^{+3} and Cr^{+6} is a function of the pH of the water and Cr^{+6} as chromate and dichromate are extremely soluble in water (Bhalerao and Sharma 2015). The recommended permissible limits for Cr concentration in water are 8 g L^{-1} for Cr^{+3} and 1 g L^{-1} for Cr^{+6}. Cr^{+6} in industrial effluents range from 2 to 5 g L^{-1} (Bhalerao and Sharma 2015). Cr in soil generally found in both Cr^{+3} and Cr^{+6} forms. The oxidants present in the soil (dissolved O_2 and MnO_2) can oxidize Cr^{+3}–Cr^{+6} (Bhalerao and Sharma 2015). Cr^{+6} can persist in soil for several years when soil is sandy or with low levels of organic matters ((Bhalerao and Sharma

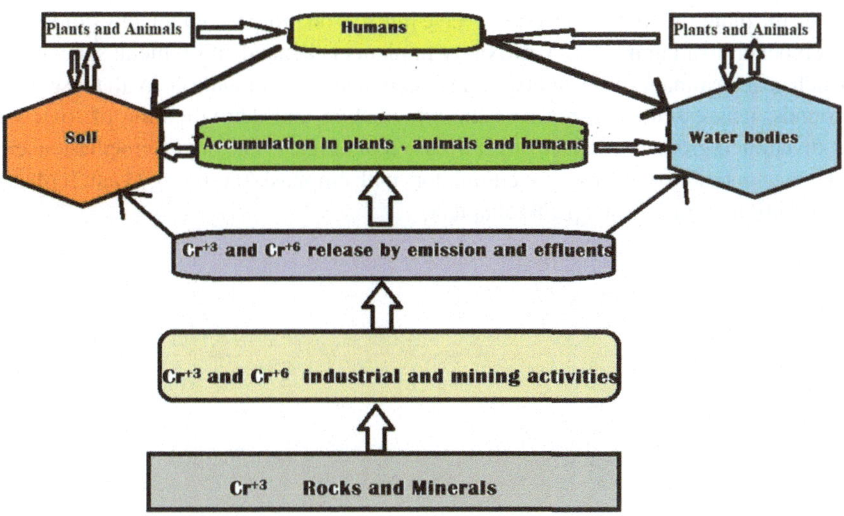

Fig. 11.1 The cycling of Cr in environment

2015). Cr^{+6} is reduced to the trivalent state (Cr^{+3}) in soil-water environment that makes Cr^{+3} less toxic to the organisms (Medda and Mondal 2017; Ao et al. 2022). The shifting of chromium from soils to living organism through water and back is a regular phenomenon. Chromium in different chemical forms reveals dissimilar effects. Cr^{+6} is highly mobile in soil and water environment. It is known to exhibit toxic impacts on living organisms leading to morpho-physiological, mutagenic and carcinogenic hazards (WHO 1988; Shanker et al. 2005; Shahid et al. 2017b; Ao et al. 2022).

11.3 Chromium in Soil-Water-Plant Relationship

The creation of living organism from non-living beings leads to the sustenance of living ones against negative situations. Plants and microbes coexist in nature, especially plants adapts to nature with acquired resistance mechanisms against biotic and abiotic stresses which ultimately becomes a nature's gift for human safety and welfare. The interaction of Cr with the plants is well-known to biologists for the last several decades till date (WHO 1988; Shanker et al. 2005; Gomes et al. 2017; Shahid et al. 2017b; Ao et al. 2022). The behaviour Cr in soil and its accumulation in different plant parts through soil-plant transfer varies with its speciation, soil physico-chemical properties and plant type (Singh et al. 2013; Bhalerao and Sharma 2015; Gomes et al. 2017; Shahid et al. 2017b; Ao et al. 2022). The pathway of Chromium in a soil-water-plant relationship and its fate in plants' cellular system is given in Fig. 11.2.

Major amount of chromium accumulation takes in plant (Shanker et al. 2005; Bulent et al. 2007; Lotfy and Mostafa 2014; Usman et al. 2020; Ao et al. 2022). Inside the plants, Cr incites several adverse impacts on morphological, physiological, and biochemical parameters. Cr toxicity in plants is dependent on metal speciation, which is determinant for its uptake, translocation and accumulation. With these back grounds, it is essential to reassess the remediation and detoxification of toxic Cr by diversified group of plants using various technologies. The phytoremediation and tolerance potential of the hyperaccumulators with emphasis on strategy I and II plants are highlighted in following sub-chapters.

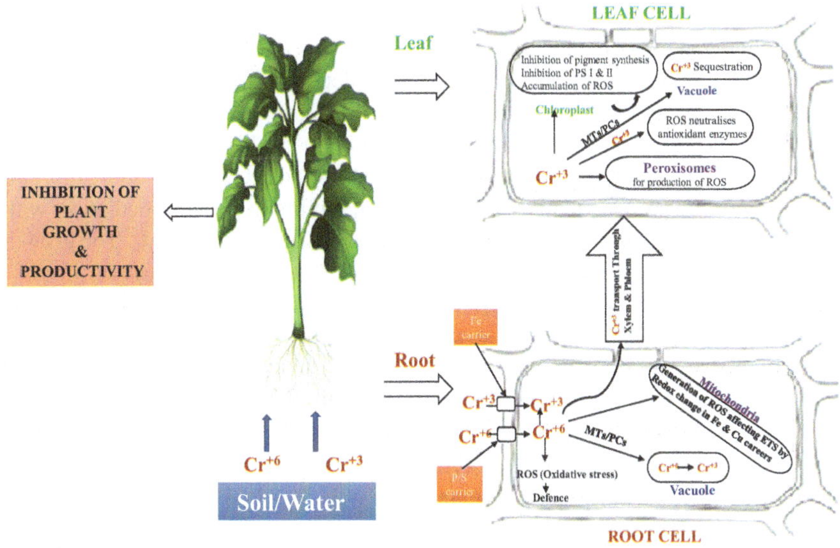

Fig. 11.2 The entry of Cr^{+6} and Cr^{+3} soil/water in to the living cells of a plant and their behaviours in plant system

11.4 Phytoremediation of Chromium Using Green Plants, Algae and Other Microbes

11.4.1 Phycoremediation

Phycoremediation is the method of removal of toxic and harmful pollutant from environment using micro and macro algae (John 2003; Danouche et al. 2021). The advantages phycoreemediation lies on its efficacy, eco-friendly approach and cost effective methodology over the conventional technology. The algae can also be commercially applied for waste water management in manmade pond eco-system for phycostabilization programme especially concerned with industries owing to its costeffectiveness and high turnover rate (Sunday et al. 2018). The capability of algae to remove and detoxify toxic metal pollutants has been developed in nature during the course progressive evolution and adaptation (Monteiro et al. 2012). This sub-chapter phycoremediation is strategically important for detoxification pollutants of waste water using algae as natural organisms. It also provides the latest views on phycoremediation approaches using micro and maco-algae for selectivity and use in biotechnology.

The phycoremediation technology includes extracellular (biosorption: nonmetabolic) and intracellular (bioaccumulation: metabolic) remediation strategies for mitigation of heavy metal toxicity (Kumar et al. 2015; Danouche et al. 2021; Kuanar et al. 2022). The extracellular process for heavy metal bioremoval is carried out by biosorption in the cell wall of the micro-algae which are achieved by physico-chemical phenomenon and production of extracellular polymeric substances (Danouche et al. 2021). Intracellular heavy metal detoxification mechanism includes bioaccumulation, bio-transformation and sub-cellular compartmentalization in vacuoles, chloroplasts and mitochondria (Kumar et al. 2015; Danouche et al. 2021). As reported by researchers (Spain et al. 2021; Ayele and Godeto 2021) *Chlorella* and *Scenedesmus* are frequently used microalgal strains in phycoremediation with varying bio-sorption efficiency for different species (Kumar et al. 2015). The growth of two algal species (*Chlorella sorokiniana* and *Scenedesmus obliquus*) in Cr^{+6} contaminated water was significantly different (Danouche et al. 2021). The algae *Spirullina plantensis* and *Spirogyra* were studied for bio-sorption of heavy metals of chromium at various concentrations. *Spirullina plantensis* showed more tolerance to the heavy metal chromium and bio-sorption than *Spirogyra sp*s. (Samuel et al. 2019). The maximum adsorption of Cr(III) ions on the cell surface of *Spirulina platensis* was recorded as 31.25 mg/g (Tatibayeva and Tazhibayeva 2021) and various other case studies on phycoremediation of Cr is given in Table 11.1.

The bioremediation capabilities of microalgae versus macroalgae have been studied comprehensively worldwide. The phycoremediation approaches using different algae visualize that the microalgal remediation potential offers enhanced biosorption advantages. Remarkable remediation ability of microalgae are being noticed by Kumar et al. (2015). They are suitable for small-scale as well as large-scale remediation purposes, particularly in contaminated waters and soils. Therefore, it is essential to emphasize on various opportunities of micro algal remediation technologies which are ecofriendly in nature.

11.4.2 Mycoremediation

Mycoremediation is a fungal-based remediation measure for pollution abatement programme. The availability of fungi in nature are less in aquatic environment due to their saprophytic character and therefore, their use for remediation of pollutants with reference to living habitats are less as compared to algae and higher plants. However, some efforts are made to use and refer some fungi for remediation of toxic pollutants (Joshi et al. 2011; Gazem and Nazareth 2013; Sharma and Malaviya 2016; Romo-Rodríguez and Gutiérrez-Corona 2019; Kullu et al. 2020; Ayele and Godeto 2021; Kuanar et al. 2022). Mycoremediation is a economic method and is often used in purpose, such as mycofiltration of waste water (Molla and Fakhru'l-Razi 2012). The hyperaccumulation activity of fungi take place via extracellular biosorption, with low intracellular uptake (Gazem and Nazareth 2013) Fungi, such as *Phanerochaete chrysosporium, Aspegillus niger, Trichoderma viride and Trichoderma longibrachiatum* have proven to be effective for Cr exclusion in aqueous medium (Joshi et al. 2011). The elimination of soil contamination by mycorrhizal fungi (AMF: arbuscular mycorrhizal fungi) is called mycorrhizo-remediation. This method is helpful for removal of Cr pollutants from contaminated soils using AM

Table 11.1 The means of phycoremediation of Cr by some mciro and macro-algae

Algae	Source of Cr	Remediation of Cr	References
Chlorella vulgaris	350 μg l^{-1} Cr^{+6} in aqueous solution	*Bioconcentration of Cr*: 399.6 μg g^{-1} d. wt	Regaldo et al. (2009)
Chlorella vulgaris	1 mg l^{-1} Cr^{+6}in aqueous solution	*Bioremoval*: 23.06 μg/100 ml *Bio-absorption*: 19.8 μg/ 100 ml *Bio-adsorption*: 3.26 μg/ 100 ml	Nacorda et al. (2010)
Scenedesmus acutus	$CrCl_2$: 20 mg l^{-1} in aqueous solution	Bioremoval efficiency: 36%	Travieso et al. (1999)
Chlorella vulgaris	$CrCl_2$: 1 mg l^{-1} in aqueous solution	Bioremoval efficiency: 48%	
Pterocladia capillacea (Marine Red Algae)	Cr^{3+}waste water	*Sorption*: Bioremoval efficiency: 20–100%	El Nemr et al. (2015)
Spirulina maxima	Cr^{3+}waste water	*Biosorption*: Bioremoval efficiency: 77%	Singh et al. (2016)
Chitosan algal biomass (*Cladophora*sp.)	Cr^{3+} Micro-beads	*Adsorbent*: Bioremoval efficiency: 68%	Sargın et al. (2016)
Chlorella miniata	Cr^{+3} in aqueous solutions	*Biosorption*: Bioremoval efficiency: 85%	Anastopoulos and Kyzas (2015)
Scenedesmus quadricauda	Cr^{+6} in Aqueous solutions	*Biosorption*: Bioremoval efficiency: 60%	
Dunalielaa salina	Cr^{+2} in Aqueous solutions	*Biosorption*: Bioremoval efficiency: 74%	
Ulva lactuca	Cr^{+6} in Aqueous solutions	*Biosorption*: Bioremoval efficiency: 96%	
Chlorella vulgaris	Cr^{+6} in waste water	*Adsorption*: 3.5 mg g^{-1}	Nourbakhsh et al. (1994)
Cladophora crispate	Cr^{+6} in waste water	*Adsorption*: 3 mg g^{-1}	
Kappaphycus alvarezii	Cr^{+6} in waste water	*Adsorption* : 0.86 mg g^{-1}	Kang et al. (2011)
Pachymeniopis sp.	Cr^{+6} in waste water	*Adsorption* : 225 mg g^{-1}	Lee et al. (2000)
Spirulina sp.	Cr^{+6} in waste water	*Adsorption* : 90.91 mg g^{-1}	Rezaei (2016)
Chlorella miniata (Dead)	Cr^{+3}in Aqueous solutions(pH 3-4.5)	Metal uptake (mg/g): 14–41	Han et al. (2006)
Sargassum hornschuchii	Cr^{+6} 25 mg L^{-1}	99.36% removal	Fakhry et al. (2013)
Planothidium lanceolatum (Diatom)	Cr^{+6}(20 mg l^{-1}) in aqueous solutions	Cr uptake: 93.45 mg g $diatoms^{-1}$	Shibi et al. (2012)
Chlorella vulgaris (Non-living)	Cr^{+6} (200 ppm) in aqueous solutions	Cr uptake: 23.6 mg g^{-1}	Aksu and Kutsal (1990)
Dunaliella sp.(Non-living)	Cr^{+6} (250–300 mg l^{-1}) in aqueous saline solutions	Cr uptake: 45.5–58.3 mg g^{-1}	Dönmez and Aksu (2002)

(continued)

Table 11.1 (continued)

Algae	Source of Cr	Remediation of Cr	References
Chlorella vulgaris (Non-living)	Cr^{+6} (1 gm/l^{-1}) in aqueous solutions	Cr uptake: 23 mg g^{-1}	Dönmez et al. (1999)
Scenedesmus obliquus (Non-living)	Cr^{+6} (25–250 mg l^{-1}) in aqueous solutions	Cr uptake: 15.6 mg g^{-1}	
Synechocystis sp. (Non-living)	Cr^{+6} (25–250 mg l^{-1}) in aqueous solutions	Cr uptake: 19.2 mg g^{-1}	
Chlamydomonas angulosa (Non-living)	Cr^{+6} in waste water	Cr uptake: 5.32 mg g^{-1} dry wt	Dwivedi et al. (2010)
Oscillatoria nigra (Non-living)	Cr^{+6} in waste water	Cr uptake: 1.86 mg g^{-1} dry wt	
Oscillatoria tenuis (Non-living)	Cr^{+6} in waste water	Cr uptake: 7.35 mg g^{-1} dry wt	
Phormidium bohneri (Non-living)	Cr^{+6} in waste water	Cr uptake: 8.55 mg g^{-1} dry wt	
Ulothrix tenuissima (Non-living)	Cr^{+6} in waste water	Cr uptake: 4.56 mg g^{-1} dry wt	
Scenedesmus sp.	Cr^{+6} (10 mg l^{-1}) in aqueous solutions	Cr^{+6} removal: 92.89%	Pradhan et al. (2019)
Pseudopediastrum boryanum	Cr^{+6} (10 mg l^{-1}) in aqueous solutions	Cr^{+6} removal: 70%	Sutkowy and Kłosowski (2018)
Chlorella colonials	Cr^{+6} (6.07mg/L) in aqueous solutions	Cr^{+6} removal: 97.8%	Jaafari and Yaghmaeian (2019)
Chlorella sorokiniana	Cr^{+6} (100 mg/L) in aqueous solutions	Cr^{+6} removal: 99.68%	Husien et al. (2019)
Chlorella vulagris	Cr^{+6} (100 ppm) in aqueous solutions	Cr^{+6} removal: 99.75%	Indhumathi et al. (2014)
Spirulina platensis	Cr^{+6} (150 ppm) in aqueous solutions	Cr^{+6}removal (Adsorption): 45.5mg/g	Nithya et al. (2019)
Chlamydomonas sp.	Cr^{+6} (250 ppm) in aqueous solutions	Cr^{+6} removal: 91.%	Ayele et al. (2021)
Polysiphonia urceolata	Cr^{+6} (250 ppm) in aqueous solutions	Cr^{+6} removal: 170.6 mg/g	Li et al. (2015)
Chondrus ocellatus	Cr^{+6} (250 ppm) in aqueous solutions	Cr^{+6} removal: 113.4 mg/g	
*Oscillatoria sp.*H1 (Living)	Cr^{+6} (200 ppm) in aqueous solutions	Cr^{+6} removal: 20.82 mg/g (98% efficiency)	Katircioğlu et al. (2012)
Nostoc sp	Cr^{+3} (100 ppm) in aqueous solutions	Cr^{+3} removal: 29 mg/g dry algae	Coronel and Varela (2015)

fungi (Kullu et al. 2020; Soni et al. 2022). Table 11.2 includes the remediation Chromium from water and soil as reported below.

The bioremediation of chromium by fungal and microbial consortia to minimize the contaminants from polluted soil and water is an attracting field of integrated phytotechnology. The biosorption capacity of biosorbents is the possible way facilitated by the metabolism-independent metal binding to the cell walls for which external surface is the only site present in the case of nonliving biomass. The biosorption process is improved by the integrated use of microbial consortia, biomass quantity, and other environmental factors like pH, temperature, contact time and metal concentration (Ayele and Godedo 2021). The above approaches might be useful in designing new methods for bioremediation of chromium from industrial and mine waste soil/water. The use of microbial consortia with immobilized bacterial, fungal/algal strains coupled with bio-surfactant will be helpful in future for Cr bioremediation processes which may prove to be economic, eco-friendly and sustainable.

11.4.3 Chelate-Based Remediation of Chromium by Plants & Phytosiderophores

The word chelate is synonymously called as ligand or chelator which originated from the Greek word 'Chele'. Chele means lobster's claw and the metal nutrient ion in soil is encircled by larger organic molecule or the claw. Chelates are known to render insoluble cations to soluble, thus they are available to plants in the root-solution-soil system (Lindsay 1974). Now a day, the chelated fertilizer is mostly applied to the soil to improve the bioavailability of micronutrients like Fe, Mn, Zn etc. for improving crop production. Chelants applied for heavy metal remediation can be divided into two group. One is amino-poly-carboxylic acids (APCAs), such as EDTA (ethylene diamine tetraacetic acid), EGTA (ethylene bis[oxyethylenetrinitrilo] tetraacetic acid, DTPA (diethylene triamino pentaacetic acid, HEDTA (hydroxyl ethylene diamine tetraaceticacid, EDDHA [ethylenediamine-N,N'bis(o-hydroxyphenyl) acetic acid] etc. which are synthetic ones. Natural APCAs such as NTA (nitrilo triacetic acid) and EDDS (ethylene diamine disuccinate) are also used for phytoextraction of heavy metals from soil. The other category are natural chelators with low molecular weight organic acids such as citric acid, oxalic acid, and tartaric acid, etc.

Chelate-based phytoremediation is considered as one of the efficient techniques for the remediation of heavy metal-contaminated soil and water (Salt et al. 1998; Roskova et al. 2022; Ranđelović et al. 2022). However, literature pertaining to the use of chelators for remediation of Cr contaminated soil/water are limited (Table 11.3) and complex due to the oxidation status of Cr in soil/water. In addition, chelate-based phytoremediation is still prone to several constraints and disadvantages. For example, leaching of mobilized metals from chelate-treated soils accompanied with

Table 11.2 The means of mycoremediation of Cr from soil and water

Fungi	Source of Cr	Remediation of Cr	References
Trichoderma longibrachiatum *Aspergillus niger* *Phanerochaete chrysosporium* *Trichoderma viride*	Cr^{+6} ($K_2Cr_2O_7$) 12.5 ppm in aqueous solution	Cr uptake: 0.55 mg g^{-1} Cr uptake: 0.05 mg g^{-1} Cr uptake: 0.1 mg g^{-1} Cr uptake: 0.03 mg g^{-1}	Joshi et al. (2011)
• Fungal strain *Aspergillus niger* (FIST1) followed by bacterial srain *Acinetobacter* sp. (IST3) • Bacterial srain *Acinetobacter* sp. *followed by fungal strain Aspergillus niger* (FIST1) (IST3	Cr^{+3}[$Cr_2(SO_4)_3$]: in soil sample of tannery effluent Cr^{+3}[$Cr_2(SO_4)_3$]: in soil sample of tannesry effluent	Cr removal: 90% after 15 days Cr removal: 64.7% after 15 days	Thakur and Srivatava (2011)
Gloeophyllm sepiarium (Brown rot fungus) *Penicillium* sp. IA-01 strain	Cr^{+6} contaminated soil Cr^{+6} (50 mg/l) in aqueous solution	Cr^{+6} removal: 94% Cr^{+6} removal: 91%	Achal et al. (2011) Acosta-Rodríguez et al. (2015)
Trichoderma viride	Cr^{+6} in bioreactor	Cr^{+6} removal: 60–94.3%	Morales-Barrera and Cristiani-Urbina (2006)
Carndida sp.	Cr^{+6} in bioreactor	Cr^{+6} removal: 100%	Guillén-Jiménez et al. (2009)
Aspergillus niger	Cr^{+3} in bioreactor	Cr^{+6} removal: 88%	Sepehr et al. (2012)
Candida tropicalis	Cr^{+6} in bioreactor	Cr^{+6} removal: 72.2%	Bahafid et al. (2013)
Bacterial-Fungal consortium (Raoultella sp., Citrobacter sp., Klebsiella sp., Salmonella sp., Achromobacter sp. and Kerstersia sp. and Pichia jadinii)	Cr^{+6} in bioreactor	Cr^{+6} removal: 100%	Tekerlekopoulou et al. (2011)
Aspergillus tubingensis	Cr^{+6} in bioreactor	Cr^{+6} removal: 100%	Coreño-Alonso et al. (2014)
Hypocrea tawa	Cr^{+6} in bioreactor	Cr^{+6} removal: 100%	Morales-Barrera and Cristiani-Urbina (2015)

(continued)

Table 11.2 (continued)

Fungi	Source of Cr	Remediation of Cr	References
Consortium (Cladosporium perangustum, Penicillium commune, Paecilomyces lilacinus, Fusarium equiseti)	Cr^{+6} in bioreactor	Cr^{+6} removal: 99.9%	Sharma and Malaviya (2016)
Wickerhamomyces anomalus	Cr^{+6} in bioreactor	Cr^{+6} removal: 100%	Fernández et al. (2017)
Aspergillus Niger *Aspergillus flavus* *Aspergillus fumigatus* *Aspergillus nidulans* *Aspergillus heteromorphus* *Aspergillus foetidus* *Aspergillus viridinutans*	Cr^{+6} (18.125mg/L) in Tannery waste water	Cr^{+6} removal: 96.3 % Cr^{+6} removal: 92.8% Cr^{+6} removal: 90.1% Cr^{+6} removal: 86.7% Cr^{+6} removal: 83.7% Cr^{+6} removal: 78.6% Cr^{+6} removal: 74.4%	Sivakumar (2016)
Aspergillus lentluls *Aspergillus terreus*	Cr^{+6} (100 mg/L) in mixed waste water (Metal & dye)	Cr^{+6} removal: 83.11% Cr^{+6} removal: 95.57%	Mishra and Malik (2014)
Consortium of *Aspergillus lentulus, Aspergillus terreus*, and *Rhizopus oryzae*	Cr^{+6} (100 mg/L) in mixed waste water(Metal & dye)	Cr^{+6} removal: 100%	Mishra and Malik (2014)
Rhizopus oryzae	Cr^{+6} (400 ppm) in aqueous solution	Cr^{+6} removal: 91.15%	Sukumar (2010)
Pleurotus ostreatus	Cr^{+6} (50 ppm) in aqueous solution	Cr^{+6} removal: 100%	da Rocha Ferreira et al. (2019)
Rhizophagus irregularis (AM Fungi) in rhizosheric association of *Brachiaria mutica* (Paragrass)	Cr^{+6}(90 mg kg^{-1} soil)	Cr^{+6} uptake: 60 mg Cr kg. dry wt^{-1} of paragrass plant	Kullu et al. (2020)

soil macronutrients are possibilities. Low biodegradability of the applied chelates are also able to cause soil toxicity due to low biodegradation (Ranđelović et al. 2022).

Based on the experimental studies on Cr reviewed (Table 11.3), there is further need of research on chelate-induced phytoremediation for exploring new mobilizing agents. Especially application of natural organic chelators in consortia with biodegrading microbes having siderophoric activity will be an effective, economical and environmentally safe remediation approach (Ranđelović et al. 2022).

Root exudates of some plants contain natural chelates. The chelated micronutrient adjacent to the root surface releases the nutrient to the root. Phytosiderophore (phyto: plant; siderophore: iron carrier) produced by members of Poaceae family can

Table 11.3 Chelate-assisted phytoremediation of Cr form contaminated soil and water

Plant Species	Chelators used for preparation of complex with Cr	Cr-Chelated complex	Cr Shoot uptake	References
Phaseolus vulgaris	DTPA/EDTA /EDDHA with Cr in equimolar ratio [Supplied $^{51}Cr^{+3}$/$^{51}Cr^{+6}$, 111 kBq (34 ng l^{-1})]	^{51}Cr-DTPA ^{51}Cr-EDTA ^{51}Cr-EDDHA $^{51}Cr^{+3}$ $^{51}Cr^{+6}$ [(370kBq(0.11 ng) ^{51}Cr] in nutrient culture solution	Bq 25.31 gm^{-1} dr.wt Bq15.8031 gm^{-1} dr.wt Bq10.8831/gm.dr.wt Bq 6.88 gm^{-1} dr.wt Bq 5.25 gm^{-1} dr.wt	Athalye et al. (1995)
Phaseolus vulgaris	EDTA/DTPA/EDDHA with Cr^{+6}/Cr^{+3} in equimolar ratio	^{51}Cr-EDTA ^{51}Cr-DTPA ^{51}Cr-EDDHA $^{51}Cr^{+3}$ $^{51}Cr^{+6}$ [2590 kBq (0.76 ng)] ^{51}Cr in ultisol soil	Bq 2958.9 gm^{-1} dr.wt Bq 2467.5 gm^{-1} dr.wt Bq 247.2 gm^{-1} dr.wt Bq 18.9 gm^{-1} dr.wt Bq 17.8 gm^{-1} dr.wt	Athalye et al. (1995)
Phaseolus vulgaris	EDTA/DTPA/ EDDHA with Cr^{+6}/Cr^{+3} in equimolar ratio	^{51}Cr-EDTA ^{51}Cr-DTPA ^{51}Cr-EDDHA $^{51}Cr^{+3}$ $^{51}Cr^{+6}$ [2590 kBq (0.76 ng)] ^{51}Cr in vertisol soil	Bq 2855.7 gm^{-1} dr.wt Bq 1376.0 gm^{-1} dr.wt Bq 439.9gm^{-1} dr.wt Bq 327.1 gm^{-1} dr.wt Bq 246.4 gm^{-1} dr.wt	Athalye et al. (1995)
Triticum vulgare	Cr^{+3}only (2μg/ ml) Cr^{+3}-Oxalic acid conc. ratio(1:1) Cr^{+3}-Malic acid conc. ratio(1:1)	Cr^{+3}(2μg/ml) Cr^{+3}-Oxalic acid Cr^{+3}-Malic acid	24.2 μg gm^{-1} dry wt 90.1 μg gm^{-1} dry wt 60.4 μg gm^{-1} dry wt	Srivastava et al. (1999)
Zea mays cv.alice	Cr^{+3}(with a 10% molar excess of EDTA $Cr^{+3}$with a 10% molar excess of DMA (2′ Deoxy-mugineic acid)	Cr^{+3} (1μM in nutrient culture solution)-EDTA Cr^{+3} (1μM in nutrient culture solution)-DMA	0.86 μg, gm^{-1} dry wt 0.76 μg, gm^{-1} dry wt	Bulent et al. (2007)

(continued)

Table 11.3 (continued)

Plant Species	Chelators used for preparation of complex with Cr	Cr-Chelated complex	Cr Shoot uptake	References
Oryza sativa	EDTA with Cr^{+6} in equimolar ratio DTPA with Cr^{+6} in equimolar ratio EDDHA with Cr^{+6} in equimolar ratio Cr^{+6} only	Cr^{+6} (10μM in nutrient culture solution)-EDTA Cr^{+6} (10μM in nutrient culture solution)-DTPA Cr^{+6} (10μM in nutrient culture solution)-EDDHA Cr^{+6} (10μM in nutrient culture solution)	12.712 mg kg^{-1} d wt 32.236 mg kg^{-1} d wt 15.469 mg kg^{-1} d wt 27.356 mg kg^{-1} d wt	Mohanty and Patra (2012b)
Triticum aestivum	EDTA with Cr^{+6} in equimolar ratio DTPA with Cr^{+6} in equimolar ratio	Cr^{+6} (10μM in nutrient culture solution)-EDTA	12.712 mg kg^{-1} d wt	Mohanty and Patra (2012b)
	EDDHA with Cr^{+6} in equimolar ratio	Cr^{+6} (10μM in nutrient culture solution)-DTPA	25.469 mg kg^{-1} d wt	
	Cr^{+6} only	Cr^{+6} (10μM in nutrient culture solution)-EDDHA	22.236 mg kg^{-1} d wt	
		Cr^{+6} (10μM in nutrient culture solution)	22.166 mg kg^{-1} d wt	
Vigna radaiata	EDTA with Cr^{+6} in equimolar ratio DTPA with Cr^{+6} in equimolar ratio EDDHA with Cr^{+6} in equimolar ratio Cr^{+6} only	Cr^{+6} (10μM in nutrient culture solution)-EDTA Cr^{+6} (10μM in nutrient culture solution)-DTPA Cr^{+6} (10μM in nutrient culture solution)-EDDHA Cr^{+6} (10μM in nutrient culture solution)	1.0 mg kg^{-1} dry wt 4.7 mg kg^{-1} dry wt 4.9 mg kg^{-1} dry wt 2.5 mg kg^{-1} dry wt	Mohanty and Patra (2012b)

(continued)

Table 11.3 (continued)

Plant Species	Chelators used for preparation of complex with Cr	Cr-Chelated complex	Cr Shoot uptake	References
Andrographis paniculata	Fe-ore tailing containing Cr of 22.7 mg kg^{-1} with EDTA (1.5 mmols kg^{-1}) Fe-ore tailing containing Cr of 22.7 mg kg^{-1} with CA (1.5 mmols kg^{-1})	22.7 mg kg^{-1} (total Cr) in Fe-ore tailing with EDTA(1.5 mmols kg^{-1}) 22.7 mg kg^{-1} (total Cr) in Fe-ore tailing with CA (1.5 mmols kg^{-1})	17.23 μg gm^{-1} dry wt of Cr 13.18 of Cr μg gm^{-1} dry wt	Chaturvedi et al. (2015) Chaturvedi et al. (2015)
Cymbopogon flexuosus	Cr^{+6} only Cr^{+6}-EDTA Cr^{+6}-DTPA Cr^{+6}-Citric acid Cr^{+6-}Salicylicacid Cr+^{6}-Fe- EDTA **Cr^{+6} Content**: (50 mg Cr^{+6} kg^{-1} of soil with respective chelators in 1:1 ratio)	Cr^{+6} only Cr^{+6}-EDTA Cr^{+6}-DTPA Cr^{+6}-Citric acid Cr^{+6}-Salicylic acid Cr+^{6}-Fe-EDTA **Cr^{+6} Content**: (50 mg Cr^{+6} kg^{-1} of soil with respective chelators in 1:1 ratio)	20 μg kg^{-1} dry wt 22 μg kg^{-1} dry wt 25 μg kg^{-1} dry wt 30 μg kg^{-1} dry wt 21 μg kg^{-1} dry wt 28 μg kg^{-1} dry wt	Patra et al. (2018b)
Cirsium Vulgare	30 mg kg^{-1} of chromium (CrO_3) in soil with 6mmol kg^{-1} EDTA	Cr (CrO_3)-EDTA	8.23 mg kg^{-1} of plant	Dökmeci and Adiloğlu (2020)

grow in low-iron stress conditions (Sugiura and Tanaka 1981). Some graminaceous species and some non-graminaceous species with phytosiderophores were grouped as strategy-II and strategy-I plants respectively (Takagi 1976) which can acquires Fe by the rerelease of phytosiderophores and subsequent uptake of Fe (Roemheld and Marschner 1986). The possible phytosiderophore mediated Cr^{+3} uptake was verified using Fe-deficient maize plants supplied with Cr^{+3}-EDTA and Cr^{+3}-DMA (2′-deoxymugineic acid) in both long (Cr^{+3}) and short term ($^{51}Cr^{+3}$) experiments. It was concluded that the Fe deficiency-induced phytosiderophore release in graminaceous plant species does not further increase Cr uptake as compared to that by Cr^{+3}-EDTA. However, DMA (Phytosiderohore) increases Cr^{+3} solubility as that of EDTA, but do not additionally contribute to Cr^{+3} uptake in maize plant (Bulent et al. 2007). On the otherhand, the phytosiderophore mediated metal uptake for Fe, Zn, Cd, Pb and Ni have been established (Roemheld and Marschner 1986; von Wiren et al. 1996; Gupta and Singh 2017).

11.4.4 Phytoremediation Techniques for Chromium Removal

Phytoremediation is a green approach of using plants or hyperaccumulators to get rid of broad variety of harmful heavy metals from contaminated soil, sediments, water etc. (Sharma et al. 2016). Many scientists noticed that the root system of plant facilitates the interaction of plants with the contaminants. Depending upon the level of contamination and the area of metal accumulation, different phytoremediation techniques like rhizofiltration phytoextraction, phytostabilisation, and Rhizodegradation are used (Tangahu et al. 2011). The techniques are shown in Fig. 11.3 and the details provided in Table 11.4.

11.4.4.1 Rhizofiltration

Rhizofiltration is one of the main mechanisms that is used in phytoremediation (Fulekar and Pathak 2012).

Rhizofiltration can be divided into two categories: Ex-situ rhizofiltration of heavy metals where contaminated water is diverted into a huge array of plants for rhizofiltration process. Some of the terrestrial plants viz. sunflower species, *Jatropha curcas* and brassica species are efficient rhizofiltrators (Kamusoko and Jingura 2017).

11.4.4.2 Phytoextraction

Absorbtion of organic and inorganic contaminants by the plant sprigs though the help of natural ligands viz. metallothioneins and phytochelatins (Sharma et al. 2016; Cobbett and Goldsbrough 2002). Limitations to this technology are described in

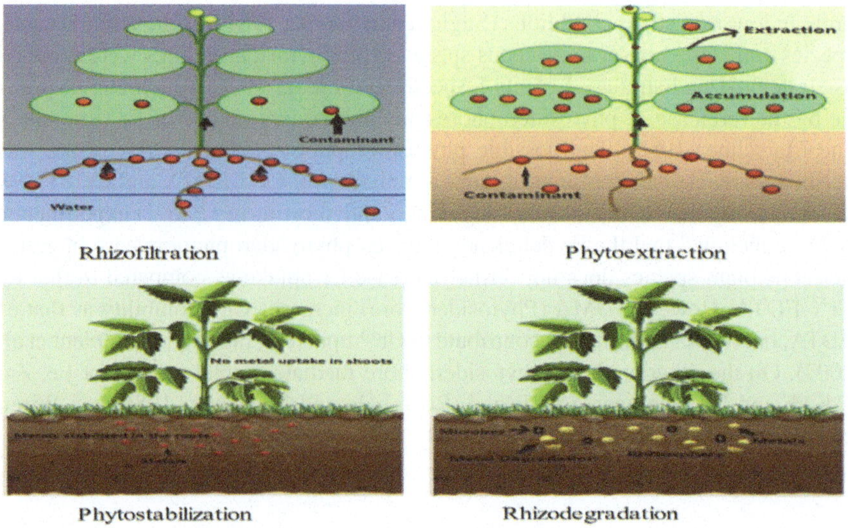

Fig. 11.3 Different techniques of phytoremediation. Adopted from Huang et al. (2004)

Table 11.4 (Li et al. 2003). The advantages of using heavy metal hyperaccumulating plants for detoxification of metals from contaminated sites include lower costs, public acceptance, wide range applicability, production of recyclable metal-rich plant residue and limited environmental disturbance. Hyperaccumulator species are also used to increase crop productivity in suboptimal metal contaminated soil (Hall 2002; McGrath and Zhao 2003).

11.4.4.3 Phytostabilization

Phytostabilization reduces the mobility, immobilization and flexibility of heavy metals through reducing dust by lowering the solubility of the contaminants, by decreasing soil erosion, and restricting heavy metal bioavailability, sedimentation, sorption, reduction or complexation of heavy metal valences (Wu et al. 2011). The primary mechanism of phytostabilization is to physically stabilize the heavy metal ions in the soil through root exudates, rhizospheric microbes, cell wall binding of heavy metal ions, complexation of heavy metal ions with the root mucilage, chelation of metal ions by metallothioneins and phytochelatins, and sequestration of metal ions into the vacuoles (Shackira and Puthur 2019). The higher accumulation of heavy metals in root tissue reduces the mobility of toxic metals in sediment (Shackira and Puthur 2019).

Table 11.4 Process, mechanism of action, limitations and application of different phytoremediation techniques

Technique	Process	Mechanism of action	Limitation	Application
Rhizofiltration	To concentrate, absorb, and precipitate harmful metals from contaminated effluents through massive root system with extensive surface area for absorbing and concentrating both water and nutrients along with heavy metals.	Physical and biochemical effects of plant roots (i) Chemcals synthesized by plant roots that helps in heavy metal uptake in plants (ii) Changes in rhizosphere and pH of the root environment causing heavy metal precipitation onto root system (iii) Organic acid and phenolic root exudates alter heavy metals speciation, uptake and acidify the medium and increase heavy metal transport and bioavailability.	(i) Optimum pH for maximum metal uptake (ii) Regulated flow rate and concentration of the influent (iii) Plant disposal and periodic harvesting	(i) Treatment of industrial and residential effluents, acid mine drainage, diluted sludges, agricultural runoffs, and radionuclide-contaminated solutions (ii) Efficiently remove toxic heavy metals viz. Cr^{6+}, Ni^{2+}, Cu^{2+}, Cd^{2+}, Pb^{2+}, and Zn^{2+} from polluted water
Phytoextraction	Suitable plants with higher metal absorbing ability i.e., hyperaccumulator species	(i) Root metal uptake, roots to shoot translocation, complexation with chelating molecules and vacuolar compartmentalization (Hall 2002; McGrath and Zhao 2003) (ii) Presence of metallothioneins and phytochelatins acting as meatl binding ligands in plants	(i) Identification and generation of new hypperaccumulators for advanced soil, crop and biomass management practices (Li et al. 2003) (ii) Lengthened clean-up process, limited number of target heavy metals, restricted root depth, elevated heavy metal concentration in soil and lack of management skill	Phytoextraction can be exploited in heavy metal-contaminated soils or mines to avoid complex and costly traditional mining technology

(continued)

Table 11.4 (continued)

Technique	Process	Mechanism of action	Limitation	Application
Phytostabilization	Heavy metal immobilization in soil may lower its bioavailability	Physically stabilize the heavy metal ions in the soil	(i) Regularly checking of root tissues, root exudates, contaminants, as well as soil amendments to prevent increased leaching solubility (ii) Only applicable to metals at shallow contaminated areas	Establishment of an insistent plant cover that restricts *contamination* through soil erosion, heavy metal leaching, water percolation, and wind dispersion of toxic dusts.
Rhizodegradation	Breaking down of organic or inorganic contaminants in soil around the root zone or rhizosphere through enhanced microbial activity	Enhance microbial activity by producing root exudates that release certain elements such as carbon and nitrogen and also compounds like sugars, organic acids, growth factors, amino acids, fatty acids, enzymes and acetates. Facilitates nutrient rich environment for growth of indigenous microbial population	(i) Time-consuming process, with the formation of toxic intermediates (ii) Other constraints are limited plant growth rate, low root depth, root contact, plant age, soil chemistry and contamination level	Such practice enhances the soil quality having functional and improved soil ecosystem

11.4.4.4 Rhizodegradation

In this method rhizosphere is linked with the symbiotic soil microbes that are the key players in rhizodegradation to decompose the toxic contaminants. Rhizodegradation is also referred to as phytostimulation (Tangahu et al. 2011). In rhizodegradation, the efficient conversion of heavy metal ions in the rhizospheric region can be contributed to the increased rhizospheric activity (Jacob et al. 2018). Rhizospheric microbes have produce several chelating agents like gluconic acids, siderophores, oxalic acids, biosurfactants that play a major role in heavy metal transformation and immobilization (Gadd 2010). Rhizodegradation is limited for remote areas where human contact is less as well as sites with shallow contamination (Naeem et al. 2020). The most valuable benefit of rhizodegradation method is broad range applicability with minimal environmental disturbance and the dissipation of the toxic contaminants in their original environment (Tangahu et al. 2011).

11.4.5 Genetically Engineered Microorganisms and Phytoremediation

The genetically modified microbes are potential and effective to enhance the efficiency of phytoremediation techniques (Fig. 11.4).

The various bacterial genera viz. *Acidovorax, Bacillus, Alcaligenes, Mycobacterium, Pseudomonas, Paenibacillus* and Rhodococcus has been widely used in the

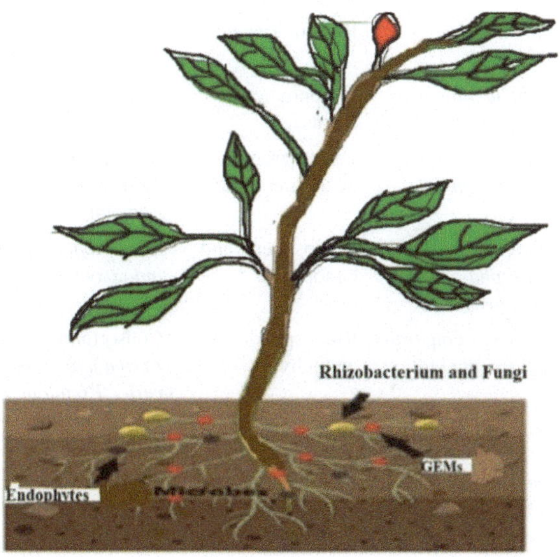

Fig. 11.4 Genetically engineered microbes for phytoremediation

phytoremediation process (Dixit et al. 2015). There are generally four major pathways to produce genetically engineered microbes for phytoremediation technologies, that includes enzyme with high specificity and affinity, pathway design and regulation, development of bioprocess, improvement in tracking and control through bioreporter applications for toxicity reduction. Application of genetically modified endophytic and rhizospheric bacteria in association with plants is an exciting emerging technology for water treatment in contaminated industrial sites (Divya and Deepak Kumar 2011; Pandotra et al. 2018; Chakraborty and Das 2016; Hussain et al. 2018). High expression safe target gene; contaminant resistant strain; and strains surviving in only few unique plant rhizospheres should be chosen for recombination (Huang et al. 2004).Various biosensors such as bacteria like *Moreaxella* sp. and *Escherichia coli* which express phytochelatin showed 25 times higher metal bioaccumulation than that of wild-type strains (Dixit et al. 2015). The major delimitation is sustainability of recombinant microbial population in soil. Moreover, the limitations of molecular approaches have been restricted to bacterial strains like *Escherichia coli, Pseudomonas putida, Bacillus subtilis* etc. Other engineered microbes are yet to be explored for their phytoremediation ability (Dixit et al. 2015). A list of microbes used in genetic engineering for phytoremediation purpose are listed in Table 11.5.

Table 11.5 Genetically modified microbes used in phytoremediation

Plant	GEM	Nature of bacteria	References
Canna Indica and *Iris pseudacorus*	Proteobacteria	Nitrosomonadaceae	Wu et al. (2018)
Cymbidium faberi	Ammonifying bacterial strains	Genetically engineered bacterial strain	Zhao et al. (2014)
Bambusoideae	Biofilm Reactor	*Protozoa* and *Metazoa*	Cao et al. (2012)
Typha domingensis	Endophytic Bacteria	*Klebsiella* sp., *Pseudomonas* sp. and *Acinetobacter* sp.	Ijaz et al. (2016)
Brachiara mutica and *Phragmites australis*	Hydrocarbon degrading bacteria	*Bacillus subtilis, Klebsiella sp., Acinetobacter Junii* and *Acinetobacter* sp.	Rehman et al. (2018)
Phragmites australis and *Typha domingensis*	Rhizospheric and endophytes	*Acinetobacter junii, Pseudomonas indoloxydans*, and *Rhodococcus* sp.	Tara et al. (2019)
Phragmites australis and *Brachia mutica*	Rhizospheric and endophytes	Consortium of five strains namely *Aeromonas salmonicida, Bacillus cerus, Pseudomonas indoloxydans, Pseudomonas gessardii*, and *Rhodococcus* sp.	Shahid et al. (2019b)
Typha domingensis and *Leptochloa fusca*	Rhizospheric and endophytes	*Bacillus cerus, Cyperus laevigatus, Aeromonas salmonicida* and *Pseudomonas gessardii*	Shahid et al. (2019a, c)

11.4.6 Phytoremediation Approaches Using Chromium Tolerant Plants, Hyperaccumulators and Aquatic Macrophytes

11.4.6.1 Chromium Tolerant Plants

Effective removal of chromium from contaminated regions by plants for the phytoremediation process are listed Table 11.6. Use of Cr accumulators and translocators, around the industrial effluent sites to reduce the excess chromium are in practice. Rhizospheric microflora efficiently plays a vital role in chromium detoxification (Wu et al. 2018; Sharma 2021). Trigerring of Reactive Oxygen Species (ROS), signalling and release of defense proteins viz. metallothionine (MTs), phytochelatins (PCs), and Glutathione-S Transferases (GSTs) causes phytosequestration and compartmentalization of metal that ultimately accelerates the bio-accumulation potential of the plants (Shanker et al. 2004; Wu et al. 2018).

Plants have introduced various mechanisms to avoid post accumulation impacts of chromium ions in cells or intracellular compartments and finally prevents their entry into intracellular compartments such as chloroplast and mitochondria (Sinha et al. 2018). A study on *Leersia hexandra* demonstrated accumulation and sequestratiion of hexavalent chromium inside the root cell wall and leaf vacuoles. Increased production of anthocyanin, oxalic acid and carotene is another strategy of tolerance (Srivastava et al. 2021).

11.4.6.2 Hyperaccumulating Plants

Hyperaccumulation is an environmentally imperishable, efficient and cost-effective strategy of phytoremediation to combat the toxicity of Chromium. Heavy metal hyperaccumulations using high affinity chelatorsis viz. organic acids, amino acids and peptides equesters them into the vacuole. Elevated metal mobilization in the rhizospheric region by organic acids; absorption through transporters and xylem transport in the shoot through xylem loading, and ultimate detoxification by chelation or compartmentalization within the vacuoles aree important factors for hyperaccumulation (Shahid et al. 2017a).

Lonicera japonica, has been traditionally used as drug in China and it can be utilized in the phytoremediation process as chromium hyperaccumulating plant. Eze et al. (2018) showed that the chromium removal potential of *Vigna unguiculata* was higher than *Arachis hypogea*. Basically, *Vigna unguiculata* can uptake and accumulate hexavalent chromium in their roots cells. Nayak et al. (2018) demonstrated that the hyperaccumulating ability of *Vetiveria zizanoides* could be stimulated by introducing *Bacillus cereus* T1B3 strain as plant growth promoting bacteria increase the phytoremediation ability. So also, *Colocasia esculenta* effectively reduce hexavalent chromium level in wetlands due high accumulation in their roots (Men and Ghazi 2018).

Table 11.6 List of some chromium tolerant plant species, their habitats and tolerance mechanisms

Family	Plant	Habitat	Tolerance mechanism	References
Aizoaceae	*Mesembryanthemum crystallinum* L.	Large, mat-forming annual with sprawling stems	Phytoextraction	Śliwa-Cebula et al. (2020)
Amaranthaceae	*Gomphrena celosoides* Mart.	Perennial herb	Increased production of proline and antioxidant enzyme activities	Adejumo et al. (2019)
Apocynaceae	*Calotropis procera* (Aiton) W.T. Aiton	Large shrub or small tree	Superoxide dismutase (SOD), catalase (CAT), and glutathione reductase (GR) shows in creased activity	Usman et al. (2020)
Araceae	*Colocasia esculenta* (L.) Schott	Fast growing & herbaceous	High-Cr bioaccumulation	Men and Ghaz (2018)
Araceae	*Lemna minor* L.	Free floating aquatic plants	High anti-oxidant activity, Phytoextraction	Sallah-Ud-Din et al. (2017)
Araceae	*Lemna minuta* Kunth	Small aquatic floating plant	Enhanced antioxidant activity	Paisio et al. 2018
Araceae	*Pistia stratiotes* L.	Aquatic plant	Antioxidant activity and accumulation was high	Mondal and Nayek (2020)
Asteraceae	*Gynura pseudochina* (L.) DC.	Terrestrial	Cr VI reduction	Mongkhonsin et al. (2011)
Brassicaceae	*Brassica napus* L.	Annual or biennial herb	Gentle Remediation Options (GROs)	Tauqeer et al. (2019)
Callitraceae	*Callitriche cophocarpa* Sendtn.	Aquatic macrophyte	Cr(VI) reduction	Augustynowicz et al. (2014)
Commeliniaceae	*Tradescantia pallida* (Rose) D. R. Hunt	Succulent perennial herb	Increased anti-oxidant activity	Sinha et al. (2014)
Fabaceae	*Medicago sativa* L.	Perennial flowering plant	High proline and GST accumulation	Christou et al. (2020)

(continued)

Table 11.6 (continued)

Family	Plant	Habitat	Tolerance mechanism	References
Fabaceae	*Medicago truncatula* Gaertn.	Small annual legume	Regulating the sulphur transport and metabolism	Wu et al. (2011)
Fabaceae	*Sesbania sesban* (L.) Merr.	Fast-growing, perennial legume tree	Phyto-stabilizer	Patra et al. (2020a)

A comparative study on the chromium accumulation potential of *Cannabis, Parthenium, Euphorbia and Rumex* showed that *Parthenium* and *Cannabis* have higher accumulation capability than *Euphorbia and Rumex* (Sajad et al. 2020). The study also revealed that *Cannabis sativa* and *Allium griffithianum* accumulated an increased concentration of chromium to thrive in high chromium contaminated regions and showed that they are potential hyperaccumulators. Singh et al. (2020) found that *Spirodela polyrrhiza* (L.) efficiently accumulated a high chromium from tannery effluent and can be utilized for waste water management. For instance, plants like *Hydrocotyle umbellata* L. and *Canna indica* L. have a great role in remediating hexavalent chromium in a contaminated waste water as well as in constructed wetland system. Since, *Canna indica* is a potent hyperaccumulator plant, it has a significant role in phytoremediation (Taufikurahman et al. 2019). A detailed study which has been carried out by Rajendran et al. (2019) resulted in higher accumulation efficiency of *Chrysopogon zizanioides* L. under toxic chromium stress condition by evolving mechanisms such as detoxification, elevation of antioxidant enzymes, production of high level of photosynthetic pigments, as well as malondialdehyde level (Rajendran et al. 2019).

Levizou et al. carried out a pot experiment in which they found that *Origanum vulgare* has a prominent potential to bioaccumulate chromium in both aerial and root system when exposed to chromium. The hybrid of *Pennisetum americanus* L. and *Pennisetum purpureum* can also be utilized for chromium phytostabilization (Ram et al. 2019). *Solanum viarum* also found to accumulate high chromium, and are advantageous in phytoremediation and may be applied for the treatment of chromium contaminated sites (Afonso et al. 2019). Patra et al. (2020b) the reported the efficiency of *Cassia tora* for phytoremediation of chromium from contaminated sites due to its increased bioaccumulation activity and high Cr tolerance potential. *Arundo donax* L. has shown tolerance and hyperaccumulating ability for chromium. The highest chromium concentrations were found in the shoots of Vernonia cinerea (5500 mg/kg dry matter) and roots of *Diectomis fastigiata* (2371 mg/kg dry matter) depicting their high chromium phytoremediation ability (Mohanty and Patra 2020). *Gomphrena celosoides* showed high antioxidant enzyme and proline activities with increased accumulation (Adejumo et al. 2019). *Melia azedarach* L. is identified as chromium and cadmium hyperaccumuator. It can cause increase in activity of non-enzymatic antioxidants viz. glutathione, soluble protein, and proline in the plant for heavy metal

Table 11.7 List of hyperaccumulating plant species with their family and habitats

Plant	Family	Habitat	References
Allium griffithianum Boiss.	Amaryllidaceae	Perennial herb	Sajad et al. (2020)
Spirodela polyrrhiza (L.)	Araceae	Aquatic weed	Singh and Malaviya (2019)
Hydrocotyle umbellata L.	Araliaceae	Creeping, aquatic herb	Taufikurahman et al. (2019)
Cirsium vulgare (Savi)	Asteraceae	Annual or biennial, herbaceous plant	Dökmeci and Adiloğlu (2020)
Dicoma niccolifera Wild		Terrestial	Banach et al. (2012)
Helianthus annuus L.		Annual forb	Ranieri et al. (2013)
Parthenium hysterophorus L.		Annual, erect, herbaceous	Ullah et al. (2019)
Vernonia cinerea (L.)		Perennial herb	Mohanty and Patra (2020)
Cannabis sativa L.	Cannabaceae	Annual, herbaceous, flowering	Ullah et al. (2019)
Canna indica L.	Cannaceae	Long-lived, perennial herb	Taufikurahman et al. (2019)
Ipomoea aquatica Forssk.	Convolvulaceae	Semi-aquatic, tropical plant	Haokip and Gupta (2021)
Euphorbia helioscopia L.	Euphorbiaceae	Desert, herbaceous spurge	Ullah et al. (2019)

detoxification (Yan et al. 2020). *Calotropis procera* when used for phytoremediation of Cr showed a very high Cr bioaccumulation potential with increased the activities of glutathione reductase, catalase and superoxide dismutase (Srivastava et al. 2021). The list of different hyperaccumulators are given in Table 11.7

11.4.6.3 Aquatic Macrophytes

Aquatic macrophytes is believed to be one of the best contender that can be utilized for phytoremediation technique. Various aquatic species possess anti-oxidant activity which enhance their ability to tolerate and accumulate high quantities of heavy metals from waste water. Some of them are duckweeds (Lemna sp., Spirodella sp.), water hyacinth (Eichhornia sp.), small water fern (Azolla sp.), and water lettuce (Pistia sp.) which reduced Cr (VI) content from waste water (Okunowo and Ogunkanmi 2010; Akhtar et al. 2017). Water hyacinth due to its prolific growth rate is being widely used for phytoremediation of Chromium (Xia and Ma 2006). Chromium, is a non-essential micronutrient and it is hazardous for plant growth and metabolism. Chromium toxicity growth and metabolism of plants, induces stunted growth, delayed germination, and reduced biosynthesis (Zayed et al. 1998; Srivastava et al. 2021). Excessive

Table 11.8 List of aquatic macrophytes used for heavy metal phytoremediation

Common name	Scientific name	Heavy metals	Mechanism	References
Reed	*Phragmites australis*; *Phragmites karka*	Fe, Ni, Mn, Zn, Pb, Cd, Co, Cu	Accumulation	Aslam et al. (2007)
Water fern, water velvet	*Azolla caroliniana*, *Azolla pinnata*	Hg, Cr		Arora et al. (2006)
Pond weed/curly leaf pond weed	*Potamogeton natans*; *Potamogeton crispus*	Co, Cu, Zn, Mn, Cd, Pb		Ali et al. (1999)
Umbrella plant	*Cyperus alternifolius*	Cu, Cd, Pb, Cr, Zn, Ni, Fe		Qian et al. (1999)
Smart weed	*Polygonum hydropiper*	Ni, Cu, Zn, Cr, Pb, Cd		Qian et al. (1999)
Smooth cordgrass	*Spartina alterniflora*	Cu, Fe, Ni, Zn		Qian et al. (1999)
Water lettuce	*Pistia stratiotes*	Cr, Ni		Skinner et al. (2007)
Duckweed	*Lemna minor*	Cd, Pb, Ni	Bioaccumulation	Wang et al. (2002)
Duckweed	*Wolffia globosa*	Cr, Cd	Biosorption, Accumulation	Boonyapookana et al. (2002)
Water hyacinth	*Eichhornia crassipes*	Cd, Pb, Hg, Fe, Cu	Adsorption, Accumulation	Mishra et al. (2007)
Reed canarygrass	*Phalaris arundinacea*	Ni, As, Cd, Pb	Phytostabilization, bioaccumulation	Vymazal et al. (2007)
Salt marsh bulrush	*Scirpus robustus*	Cd, Zn, Ni, Pb, Cr	Accumulation	de Souza et al. (1999)
Rabbitfoot grass	*Polypogon monspeliensis*	Cd, Cr, Ni	Biosorption	de Souza et al. (1999)
Zebra rush	*Scirpus tabernaemontani*	As, Pb, Hg	Biosorption	Skinner et al. (2007)
Water zinnia	*Wedelia trilobata*	Pb, Ni, Cu, Cr	Biosorption, accumulation	Qian et al. (1999)

Adopted from Rai (2008)

Cr bioaccumulation in plants can cause its death by diisturbing physiological and biochemical processes (Table 11.8).

Aquaremediation is one of the most suitable approach for removal of Cr(VI) ions as compared to conventional approach. There are typically seven species of water hyacinth in which *Eichhornia crassipes* is one of the most explored aquatic macrophyte high nitrogen content and biomass used for phytoremediation. Researches (Chanakya et al. 1993; Singhal and Rai 2003) reports that it also provides high range of tolerance activity against wide variety of toxic contaminants including heavy metal with high nutrient absorption ability. It can treat about 1 million L domestic sewage per day and reduces the chemical oxygen demand (COD) and biological oxygen demand (BOD) up to 71% and 89%, respectively (Saha et al. 2018).

Due to its fast-growing ability, *Ipomoea aquatica* could effectively remove hexavalent chromium within a relatively short duration. *Ipomonea aquatica* could be utilized to efficiently remediate highly concentrated chromium contaminated wastewater (Chen et al. 2010). Moreover, Hakeem et al. (2014) proved that the tolerant, fast-growing, hyperaccumulating *Ipomea aquatica* plant possess translocation and bio-accumulation factor >1 and hence showed promising phytoremediation activity against chromium. One of the study reported the implementation of *Lemna minuta* Kunth for the spontaneous reduction of hexavalent chromium depicting its usage in the treatment of industrial effluents (Srivastava et al. 2021).

The use of aquatic vascular plants for the treatment of industrial effluents before discharge could be a novel approach in phytoremediation of heavy metals. Aquatic macrophytes mostly remediate the contaminant through the mechanism of rhizofiltration. The process includes plants under hydroponic treatment can be transported into metal polluted waters where they absorb and accumulate the heavy metals in their roots and shoots (Salt et al. 1995; Zhu et al. 1999). Root exudates and changes in rhizosphere pH may cause metal precipitation on root surfaces. When the plants are fully saturated with the metal contaminants, roots or whole plants are harvested for extraction (Zhu et al. 1999). Water hyacinths are responsible for conducting phytoremediation via several mechanisms such as concentration, root biosorption and metabolic degradation.

Eichhornia crassipes is a biological indicator of metal pollution (Zaranyika and Ndapwadza 1995). Considering its phytoremediation efficacy, *Eichhornia* crassipes is a promising plant species that is most oftenly used for remediation of natural water system as well as wastewater contaminated with low levels of Cr and other heavy metals (Odjegba and Fasidi 2007). An amalgamation of algae and Eichhornia in reducing cadmium content of wastewater and its filtration has been well demonstrated by Tripathi and Shukla (1991). Soltan and Rashed (2003) introduced water hyacinth with various heavy metals and Cr, and suggested that accumulation of heavy metals was higher in the roots than in the aerial parts. (Zhao and Duncan 1998) investigated reduction of hexavalent chromium and extraction of nickel from aqueous environment. In addition, *Azolla caroliniana* reduced hexavalent chromium and lead from wastewater, by accumulating in its biomass, so it may be used as a potent Cr and Pb bioaccumulator (Bennicelli et al. 2004).

Spirodela polyrhiza, Hydrodictyon reticulatum, Chara coralline, Ceratophyllum demersum, Bacopa monnieri, Vallisneria spiralis, Alternanthera sessilis and *Hygrorrhiza aristata. Potamogeton lucens, Salvinia herzegoi* and *Eichhornia crassipes* were also proven to be magnificient biosorbents for Fe, Mn, Cu, Cr, Cd, and Pb under experimental conditions. Table listed all such plants used in Cr phytoremediation (Rai 2008).

11.4.7 Biosorption for Effective Removal of Chromium from Wastewater by Bioadsorbents

Biosorption is one of the emerging waste management mechanism (Fig. 11.5) attained by plants. It exploits the bioefficacy organic matters for accumulation of heavy metals from waste water system. The technique employs metabolically mediated or physico-chemical pathways of uptake (Ali Redha 2020).

Many researches clearly stated that both living and dead aquatic plants are potent to remove heavy metals. They inhabit different aquatic environments like streams, marginal zones of the lakes, riverine, drainage systems and wetlands, examples: *Salvinia herzogoi, Myriophyllum spicatum, Ceratophyllum demersum, Potamogeton lucens,*

Biosorption utilizes inexpensive raw materials that are abundant in nature and is an environmental friendly process (Fiol et al. 2008; Gupta et al. 2010, Pradhan et al. 2017). Various studies indicated that several materials that are used for biosorption mechanism depend on the nature of the substrate, contact time, material dosage, solution pH, temperature, etc. (Farooq et al. 2010). The potential of biosorption of a metal ion by a bioadsorbent is calculated by following formula (Wei et al. 2016).

$$q_e = (C_0 - C_e)\, V/W$$

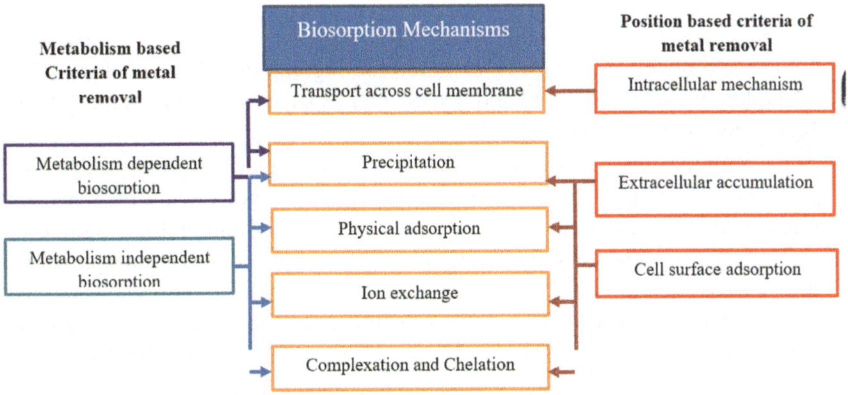

Fig. 11.5 Classification of biosorption mechanisms

where,

q_e	biosorption capacity (mg of metal/g of biosorbent),
C_0 and C_e	the initial and equilibrium concentration of metal ion solution (in mg/L)
V	volume of metal ion solution (L), and
W	quantity of biosorbent material dose used (g).

The higher the bioadsorption capacity, the greater the amount of heavy metal ions that a bioadsorbent can bioadsorb.

The use of plant based biosorbent for the removal of heavy metals from waste water can be attributed to the activity of hydroxyl, carboxyl and carbonyl groups that facilitates metal binding (Bilal et al. 2018; Noli et al. 2019).

Heavy metals are taken up by the bioadsorbent through various mechanisms, several studies signify that different mechanisms of bioadsorption are based on the classification criteria and type of biosorbent.

11.5 Conclusion and Future Perspectives

Chromium is known as a toxic environmental pollutant generated by natural and anthropogenic sources. The stress imposed on living organisms by Cr exposure is responsible for production of oxygen free radicals which degrade essential biomolecules and causes nuclear damage. The book chapter illustrates the sources of environmental chromium contamination and wide information about current phytoremediation processes to mitigate toxic environmental pollution from contaminated soil and water. The eco-friendly phytoremediation remediation approaches for cleaning chromium contaminated soil and water utilizing living and dead organisms of algae, fungi, and higher plants are discussed. To achieve this, different phytoremediation practices with integrated phytotechnological approaches have been described vividly. Based on the facts and figures summarized in this book chapter, the next future strategies are described below:

1. Novel hyperaccumulator plants with good adsorption and absorption capabilities for use at polluted sites must be explored for ongoing phytoremediation efforts. The tolerant native terrestrial and aquatic plant species need to be screened with better phytoremediation efficacy.
2. The continuous monitoring and remediation of heavy metals from groundwater and soil are important which must be based on advance integrated abatement technologies like physical, chemical, physico-chemical and biological remediation approaches. The integrated green phtotechnological remediation approaches are gaining attention involving different methods and different organisms. Integration of different processes needs careful understanding with adequate research trials before implementing integrated phytotechnological approaches for remediation of Cr from contaminated soil and water. The integrated technologies should be experimentally feasible for large range applications, economically viable

which are expected to be relatively efficient than the individual processes. To the best of our understandings, different integrated phytotechnologcal remediation options are important in future research for removal of Cr from contaminated environment.

References

Achal V, Kumari D, Pan X (2011) Bioremediation of chromium contaminated soil by a brown-rot fungus, *Gloeophyllum sepiarium*. Res J Microbiol 6:166–171. https://doi.org/10.3923/jm.2011.166.171

Acosta-Rodríguez I, Arévalo-Rangel DL, Cárdenas-González JF, Moctezuma-Zárate MG, Martínez-Juárez VM (2015) Hexavalent Chromium (VI) Removal by *Penicillium sp.* IA-01. In: Shiomi N (ed) Advances in bioremediation of wastewater and polluted soil, Chapter 8. Intech Open, pp 166–192. https://doi.org/10.5772/60660

Adejumo SA, Tiwari S, Thul S, Sarangi BK (2019) Evaluation of lead and chromium tolerance and accumulation level in *Gomphrena celosoides*: a novel metal accumulator from lead acid battery waste contaminated site in Nigeria. Int J Phytoremed 21(13):1341–1355. https://doi.org/10.1080/15226514.2019.1633258

Afonso TF, Demarco CF, Pieniz S, Camargo FAO, Quadro MS, Andreazza R (2019) Potential of Solanum viarum Dunal in use for phytoremediation of heavy metals to mining areas, southern Brazil. Environ Sci Poll Res 26(23):24132–24142. https://doi.org/10.1007/s11356-019-05460-z

Akhtar ABT, Yasar A, Ali R, Irfan R (2017) Phytoremediation using aquatic macrophytes. In: Ansari AA, Gill SS, Gill R, Lanza GR, Newman L (eds) Phytoremediation. Springer International Publishing, pp 259–276. https://doi.org/10.1007/978-3-319-52381-1_8

Aksu Z, Kutsal T (1990) A comparative study for biosorption characteristics of heavy metal ions with *Chlorella vulgaris*. Environ Technol 11(10):979–987. https://doi.org/10.1080/09593339009384950

Ali MB, Tripathi RD, Rai UN, Pal A, Singh SP (1999) Physico-chemical characteristics and pollution level of lake Nainital (U.P., India): role of macrophytes and phytoplankton in biomonitoring and phytoremediation of toxic metal ions. Chemosphere 39(12):2171–2182. https://doi.org/10.1016/S0045-6535(99)00096-X

Ali Redha A (2020) Removal of heavy metals from aqueous media by biosorption. Arab J Basic Appl Sci 27(1):183–193. https://doi.org/10.1080/25765299.2020.1756177

Anastopoulos I, Kyzas GZ (2015) Progress in batch biosorption of heavy metals onto algae. J Mol Liq 209:77–86

Ao M, Chen X, Deng T, Sun S, Yetao T, Morel JL, Qiu R, Wang S (2022) Chromium biogeochemical behaviour in soil-plant systems and remediation strategies: a critical review. J Hazard Mater 424:127233

Arora A, Saxena S, Sharma DK (2006) Tolerance and phytoaccumulation of Chromium by three Azolla species. World J Microbiol Biotech 22(2):97–100. https://doi.org/10.1007/s11274-005-9000-9

Aslam MM, Malik M, Baig MA, Qazi IA, Iqbal J (2007) Treatment performances of compost-based and gravel-based vertical flow wetlands operated identically for refinery wastewater treatment in Pakistan. Ecol Eng 30(1):34–42. https://doi.org/10.1016/j.ecoleng.2007.01.002

Athalye VV, Ramachandran V, D'Souza TJ (1995) Influence of chelating agents on plant uptake of 51Cr, 210Pb, and 210Po. Environ Pollut 89:47–53

Augustynowicz J, Wróbel P, Płachno BJ, Tylko G, Gajewski Z, Węgrzynek D (2014) Chromium distribution in shoots of macrophyte Callitriche cophocarpa Sendtn. Planta 239(6):1233–1242. https://doi.org/10.1007/s00425-014-2047-9

Ayele A, Godeto YG (2021) Bioremediation of chromium by microorganisms and its mechanisms related to functional groups. J Chem 2021:1–21

Ayele A, Suresh A, Benor S, Konwarh R (2021) Optimization of chromium (VI) removal by indigenous microalga (*Chlamydomonas sp.)* based biosorbent using response surface methodology. Water Environ Res 93(8):1276–1288. https://doi.org/10.1002/wer.1510

Bahafid W, Tahri-Joutey JN, Sayel H, Boularab I, Ghachtouli NE (2013) Bioaugmentation of chromium-polluted soil microcosms with *Candida tropicalis* diminishes phytoavailable chromium. J Appl Microbiol 115(3):727–734. https://doi.org/10.1111/jam.12282

Banach AM, Banach K, Stepniewska Z (2012) Phytoremediation as a promising technology for water and soil purification: Azolla caroliniana Willd. as a case study. Acta Agrophysica 19(2)

Bennicelli R, Stępniewska Z, Banach A, Szajnocha K, Ostrowski J (2004) The ability of Azolla caroliniana to remove heavy metals (Hg(II), Cr(III), Cr(VI)) from municipal waste water. Chemosphere 55(1):141–146. https://doi.org/10.1016/j.chemosphere.2003.11.015

Bhalerao SA, Sharma AS (2015) Chromium: as an environmental pollutant. Int J Curr Microbio Appl Sci 4:732–746

Bilal M, Rasheed T, Sosa-Hernández J, Raza A, Nabeel F, Iqbal H (2018) Biosorption: an interplay between Marine Algae and potentially toxic elements—A review. Marine Drugs 16(2):65. https://doi.org/10.3390/md16020065

Boonyapookana B, Upatham ES, Kruatrachue M, Pokethitiyook P, Singhakaew S (2002) Phytoaccumulation and phytotoxicity of cadmium and chromium in duckweed *Wolffia globosa*. Int J Phytoremed 4(2):87–100. https://doi.org/10.1080/15226510208500075

Bulent A, Taner DR, Tolga T, Sertac Y, Celik T, Ferruh Z, Haluk O (2007) Bilateral testicular germ cell tumors in Turkey: increase in incidence in last decade and evaluation of risk factors in 30 patients. J Urol 178(1):129–133

Cao Y, Chan F, Chui YH, Xiao H (2012) Characterization of flax fibres modified by alkaline, enzyme, and steam-heat treatments. BioResources 7(3):4109–4121

Chakraborty J, Das S (2016) Molecular perspectives and recent advances in microbial remediation of persistent organic pollutants. Enviorn Sci Pollut Res 23(17):16883–16903. https://doi.org/10.1007/s11356-016-6887-7

Chanakya HN, Borgaonkar S, Meena G, Jagadish KS (1993) Solid-phase biogas production with garbage or water hyacinth. Biores Tech 46(3):227–231. https://doi.org/10.1016/0960-8524(93)90125-U

Chaturvedi N, Dhal N, Patra HK (2015) EDTA and Citric acid mediated phytoextraction of heavy metals from iron ore tailings using *Andrographis paniculata*: a comparative study. Intl J Min Reclama Env 29(1):33–46. https://doi.org/10.1080/17480930.2014.955328

Chen J-C, Wang K-S, Chen H, Lu C-Y, Huang L-C, Li H-C, Peng T-H, Chang S-H (2010) Phytoremediation of Cr(III) by Ipomonea aquatica (water spinach) from water in the presence of EDTA and chloride: effects of Cr speciation. Biores Tech 101(9):3033–3039. https://doi.org/10.1016/j.biortech.2009.12.041

Christou A, Georgiadou EC, Zissimos AM, Christoforou IC, Christofi C, Neocleous D, Dalias P, Torrado SOCA, Argyraki A, Fotopoulos V (2020) Hexavalent chromium leads to differential hormetic or damaging effects in alfalfa (Medicago sativa L.) plants in a concentration-dependent manner by regulating nitro-oxidative and proline metabolism. Environ Pollut 267:115379. https://doi.org/10.1016/j.envpol.2020.115379

Cobbett C, Goldsbrough P (2002) Phytochelatins and metallotheonine: roles in heavy metal detoxification and homeostasis. Ann Rev Plant Biol 53(1):159–182. https://doi.org/10.1146/annurev.arplant.53.100301.135154

Coreño-Alonso A, Solé A, Diestra E, Esteve I, Gutiérrez-Corona JF, Reyna López GE, Fernández FJ, Tomasini A (2014) Mechanisms of interaction of chromium with *Aspergillus niger* var tubingensis strain Ed8. Bioresour Technol 158:188–192. https://doi.org/10.1016/j.biortech.2014.02.036

Coronel E, Varela M (2015) Kinetics of chromium (iii) uptake by *Nostoc sp*. Algae Rev Boliviana de Qu_ımica 32(3):60–64 da Rocha Ferreira GL, Vendruscolo F, Antoniosi Filho NR (2019) Biosorption of hexavalent chromium by Pleurotus ostreatus. Heliyon 5(3):e01450. https://doi.org/10.1016/j.heliyon.2019.e01450

Danouche M, Ghachtouli NL, Hicham El Arroussi HE (2021) Phycoremediation mechanisms of heavy metals using living green microalgae: physicochemical and molecular approaches for enhancingselectivity and removal capacity. Heliyon 7:e07609. https://doi.org/10.1016/j.heliyon.2021.e07609

de Souza MP, Huang CP, Chee N, Terry N (1999) Rhizosphere bacteria enhance the accumulation of selenium and mercury in wetland plants. Planta 209:259–263

Divya B, Deepak Kumar M (2011) Plant–microbe interaction with enhanced bioremediation. Res J Biotech 6:4

Dixit R, Wasiullah X, Malaviya D, Pandiyan K, Singh UB, Sahu A, Shukla R, Singh BP, Rai JP, Sharma PK, Lade H (2015a) Bioremediation of heavy metals from soil and aquatic environment: an overview of principles and criteria of fundamental processes. Sustainability7(2):2189–2212. https://doi.org/10.3390/su7022189

Dökmeci AH, Adiloğlu S (2020) The phytoremediation of chromium from soil using cirsium vulgare and the health effects. Biosci Biotech Res Asia 17(03):535–541. https://doi.org/10.13005/bbra/2857

Dönmez GC, Aksu Z, Öztürk A, Kutsal T (1999) A comparative study on heavy metal biosorption characteristics of some algae. Process Biochem 34:885–892

Dönmez G, Aksu Z (2002) Removal of chromium(VI) from saline wastewaters by *Dunaliella* species. Process Biochem 38:751–762

Dwivedi S, Srivastava S, Mishra S, Kumar A, Tripathi RD, Rai UN, Dave R, Tripathi P, Charkrabarty D, Trivedi PK (2010) Characterization of native micro-algal strains for their chromium bioaccumulation potential: phytoplankton response in polluted habitats. J. Hazard Mater 173:95–101

El Nemr A, El-Sikaily A, Khaled A, Abdelwahab O (2015) Removal of toxic chromium from aqueous solution, wastewater and saline water by marine red alga *Pterocladia capillacea* and its activated carbon. Arab J Chem 8:105–117

Eze CN, Odoh CK, Eze EA, Orjiakor PI, Enemuor SC, Okobo UJ (2018) Chromium (III) and its effects on soil microbial activities and phytoremediation potentials of Arachis hypogea and Vigna unguiculata. Afr J Biotechnol 17(38):1207–1214. https://doi.org/10.5897/AJB10.5897/AJB2018.16566

Fakhry EM, Ismail S, Noaman ND, Shafik MA, Wafaa M, Menesi WM (2013) Adsorptive removal of chromium by dried brown alga *Sargassum hornschuchii* Egyptian J Phycol 14:1–14

Farooq U, Kozinski JA, Khan MA, Athar M (2010) Biosorption of heavy metal ions using wheat based biosorbents—A review of the recent literature. Biores Technol 101(14):5043–5053. https://doi.org/10.1016/j.biortech.2010.02.030

Fernández PM, Cruz EL, Viñarta SC, Castellanos de Figueroa LI (2017) Optimization of culture conditions for growth associated with Cr(VI) removal by *Wickerhamomyces anomalus* M10. Bull Environ Contam Toxicol 98(3):400–406. https://doi.org/10.1007/s00128-016-1958-5

Fiol N, Escudero C, Villaescusa I (2008) Chromium sorption and Cr(VI) reduction to Cr(III) by grape stalks and yohimbe bark. Biores Tech 99(11):5030–5036. https://doi.org/10.1016/j.biortech.2007.09.007

Fulekar MH, Pathak B (2012) Rhizofiltration: a green technology for remediation of heavy metals 2:7

Gadd GM (2010) Metals, minerals and microbes: geomicrobiology and bioremediation. Microbiology 156(3):609–643. https://doi.org/10.1099/mic.0.037143-0

Gazem MA, Nazareth S (2013) Sorption of lead and copper from an aqueous phase system bymarine-derived Aspergillus species. Ann Microbiol 63(2):503–511. https://doi.org/10.1007/s13213-012-0495-7

Gomes MAC, Hauser-Davis RA, Suzuki MS, Vitória AP (2017) Plant chromium uptake and transport, physiological effects and recent advances in molecular investigations. Ecotoxicol Environ Saf 140:55–64

Greenwood NN, Earnshaw A (1997) Chemistry of the elements, 2nd edn. Butterworth-Heinemann. ISBN 978-0-08-037941-8

Guillén-Jiménez FDM, Netzahuatl-Muñoz AR, Morales-Barrera L, Cristiani-Urbina E (2009) Hexavalent chromium removal by *Candida* sp. in a concentric draft-tube airlift bioreactor. Water Air Soil Pollut 204(1):43. https://doi.org/10.1007/s11270-009-0024-x

Gupta VK, Rastogi A, Nayak A (2010) Biosorption of nickel onto treated alga (*Oedogonium hatei*): application of isotherm and kinetic models. J Colloid Interface Sci 342(2):533–539. https://doi.org/10.1016/j.jcis.2009.10.074

Gupta CK, Singh B (2017) Uninhibited biosynthesis and release of phytosiderophores in the presence of heavy metal (HM) favors HM remediation. Environ Sci Pollut Res 24:9407–9416. https://doi.org/10.1007/s11356-017-8636-y

Hakeem K, Sabir M, Ozturk M, Mermut AR (eds) (2014) Soil remediation and plants: prospects and challenges. Academic Press, Cambridge, MA, USA

Hall JL (2002) Cellular mechanisms for heavy metal detoxification and tolerance. J Exp Bot 53(366):1–11. https://doi.org/10.1093/jexbot/53.366.1

Han X, Wong YS, Tam NFY (2006) Surface complexation mechanism and modeling in Cr(III) biosorption by a microalgal isolate *Chlorella miniata*. J Colloid Interface Sci 303(2):365–371. https://doi.org/10.1016/j.jcis.2006.08.028

Hansel CM, Ferdelman TG, Tebo BM (2015) Cryptic cross-linkages among biogeochemical cycles: novel insights from reactive intermediates. Elements 11(40):414. https://doi.org/10.2113/gselements.11.6.409

Haokip N, Gupta A (2021) Phytoremediation of chromium and manganese by *Ipomoea aquatica* Forssk. From aqueous medium containing chromium-manganese mixtures in microcosms and mesocosms. Water Environ J 35(3):884–891. https://doi.org/10.1111/wej.12676

Holleman AF, Wiber E, Wiberg N (1985) "Chromium". Lehrbuch der Anorganischen Chemie (in German), Walter de Gruyter, pp 1081–1095. ISBN 978-3-11-007511-3

Huang X-D, El-Alawi Y, Penrose DM, Glick BR, Greenberg BM (2004) Responses of three grass species to creosote during phytoremediation. Environ Pollut 130(3):453–463. https://doi.org/10.1016/j.envpol.2003.12.018

Husien S, Labena A, El-Belely EF, Mahmoud HM, Hamouda AS (2019) Absorption of hexavalent chromium by green micro algae *Chlorella sorokiniana*: live planktonic cells. Water Pract Technol 14:515–529. https://doi.org/10.2166/wpt.2019.034

Hussain I, Aleti G, Naidu R, Puschenreiter M, Mahmood Q, Rahman MM, Wang F, Shaheen S, Syed JH, Reichenauer TG (2018) Microbe and plant assisted-remediation of organic xenobiotics and its enhancement by genetically modified organisms and recombinant technology: a review. Sci Total Environ 628–629:1582–1599. https://doi.org/10.1016/j.scitotenv.2018.02.037

Ijaz A, Imran A, Anwar ul Haq M, Khan QM, Afzal M (2016) Phytoremediation: recent advances in plant-endophytic synergistic interactions. Plant Soil 405(1–2):179–195. https://doi.org/10.1007/s11104-015-2606-2

Indhumathi P, Syed Shabudeen PS, Shoba US, Saraswathy CP (2014) The removal of chromium from aqueous solution by using green micro algae. J Chem Pharm Res 6(6):799–808

Jaafari J, Yaghmaeian K (2019) Optimization of heavy metal biosorption onto freshwater algae (*Chlorella coloniales*) using response surface methodology (RSM). Chemosphere 217:447–455

Jacob JM, Karthik C, Saratale RG, Kumar SS, Prabakar D, Kadirvelu K, Pugazhendhi A (2018) Biological approaches to tackle heavy metal pollution: a survey of literature. J Environ Manag 217:56–70. https://doi.org/10.1016/j.jenvman.2018.03.077

John J (2003) Phycoremediation: algae as tools for remediation of mine-void wetlands. Modern trends in applied aquatic ecology. Springer, Boston, MA, pp 133–147

Joshi PK, Swarup A, Maheshwari S, Kumar R, Singh N (2011) Bioremediation of heavy metalsin liquid media through fungi isolated from contaminated sources. Indian J Microbiol 51(4):482–7. https://doi.org/10.1007/s12088-011-0110-9

Jutsz AM, Gnid A (2015) Mechanisms of stress avoidance and tolerance by plants used in phytoremediation of heavy metals. Arch Environ Protect 41:104–114. https://doi.org/10.1515/aep-2015-0045

Kamusoko R, Jingura RM (2017) Utility of Jatropha for phytoremediation of heavy metals and emerging contaminants of water resources: a review. CLEAN—Soil Air Water 45(11):1700444. https://doi.org/10.1002/clen.201700444

Kang OL, Ramli N, Said M et al (2011) *Kappaphycus alvarezii* waste biomass. A potential biosorbent for the chromium ion removal. J Environ Sci 23:918–922

Katircioğlu H, Aslim B, Tunceli A (2012) Chromium (VI) biosorption from aqueous solutions by free and immobilized biomass of *Oscillatoria sp.* H1 isolated from freshwater. ISIJ Int 52(7):1173–1178

Katz SA, Salem H (1994) The biological and environmental chemistry of chromium. VCH Publishers, New York

Kuanar A, Kabi SK, Rath M, Dhal NK, Bhuyan R, Das S, Kar D (2022) A comparative review on bioremediation of chromium by bacterial, fungal, algal and microbial consortia. Geomicrobiol J 39(6):515–530. https://doi.org/10.1080/01490451.2022.2035019

Kullu B, Patra DK, Acharya S, Pradhan C, Patra HK (2020) AM fungi mediated bioaccumulation of hexavalent chromium in Brachiaria mutica-a mycorrhizal phytoremediation approach. Chemosphere 258:127337. https://doi.org/10.1016/j.chemosphere.2020.127337

Kumar KS, Dahms HU, Won EJ, Lee JS, Shin KH (2015) Microalgae—A promising tool for heavy metal remediation. Ecotoxicol Environ Saf 113:329–352. https://doi.org/10.1016/j.ecoenv.2014.12.019

Lee DC, Park CJ, Yang JE, Jeong YH, Rhee HI (2000) Screening of hexavalent chromium biosorbent from marine algae. Appl Microbiol Biotechnol 54(3):445–448. https://doi.org/10.1007/s002530000387

Li Y-M, Chaney R, Brewer E, Roseberg R, Angle JS, Baker A, Reeves R, Nelkin J (2003) Development of a technology for commercial phytoextraction of nickel: economic and technical considerations. Plant Soil 249(1):107–115. https://doi.org/10.1023/A:1022527330401

Li YX, Wang Y, Zhao FJ (2015) Kinetic and equilibrium studies of chromium (VI) biosorption by spent macroalgae *Polysiphonia urceolata* and *Chondrus ocellatus*. Biotechnol Biotechnol Equip 29(3):498–505. https://doi.org/10.1080/13102818.2015.1011374

Lindsay WL (1974) Role of chelation in micronutrient availability. In: Carson FW (ed) Plant roots and environment. Univ. Press of Virginia. Charlottesville, VA, USA, pp 507–524

Lotfy S, Mostafa A (2014) Phytoremediation of contaminated soil with cobalt and chromium. J Geochem Explorat 144:367–373. https://doi.org/10.1016/j.gexplo.2013.07.003

McGrath SP, Zhao F-J (2003) Phytoextraction of metals and metalloids from contaminated soils. Curr Opin Biotech 14(3):277–282. https://doi.org/10.1016/S0958-1669(03)00060-0

Medda S, Mondal NK (2017) Chromium toxicity and ultrastructural deformation of Cicer arietinum with special reference of root elongation and coleoptile growth. Ann Agrarian Sci 15(3):396–401. https://doi.org/10.1016/j.aasci.2017.05.022

Men CK, Ghazi RM (2018) Phytoremediation of chromium (VI) using Colocasia esculenta in laboratory scale constructed wetlands. J Trop Resour Sustain Sci (JTRSS) 6(1):45-49. https://doi.org/10.47253/jtrss.v6i1.727

Mishra A, Malik A (2014) Novel fungal consortium for bioremediation of metals and dyes from mixed waste stream. Bioresour Technol 171:217–226. https://doi.org/10.1016/j.biortech.2014.08.047

Mishra KK, Rai UN, Prakash O (2007) Bioconcentration and phytotoxicity of Cd in Eichhornia crassipes. Environ Monit Assess 130(1–3):237–243. https://doi.org/10.1007/s10661-006-9392-5

Mohanty M, Patra HK (2011) Attenuation of chromium toxicity by bioremediation technology. Rev Environ Contam Toxicol 210:1–34. https://doi.org/10.1007/978-1-4419-7615-41

Mohanty M, Patra HK (2012a) Phytoremediation potential of Paragrass—An in situ approach for chromium contaminated soil. Int J Phytoremediation 14:796–805. https://doi.org/10.1080/15226514.2011.619595

Mohanty M, Patra HK (2012b) Effect of chelate assisted hexavalent chromium on physiological changes, biochemical alterations and Cr bioavailability in crop plants—An in vitro phytoremediation approach. Bioremediation J 16(3):147–155. https://doi.org/10.1080/10889868.2012.687414

Mohanty M, Patra HK (2020) Phytoassessment of in situ weed diversity for their chromium distribution pattern and accumulation indices of abundant weeds at South Kaliapani chromite mining area with their phytoremediation prospective. Ecotoxic Env Saf 194:110399. https://doi.org/10.1016/j.ecoenv.2020.110399

Molla AH, Fakhru'l-Razi A (2012) Mycoremediation-a prospective environmental friendly technique of bio-separation and dewatering of domestic waste water sludge. Environ Sci Pollut Res Int 19(5):1612–9. https://doi.org/10.1007/s11356-011-0676-0

Mondal NK, Nayek P (2020) Hexavalent chromium accumulation kinetics and physiological responses exhibited by Eichhornia sp. and Pistia sp. Int J Environ Sci Technol 17(3):1397–1410. https://doi.org/10.1007/s13762-019-02418-z

Mongkhonsin B, Nakbanpote W, Nakai I, Hokura A, Jearanaikoon N (2011) Distribution and speciation of chromium accumulated in Gynura pseudochina (L.) DC. Environ Exp Bot 74:56–64. https://doi.org/10.1016/j.envexpbot.2011.04.018

Monteiro CM, Castro PML, Malcata FX, Instituto I, Carlos A, Campos O, Maia, Monteiro C, Castro PML, Malcata F (2012). Metal uptake by Microalgae: underlying mechanisms and practical applications. Biotechnol Prog 28(2):299–311

Morales-Barrera L, Cristiani-Urbina E (2006) Removal of hexavalent chromium by *Trichoderma viride* in an airlift bioreactor. Enzyme Microb Technol 40(1):107–113. https://doi.org/10.1016/j.enzmictec.2005.10.044

Morales-Barrera L, Cristiani-Urbina E (2015) Bioreduction of hexavalent chromium by *Hypocrea tawa* in a concentric draft-tube airlift bioreactor. J Environ Biotechnol Res 1(1):37–44

Nacorda JOO, Martinez-Goss MR, Nerissa K, Torreta NK (2010) Bioremoval and bioreduction of chromium(VI) by the Green Microalga, *Chlorella vulgaris* Beij, Isolated from Laguna de Bay, Philippines. Philipp J Sci 139:181–188

Naeem N, Tabassum I, Majeed A, Amjad Khan M, Shahbaz S (2020) Review article on phytoremediation and other remediation technologies of soil contaminated with heavy metals. Acta Sci Agr 4(3):01–05

Nayak AK, Panda SS, Basu A, Dhal NK (2018) Enhancement of toxic Cr(VI), Fe, and other heavy metals phytoremediation by the synergistic combination of native *Bacillus cereus* strain and *Vetiveria zizanioides* L. Int J Phytoremed 20(7):682–691

Nithya K, Sathish A, Pradeep K. Kiran, Baalaji S (2019) Algal biomass waste residues of *Spirulina platensis* for chromium adsorption and modelling studies. J Environ Chem Eng 7(5):103273

Noli F, Kapashi E, Kapnisti M (2019) Biosorption of uranium and cadmium using sorbents based on Aloe vera wastes. J Environ Chem Eng 7(2):102985. https://doi.org/10.1016/j.jece.2019.102985

Nourbakhsh M, Sag Y, Ozer D, Aksu Z, Çağlar A (1994) A comparative study of various biosorbents for removal of chromium(VI) ions from industrial waste-water. Process Biochem 29:1–5

Odjegba VJ, Fasidi IO (2007) Phytoremediation of heavy metals by Eichhornia crassipes. Environmentalist 27(3):349–355. https://doi.org/10.1007/s10669-007-9047-2

Okunowo W, Ogunkanmi L (2010) Phytoremediation potential of some heavy metals by water hyacinth. Int J Biol Chem Sci 4(2). https://doi.org/10.4314/ijbcs.v4i2.58121

Paisio CE, Fernandez M, González PS, Talano MA, Medina MI, Agostini E (2018) Simultaneous phytoremediation of chromium and phenol by Lemna minuta Kunth: a promising biotechnological tool. Int J Environ Sci Technol 15(1):37–48

Pandotra P, Raina M, Salgotra RK, Ali S, Mir ZA, Bhat JA, Tyagi A, Upadhahy D (2018) Plant-bacterial partnership: a major pollutants remediation approach. In: Oves M, Zain Khan M, Ismail IMI (eds) Modern age environmental problems and their remediation. Springer International Publishing, pp 169–200. https://doi.org/10.1007/978-3-319-64501-8_10

Patra DK, Pradhan C, Patra HK (2018a) An in situ study of growth of Lemongrass: *Cymbopogon flexuosus* (Nees ex Steud.) W. Watson on varying concentration of Chromium (Cr^{+6}) on soil and its bioaccumulation: perspectives on phytoremediation potential and phytostabilisation of chromium toxicity. Chemosphere 193:793–799

Patra DK, Pradhan C, Patra HK (2018b) Chelate based phytoremediation study for attenuation of chromium toxicity stress using lemongrass: *Cymbopogon flexuosus* (nees ex steud.) W. Watson. Int J Phytoremediation 20(13):1324–1329. https://doi.org/10.1080/15226514.2018.1488812

Patra DK, Pradhan C, Patra HK (2019) Chromium bioaccumulation, oxidative stress metabolism and oil content in lemon grass *Cymbopogon flexuosus* (Nees ex Steud.) W. Watson grown in chromium rich over burden soil of Sukinda chromite mine, India. Chemosphere 218:1082–1088

Patra DK, Pradhan C, Kumar J, Patra HK (2020a) Assessment of chromium phytotoxicity, phytoremediation and tolerance potential of Sesbania sesban and Brachiaria mutica grown on chromite mine overburden dumps and garden soil. Chemosphere 252:126553. https://doi.org/10.1016/j.chemosphere.2020.126553

Patra DK, Grahacharya A, Pradhan C, Patra HK (2020b) Phytoremediation potential of coffee pod (*Cassia tora*): an in situ approach for attenuation of chromium from overburden soil of Sukinda Chromite Mine, India. Environ Progr Sustain Energy e13510

Patra DK, Acharya S, Pradhan C, Patra HK (2021) Poaceae plants as potential phytoremediators of heavy metals and eco-restoration in contaminated mining sites. Environ Technol Innov 21:101293

Pradhan D, Devi N, Sukla LB (2017) Biosorption of hexavalent chromium using biomass of microalgae scenedesmus SP. Int J Eng Technol 7(3):558–563

Pradhan D, Sukla LB, Mishra BB, Devi N (2019) Biosorption for removal of hexavalent chromium using microalgae *Scenedesmus* sp. J Cleaner Prod 209:617–629

Qian J, Zayed A, Zhu Y, Yu M, Terry N (1999) Phytoaccumulation of trace elements by Wetland Plants: III. Uptake and accumulation of ten trace elements by twelve plant species. J Environ Qual 28(5):1448–1455. https://doi.org/10.2134/jeq1999.00472425002800050009x

Rai PK (2008) Heavy metal pollution in aquatic ecosystems and its phytoremediation using wetland plants: an ecosustainable approach. Int J Phytoremed 10(2):133–160. https://doi.org/10.1080/15226510801913918

Rajendran M, An W, Li W, Perumal V, Wu C, Sahi SV, Sarkar SK (2019) Chromium detoxification mechanism induced growth and antioxidant responses in vetiver (Chrysopogon zizanioides(L.) Roberty). J Central South Univ 26(2):489–500. https://doi.org/10.1007/s11771-019-4021-y

Ram BK, Han Y, Yang G, Ling Q, Dong F (2019) Effect of hexavalent chromium [Cr(VI)] on phytoremediation potential and biochemical response of hybrid napier grass with and without EDTA application. Plants8(11):515. https://doi.org/10.3390/plants8110515

Ranđelović D, Jakovljević K, Tijana Zeremski T (2022) Chelate-assisted phytoremediation In: Pandey VC (ed) Assisted phytoremediation (Chapter 5). Elsevier, Science Direct, pp 131–154. ISBN: 978-0-12-822893-7. https://doi.org/10.1016/B978-0-12-822893-7.00004-5

Ranieri E, Fratino U, Petruzzelli D, Borges AC (2013) A comparison between phragmites australis and Helianthus annuus in chromium phytoextraction. Water Air Soil Pollut 224(3):1465. https://doi.org/10.1007/s11270-013-1465-9

Regaldo L, Gagneten AM, Troiani H (2009) Accumulation of chromium and interaction with other elements in *Chlorella vulgaris* (Cloroficeae) and *Daphnia magna* (Crustacea, Cladocera). J Environ Biol 30(2):213–216. https://www.jeb.co.in, https://www.info@jeb.co.in

Rehman K, Imran A, Amin I, Afzal M (2018) Inoculation with bacteria in floating treatment wetlands positively modulates the phytoremediation of oil field wastewater. J Haz Mater 349:242–251. https://doi.org/10.1016/j.jhazmat.2018.02.013

Rezaei H (2016) Biosorption of chromium by using *Spirulina Sp.* Arabian J Chem 9:846–853

Roemheld V, Marschner H (1986) Evidence for a specific uptake system for Iron phytosiderophores in roots of grasses. Plant Physiol 80:175–180

Romo-Rodríguez P, Gutiérrez-Corona JF (2019) Fungal processes of interaction with chromium. In: Tomasini A, León-Santiesteban HH (eds) Fungal bioremediation: fundamentals and applications, Chapter 6. CRC Press, Taylor & Francis Group, pp 173–185. http://www.taylorandfrancis.com, http://www.crcpress.com

Roskova Z, Skarohlid R, McGachy L (2022) Siderophores: an alternative bioremediation strategy? Sci Total Environ 819:153144. https://doi.org/10.1016/j.scitotenv.20

Saha P, Mondal A, Sarkar S (2018) Phytoremediation of cyanide containing steel industrial wastewater by Eichhornia crassipes. Int J Phytorem 20(12):1205–1214. https://doi.org/10.1080/15226514.2017.1375898

Sajad MA, Khan MS, Bahadur S, Naeem A, Ali H, Batool F, Shuaib M, Khan MAS, Batool S (2020) Evaluation of chromium phytoremediation potential of some plant species of Dir Lower, Khyber Pakhtunkhwa, Pakistan. Acta Ecol Sinica 40(2):158–165. https://doi.org/10.1016/j.chnaes.2019.12.002

Sallah-Ud-Din R, Farid M, Saeed R, Ali S, Rizwan M, Tauqeer HM, Bukhari SAH (2017) Citric acid enhanced the antioxidant defense system and chromium uptake by Lemna minor L. grown in hydroponics under Cr stress. Environ Sci Pollut Res 24(21):17669–17678. https://doi.org/10.1007/s11356-017-9290-0

Salt DE, Blaylock M, Kumar NPBA, Dushenkov V, Ensley BD, Chet I, Raskin I (1995) Phytoremediation: a novel strategy for the removal of toxic metals from the environment using plants. Nat Biotechnol 13(5):468–474. https://doi.org/10.1038/nbt0595-468

Salt DE, Smith RD, Raskin I (1998) Phytoremediation. Annu Rev Plant Physiol Plant Mol Biol 49:643–668

Samuel RT, Menon LP, Bhavana V, Sathya P (2019) Studies on phycoremediation of chlorpyrifos and heavy metal chromium using algae. Int J Pharm Biol Sci 9:64–71

Sargın I, Arslan G, Kaya M (2016) Efficiency of chitosan-algal biomass composite microbeads at heavy metal removal. React Funct Polym 98:38–47

Sepehr MN, Nasseri S, Zarrabi M, Samarghandi MR, Amrane A (2012) Removal of Cr(III) from tanning effluent by *Aspergillus niger* in airlift bioreactor. Sep Purif Technol 96:256–262. https://doi.org/10.1016/j.seppur.2012.06.013

Shackira AM, Puthur JT (2019) Phytostabilization of heavy metals: understanding of principles and practices. In: Srivastava S, Srivastava AK, Suprasanna P (eds) Plant-metal interactions. Springer International Publishing, pp 263–282. https://doi.org/10.1007/978-3-030-20732-8_13

Shahid M, Dumat C, Khalid S, Niazi NK, Antunes PM (2017a) Cadmium bioavailability, uptake, toxicity and detoxification in soil-plant system. In: Reviews of environmental contamination and toxicology, vol 241. Springer International Publishing, Cham, pp 73–137

Shahid M, Shamshad S, Rafiq M, Khalid S, Bibi I, Niazi NK, Dumat C, Rashid MI (2017b) Chromium speciation, bioavailability, uptake, toxicity and detoxification in soil-plant system: a review. Chemosphere 178:513–533

Shahid UB, Bicer Y, Ahzi S, Abdala A (2019a) Thermodynamic assessment of an integrated renewable energy multigeneration system including ammonia as hydrogen carrier and phase change material energy storage. Energy Convers Manage 198:111809

Shahid MJ, Arslan M, Siddique M, Ali S, Tahseen R, Afzal M (2019b) Potentialities of floating wetlands for the treatment of polluted water of river Ravi, Pakistan. Ecolog Eng 133:167–176. https://doi.org/10.1016/j.ecoleng.2019.04.022

Shahid MJ, Tahseen R, Siddique M, Ali S, Iqbal S, Afzal M (2019c) Remediation of polluted river water by floating treatment wetlands. Water Supply 19(3):967–977

Shanker AK, Cervantes C, Loza-Tavera H, Avudainayagam S (2005) Chromium toxicity in plants. Environ Int 31:739–753

Shanker A, Djanaguiraman M, Sudhagar R, Chandrashekar C, Pathmanabhan G (2004) Differential antioxidative response of ascorbate glutathione pathway enzymes and metabolites to chromium speciation stress in green gram ((L.) R.Wilczek. Cv CO4) roots. Plant Sci 166(4):1035–1043. https://doi.org/10.1016/j.plantsci.2003.12.015

Sharma S, Malaviya P (2016) Bioremediation of tannery wastewater by *chromium* resistant novel fungal consortium. Ecol Eng 91:419–425. https://doi.org/10.1016/j.ecoleng.2016.03.005

Sharma S, Rana S, Thakkar A, Baldi A, Murthy RSR, Sharma RK (2016) Physical, chemical and phytoremediation technique for removal of heavy metals. J Heavy Metal Toxicity Dis 1(2). https://doi.org/10.21767/2473-6457.100010

Sharma P (2021) Efficiency of bacteria and bacterial assisted phytoremediation of heavy metals: an update. Biores Technol 328:124835. https://doi.org/10.1016/j.biortech.2021.124835

Shibi K, Cherifi O, El gharmali A, Oudra B, Aziz F (2012) Accumulation and toxicological effects of cadmium, copper and zinc on the growth and photosynthesis of the freshwater diatom *Planothidium lanceolatum* (Brébisson) Lange-Bertalot: a laboratory study. J Mater Environ Sci 3(3):497–506

Singh HP, Mahajan P, Kaur S, Batish DR, Kohli RK (2013) Hromium toxicity and tolerance in plants. Environ Chem Lett 11:229–254. https://doi.org/10.1007/s10311-013-0407-5

Singh N, Raghubanshi A, Upadhyay A, Rai U (2016) Arsenic and other heavy metal accumulation in plants and algae growing naturally in contaminated area of West Bengal, India. Ecotoxicol Environ Saf 130:224–233

Singh A, Malaviya P (2019) Chromium phytoaccumulation and its impact on growth and photosynthetic pigments of Spirodela polyrrhiza (L.) Schleid. On exposure to tannery effluent. Environ Sustain 2(2):157–166. https://doi.org/10.1007/s42398-019-00062-4

Singh H, Kumar D, Soni V (2020) Copper and mercury induced oxidative stresses and antioxidant responses of Spirodela polyrhiza (L.) Schleid. Biochem Biophys Rep 23:100781. https://doi.org/10.1016/j.bbrep.2020.100781

Singhal V, Rai JPN (2003) Biogas production from water hyacinth and channel grass used for phytoremediation of industrial effluents. Bioresour Technol 86(3):221–225. https://doi.org/10.1016/S0960-8524(02)00178-5

Sinha V, Pakshirajan K, Chaturvedi R (2014) Chromium(VI) accumulation and tolerance by tradescantia pallida: biochemical and antioxidant study. Appl Biochem Biotechnol 173(8):2297–2306. https://doi.org/10.1007/s12010-014-1035-7

Sinha V, Pakshirajan K, Chaturvedi R (2018) Chromium tolerance, bioaccumulation and localization in plants: an overview. J Environ Manag 206:715–730. https://doi.org/10.1016/j.jenvman.2017.10.033

Sivakumar D (2016) Biosorption of hexavalent chromium in atannery industry wastewater using fungi species. Glob J Environ Sci Manag 2(2):105–124. https://doi.org/10.7508/gjesm.2016.02.002

Skinner K, Wright N, Porter-Goff E (2007) Mercury uptake and accumulation by four species of aquatic plants. Environ Pollut 145(1):234–237. https://doi.org/10.1016/j.envpol.2006.03.017

Śliwa-Cebula M, Kaszycki P, Kaczmarczyk A, Nosek M, Lis-Krzyścin A, Miszalski Z (2020) The common ice plant (Mesembryanthemum crystallinum L.)–Phytoremediation potential for cadmium and chromate-contaminated soils. Plants 9(9):1230. https://doi.org/10.3390/plants9091230

Soltan ME, Rashed MN (2003) Laboratory study on the survival of water hyacinth under several conditions of heavy metal concentrations. Adv Environ Res 7(2):321–334. https://doi.org/10.1016/S1093-0191(02)00002-3

Soni SK, Singh R, Tiwari S (2022) Management of chromium(VI)-contaminated soils through synergisticapplication of vermicompost, chromate reducing rhizobacteriaand Arbuscular mycorrhizal fungi (AMF) reduced plant toxicityand improved yield attributes in *Ocimum basilicum* L. Arch Microbiol 204:614

Spain O, Plohn M, Funk C (2021) The cell wall of green microalgae and its role in heavy metal removal. Physiol Plantarum 1–10

Srivastava S, Prakash S, Srivastava MM (1999) Chromium mobilization and plant availability—The impact of organic complexing ligands. Plant Soil 212(2):201–206

Srivastava D, Tiwari M, Dutta P, Singh P, Chawda K, Kumari M, Chakrabarty D (2021) Chromium stress in plants: toxicity, tolerance and phytoremediation. Sustainability 13(9):4629
Sugiura Y, Tanaka H (1981) Structure, properties and transport mechanism of Iron(III) complex of mugineic acid, a possible phytosiderophore. J Am Chem Soc 103:6979–6982
Sukumar M (2010) Reduction of hexavalent chromium by rhizopus oryzae. Afr J Environ Sci Technol 4:412–418. http://www.academicjournals.org/AJEST ISSN 1991-637X © Academic Journals
Sunday ER, Uyi OJ, Caleb OO (2018) Phycoremediation: an eco-solution to environmental protection and sustainable remediation. J Chem Environ Biol Eng 2:5–10
Sutkowy M, Kłosowski G (2018) Use of the coenobial green algae pseudopediastrum boryanum (chlorophyceae) to remove hexavalent chromium from contaminated aquatic ecosystems and industrial wastewaters. Water 10(6):712
Takagi S (1976) Naturally occurring iron-chelating compounds in oat and rice root washings. Soil Sco Plant Nutr 232:423–433
Tangahu BV, Sheikh Abdullah SR, Basri H, Idris M, Anuar N, Mukhlisin M (2011) A review on heavy metals (As, Pb, and Hg) uptake by plants through phytoremediation. Int J Chem Eng 2011:1–31
Tara N, Arslan M, Hussain Z, Iqbal M, Khan QM, Afzal M (2019) On-site performance of floating treatment wetland macrocosms augmented with dye-degrading bacteria for the remediation of textile industry wastewater. J Clean Prod 217:541–548. https://doi.org/10.1016/j.jclepro.2019.01.258
Tattiayeva Z, Tattiayeva S (2021) Analysis of Cr(III) ions adsorption on the surface of algae: implications for the removal of heavy metal ions from water. Eastern-Euro J Enterp Technol 4/10(112):14–23
Taufikurahman T, Pradisa MAS, Amalia SG, Hutahaean GEM (2019) Phytoremediation of chromium (Cr) using Typha angustifolia L., Canna indica L. and Hydrocotyle umbellata L. in surface flow system of constructed wetland. IOP Conf Ser: Earth Environ Sci 308(1):012020
Tauqeer HM, Ur-Rahman M, Hussain S, Abbas F, Iqbal M (2019) The potential of an energy crop "Conocarpus erectus" for lead phytoextraction and phytostabilization of chromium, nickel, and cadmium: An excellent option for the management of multi-metal contaminated soils. Ecotoxicol Environ Saf 173:273–284
Tekerlekopoulou AG, Tsiflikiotou M, Akritidou L, Viennas A, Tsiamis GP, Thakur IS, Srivastava S (2011) Bioremediation and bioconversion of chromium and pentachlorophenol in tannery effluent by microorganisms. Int J Technol 3:224–233
Thakur IS, Srivastava S (2011) Bioremediation and bioconversion of chromium and pentachlorophenol in tannery effluent by microorganisms. Civ Eng 2(3)
Travieso and Canizares -Villanueva, R. O., Borja, R., Benítez, F., Domínguez-Bocanegra, A. R., Dupeyrón, R., Valiente, Y.V. , 1999Travieso L, Canizares-Villanueva RO, Borja R, Benítez F, Domínguez-Bocanegra AR, Dupeyrón R, Valiente YV (1999) Heavy metal removal by microalgae. Bull Environ Contam Toxicol 62:144–151
Tripathi BD, Shukla SC (1991) Biological treatment of wastewater by selected aquatic plants. Environ Pollut 69(1):69–78
Ullah R, Hadi F, Ahmad S, Jan AU, Rongliang Q (2019) Phytoremediation of lead and chromium contaminated soil improves with the endogenous phenolics and proline production in parthenium, cannabis, Euphorbia, and Rumex species. Water Air Soil Pollut 230(2):40
Usman K, Al Jabri H, Abu-Dieyeh MH, Alsafran MHSA (2020) Comparative assessment of toxic metals bioaccumulation and the mechanisms of chromium (Cr) tolerance and uptake in *Calotropis procera*. Front Plant Sci 11. https://doi.org/10.3389/fpls.2020.00883
von Wiren N, Marschner H, Romheld V (1996) Root of iron-efficient maize also absorbs phytosiderophore chelated zinc. Plant Physiol 106:71–77
Vymazal J, Švehla J, Kröpfelová L, Chrastný V (2007) Trace metals in phragmites australis and phalaris arundinacea growing in constructed and natural wetlands. Sci Total Environ 380(1–3):154–162. https://doi.org/10.1016/j.scitotenv.2007.01.057

Wang Q, Cui Y, Dong Y (2002) Phytoremediation of polluted waters potentials and prospects of wetland plants. Acta Biotechnol 22(1–2):199–208
Wei W, Wang Q, Li A, Yang J, Ma F, Pi S, Wu D (2016) Biosorption of Pb (II) from aqueous solution by extracellular polymeric substances extracted from Klebsiella sp. J1: adsorption behavior and mechanism assessment. Sci Rep 6(1):31575. https://doi.org/10.1038/srep31575
World Health Organisation (1988) Chromium: environmental health criteria 61:1–241
Wu Q, Wang S, Thangavel P, Li Q, Zheng H, Bai J, Qiu R (2011) Phytostabilization potential of *Jatropha Curcas* L. in polymetallic acid mine tailings. Int J Phytoremed 13(8):788–804
Wu S, Hu Y, Zhang X, Sun Y, Wu Z, Li T, Lv J, Li J, Zhang J, Zheng L, Huang L, Chen B (2018) Chromium detoxification in arbuscular mycorrhizal symbiosis mediated by sulfur uptake and metabolism. Enviorn Exp Bot 147:43–52
Xia H, Ma X (2006) Phytoremediation of ethion by water hyacinth (Eichhornia crassipes) from water. Bioresour Tech 97(8):1050–1054
Yan X, Wang J, Song H, Peng Y, Zuo S, Gao T, Duan X, Qin D, Dong J (2020) Evaluation of the phytoremediation potential of dominant plant species growing in a chromium salt–producing factory wasteland, China. Enviorn Sci Pollut Res 27(7):7657–7671
Zaranyika MF, Ndapwadza T (1995) Uptake of Ni, Zn, Fe, Co, Cr, Pb, Cu and Cd by water hyacinth (eichhornia crassipes) in mukuvisi and manyame rivers, Zimbabwe. J Environ Sci Health Part A: Environ Sci Eng Toxicol 30(1):157–169
Zayed AM, Terry N (2003) Chromium in the environment: factor affecting biological remediation. Plant and Soil 249:139–156
Zayed A, Lytle CM, Qian J-H, Terry N (1998) Chromium accumulation, translocation and chemical speciation in vegetable crops. Planta 206(2):293–299
Zhadra T, Tazhibayeva SM, Kujawski W, Zayadan B, Musabekov K, Adilbekova A (2021) Analysis of Cr (III) ions adsorption on the surface of algae: implications for the removal of heavy metal ions from water. East-Eur J Enterp Technol 4(10(112)):14–23. https://doi.org/10.15587/1729-4061.2021.237532
Zhao M, Duncan JR (1998) No title found. Biotechnol Lett 20(1):37–39
Zhao XL, Yang JZ, Liu S, Chen CL, Zhu HY, Cao JX (2014) The colonization patterns of different fungi on roots of Cymbidium hybridum plantlets and their respective inoculation effects on growth and nutrient uptake of orchid plantlets. World J Microbiol Biotech 30:1993–2003
Zhu YL, Pilon-Smits EAH, Tarun AS, Weber SU, Jouanin L, Terry N (1999) Cadmium tolerance and accumulation in Indian mustard is enhanced by overexpressing γ-glutamylcysteine synthetase. Plant Physiol 121(4):1169–1177

Chapter 12
Toxicity of Rhizospheric Chromium Contaminated Soil and Its Phytoremediation

Pratyush Kumar Das, Bikash Kumar Das, Bidyut Prava Das, and Patitapaban Dash

Abstract The chromium is a common ingredient of industrial products for providing tensile strength, corrosion inhibition and shining ability to metals. The overuse of chromium during industrial production is one of the factors responsible for rhizospheric soil chromium contamination and phytotoxicity. Billion dollars of chromite resources are present across the world. The chromite mining and release of hexavalent chromium from industrial refuges, augment the risk associated with rhizospheric soil chromium contamination. The hexavalent chromium is recognized by USEPA, as a hazardous metal. Selection of hyper-accumulators for operation of phytoremediation is a possible solution for this burning environmental problem. The hyperaccumulator, associated soil biota and available chromium, interactions in rhizospheric soil decides the fate of phytoremediation. The disposal of hyperaccumulators biomass used during phyto-remediation may have dire consequences but found to be sustainable, economical, and advantageous, as compared to possible physico-chemical processes. The present approach of biomass use, during rhizospheric remediation of chromium contaminated soil is gaining acceptance over the years. For process efficiency improvisation, it is required to optimize the operating conditions, during pilot and field scale applications. The successful operation of phytoremediation using selected chromium hyperaccumulators, at pilot and field stages of application could help in promoting the detoxification of environmental components like soil and minimization of adverse impacts of chromium on public health and environment. It is a step towards up-gradation of environmental quality and protection of living society on a sustainable basis.

Keywords Chromium · Hyperaccumulators · Phytoremediation · Rhizosphere · Toxicity

P. K. Das · B. K. Das · P. Dash (✉)
Centre for Biotechnology, Siksha 'O' Anusandhan (Deemed to be University), Bhubaneswar, Odisha, India
e-mail: patitapabandash@soa.ac.in

B. P. Das
Department of Botany, Sailabala Women's Autonomous College, Cuttack, Odisha, India

© The Author(s), under exclusive license to Springer Nature Switzerland AG 2023

N. Kumar et al. (eds.), *Chromium in Plants and Environment*, Environmental Science and Engineering, https://doi.org/10.1007/978-3-031-44029-8_12

12.1 Introduction

The application of chromium (Cr) individually or in combination with other heavy metals, like nickel-based alloys, improves the strength and corrosion resistivity of manufactured steel products. It is also commonly used as an ingredient during commercial activities like metal plating, leather tanning, wood keeping, painting, dyeing and chemicals manufacturing. The excessive use of Cr with industrialization and urbanization is one of the prime factors behind degradation of environment by Cr rich effluents, sludge and solid wastes. The contamination of soil profile has increased the human health risk around the mining and industrial sites. The soil toxicity of Cr contaminated sites is correlated with the proportionate distribution of Cr(VI) (hexavalent chromium), Cr(III) (trivalent chromium), and TCr (total chromium), in its structural horizons.

The contaminated soils, rich in Cr(VI) are extremely toxic and removal of toxicity is possible by enriching its rhizospheric segments with organic carbons, during phytoremediation. It may get channelized by the possible reduction of hydrophilic Cr(VI)–Cr(III), with the decrease in its stability and water solubility (USEPA 1998; Zayed and Terry 2003). The soil Cr kinetics during phytoremediation is possibly modulated by the locally involved abiotic and biotic components of the soil environment (Eco-USA 2001).

The detoxification of Cr contaminated soils can be possible with the application of physical and chemical principles, but unlike biological principles, are responsible for secondary environmental pollution, at many instances. The physico-chemical techniques like soil flushing, solidification, stabilization, vitrification, redox reactions, excavation and off-site disposal were attempted earlier with different degrees of success, but not free from disadvantages (USEPA 1993). These techniques are either proved to be costly or inappropriate for successful detoxification of Cr contaminated soils. The operation of phytoremediation is a viable option under the present context for successful detoxification of Cr contaminated industrial and mining sites.

12.2 Speciation of Chromium and Toxicity

In its natural state, Cr is a hard silvery metal, ranked as the 17th top most hazardous substance (USEPA 1999, 1998). The two stable forms of this toxic metal are Cr(III) and Cr(VI). The intermediate unstable forms like Cr(IV) and Cr(V) are formed during conversion of Cr(VI) and Cr(III), following redox reactions. During intracellular reduction in tissues of living organisms, the concentration of TCr may be same as the concentration of Cr(III), if, all Cr(VI) gets reduced to Cr(III), in the system,

The hazardous Cr(VI) is extremely toxic to biological cells, beyond threshold limits. Its high toxicity is more pronounced with increase in solubility, permeability and mobility, as compared to those under trivalent conditions (Das et al. 2021a, b). It may be due to the variation in configuration and confirmation of specified chemical species. The Cr(III) is mostly non-toxic within the threshold limits and much required for living cells, as a trace dietary supplement (Panda and Choudhury 2005; Nematshahi et al. 2012).

12.3 Hexavalent Chromium as a Toxic Heavy Metal

The Cr(VI) is a commercially useful heavy metal, required during industrial production. Besides its tensile strength and corrosion resistance, some other features attract its presence, as an adjunct during industrial processing. The ability of Cr salts to change colour at different concentrations is another aspect for its consideration as a colouring agent, during industrial requirements (Augustynowicz et al. 2020). The production and post-production processes released wastes, rich in Cr(VI) to surroundings and responsible for occurrence of soil pollution, directly or indirectly. The enrichment of soils with Cr(VI), leads to expression of its adverse effects on components like resident biota. The excessive accumulation of Cr(VI) in living cells, sourced from contaminated soils, expresses its toxicity in affected cells.

The industrial effluents, from metals finishing, leather tanning, cement production and similar processes cause Cr(VI) based water pollution. Besides surface water pollution, the Cr based contamination of bore well water is an example of groundwater pollution (Zaidi et al. 2014). The soil pollution, directly from Cr rich industrial wastes or indirectly through contaminated surface water or harvested groundwater shows wide range of variation in Cr(VI) contamination. The spectrum of Cr rich wastes from industries, mines and urban sectors are released in solid, liquid or gaseous phases. The Cr(VI) from these wastes, directly or indirectly, channelized into the soil and responsible for wide range of Cr(VI) led soil pollution.

The post-contamination changes caused by Cr(VI), includes, irreversible alterations in the genomic constituents, errors at the levels of transcription, translation and post-translation, anomalous cell division, and activity of proteins inside exposed tissues, and subsequently, direct or indirect interruption of the cellular development (Das et al. 2017, 2018).

12.4 Sources of Chromium Release to Rhizospheric Soil

The main source of Cr required for industrial purposes is chromites. It is one of the main reasons behind wide scale geological activities and chromite mining. The anthropogenic causes are not the only route of soil Cr pollution, as has been caused by multiple natural sources, also. Naturally, it is caused by sources like volcanic eruptions, soil erosion, rocks disintegration, sands and dusts dispersion by blowing of wind (Memon and Schröder 2009; Apte et al. 2006; Das et al. 2022a, b, c). The impacts from anthropogenic causes of soil Cr pollution is more than the impacts of pollution caused by natural phenomena (Fig. 12.1).

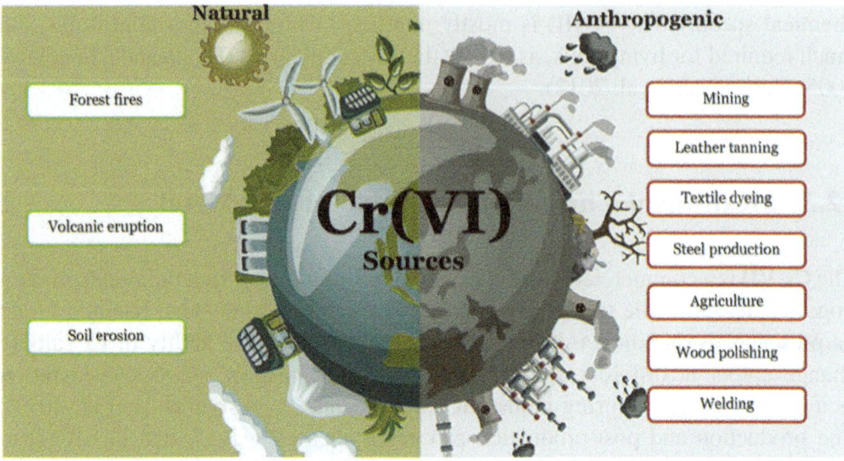

Fig. 12.1 Sources of release of Cr(VI) as a soil pollutant

The Cr as a pollutant is directly or indirectly, expressing its adverse impacts on the exposed environmental components. The activities like tanning of leather, electroplating of metals, processing of timber, dyeing of textiles, smoking of tobacco, leaching of toxicants from improper sanitary landfills, refining of ferrochrome ores, production of cement and stainless steel are noteworthy examples from anthropogenic sources of soil pollution by Cr. It imparts hazardous effects on components of ecosystems (Das et al. 2021a; Saha et al. 2011; Guidotti et al. 2015). Even, the application of phosphate fertilizers can be able to cause Cr based soil pollution, as 30–3000 mg kg^{-1} of Cr was found in it (Singh et al. 2013).

12.5 Mechanism of Rhizospheric Soil Chromium Toxicity

The chromium toxicity at the soil rhizosphere is governed by few abiotic and biotic factors prevalent at the site. This toxicity resulted due to the interactions of relative proportion of chromium species and the associated environmental components present there. The Cr forms like Cr(VI) and Cr(III) are stable and have attended high residence times as compared to its unstable forms. At a point of time, the relative proportion of Cr species present in the rhizospheric soil systems are determined by the residence times of available chromium forms.

The environmental factors determining the rhizospheric soil chromium toxicity are abiotic and biotic in nature. The abiotic factors influencing the rhizospheric soil Cr toxicity levels include texture, pH, precipitation, redox potential and nutrient status of the soil systems, and the biotic factors include soil microorganisms, organic carbon contents of the soil system.

12.5.1 Abiotic Factors Determining Rhizospheric Soil Chromium Toxicity

Texture of soil

It is determined by particle size, porosity, and water holding capacity of soil. These characters determine the type, concentration, affinity and leaching ability of the Cr species and ultimately its toxicity level in the rhizospheric soil.

Soil pH

It determines the inter-conversion of Cr(VI) and Cr(III), at a point of time in soil. Mostly, the presence of Cr(VI), makes the soil more acidic by decreasing its pH due to the induction of deprotonation.

Precipitation at the site

It is required for determining soil Cr toxicity, as Cr(VI) is hydrophilic. Its concentration is influenced by the fluctuation in soil Cr dilution coefficient.

Soil Redox potential

Specific chromium forms play significant role in determining net soil Cr toxicity. The redox potential determines the presence of specified chromium forms and fluctuations in the relative proportion of those forms.

Soil nutrient status

The sequestration of chromium from soil to flora by living cells reduces its concentration and toxicity in rhizospheric soil. The chromium is not an essential element for plants growth and survival. Specific channels are absent in plants for chromium absorption and translocation. The soil nutrient status is an important factor, as chromium species utilizes the path of specific nutrients for absorption and translocation in plants. It follows the path of nutrients sharing similarities with the structure of chromium species during the absorption and translocation in plants.

12.5.2 Biotic Factors Determining Rhizospheric Soil Chromium Toxicity

Soil microorganisms

The microbial populations present in soil, helps the plants during adsorption, absorption and translocation of Cr species from soil. It ultimately reduces the Cr toxicity in that soil. The microorganisms like species of algal, fungal and bacteria population are quite useful for inducing phytoremediation, during soil Cr detoxification. Besides, production of Cr reductase by specified microorganisms, helps in the net reduction in Cr toxicity, at rhizospheric soil systems.

Soil organic carbon content

It is required for minimization of Cr toxicity in soil. It helps in soil Cr toxicity reduction, following a series of protonation and deprotonation reactions.

12.6 Focus on Soil Chromium Toxicity in India

About 2% of the world's chromium resources come from Indian chromite reserves. The Sukinda mines playing pivotal role in chromite distribution map, as it acquires 97% of India's chromite deposits (Mishra and Sahu 2013). The main chromite reserves are located in the states of Odisha (Sukinda), Karnataka (Nuggihalli), Maharashtra (Nagpur and Sindhudurg), Jharkhand (Jojohau), Andhra Pradesh (Jannaram), and Tamil Nadu (Namakkal and Thiruchengoddu). The Cr pollution from chromite mining, leaching, effluents discharge, improper Cr waste disposal, mine-tailing infiltration, and other growing industrial operations, primarily contribute to pollution in India (Prasad et al. 2021).

The Sukinda Valley, one of the top ten polluted locations on earth, is well known for major chromite reservoirs of India. It generates a substantial chunk of mining waste, which worsens the health of those exposed to it and causes severe environmental problems (Yadav et al. 2018). The chromite mining activities have ruined the topography, soil and water resources of the site and threatened the associated public health. The Cr emission from these mines to environmental components ranges in between 10 and 4000 mg Kg^{-1} (Vijayana and Nikos 2010). In Sukinda, the surface and groundwater have Cr(VI) levels much above the threshold limit, 0.05 mg L^{-1}. The Blacksmith Institute's (2007), found an alarming level of Cr(VI) in the surface water of mining area. In the Damsala nala, Cr(VI) concentrations were ranging in between 0.018 and 0.172 mg L^{-1}, throughout the summer. In contrast, it exceeded the threshold limit (0.05 mg L^{-1}) for B and C category surface water at village Ostapa, reaching up to 0.201 mg L^{-1}, during monsoon season. The Cr(VI) and TCr concentration varied from 12–311 mg Kg^{-1} to 3589–14,486 mg Kg^{-1}, respectively (Mishra et al. 2009). The adverse effects of chromite mining are observed more, within 1 km radius from the centre of mining and industrial activities. The adjacent villages are not even free from its adverse effects. The mortality rate is 86.42% in adjacent villages, due to diseases associated with chromite mining activities. The acute pollution and health risks associated with Cr(VI) result in irreversible damage to the exposed organisms (Gupta et al. 2019).

The tanneries in India use chrome tanning methods. It is a leading contributor to soil pollution specifically in states having numerous leather tanning industries. These industries release 2000–3000 tonnes of Cr per annum, thereby contaminating soil and water bodies. The states like Tamil Nadu, Gujarat, Uttar Pradesh, and West Bengal are home to majority of these industries. The tanneries generate almost 1500 metric tonnes of chromium sulphate per annum, as trash (Down to Earth 2005). The untreated effluents have Cr concentrations up to 2000–5000 mg L^{-1} and being released to nearby lakes, rivers, and streams (Dhal et al. 2013). The tanneries discharged significant amount of untreated effluents into the river Ganges (Mohan et al. 2011). The Cr(VI)-containing sludge is potentially toxic and is anticipated to have detrimental impacts on human health, when it seeps into groundwater, subsoil, and rivers. This sludge releases hazardous pollutants and volatile methane into the environment and occasionally catches fire during summer. The assessment of water quality of Kanpur revealed, groundwater with Cr(VI) content of 6.2 mg L^{-1} against the threshold limit of 0.05 mg L^{-1}. It was observed that a steady increase in the Cr concentration from upstream (0.039 ± 0.02 mg L^{-1}) to downstream (4.47 ± 1.85 mg L^{-1}) of the river, with summer being the optimal season and declining during the monsoon period (Khatoon et al. 2013). It may occur due to the increase in dilution factor during monsoon period.

Tonnes of garbage containing Cr have been piling up within the shuttered offices of an industrial complex, at Ranipet, for almost three decades. Besides, the high level of soil pollution may be due to the presence of hundreds of tanneries and small chemical companies (Rao et al. 2013). The environmental experts believe that, within a 30 km radius the groundwater has already been poisoned by Cr wastes. The study of soil and groundwater qualities at Ranipet was done in 2016. It revealed serious contamination of those environmental components. As per the Geological Survey of India, Cr(VI) contamination has a southward spread up to 2–2.5 kms. The assessment of Thandalam and Manianpattu lakes confirmed heavy contamination with Cr and thus making the water unfit for human use (Madhavan 2020). The TCCL was responsible for the production of chromium sulphate, sodium bichromate, and sodium sulphate tanning powder. The TCCL factory was shut down for environmental issues, including soil and water pollution, in 1996. A serious health risk is posed, by the estimated 1,50,000 tonnes of Cr-containing wastes, dumped there. The irresponsible dumping of waste containing Cr(VI) over a long period of time has resulted in the accumulation of trash to a height of 3–5 m over 2–4 ha. During the rainy season, Cr(VI) leachate infiltrates through the subsurface, hence, affecting the groundwater quality (CPCB 2016).

According to an assessment report, an industrial unit released, about 77,000 tonnes of hazardous Cr wastes into the environment at Gujarat (Rao et al. 2009). The area of the abandoned industrial unit is heavily contaminated with chromate salts covering an area of 15,000 square feet. The unauthorized Cr waste dumping sites are located along roadways close to the factory. Workers, exposed to Cr were shown serious health effects, including yellow discoloration of the affected parts.

The cement manufacturing industries, breaking down asbestos, catalytic converter emissions and other solid organic wastes are other sources of Cr contamination. As it enters into the living organisms, it becomes the part of the food chain. Its concentration rises in tissues and eventually biomagnified in top order organisms (Mitra et al. 2017). The countries like India, Bangladesh, and Pakistan are making protein concentrates as a feed for fish and poultry from tannery wastes (skin). The high Cr content (0.3–0.4%, dry weight) in these products could be dangerous for the public health due to biomagnifications. It is a possibility that, 1 metric tonne (dry weight) of excreta, from those contaminated poultry, might expose the environment to a Cr burden of 2.94 kg. (Hossain et al. 2017). At higher concentrations, Cr is noxious

Table 12.1 Reported soil and water contamination of few Indian cities by chromium

Region	Contaminated environmental component	Causes of chromium release	Chromium concentration (in ppm)	References
Nauriyakhera (Kanpur)	Groundwater	Textile effluents, tannery effluents, chromium-rich wastes, dumpsites	16.30	Singh et al. (2009)
Pernampattu, Madhnur, Alangayam, Natrampalli (Vellore)	Groundwater	Tannery effluents	0.04	Kanagaraj and Elango (2019)
Maheshwaram watershed (Hyderabad)	Groundwater	Urban wastes, irrational waste disposal	0.011–0.418	Purushotham et al. (2013)
River Yamuna (Delhi stretch)	Surface water	Human interference	0.002–1.98	Bhardwaj et al. (2017)
Ashtamudi wetland (Kollam)	Surface water	Dumping of municipal wastes, wastes from fishing harbor, oil spillage	1.1–0.08	Karim and Williams (2015)
Ropar wetland (Amritsar)	Soil	Human interference	0.29–10.30	Sharma et al. (2018)
Ranaghat–Fulia–Shantipur area (Nadia)	Surface water	Textile effluents	0.0–4.9	Sanyal et al. (2015)

to plants and negatively impacts a variety of biological processes. In some cases, it may lead to the destruction of the entire population (Dotaniya et al. 2014). The level of Cr contamination of environmental components is variable and it depends upon the sources of Cr release (Table 12.1).

12.7 Overview of Toxic Effects of Soil Chromium Contamination

The chromium gets accumulated in soil slowly but in it maintains a long residence time. It leads to Cr based soil pollution, a burning environmental problem. The soils irrigated with sewage sludge and effluents, accumulates Cr(VI) in its surface layer (Abdel-Sabour 2007).

In the geogenic processes of chromite oxidation, the microbes interact with mafic and ultramafic rocks at the same time. It releases Cr(VI) in our natural environment.

The Cr(VI) shows opposite physical and chemical characteristics in soil colloids and has a strong affinity towards negative charge (Tumolo et al. 2020) with the pH ranges in between 4 and 8. The Cr is present in soil in low concentrations, but it may get increased in it with certain natural and manmade activities. Generally, Cr(VI) is highly reactive and toxic, as compared to other Cr forms, due to its hydrophilic structure with high oxidation state. Its small concentration in soil may be due to the result of conversion of natural Cr(III) by oxidation. In contrast, larger concentration of Cr(VI) in soil may be due to the Cr(VI) pollution or the conversion of Cr(III) by oxidation. The combustion of fossil fuels, mining, smelting of ores, amendment of sludge to soil, application of fertilizer and chemical agricultural practices are examples of major causes of soil Cr contaminations. When Cr added to sewage sludge, it may change its form but present in soil for an extended period and available to plants for many years (Dhal et al. 2013).

The Cr polluted soil samples collected from a depth of 30 cm shows variation in the levels of different forms of Cr. As an example, out of 8 km^2 sampled area, almost, 0.9 km^2 was observed to be polluted with Cr, with a high concentration up to 12 960 mg Kg^{-1} (Ayari et al. 2010). The Cr(VI) at moderate to high concentrations has been found to affect plant growth and physiology. The wilting and discoloration of leaves have been observed initially in plants during exposure to Cr toxicity (ANRCP 1998). The 0.5 ppm Cr(VI) concentration in aquatic conditions or 5 ppm of its concentration in soils, can impart phytotoxicity (Fendorf 1995).

12.8 Possible Techniques for Remediation of Soil Chromium Contamination

The widespread use of Cr in industrial installations and its extensive extraction at mining sites, pollutes the soil matrices to a larger extent. Soil being an intrinsic part of the environment is strongly related to determination of environmental health. The toxic Cr(VI) exists in soil with pH ranging from 7 to >7, as highly, moderately or sparsely soluble salts or anions, like chromate (CrO_4^{2-}). Remediation of Cr(VI) contaminated soil is extremely important to protect the public health from its adverse effects. However, the complexity of chromium compounds makes the remediation process really challenging. The stable forms of chromium are capable of conversion among stable and unstable forms in nature, due to redox reactions. It is making the soil system complicated to determine as Cr contaminated soils are hazardous or not (James 1996).

Cr is a hypertoxic and carcinogenic agent, capable of accumulation and transfer through food webs, affecting human health (Deb et al. 2022; Ding et al. 2021). Researchers worldwide are keen towards studying Cr pollution and to devise potential harmless techniques to manage the same (Zhang et al. 2021). Most of the remediation techniques, target the conversion of Cr(VI) to its least toxic and stable form in soil

(Yang et al. 2021). The remediation of Cr polluted soil can be done through physico-chemical and/or biological methods.

12.8.1 Physico-Chemical Methods for Remediation of Cr(VI) Contaminated Soils

The commonly used physico-chemical methods are land filling, soil washing, stabilization, vitrification, and chemical reduction. These methods can be used at the site of pollution or away from it, depending on the locality and contaminants load on soil.

Landfilling

Landfilling, also known as "dig and haul" is the most simple of the remediation techniques. This technique is used to remove, soil pollutants from its actual site, to a secure landfill that has been engineered with impermeable walls, drains for leachates, and other facilities. The landfill area generally located far away from urban areas and generally in isolated places. This makes the transport of the contaminated soil to the landfill site, a very costly affair. Moreover, the technique of landfilling is possible for contaminated soil over a small area. Cr(VI) pollution arising out of large sites like mines cannot be remediated using this technique.

Soil washing

Soil washing is another option that uses an aqueous solution to separate contaminants like Cr(VI) adsorbed onto the soil particles. The washing solution generally mobilizes heavy metals by making changes to soil's ionic strength, pH, complexation, oxidizing and reducing abilities (Beiyuan et al. 2017). An array of acids, alkalis and other chemicals are used in formulation of a washing solution. Despite washing, some metals and leachates tend to present in soils (Zhai et al. 2018). This technique is also not feasible for large contaminated sites.

Vitrification

The process of vitrification uses thermal energy in order to melt the soil so as to bring physical or chemical stabilization. Heavy metals like Cr in the soil are isolated in glass material wherein they remain chemically bound (Shao et al. 2022; Shu et al. 2020). This process in particular is highly energy demanding and therefore a costly option.

Stabilization

Stabilization of Cr(VI), a toxic metal in contaminated soils is mostly carried out by the use of types of stabilizing agents. The stabilizing agents react with heavy metals through a sequence of reactions like adsorption using suitable adsorbents, precipitation, and reduction using requisite redox agents, to reduce the mobility, toxicity, or biological efficiency of contaminants (Xu et al. 2021; Mei et al. 2022).

Adsorption

It is a common technique, used for decontamination of soils polluted by metals (Wadhawan et al. 2020). Agricultural residues and charged carbons from organic sources are excellent adsorbents, having metal removal ability from soil. Hence, it is being used as natural adsorbents (Cheng et al. 2019). Chitosan, a natural polymeric nanoparticle has high adsorption properties, may be attributed to the reactive amino and hydroxyl groups present as functional groups on it. Recent uses of chitosan in nanofiltration of metal contaminants from contaminated soil have been done, successfully (Wadhawan et al. 2020). Biochar is also considered to be a good adsorbent on the basis of its economy, easy availability, and optimum water retention capacity. Biochar also helps in cycling of nutrients during crop growth, along with reduced uptake of heavy metals from soils by engaged plants (Fu et al. 2021; Kavitha et al. 2018). A dose of 10 g Kg^{-1} of biochar-nZVI was found to successfully remove 86.55% Cr(VI) from polluted soil along with increasing the abundance and diversity of indigenous bacterial species (Yang et al. 2022). In a recent study, double hydroxides of magnesium (MgAl) and calcium (CaAl), calcined with temperature variation, to immobilize Cr(VI) in soil. The use of MgAl at 500 °C and CaAl at 900 °C were found to adsorb Cr at a rate of 13.89 mg g^{-1} and 33.78 mg g^{-1}, respectively. It indicates that, the double hydroxides could stabilize Cr(VI) better in soil and thus prevent its movement from soil to plants (Zhao et al. 2021). The use of appropriate adsorbents for remediation of Cr contaminated soil is not a long-term solution, as heavy metals like Cr(VI) will eventually undergo decomplexation over time and leads to the release of several secondary pollutants (Lin et al. 2022).

Precipitation

The process of precipitation, makes use of certain chemicals, referred to as 'precipitants'. These chemicals have the ability to react with heavy metals, to form insoluble complexes. Soil pH and metal concentration are two major factors to determine success of the process. Cr is more soluble and mobile at low pH and can be precipitated by increasing the pH of soil matrix. Sludge rich in Cr generated from industries is first digested, followed by addition of specific salts and hydroxides (Pham et al. 2019). However, precipitation does not work out all alone. It needs certain secondary techniques, like exchange of ions, adsorption or both in sequence, for complete metal removal.

Chemical Reduction

Chemical reduction makes use of chemicals to reduce the metal from its toxic to least toxic form. This process is generally used to reduce Cr(VI) to least toxic, Cr(III). Several industries generate huge amount of Cr(VI) rich effluents, and commonly treat them by the process of chemical reduction. The chemicals like ferrous sulphate, sodium bisulphite, sulphur dioxide, and ferrous ammonium sulphates are some of the reluctant, used for the reduction of toxic Cr(VI), in industries. Soil mixing equipment, injection wells are some of methods to introduce reducing agents to sub-surface soils, at metal polluted sites. The drawbacks of the process include, occurrence of several side reactions, making the soil Cr(VI) treatment an arduous task (Higgins et al. 1997).

12.8.2 Biological Approaches for Remediation of Cr Contaminated Soil

The technique of bioremediation uses organisms like microbes for microbial remediation and plants for phytoremediation. These organisms, remove toxic metal contaminants, like Cr(VI) from soil (Leong et al. 2019; Khoo et al. 2021). A major advantage of bioremediation lies in the fact that in certain cases the remediation of the environment can be easily carried out without the need for any human intervention. Bioremediation can be carried out as in-situ (on-site) and ex-situ (off-site), for detoxification of Cr contaminated soils. The in-situ technique involves processes such as Biosparging (Hussain et al. 2021), Bioventing (Anekwe and Isa 2021), Bacterial remediation (Dhaliwal et al. 2020), Fungal remediation (Srivastava et al. 2015), and Phytoremediation (Lakkireddy and Kües 2017). Similarly, the ex-situ techniques include Land farming (Mosa et al. 2016), Composting (Dhaliwal et al. 2020), and Bio-piling (Gogoi et al. 2021). The ex-situ mode of remediation involves excavation of soil from polluted sites and its shifting to an off-site condition for treatment of pollutants. At the post-treatment stage the disposal of treated wastes have to be done at some pre-approved sites. It makes the whole process more tedious and expensive (Fasani et al. 2018).

12.9 Phytoremediation as a Technique for Soil Chromium Remediation: Opportunities and Challenges

The soil Cr remediation is an arduous task and needs proper attention, keeping in mind the several lethal impacts it poses on environment as well as on its components. There are several methods, being employed worldwide for the remediation of Cr contaminated soil and water, as discussed in the previous section. However, it is quite important to select an appropriate method that is not only feasible, cost-effective, but also environmentally sustainable. Phytoremediation is one such environmentally sustainable technique that makes use of flora, soil conditioners, and rhizospheric microbes to reduce the toxicity of environmental contaminants (Das 2018). Plants like hyperaccumulators, can withstand and accumulate high concentrations of soil Cr(VI) during remediation (Das et al. 2017).

Phytoremediation as a whole is comprised of many techniques which includes phytoextraction (Ali et al. 2013), phytostabilization (Lone et al. 2008), phytodegradation (Pilon-Smits 2005), phytostimulation (Dzantor 2007), phytovolatilization

(Limmer and Burken 2016), rhizofiltration, and phytodesalination (Ali et al. 2013). Reduction of soil Cr(VI) through phytoremediation, mainly employs stabilization of metal in rhizospheric soil and/or its translocation to aerial plant biomass. Plants generally utilize the xylem tissues to translocate Cr(VI) from contaminated soils into their tissues. Plants generally uptake Cr(VI) through their roots by either the apoplastic or the symplastic systems of transport. Apoplastic transport of Cr(VI) is an energy independent pathway and occurs using intercellular spaces in roots. However, the symplastic pathway is energy dependent and takes into account the involvement of sulphate or phosphate ion channels/carriers (Chaudhary et al. 2018). The non-hyperaccumulators among plants tend to accumulate heavy metals in vacuoles of roots whereas, the hyperaccumulators among plants transfer metals from roots to different portions of shoots through vessels of xylem using symplastic pathway (Chandra et al. 2017). Several membrane proteins facilitate, the transfer of the metal through the tissues of xylems (Chaudhary et al. 2018). The metal moves to aerial parts of plants, especially to foliar tissues for accumulation. Once inside those tissues, the Cr is sequestered to vacuoles present in foliar cells, with the action of several transporter proteins.

The detoxification of Cr contaminated soil systems is possible using physico-chemical and biological principles. Analysis of these methods indicates that, phytoremediation is an economical and sustainable technique for detoxification of Cr contaminated rhizospheric soil (Schnoor 1997; USEPA 2000). Thc plant species engaged for phytoremediation purpose is decided by its ability to detoxify the Cr polluted soil systems, in an economical, optimal and sustainable manner. The quantity of Cr uptake by those plants, engaged for phytoremediation, is variable and species specific.

The roots and stems of cruciferous plants like, *Brassica juncea* has the ability for efficient accumulation of toxic Cr(VI) from polluted soil or aquatic systems (Salt et al. 1997). Besides *B. juncea*, other crucifers like *B. nigra*, *B. oleracea*, *B. campestris*, *B. carinata*, and *B. napus* have the ability for efficient metals accumulation (Kumar et al. 1995). Similarly, aquatic species like *Eichhornia crassipes* are useful for remediation of oxidation pond designed for the loading of discharges from industrial units. In an earlier study, the *Eichhornia crassipes*, were allowed to grow on Cr(VI) polluted systems, but it shows the accumulation Cr(III) in tissues of roots and stems (Lytle et al. 1998). The *Eichhornia crassipes* was not only a hyperaccumulator of Cr, but can accumulate cadmium to a certain extent. It shows poor accumulation of arsenic and nickel under polluted conditions (Zhu et al. 1999). The preference for bioaccumulation of heavy metals is species specific. The plant species *Helianthus annuus* accumulates heavy metals in the order of cadmium > nickel > chromium which is reverse to the trend shown by *Brassica juncea* (Zavoda et al. 2001). It is an indication towards transformation of Cr species with the reduction of its toxicity level during phytoremediation. All the plant species are not equally capable of heavy metals remediation. The plant species shows gradation, with respect to heavy metal remediation.

The roots are the main region for Cr accumulation in plants (Das et al. 2022a, b, c). The X-ray absorption spectroscopy revealed that, Cr(VI) entered into the roots of *Prosopis* sp. was completely reduced to Cr(III) during its movement from root to leaves and being present in Cr(III) in foliar biomass (Aldrich et al. 2003). The interest concentrates on the point of conversion of Cr(VI)–Cr(III) during phytoremediation. An earlier study reflected that, the Cr(VI) was converted to Cr(III) during its presence in lateral roots of plants engaged for phytoremediation, and then the Cr (III) was moved into the foliar tissues (Lytle et al. 1998).

A specific plant species is not equally capable of reducing the toxicity of a number of heavy metals. The *Brassica sp* is capable of effective extraction of Cr from soil as compared to the extraction of other heavy metals like zinc, cadmium, copper and nickel present in soil systems (Kumar et al. 1995).

The chelators induced bioaccumulation of heavy metals is not so encouraging. The chelating agents can induce the accumulated metal concentration in plants but the overall metal detoxification by plants from soil, decreased significantly. The metal detoxification in soil decreases due to the tissue necrosis of plants engaged for phytoremediation purpose (Chen and Cutright 2001).

The dicotyledonous plants are more suitable for phytoremediation as compared to monocotyledonous plants, as these two groups of plants have differences in the structure and composition of root systems. The dicotyledonous plants with network of taproot system are preferable for phytoremediation as compared to the fibrous root systems of monocotyledonous plants. The extensive network of celluloses and hemicelluloses in dicotyledonous taproot system, provides more polar hydroxyl groups, required for the transport of Cr from soil to cellular systems of flora present in rhizospheric soil. Presence of these polar OH^- groups helps in the lysis of water and formation of bonds between Cr and water. It may be the possible reason for more solubility and mobility of Cr(VI) inside the plants. It is supported by the earlier findings like, the buckwheat shows more Cr absorption by roots and its translocation from roots to the stems and leaves, as compared to monocot plants like corn and barley (ANRCP 1998; Das et al. 2022a).

The gradient of bioaccumulation of Cr species across plant parts may be attributed to the difference in chemical structures of those plant parts. The roots show relatively high polysaccharides fractions as compared to stems and leaves. The roots and stems have rich OH^- fractions as compared to leaves rich in proteins. It may be a possible explanation for better uptake and bioaccumulation of total chromium (TCr) in roots as compared leaves. It is strengthened by the outcomes of the study on bioaccumulation of TCr, in tissues of *Larrea tridentate* (Gardea-Torresdey et al. 1998).

The phytoremediation efficiency, not only depending on plants as hyperaccumulators, but also on parameters like, characteristics of soil, metals, and microorganisms present in the rhizospheric systems (Das et al. 2018). The multiple factors required for successful operation of phytoremediation of Cr from contaminated soils are described in brief in the succeeding sections.

12.9.1 Selection of Appropriate Plants

Selection of suitable plant species as hyperaccumulators of soil Cr is the initial and crucial step for the smooth functioning of phytoremediation. It is better to select plant species for this purpose, on the basis of certain distinctive features acquired by those species (Fig. 12.2).

The term hyperaccumulator was first used for plants that can retain/tolerate >1000 mg Kg^{-1}, dry weight of heavy metals like nickel in their tissues. Plants growing in natural environment and dry vegetation with hyperaccumulation ability of 300 mg Kg^{-1} of Cr can be labelled as Cr-hyperaccumulating plants (Farooqi et al. 2022). An indigenous plant species is more preferable over other plants due to fewer requirements of management and easy acclimatization to the soil profile, native climate, and seasonal fluctuations. Besides the indigenous plants, there are certain exotic species that can outperform others in terms of accumulation. Plants with high biomass yield, tolerance to extreme climatic conditions, tolerance to variations in soil chemical profiles, and deep and branched network of roots can be considered as an effective phytoremedial species for decontamination of soil Cr(VI) (Sarma 2011). Higher biomass yield means higher capacity to retain Cr(VI) in its biomass. Similarly, tolerance to varied climatic conditions and soil chemical profiles ensures that the plant can easily be grown and thrive in most part of the globe, thus not making its use restricted to a particular geographical area. Deep and branched network of roots will ensure maximum coverage and contact with the soil, thereby making the accumulation process much efficient.

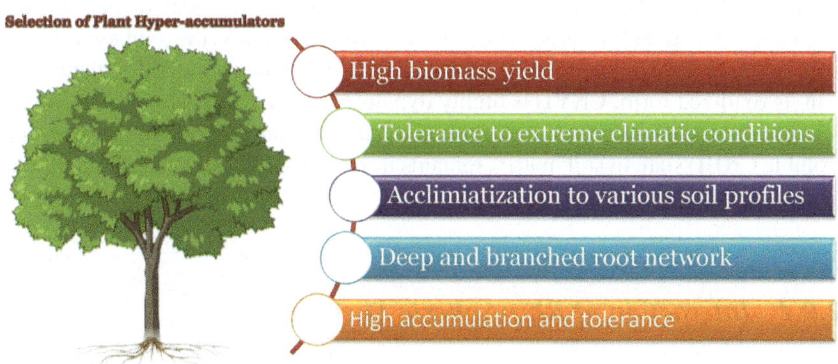

Fig. 12.2 Important factors for selecting a hyper-accumulator plant

12.9.2 Characteristics of the Rhizospheric Soil

The soil chemical features at rhizosphere, like organic carbon content, pH, and texture, to certain extent determine the local availability of metals for phytoremediation (Shah and Daverey 2020). At a low pH of <5, the Cr(VI) is present in soil as oxyanion, like CrO_4^{-2} and remained there in immobilized condition. With an increase in pH (>5), the toxic heavy metal becomes highly mobile and available in soil for absorption by plants root system. Organic amendment in soil rich in Cr(VI) results in reducing the soil pH. It forms a stable organo-metallic complex and exhibits reduced toxicity (Das et al. 2021a). Recent studies have suggested organic acids supplementation as a source of protons that may help in inducing the detoxification Cr(VI) based polluted soils (Das et al. 2021a).

The texture of soil is one of the determining factors, to estimate the efficiency of phytoextraction (Złochet al. 2017). The small sized soil particles have higher concentration of metals like Cr(VI), as compared to coarse particles. The more reactive surface area of small sized soil particles may be a factor for enhanced translocation of metals into plants, from those soils.

12.9.3 Concentration and Nature of the Metal (Cr(VI))

The concentration of metals like Cr(VI) in soil, is another factor to determine the extent of phytoremediation under the stipulated conditions. The upsurge in concentration of soil Cr(VI), beyond the metal accumulation limit of selected plant species, would not work out in favour of remediation process. Similarly, the oxidation state of Cr in soil is directly linked with the phytoavailability of heavy metal for plants. The Cr in its oxidized form, Cr(VI) is highly dynamic and can be easily moved into the root systems of hyperaccumulators from soil matrix. Contrary to it, the reduced and insoluble Cr(III) state in soil matrix, remains immobilized in soil and unavailable for plants to get absorbed.

12.9.4 Interaction of Rhizospheric Microbes

Rhizospheric microbes, also referred to as 'plant growth promoting rhizobacteria' (PGPRs) positively influence the phytoremediation of chromium (Fig. 12.3). The PGPRs are capable of producing several organic acids that can reduce the soil pH, thus enhancing the heavy metals bioavailability for phytoextraction (Yang et al. 2018). The rhizospheric microbes involved in the redox reactions of metals, present in soil, by producing certain oxidising or reducing agents. These agents stabilize the metals in soil or transform them into less toxic forms (Ma et al. 2016).

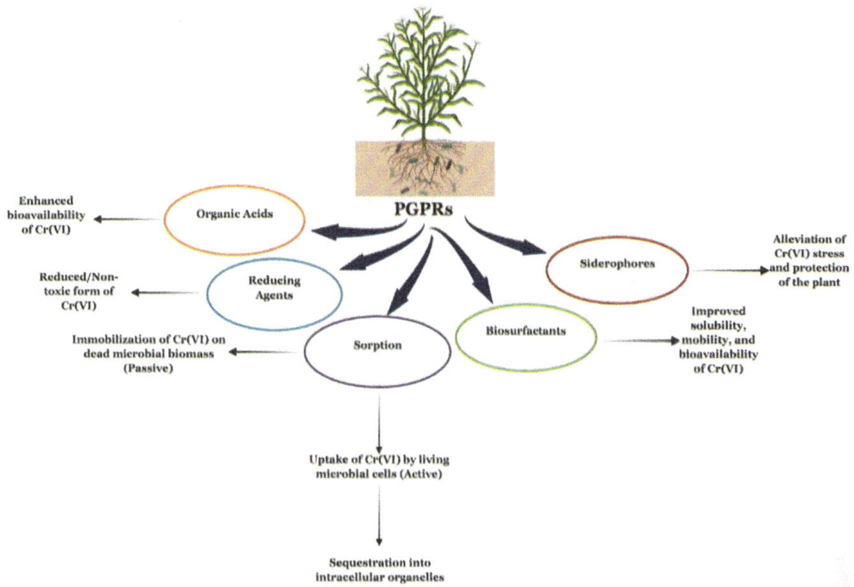

Fig. 12.3 Role of PGPRs in the phytoremediation of Cr(VI)

Specific microbes have an inherent capability for biosorption of heavy metals from soil, following passive or active mechanism, and thereby help in phytoremediation. Passive sorption immobilizes heavy metals by attaching to functional groups present on the surface of dead microbes. In active sorption process, the heavy metals are trapped by the living microbial cells. These metals then sequestered within the intracellular organelles by binding with metallothioneins, present in cells (Das et al. 2021a).

Some microorganisms have the ability to produce amphiphilic compounds, known as 'biosurfactants'. These compounds can make desorption of heavy metals from soil. They also improve the solvation and dynamism of metals, thus making them available for intake by plant hyperaccumulators (Lal et al. 2018). Rhizospheric microbes can also produce 'siderophores' (compounds having low molecular weight), capable of protecting plants under heavy metal stress conditions. The siderophores are basically iron chelators and thus alleviating the biosynthesis of chlorophyll pigments with healthy growth of the plants under metal stress conditions (Ahemad 2015). Many rhizospheric microbes can be utilized for Cr(VI) remediation in soil (Table 12.2).

Table 12.2 Application of rhizospheric microbes for Cr(VI) remediation in soil

Microorganism	Mechanism	Remediation effect	References
Bacillus subtilis MAI3	Production of antioxidants and reductase enzymes	Reduction of Cr(VI)–Cr(III), promoted growth and production of photosynthetic pigments in soybean	Wani et al. (2018)
Paenibacilus konsidensis SK3	Rhizospheric interactions, lowering of pH, secretion of organic acids in soil	Decrease in soil Cr(VI) and increase in phytoremedial ability of *Pongamia pinnata*	Das et al. (2022b)
Klebsiella sp. CPSB4	Secretion of organic acids for phosphate solubilization	Plant growth promotion, 95% reduction of soil Cr(VI)	Gupta et al. (2018)
Cellulosimicrobium cellulans $KUCr_3$	Production of IAA, and solubilization of phosphate	Reduction of Cr(VI)–Cr(III), promoted growth of chilly plant and reduced the uptake of Cr	Chatterjee et al. (2009)
Microbacterium sp.	Cr(VI) reduction	Reduced Cr(VI) toxicity and improved biomass in fenugreek	Soni et al. (2023)
Sphingomonas sp.	Upregulation of antioxidant system	Increase in plant biomass, reduced translocation of Cr(VI), and reduced oxidative stress in soybean	Bilal et al. (2018)
Cellulosimicrobium funkei	Promotion of plant growth promoting substances	High Cr(VI) tolerance up to 1200 μg/ml, enhanced root length in *Phaseolus vulgaris* L	Karthik et al. (2017)

12.10 Socio-economic Aspects of Phytoremediation of Chromium Contaminated Rhizospheric Soil

The good health of common people is intricately linked with sustainable development, and rational land use practices. The soil toxicity due to Cr enrichment breached these notions and not only affecting the public health but also have some adverse impacts on the economic development of a country. The leachable Cr(VI), is a hazardous and Group I human carcinogen (IARC 1990), released from industrial activities. To obtain a sustainable economic development, it is required to take steps for detoxification of Cr based polluted soils, using phytoremediation. Primary target of phytoremediation is to restore the quality of Cr contaminated lands by restricting the Cr content of soil. It is essential to restrict the Cr toxicity of soil, caused by anthropogenic activities, as this toxicity is increasing with time. The toxicity caused by release of Cr(VI) from natural sources is highly insignificant, as compared to its release from anthropogenic sources.

The use of edible plant species for phytoremediation purpose may create havoc for the society. The consumption of those species is significantly fatal due to bioaccumulation of toxic products. As a bio-safety measure, it is better to engage non-edible plant species or weeds for the remediation of soil Cr level. Post-remediation measures require safe disposal of used plants, to prevent secondary environmental pollution and to establish sustainable development.

The phytoremediation is an economical and user friendly technique for land detoxification. It can be suitable for use by all the sections of society, at pilot and field scale levels. It can bring reclamation of hectares of land, which is essential for sustainable socio-economic development at the regional and global levels.

12.11 Conclusion

The release of chromium from multiple sources is degrading the quality of land resources. The high concentration of chromium in soil is making it unfit for productive uses. To protect the living systems and their associated environment from chromium adverse effects, it is better to use the phytoremediation to control Cr based soil pollution. It is advantageous to use phytoremediation for redressal of soil chromium toxicity, as compared to other physico-chemical techniques. It is an economical, user friendly and effective technique. The use of dicotyledonous plants for phytoremediation is preferable, as compared to the use of monocotyledonous plants for this purpose. The efficiency of soil Cr phytoremediation can be upgraded with proper regulation of hyperaccumulators rhizosphere. The active and passive mechanisms involved in this phytoremediation are step towards restoration of healthy state of degraded environment. To a certain extent, it can prevent the emerging socio-economic disruptions caused by toxic pollutants, at different levels. Further study on this aspect is essential to bring improvement of phytoremediation process.

Acknowledgements The author(s) acknowledge the infrastructural support provided by Siksha 'O' Anusandhan (Deemed to be University), Bhubaneswar, Odisha.

References

Abdel-Sabour MF (2007) Chromium in receiving environment in Egypt (an overview). Electron J Environ Agric Food Chem 6:2178–2198

Ahemad M (2015) Enhancing phytoremediation of chromium-stressed soils through plant-growth-promoting bacteria. J Genet Eng Biotechnol 13(1):51–58. https://doi.org/10.1016/j.jgeb.2015.02.001

Aldrich MV, Gardea-Torresdey JL, Peralta-Videa JR, Parsons JG (2003) Uptake and reduction of Cr (VI) to Cr (III) by mesquite (Prosopis spp.): Chromate–plant interaction in hydroponics and solid media studied using XAS. Environ Sci Technol 37(9):1859–1864

Ali H, Khan E, Sajad MA (2013) Phytoremediation of heavy metals—Concepts and applications. Chemosphere 91(7):869–881. https://doi.org/10.1016/j.chemosphere.2013.01.075

Anekwe IM, Isa YM (2021) Wastewater and bioventing treatment systems for acid mine drainage–contaminated soil. Soil Sedim Contam: Int J 30(5):518–531. https://doi.org/10.1080/15320383.2020.1863909

ANRCP (1998) Literature review of the lifetime of DOE material: Aging of plastic bonded explosives and the explosives and polymers contained therein. September 1998, https://www.osti.gov/servlets/purl/290850

Apte AD, Tare V, Bose P (2006) Extent of oxidation of Cr(III)–Cr(VI) under various conditions pertaining to natural environment. J Hazard Mater 128(2–3):164–174. https://doi.org/10.1016/j.jhazmat.2005.07.057

Augustynowicz J, Sitek E, Bryniarski T, Baran A, Ostachowicz B, Urbańska-Stopa M, Szklarczyk M (2020) The use of Callitriche cophocarpa Sendtn. for the reclamation of Cr-contaminated freshwater habitat: benefits and limitations. Environ Sci Pollut Res 27(20):25510–25522. https://doi.org/10.1007/s11356-020-08887-x

Ayari F, Hamdi H, Jedidi N, Gharbi N, Kossai R (2010). Heavy metal distribution in soil and plant in municipal solid waste compost amended plots. Int J Environ Sci Technol 7(3):465–472. https://doi.org/10.1007/BF03326156

Beiyuan J, Tsang DC, Valix M, Zhang W, Yang X, Ok YS, Li XD (2017) Selective dissolution followed by EDDS washing of an e-waste contaminated soil: extraction efficiency, fate of residual metals, and impact on soil environment. Chemosphere 166:489–496. https://doi.org/10.1016/j.chemosphere.2016.09.110

Bhardwaj R, Gupta A, Garg JK (2017) Evaluation of heavy metal contamination using environmetrics and indexing approach for River Yamuna, Delhi stretch, India. Water Sci 31(1):52–66

Bilal S, Khan AL, Shahzad R, Kim YH, Imran M, Khan MJ, Al-Harrasi A, Kim TH, Lee IJ (2018) Mechanisms of Cr(VI) resistance by endophytic Sphingomonas sp. LK11 and its Cr(VI) phytotoxic mitigating effects in soybean (Glycine max L.). Ecotoxicol Environ Saf 164:648–658. https://doi.org/10.1016/j.ecoenv.2018.08.043

Black Smith Institute Report (2007) The world's worst polluted places. A project of Blacksmith Institute, pp 16–17

Chandra R, Kumar V, Singh K (2017) Hyperaccumulator versus nonhyperaccumulator plants for environmental waste management. In: Phytoremediation of environmental pollutants. CRC Press, pp 43–80

Chatterjee S, Sau GB, Mukherjee SK (2009) Plant growth promotion by a hexavalent chromium reducing bacterial strain, Cellulosimicrobium cellulans KUCr3. World J Microbiol Biotechnol 25(10):1829–1836. https://doi.org/10.1007/s11274-009-0084-5

Chaudhary K, Agarwal S, Khan S (2018) Role of phytochelatins (PCs), metallothioneins (MTs), and heavy metal ATPase (HMA) genes in heavy metal tolerance. In: Mycoremediation and environmental sustainability. Springer, Cham, pp 39–60. https://doi.org/10.1007/978-3-319-77386-5_2

Chen H, Cutright T (2001) EDTA and HEDTA effects on Cd, Cr, and Ni uptake by Helianthus annuus. Chemosphere 45(1)21–28

Cheng C, Ren X, Wang Z, Yan C (2019) Heterogeneous impacts of renewable energy and environmental patents on CO2 emission-Evidence from the BRIICS. Sci Total Environ 668:1328–1338

CPCB (Central Pollution Control Board) (2016) Central pollution control board environmental data. http://cpcb.nic.in/Water_Quality_Data.php

Das PK, Das BP, Dash P (2017) Hexavalent chromium induced toxicity and its remediation using macrophytes. Pollut Res 36(1):92–98

Das PK (2018) Phytoremediation and nanoremediation: emerging techniques for treatment of acid mine drainage water. Defence Life Sci J 3(2):190–196. https://doi.org/10.14429/dlsj.3.11346

Das PK, Das BP, Dash P (2018) Role of plant species as hyper-accumulators in the decontamination of hexavalent chromium contaminated soil. Indian J Environ Prot 38(12):1016–1024

Das BK, Das PK, Das BP, Dash P (2021a) Green technology to limit the effects of hexavalent chromium contaminated water bodies on public health and vegetation at industrial sites. J Appl Biol Biotechnol 9(2):28–35. https://doi.org/10.7324/JABB.2021.9203

Das PK, Das BP, Dash P (2021b) Chromite mining pollution, environmental impact, toxicity and phytoremediation: a review. Environ Chem Lett 19(2):1369–1381. https://doi.org/10.1007/s10311-020-01102-w

Das PK, Das BP, Dash P (2022a) Analytical study on hexavalent chromium accumulation in plant parts of Pongamia pinnata (L.) Pierre and remediation of contaminated soil. J Appl Biol Biotechnol 10(1):22–30. https://doi.org/10.7324/JABB.2021.100103

Das PK, Das BP, Dash P (2022b) A super-tolerant bacteria strain improves phytoremediation of Cr(VI) contaminated soil with *Pongamia pinnata*. Rhizosphere 100543. https://doi.org/10.1016/j.rhisph.2022.100543

Das PK, Das BK, Das BP, Dash P (2022c) Evaluation of remediation ability of pongamia pinnata (l.) Pierre under hexavalent chromium stress soil conditions. Pollut Res 41(3):989–996. https://doi.org/10.53550/PR.2022.v41i03.033

Deb AK, Biswas B, Naidu R, Rahman MM (2022) Mechanistic insights of hexavalent chromium remediation by halloysite-supported copper nanoclusters. J Hazard Mater 421:126812. https://doi.org/10.1016/j.jhazmat.2021.126812

Dhal B, Thatoi HN, Das NN, Pandey BD (2013) Chemical and microbial remediation of hexavalent chromium from contaminated soil and mining/metallurgical solid waste: a review. J Hazard Mater 250:272–291

Dhaliwal SS, Singh J, Taneja PK, Mandal A (2020) Remediation techniques for removal of heavy metals from the soil contaminated through different sources: a review. Environ Sci Pollut Res 27(2):1319–1333. https://doi.org/10.1007/s11356-019-06967-1

Ding K, Zhou X, Hadiatullah H., Lu, Y., Zhao, G., Jia, S., Zhang, R. and Yao, Y., 2021. Removal performance and mechanisms of toxic hexavalent chromium (Cr(VI)) with ZnCl2 enhanced acidic vinegar residue biochar. J Hazard Mater 420:126551. https://doi.org/10.1016/j.jhazmat.2021.126551

Dotaniya ML, Thakur JK, Meena VD, Jajoria DK, Rathor G (2014) Chromium pollution: a threat to environment-a review. Agricult Rev 35(2)

Down to Earth (2005) Rs. 67 crore later. Down earth: science and environment online, India 13(20):36

Dzantor EK (2007) Phytoremediation: the state of rhizosphere 'engineering' for accelerated rhizodegradation of xenobiotic contaminants. J Chem Technol Biotechnol: Int Res Process Environ Clean Technol 82(3):228–232. https://doi.org/10.1002/jctb.1662

Eco-USA (2001) Toxics: chromium. http://www.eco-usa.net/toxics/chromium.html

Farooqi ZUR, Hussain MM, Ayub MA, Qadir AA, Ilic P (2022) Potentially toxic elements and phytoremediation: opportunities and challenges. Phytoremediation 19–36. https://doi.org/10.1016/B978-0-323-89874-4.00020-0

Fasani E, Manara A, Martini F, Furini A, DalCorso G (2018) The potential of genetic engineering of plants for the remediation of soils contaminated with heavy metals. Plant Cell Environ 41(5):1201–1232. https://doi.org/10.1111/pce.12963

Fendorf SE (1995) Surface reactions of chromium in soils and waters. Geoderma 67(1–2):55–71

Fu H, Ma S, Xu S, Duan R, Cheng G, Zhao P (2021) Hierarchically porous magnetic biochar as an efficient amendment for cadmium in water and soil: performance and mechanism. Chemosphere 281:130990. https://doi.org/10.1016/j.chemosphere.2021.130990

Gardea-Torresdey JL, Arenas JL, Francisco NMC, Tiemann KJ, Webb R (1998) Ability of immobilized cyanobacteria to remove metal ions from solution and demonstration of the presence of metallothionein genes in various strains. J Hazard Substan Res 1(1):2

Gogoi NM, Baroowa B, Gogoi N (2021) Ecological tools for remediation of soil pollutants. In: Bioremediation science from theory to practice. CRC Press, pp 57–78

Guidotti L, Abad SQ, Rodríguez-González P, Alonso J, Beone GM (2015) Quantification of Cr(VI) in soil samples from a contaminated area in northern Italy by isotope dilution mass spectrometry. Environ Sci Pollut Res 22(22):17569–17576. https://doi.org/10.1007/s11356-015-4963-z

Gupta P, Kumar V, Usmani Z, Rani R, Chandra A (2018) Phosphate solubilization and chromium (VI) remediation potential of Klebsiella sp. strain CPSB4 isolated from the chromium contaminated agricultural soil. Chemosphere 192:318–327. https://doi.org/10.1016/j.chemosphere.2017.10.164

Gupta N, Yadav KK, Kumar V, Kumar S, Chadd RP, Kumar A (2019) Trace elements in soil-vegetables interface: translocation, bioaccumulation, toxicity and amelioration-a review. Sci Total Environ 651:2927–2942

Higgins TE, Halloran AR, Petura JC (1997) Traditional and innovative treatment methods for Cr(VI) in soil. Soil Sedim Contam 6(6):767–797. https://doi.org/10.1080/15320389709383597

Hossain MS, Persicke M, ElSayed AI, Kalinowski J, Dietz KJ (2017) Metabolite profiling at the cellular and subcellular level reveals metabolites associated with salinity tolerance in sugar beet. J Experim Bot 68(21–22):5961–5976

Hussain K, Haris M, Qamar H, Hussain T, Ahmad G, Ansari MS, Khan AA (2021) Bioremediation of waste gases and polluted soils. In: Microbial Rejuvenation of polluted environment. Springer, Singapore, pp 111–137. https://doi.org/10.1007/978-981-15-7455-9_5

IARC (1990) Chromium, nickel, and welding, Monogr on the evaluation of carcinogenic risks to humans, vol 49. International Agency for Research on Cancer, Lyons

James BR (1996) Peer reviewed: the challenge of remediating chromium-contaminated soil. Environ Sci Technol 30(6):248A-251A

Kanagaraj G, Elango L (2019) Chromium and fluoride contamination in groundwater around leather tanning industries in southern India: Implications from stable isotopic ratio δ53Cr/δ52Cr, geochemical and geostatistical modelling. Chemosphere 220:943–953

Karim LR, Williams ES (2015) Accumulation of heavy metals in the surface water of Asthamudi Lake, Kollam, Kerala. Nat Environ Pollut Technol 14(2):431

Karthik C, Elangovan N, Kumar TS, Govindharaju S, Barathi S, Oves M, Arulselvi PI (2017) Characterization of multifarious plant growth promoting traits of rhizobacterial strain AR6 under Chromium (VI) stress. Microbiol Res 204:65–71. https://doi.org/10.1016/j.micres.2017.07.008

Kavitha B, Reddy PVL, Kim B, Lee SS, Pandey SK, Kim KH (2018) Benefits and limitations of biochar amendment in agricultural soils: a review. J Environ Manag 227:146–154. https://doi.org/10.1016/j.jenvman.2018.08.082

Khatoon N, Khan AH, Rehman M, Pathak V (2013) Correlation study for the assessment of water quality and its parameters of Ganga River, Kanpur, Uttar Pradesh, India. IOSR J Appl Chem 5(3):80–90

Khoo KS, Chia WY, Chew KW, Show PL (2021) Microalgal-bacterial consortia as future prospect in wastewater bioremediation, environmental management and bioenergy production. Indian J Microbiol 61(3):262–269. https://doi.org/10.1007/s12088-021-00924-8

Kumar PN, Dushenkov V, Motto H, Raskin I (1995) Phytoextraction: the use of plants to remove heavy metals from soils. Environ Sci Technol 29(5):1232–1238

Lakkireddy K, Kües U (2017) Bulk isolation of basidiospores from wild mushrooms by electrostatic attraction with low risk of microbial contaminations. AMB Express 7(1):1–22. https://doi.org/10.1186/s13568-017-0326-0

Lal S, Ratna S, Said OB, Kumar R (2018) Biosurfactant and exopolysaccharide-assisted rhizobacterial technique for the remediation of heavy metal contaminated soil: an advancement in metal phytoremediation technology. Environ Technol Innov 10:243–263. https://doi.org/10.1016/j.eti.2018.02.011

Leong WH, Zaine SNA, Ho YC, Uemura Y, Lam MK, Khoo KS, Kiatkittipong W, Cheng CK, Show PL, Lim JW (2019) Impact of various microalgal-bacterial populations on municipal wastewater bioremediation and its energy feasibility for lipid-based biofuel production. J Environ Manag 249:109384. https://doi.org/10.1016/j.jenvman.2019.109384

Limmer M, Burken J (2016) Phytovolatilization of organic contaminants. Environ Sci Technol 50(13):6632–6643. https://doi.org/10.1021/acs.est.5b04113

Lin H, Wang Z, Liu C, Dong Y (2022) Technologies for removing heavy metal from contaminated soils on farmland: a review. Chemosphere 135457. https://doi.org/10.1016/j.chemosphere.2022.135457

Lone MI, He ZL, Stoffella PJ, Yang XE (2008) Phytoremediation of heavy metal polluted soils and water: progresses and perspectives. J Zhejiang Univ Sci B 9(3):210–220. https://doi.org/10.1631/jzus.B0710633

Lytle CM, Lytle FW, Yang N, Qian JH, Hansen D, Zayed A, Terry N (1998) Reduction of Cr (VI) to Cr (III) by wetland plants: potential for in situ heavy metal detoxification. Environ Sci Technol 32(20):3087–3093

Ma Y, Rajkumar M, Zhang C, Freitas H (2016) Beneficial role of bacterial endophytes in heavy metal phytoremediation. J Environ Manag 174:14–25. https://doi.org/10.1016/j.jenvman.2016.02.047

Madhavan T (2020) Chromium waste remains a threat in Ranipet the Hindu. https://www.thehindu.com/news/national/tamil-nadu/chromium-waste-remains-a-threat-in-ranipet/article30898352.ece

Mei H, Huang W, Wang Y, Xu T, Zhao L, Zhang D, Luo Y, Pan X (2022) One stone two birds: Bone char as a cost-effective material for stabilizing multiple heavy metals in soil and promoting crop growth. Sci Total Environ 156163. https://doi.org/10.1016/j.scitotenv.2022.156163

Memon AR, Schröder P (2009) Implications of metal accumulation mechanisms to phytoremediation. Environ Sci Pollut Res 16(2):162–175. https://doi.org/10.1007/s11356-008-0079-z

Mishra S, Das AP, Seragadam P (2009) Microbial remediation of hexavalent chromium from chromite contaminated mines of Sukinda Valley, Orissa (India). J Environ Res Dev 3:1122–1127

Mishra H, Sahu HB (2013) Environmental scenario of chromite mining at Sukinda Valley—A review. Int J Environ Eng Manag 4:287–292

Mitra S, Sarkar A, Sen S (2017) Removal of chromium from industrial effluents using nanotechnology: a review. Nanotechnol Environ Eng 2(1):1–14

Mohan D, Rajput S, Singh VK, Steele PH, Pittman CU Jr (2011) Modeling and evaluation of chromium remediation from water using low cost bio-char, a green adsorbent. J Hazard Mater 188(1–3):319–333

Mosa KA, Saadoun I, Kumar K, Helmy M, Dhankher OP (2016) Potential biotechnological strategies for the cleanup of heavy metals and metalloids. Front Plant Sci 7:303. https://doi.org/10.3389/fpls.2016.00303

Nematshahi N, Lahouti M, Ganjeali A (2012) Accumulation of chromium and its effect on growth of (Allium cepa cv. Hybrid). Euro J Exp Biol 2(4):969–974

Panda SK, Choudhury S (2005) Chromium stress in plants. Brazilian J Plant Physiol 17:95–102. https://doi.org/10.1590/S1677-04202005000100008

Pham TD, Tran TT, Pham TT, Dao TH, Le TS (2019). Adsorption characteristics of molecular oxytetracycline onto alumina particles: the role of surface modification with an anionic surfactant. J Mol Liquids 287:110900. https://doi.org/10.1016/j.molliq.2019.110900

Pilon-Smits E (2005) Phytoremediation. Annu Rev Plant Biol 56:15

Prasad S, Yadav KK, Kumar S, Gupta N, Cabral-Pinto MM, Rezania S, Radwan N, Alam J (2021) Chromium contamination and effect on environmental health and its remediation: a sustainable approaches. J Environ Manag 285:112174

Purushotham D, Rashid M, Lone MA, Rao AN, Ahmed S, Nagaiah E, Dar FA (2013) Environmental impact assessment of air and heavy metal concentration in groundwater of Maheshwaram watershed, Ranga Reddy district, Andhra Pradesh. J Geol Soc India 81(3):385–396

Rao DP, Saxena R, Saxena V, Singh A (2009) Toxic load of tannery industries situated in Kanpur. Int J Appl Environ Sci 4(3):327–336

Rao GT, Rao VG, Ranganathan K (2013) Hydrogeochemistry and groundwater quality assessment of Ranipet industrial area, Tamil Nadu, India. J Earth Syst Sci 122(3):855–867
Saha R, Nandi R, Saha B (2011) Sources and toxicity of hexavalent chromium. J Coord Chem 64(10):1782–1806. https://doi.org/10.1080/00958972.2011.583646
Salt DE, Pickering IJ, Prince RC, Gleba D, Dushenkov S, Smith RD, Raskin I (1997) Metal accumulation by aquacultured seedlings of Indian mustard. Environ Sci Technol 31(6):1636–1644
Sanyal T, Kaviraj A, Saha S (2015) Deposition of chromium in aquatic ecosystem from effluents of handloom textile industries in Ranaghat-Fulia region of West Bengal, India. J Adv Res 6(6):995–1002
Schnoor JL, Galloway JN, Moldan B (1997) Peer Reviewed: East Central Europe: An Environment in Transition. Environ Sci Technol 31(9):412A-416A
Sarma H (2011) Metal hyperaccumulation in plants: a review focusing on phytoremediation technology. J Environ Sci Technol 4(2):118–138. https://doi.org/10.3923/jest.2011.118.138
Shah V, Daverey A (2020) Phytoremediation: a multidisciplinary approach to clean up heavy metal contaminated soil. Environ Technol Innov 18:100774. https://doi.org/10.1016/j.eti.2020.100774
Shao Y, Shao Y, Zhang, W., Zhu, Y., Dou, T., Chu, L. and Liu, Z., 2022. Preparation of municipal solid waste incineration fly ash-based ceramsite and its mechanisms of heavy metal immobilization. Waste Manag 143:54–60. https://doi.org/10.1016/j.wasman.2022.02.021
Sharma S, Nagpal AK, Kaur I (2018) Heavy metal contamination in soil, food crops and associated health risks for residents of Ropar wetland, Punjab, India and its environs. Food Chem 255:15–22
Shu X, Li Y, Huang W, Chen S, Xu C, Zhang S, Li B, Wang X, Qing Q, Lu X (2020) Rapid vitrification of uranium-contaminated soil: Effect and mechanism. Environ Pollut 263:114539. https://doi.org/10.1016/j.envpol.2020.114539
Singh RK, Sengupta B, Bali R, Shukla BP, Gurunadharao VVS, Srivatstava R (2009) Identification and mapping of chromium (VI) plume in groundwater for remediation: a case study at Kanpur, Uttar Pradesh. J Geol Soc India 74(1):49–57
Singh HP, Mahajan P, Kaur S, Batish DR, Kohli RK (2013) Chromium toxicity and tolerance in plants. Environ Chem Lett 11(3):229–254. https://doi.org/10.1007/s10311-013-0407-5
Soni SK, Kumar G, Bajpai A, Singh R, Bajapi Y, Tiwari S (2023) Hexavalent chromium-reducing plant growth-promoting rhizobacteria are utilized to bio-fortify trivalent chromium in fenugreek by promoting plant development and decreasing the toxicity of hexavalent chromium in the soil. J Trace Elements Med Biol 76:127116. https://doi.org/10.1016/j.jtemb.2022.127116
Srivastava S, Agrawal SB, Mondal MK (2015) A review on progress of heavy metal removal using adsorbents of microbial and plant origin. Environ Sci Pollut Res 22(20):15386–15415. https://doi.org/10.1007/s11356-015-5278-9
Tumolo M, Ancona V, De Paola D, Losacco D, Campanale C, Massarelli C, Uricchio VF (2020) Chromium pollution in European water, sources, health risk, and remediation strategies: an overview. Int J Environ Res Publ Health 17(15):5438. https://doi.org/10.3390/ijerph17155438
USEPA (1993) Remediation technologies screening matrix and reference guide. http://www.frtr.gov/matrix2/appde/appde07.html
USEPA (1998) Toxicological review of hexavalent chromium. In: Support of summary information on the integrated risk information system. USA, Washington D.C.
USEPA (United States Environmental Protection Agency) (1999) Integrated risk information system (IRIS) on chromium VI. National Center for Environmental Assessment, Office of Research and Development, Washington, DC
USEPA (2000) National Priorities List (NPL) Sites with fiscal year 2000, Records of Decision (RODs). USEPA-December 2000, Office of Emergency and Remedial Response, OERCLIS
Vijayana GI, Nikos M (2010) Unsafe chromium and its environmental health effects of Odisha chromite mines. In: Proceedings of the international conference on energy and environment technologies and equipment, pp 1790–5095
Wadhawan S, Jain A, Nayyar J, Mehta SK (2020) Role of nanomaterials as adsorbents in heavy metal ion removal from waste water: a review. J Water Process Eng 33:101038. https://doi.org/10.1016/j.jwpe.2019.101038

Wani PA, Wahid S, Singh R, Kehinde AM (2018) Antioxidant and chromium reductase assisted chromium (VI) reduction and Cr(III) immobilization by the rhizospheric Bacillus helps in the remediation of Cr(VI) and growth promotion of soybean crop. Rhizosphere 6:23–30. https://doi.org/10.1016/j.rhisph.2018.01.004

Xu DM, Fu RB, Wang JX, Shi YX, Guo XP (2021) Chemical stabilization remediation for heavy metals in contaminated soils on the latest decade: available stabilizing materials and associated evaluation methods-a critical review. J Clean Prod 321:128730. https://doi.org/10.1016/j.jclepro.2021.128730

Yadav KK, Gupta N, Kumar A, Reece LM, Singh N, Rezania S, Khan SA (2018) Mechanistic understanding and holistic approach of phytoremediation: a review on application and future prospects. Ecol Eng 120:274–298

Yang P, Zhou XF, Wang LL, Li QS, Zhou T, Chen YK, Zhao ZY, He BY (2018) Effect of phosphate-solubilizing bacteria on the mobility of insoluble cadmium and metabolic analysis. Int J Environ Res Publ Health 15(7):1330. https://doi.org/10.3390/ijerph15071330

Yang Z, Zhang X, Jiang Z, Li Q, Huang P, Zheng C, Liao Q, Yang W (2021) Reductive materials for remediation of hexavalent chromium contaminated soil—A review. Sci Total Environ 773:145654. https://doi.org/10.1016/j.scitotenv.2021.145654

Yang J, Tan X, Shaaban M, Cai Y, Wang B, Peng QA (2022) Remediation of Cr(VI)-contaminated soil by biochar-supported nanoscale zero-valent iron and the consequences for indigenous microbial communities. Nanomaterials 12(19):3541. https://doi.org/10.3390/nano12193541

Zaidi S, Panchal M, Parekh V, Shaikh A, Zaidi U, Desai U, Patel K, Dave K, Ansari S, Upadhyay P, Shivgotra V (2014) Ground water contamination with hexavalent chromium and its health effects: debatable and unsettled issue of regulation. In: Environmental sustainability: concepts, principles, evidences and innovations, p 300

Zavoda J, Cutright T, Szpak J, Fallon E (2001) Uptake, selectivity, and inhibition of hydroponic treatment of contaminants. J Environ Eng 127(6):502–508

Zayed AM, Terry N (2003) Chromium in the environment: factors affecting biological remediation. Plant Soil 249(1):139–156. https://doi.org/10.1023/A:1022504826342

Zhai X, Li Z, Huang B, Luo N, Huang M, Zhang Q, Zeng G, (2018) Remediation of multiple heavy metal-contaminated soil through the combination of soil washing and in situ immobilization. Sci Total Environ 635:92–99. https://doi.org/10.1016/j.scitotenv.2018.04.119

Zhang X, Gai X, Zhong Z, Bian F, Yang C, Li Y, Wen X (2021) Understanding variations in soil properties and microbial communities in bamboo plantation soils along a chromium pollution gradient. Ecotoxicol Environ Saf 222:112507. https://doi.org/10.1016/j.ecoenv.2021.112507

Zhao J, Zhang L, Zhang S, Yuan W, Fang X, Yu Q, Qiu X (2021) Remediation of chromium-contaminated soil using calcined layered double hydroxides containing different divalent metals: temperatures and mechanism. Chem Eng J 425:131405. https://doi.org/10.1016/j.cej.2021.131405

Zhu YL, Pilon-Smits EA, Tarun AS, Weber SU, Jouanin L, Terry N (1999) Cadmium tolerance and accumulation in Indian mustard is enhanced by overexpressing γ-glutamylcysteine synthetase. Plant Physiol 121(4):1169–1177

Złoch M, Kowalkowski T, Tyburski J, Hrynkiewicz K (2017) Modeling of phytoextraction efficiency of microbially stimulated *Salix dasyclados* L. in the soils with different speciation of heavy metals. Int J Phytoremed 19(12):1150–1164. https://doi.org/10.1080/15226514.2017.1328396

Chapter 13
Microbial Remediation Technologies for Chromium Removal: Mechanism, Challenges and Future Prospect

Aashna Monga, Abhay B. Fulke, Manisha D. Giripunje, and Debjani Dasgupta

Abstract Heavy metal (HM) exposure is regarded as one of the greatest environmental concerns worldwide due to their non-biodegradability, high bioaccumulation in the food chain, and most importantly, human carcinogenicity. The industrial uses of chromium (Cr) are diverse and include metallurgy, paint, leather tanning, and electroplating. Because of inadequate waste discharge regulations, toxic amounts of Cr are released into the environment, severely damaging the ecosystem. It is now understood that Cr has some advantages for humans in its trivalent oxidation form [Cr(III)] as a micronutrient. However, its hexavalent form [Cr(VI)] is a strong carcinogen and has no recognized biological functions. Over the years, a number of physico-chemical, and biobased techniques have appeared in the effort to eliminate Cr from the environment. Bioremediation of Cr have several advantages over the conventional physical and chemical treatment methods due to its low cost, environment friendly practices and sustainability. Bacteria employs several mechanisms such as biosorption, efflux, bioreduction and bioaccumulation that they possess either inherently or have acquired to counter the toxic effects of Cr with time. This chapter focuses in detail on microbial mechanisms and responses against Cr toxicity, their applications and challenges in real time applicability of these. Further, the latest strategies and solutions in developing bioremediation applications are also discussed in this chapter. Nanobioremediation, immobilization techniques and use of enhancers have immense scope in improving the bioremediation efficiency and also in metal recovery. This information will be helpful in understanding the current status of research of Cr pollution remediation and bridging the gap between lab scale findings and its real time applicability in the environment.

A. Monga · A. B. Fulke (✉)
Microbiology Division, CSIR-National Institute of Oceanography (CSIR-NIO), Regional Centre, Mumbai, Maharashtra 400053, India
e-mail: afulke@nio.org

A. Monga · D. Dasgupta
School of Biotechnology and Bioinformatics, D. Y. Patil University, Navi Mumbai, India

M. D. Giripunje
Sevadal Mahila Mahavidyalaya, Nagpur 440009, India

© The Author(s), under exclusive license to Springer Nature Switzerland AG 2023

N. Kumar et al. (eds.), *Chromium in Plants and Environment*, Environmental Science and Engineering, https://doi.org/10.1007/978-3-031-44029-8_13

Keywords Chromium · Remediation · Microbes · Removal · Bioremediation · Phytoremediation

13.1 Environmental Pollution

Natural resources are being consumed quickly due to the rapid development of society. Although heavy metals are used in a variety of industrial processes, some of them have the potential to seriously harm the environment. Heavy metals are hard to break down and have a long half-life. They will obstruct specific protein and nucleic acid processes after entering the body (Bartlett 1991; Chen and Tian 2021). One of the biggest environmental problems today is the discharge of dangerous heavy metals into wastewater from industry and human activity. Many academics and experts are paying attention to water pollution because it poses a serious threat to people, land animals, and aquatic animals and plants. This is mostly due to a growth in various industrial operations, which contribute significantly to the global production of waste and untreated water (Ayele and Godeto 2021). In addition to its undesirable side effects, industrial and technological advancements also damage and pollute the environment. Xenobiotics, poisonous, and other gases are unintentionally and intentionally released into the environment as a result of these revolutions (Verma and Kuila 2019). Water pollution induced from release of unregulated and large amounts of untreated or partially treated industrial effluents is a major threat to all life forms (Munjur et al. 2020). For instance, drinking water polluted with atorvastatin (a medication used to treat cardiovascular conditions) has indeed been related to serious health complications like myopathy, renal problems, amnesia and memory lapses, pancreatic and hepatic malfunction, etc. (Ali et al. 2019). Many other kinds of wastes from industrial, agricultural or domestic when dumped into water bodies untreated over the years can severely pollute and the contaminants can cause a variety of ailments including cutaneous, gastrointestinal, and vector-borne illnesses as well as blindness, paralysis, and renal (Chowdhary et al. 2017). Broadly, several drugs from pharmaceuticals (Ali et al. 2019), inorganic pollutants and polysaccharides from distillery industries (Chowdhary et al. 2017), organic wastes from pulp and paper industries (Zainith et al. 2019) and heavy metals that are continuously released from electroplating, chemical, metallurgy, tannery, textile industries etc. (Chowdhary et al. 2020) are of great concern as they present a very big environment challenge and threatening health of humans.

13.2 Heavy Metals

Metals with a weight greater than 5 g/cm^3 are classified as heavy metals (HMs) (Fulke et al. 2020). Chromium (Cr), cobalt (Co), copper (Cu), cadmium (Cd), arsenic (As), gallium (Ga), iron (Fe), mercury (Hg), lead (Pb), nickel (Ni), and manganese (Mn) are a few of the more well-known heavy metals (Pandey and Madhuri 2014). With high molecular weight, atomic number, and specific gravity, they include the majority of transitional metals, basic metals, some metalloids, and lanthanides (Ayele and Godeto 2021). Due to their non-biodegradability and prolonged atmospheric persistence, these toxic pollutants—which are commonly present in industrial effluents (Lian et al. 2019; Prasad et al. 2021); are detrimental even at extremely low concentrations. This makes them a significant environmental risk and one of the most challenging and complex environmental issues posing risks to both the ecosystem and public health (Kapahi and Sachdeva 2019; Monga et al. 2022a). HMs are naturally components of earth's crust that more than five times denser than water (Karimi-Maleh et al. 2021; Elgarahy et al. 2021; Cuellar et al. 2022). These elements can be found in nature in different forms such as hydroxides, acids and bases or as chemical complexes; can neither be destroyed or degraded from the environment (Cuellar et al. 2022). Due to their non-biodegradability, high bio accumulation in food chain and most of all human carcinogenicity, heavy metal exposure is considered one of the biggest environmental concerns globally (He and Chen 2014). Few metals are required by living beings to undertake certain metabolic activities, but several of these metals can be detrimental to human health at even very low concentration (Zhang et al. 2016a, b; Cuellar et al. 2022). Ideally, heavy metals when used in industries must undergo a regulated processing start from their sourcing extractions from ground deposits to their smelting and refining stages with a proper disposal of the resulting products. Instead, the heavy metal containing industrial wastes are released in the environment during each of these stages (Cuellar et al. 2022). Numerous industrial sectors, including electrochemical, pulp and paper industries, textile, metallurgies, mineral extraction, and the dye and paint chemical industries, employ various types of metals extensively for a variety of purposes (Igiri et al. 2018; Sun et al. 2019a; Ayele and Godeto 2021). Environmental deterioration is mostly brought on by unplanned industrial and urban expansion, which disregards the importance of a healthy environment. Due to these acts, heavy metal pollution has significantly increased, upsetting the natural balance (Wang et al. 2018). According to a WHO study, over 1.7 million children under the age of five die as a result of exposure to dangerous pollutants, particularly heavy metals (Xu et al. 2018). As a result, heavy metal environmental pollution is a major problem that necessitates immediate action (Pushkar et al. 2021).

13.3 Cr Contamination/Menace in India

The chromite deposits in India constitute around 2% of the world load. Of this, Odisha alone is responsible for 98% of the total chromite with 97% found in Sukinda valley (Mishra and Sahu 2013). In accordance with survey conducted by the Ministry of Environment, Forest, and Climate Change (MoEFCC), the Government of India (GOI), states—Andhra Pradesh, Maharashtra, Tamil Nadu, Karnataka and Gujarat—produce 80% of the metal-enriched toxic waste (Singh et al. 2020). Leaching and natural weathering of chromite from chromite mines into water bodies is a severe cause of concern for soil and water pollution (Das and Mishra 2009; Prasad et al. 2021). In these Indian states, Ranipet in Tamil Nadu, Kanpur in Uttar Pradesh, Sukinda valley in Odisha), and Vadodara in Gujarat were identified as having the highest levels of contamination (Mishra and Sahu 2013; Jamshed and Vamit 2017; Singh et al. 2020). For example, the Cr(VI) and total Cr levels reported pin the Ranipet were 142 mg/L and 158 mg/L, respectively (Jeyasingh et al. 2011). Cr(VI) concentrations of up to 80 mg/L have been reported from the Kanpur region (Singh et al. 2013). According to studies by the Regional Research Laboratory (RRL) of the Council for Scientific and Industrial Research (CSIR) in Sukinda Valley, 7.6 metric tonnes of excess and overloaded waste were being dumped annually with the potential to release 11.3 tonnes of Cr(VI) in the ecosystem. Orissa Voluntary Health Association (OVHA) furthermore investigated the human mortality rates in the vicinity of such mining areas and found that 86.42% of the population in nearby villages were affected due to chromite mine related disorders (Gupta et al. 2019; Prasad et al. 2021). Also, around 80% of Indian tannery industries are majorly involved in chrome tanning. Tannery industry produce about 1500 metric tonnes of chrome sulphate per year as effluent that is discharged in the environment. A report on heavy metal contamination and risk analysis in water and sediments of the Ganga River between Kanpur and Pryagraj India, was recently published by Aggarwal et al. In 2022. In most of the samples, sediment Cr levels were higher than the averages for the Indian River System (IRS) and the planet's surface rocks, which were 87 and 71 mg/L, respectively. Sediment Cr levels ranged from 31.4 to 100.2 mg/L on average (Aggarwal et al. 2022). Cr pollution in the Ganga River is also to be blamed on by the usage of paint components containing Cr use for vehicular refurbishment. Cr concentrations in some of the samples have reached an alarming level due to the lethality that it can cause to some of aquatic species in the river (Aggarwal et al. 2022). In Yamuna River, Pb and Cr levels have exceeded the WHO permissible limits and most of the samples tested were extremely contaminated and unfit for the purposes of drinking, cooking or washing (Singh Sankhla et al. 2021).

13.4 Chromium (Cr): Occurrence, Speciation, and Fate into the Environment

The French chemist Louis Vauquelin made the discovery of Cr in 1797. Due to the various colours seen in the Cr-containing substances, Cr was given the Greek term "chroma" (Barnhart 1997). The transition metal Cr has atomic number of 24, an atomic weight of 51.996 amu, and an electronic structure of 4d5s1. It is a member of group VI-B of the periodic table. With well almost all naturally occurring Cr being in the trivalent state [Cr(III)], it is also the 21st most abundant component in the Earth's crust and quite prevalent in river waters, lakes, seawater, and underground waters naturally. It is typically combined with Fe or other inorganic materials (Barnhart 1997). Chromite (Fe, Mn) is the most important ore of chromium being found in nature (Focardi et al. 2012). The zero [Cr], trivalent [Cr(III)], and hexavalent [Cr(VI)] forms are the most significant in industrial products and the environment due to their stability, even though it exists in multiple valence states (ranging from −2 and +6). (Barnhart 1997; Karthik et al. 2017). However, their chemical properties are contradictory, display differences in physicochemical characteristics, and exhibit biological reactivity, which has diverse effects on living cells (Bharagava and Mishra 2018; Sanjay et al. 2017; Pushkar et al. 2021). Some species need Cr(III), which is less harmful and functions as a supplement, for development and some metabolic pathways (Ma et al. 2019a, b). Moreover, due to its high bioavailability and dispersion rates in natural systems, Cr(VI) is more hazardous than Cr(III). The cellular membrane is quickly penetrated by Cr(VI), which can easily react with the proteins in the cytoplasm of the host cell (Bharagava and Mishra 2018; Pushkar et al. 2021). Also, Cr's ionic state is regulated by the pH and electrochemical state of the aqueous environment that it is present in. Table 13.1 lists some of the basic characteristics of different forms of Cr.

In the environment, Cr(III) is most stable and requires a considerable amount of energy to get converted into lower or higher valency states. Cr(II) is only stable in the absence of any oxidizing agent since otherwise it easily oxidizes to Cr(III) under anaerobic conditions. The Cr(III)/Cr(II) metal ion couple's negative standard potential (Eo) also supports this. On the other hand, Cr(VI) is unstable and strongly oxidizing in the presence of electron donors because of its extremely favorable redox potential in acidic solution (Eo between 1.33 and 1.38 V). The acidity drops as a result of the H^+ being used up during the reduction of $HCrO_4^-$ (Eq. 13.1), further lowering the chemical potential. When CrO_4^{2-} is reduced within a more basic solution, OH^- is produced in the face of a redox gradient (Eo = −0.13 V). When a result, Cr(III) is less stable than Cr(VI) and has a lower potential as basicity rises (pH > 4). However, in weakly/slightly acidic and weakly basic conditions, E versus pH has a steeper slope than Eq. (13.2) because di- and mono-hydroxy species are formed. A Pourbaix diagram has thus helped significantly in illustrating the pH and redox potential parameters that all the species must meet in order to be thermodynamically stable (Fig. 13.1) (Kotaś and Stasicka 2000).

$$HCrO_4^- + 7H^+ + 3e^- - Cr^{3+} + 4H_2O \tag{13.1}$$

Table 13.1 Physical properties of the various forms of Cr

Properties	Melting point (°C)	Boiling point (°C)	Solubility in water (g/L)	Density (g/cm^3)
Cr	1185	2672	Insoluble	7.14
$CrCl_3$	1152	–	Slightly soluble	2.76
K_2CrO_4	968.3	–	790	2.73
Cr_2O_3	226	4000	Insoluble	5.21
CrO_3	196	–	624	2.70

Adopted from WHO (1996), Pushkar et al. (2021)

$$CrO_2^- + 4H_2O + 3e^- - Cr(OH)_3 + 5OH^- \tag{13.2}$$

Gorny et al. (2016) extensively reviewed the existing literature concerning the redox pathways of Cr(III and VI) in aquatic habitats and their respective transposition to the surface sediments, where the speciation data is particularly limited and scarce. The main governing factors in Cr speciation in aquatic settings involve Mn(III, IV) hydroxides for Cr(III) oxidation, dissolved Fe(II) and HS-acting as Cr(VI) reducing species along with ferrous and sulfide minerals as Cr(VI) reducing phases as well as Fe(II) bearing minerals. Nonetheless, the redox conversion of Cr(VI)–Cr(III) is also a result of microbial action, and this conversion occurs either through detoxifying or dissimilatory reductions. The indirect conversion/oxidation of Cr(III)–Cr(VI) is also known to occur by Mn(II) oxidizing bacteria, though the mechanisms are not clearly identified yet. Moreover, Mn(II) and ammonium ions are not known to encourage reduction of Cr(VI). After it is reduced to Cr(III), the mobility of Cr(III) ions in the sediment fractions gets very restrictive and is only regulated by precipitation and sorption mechanisms (Gorny et al. 2016).

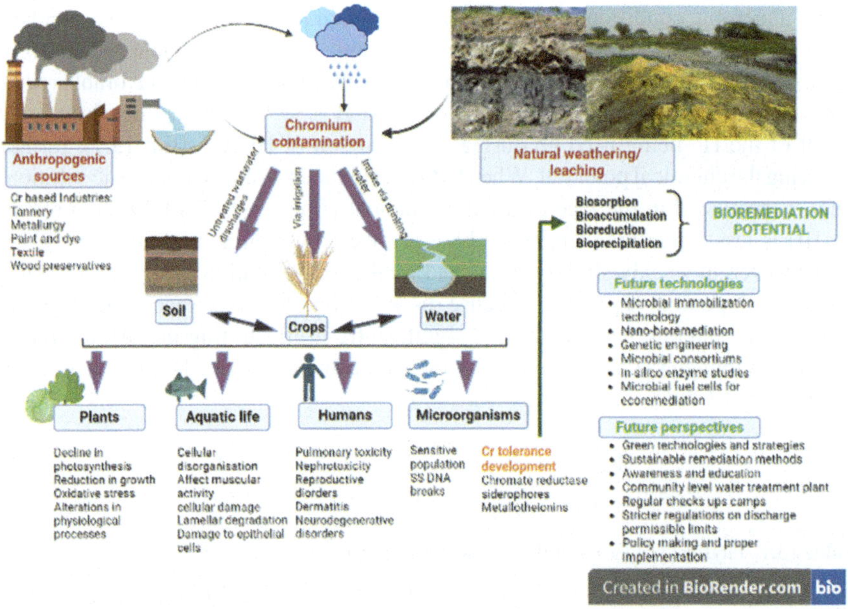

Fig. 13.1 An overview of environmental Cr contamination due to natural and anthropogenic sources; its effects on plants, aquatic life, microorganisms, and humans; future remediation technologies and perspectives

13.5 Essentiality of Cr(III)

With completely distinct reactivity in its two most abundant oxidation states, Cr(III) and Cr(VI), Cr stands out among other HMs as an intriguing exception (Genchi et al. 2021; Monga et al. 2022a, b). While Cr(VI) has a far higher bioavailability than Cr(III) due to its high solubility in water and transmembrane permeation, Cr(III) is less hazardous because it cannot easily pass through cell membranes. As consequence of this, Cr(VI) disseminates easily away from the innate site of contamination and is highly toxic even at low concentrations (Gorny et al. 2016; Nakkeeran et al. 2018). The Cr(VI) species being principally dominant in natural aquifers while the Cr(III) species being widespread in municipal wastewater rich in organics (Cheung and Gu 2007). The redox potential of Cr affects both its knietics and its dynamics (Moffat et al. 2018; Monga et al. 2022a, b). From being a vital trace element to a physiologically inert element (metallic Cr) to a strong endocrine disruptor to being genotoxic and carcinogenic, Cr exhibits a variety of traits (DesMarais and Costa 2019). It is now understood that the Cr(III) is necessary for both normal human and animal development. It has been identified as a pharmacologically active element due to its significance in the maintenance of nucleic acid (NA) structural integrity as well as glucose and lipid metabolism (Zayed and Terry 2003; Vincent 2017). Absorbed from dietary Cr, Cr(III) is now known to be a constituent of glucose tolerance factor (GTF) which is responsible for glucose clearance from the blood via an insulin stimulating mechanism. Also, Cr(III) ions contribute to the activation of the insulin receptor tyrosine kinase, which increases and enhances the insulin action threefold. Therefore, a lack of Cr(III) ions might lead to ailments and weight loss related with carbohydrates (Monga et al. 2022a). IARC (International Agency for Research on Cancer) in 1990 classified Cr(VI) as a class 1 carcinogen. Due to human activities, Cr(VI) is now widely spread in the environment and acts using complex mechanisms of generating reactive oxygen species leading to oxidative stress, epigenetic changes, chromosomal and DNA aberrations, and mutagenesis (Genchi et al. 2021; Monga et al. 2022a, b).

13.6 Origins of Cr Pollution

13.6.1 Natural Sources

The Earth's crust naturally contains Cr (Srivastav et al. 2018). It can be released naturally, primarily in Cr(III) and Cr(VI) form, from sources of Cr by processes of weathering (McCartor and Becker 2010; Stambulska et al. 2018; Prasad et al. 2021) (Fig. 13.2). According to Oze et al. (2007), the Earth's crust has Cr concentrations of over 200 mg kg^{-1} in ultramafic (ultrabasic) rock formations and ophiolite serpentinites structures, which make up about 1% of the landscape of the terrestrial environment, mostly found in the densely populated Mediterranean and Pacific regions

(Prasad et al. 2021). The sole valence state identified in the serpentine soil solids is Cr(III), however Cr(VI) has been found in the serpentine soil solutions from New Caledonia and California at quantities <30 M. The presence of Cr-spinels, specifically chromite and Cr-magnetite, has a direct impact on the concentration and range of Cr levels in serpentine sediments. However, oxidation of Cr(III) from Cr-spinels by high-valent Mn oxides or other potent oxidants has been found as a potential source of Cr(VI) in serpentine soil solutions. These phases are weather resistant and are maintained in the soil ecosystem (Oze et al. 2004). Chromite ore bodies could generate toxic Cr(VI) levels from inert chromites and contribute towards Cr pollution in waterbodies as shown in a study made on chromite bearing oxidized rocks in Orissa, India (Godgul and Sahu 1995). Cr(VI) is a toxin typically originating from anthropogenic activity (Bartlett and James 1988). However, both ground and surface waters from California, Italy, and Mexico have recorded naturally existing aqueous Cr(VI) concentrations up to 73 g/L, values exceeding the WHO's limit for drinking water of 50 g of Cr(VI) per litre, or 960 nM Cr(VI) (Oze et al. 2007). In the presence of birnessite, an ubiquitous manganese rock, Oze et al. (2007) observed rapid dissolution of chromite and subsequent conversion of Cr(III) to aqueous Cr(VI), explaining the production of Cr(VI) by a Cr(III)-bearing material regarded to be geochemically inert. Natural events may cause the Cr(III) in ultramafic- and serpentinite-derived soils and sediments to be oxidised and absorbed, resulting in dangerously high amounts of Cr(VI) in both surface and groundwater (Oze et al. 2007).

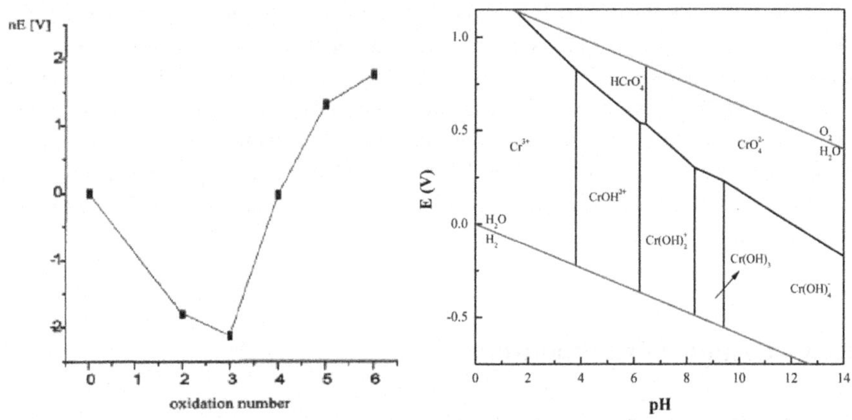

Fig. 13.2 Left: The Cr Species Frost diagram in an acidic condition. Right: A schematic Pourbaix diagram for the dominant Cr species in dilute, aerated aqueous solutions in the absence of any other agents for complexing Cr, except water or OH^-. Adapted from Kotaś and Stasicka (2000)

13.6.2 Human Activities

The global Cr reservoir is impacted by anthropogenic and natural events (Coetzee et al. 2020; Singh et al. 2020; Guo et al. 2021). The human population has been exposed to Cr through pollution exposure or drinking contaminated water (DesMarais and Costa 2019) (Fig. 13.2). Due of Cr's negative impacts on health, many laws have been put in place for monitoring and discharge. Over 100 locations release Cr, according to a Pure Earth survey from 2019, putting approximately 1.5 million individuals at risk of exposures to Cr and other contaminants (Singh et al. 2020). Effluent and sludge dumped from industrial facilities like chrome plating, metal polishing, leather tanning, and textiles are the principal sources of Cr(VI) pollution (He and Li 2020; Prasad et al. 2021; Jobby et al. 2018). Table 13.2 mentions the various industrial effluent and wastewater sources and the Cr content they usually carry. These industries contribute significantly to Cr(VI) toxicities and contamination in water and soil (Lian et al. 2019; Prasad et al. 2021). This hampers plant growth, agriculture, animal health, damaging human health eventually (Mitra et al. 2017). The USEPA and the European Union (EU) currently advise that the acceptance limit for surface wastewater should be less than 0.05 mg/L, with the total concentration of Cr [Cr(III), Cr(VI), and other forms] use around to below 2 mg/L (Labied et al. 2018; Ukhurebor et al. 2021; Monga et al. 2022a). A threshold of 0.05 mg/L of Cr in drinkable water and 0.1 mg/L for industrial effluent emission into groundwater have been set by the Central Control Board in India. Further to set criteria for controlled Cr emissions under the Clean Air Act of 1990, the USEPA increased the threshold to 0.1 mg/L (USEPA 2010). Governments and individuals all over the world are still very concerned about the presence of Cr in both natural and artificial ecosystems (Singh et al. 2020; Chen and Tian 2021). In Mexico, for instance, there are 769 tonnes of Cr(VI) waste being produced annually (Cuellar et al. 2022).

Table 13.2 Type of Cr(VI) salts used in various industries

Cr used in industries	Chemical forms
Chrome plating	Barium chromate, zinc chromate, Strontium chromate, sodium chromate
Leather tanneries	Ammonium dichromate
Wood chemical additives/preservatives	Chromium trioxide
Stainless steel factories	Potassium chromate, ammonium dichromate, potassium dichromate
Paints and pigments	Barium chromate, calcium chromate, lead chromate, zinc chromate, potassium dichromate

Adopted from Prasad et al. (2021)

13.7 Toxicity of Cr(VI)

13.7.1 Humans

The main toxic heavy metals- Pb, Cd, Hg and Cr stand out (Ozden et al. 2018; Cuellar et al. 2022) with the latter being a fascinating case due to its entirely different reactivities in its two most prevalent oxidation valence states of Cr(III) and Cr(VI) (Genchi et al. 2021; Monga et al. 2022a, b). The EPA has included Cr(VI) in the list of toxic substances (USEPA 2014) and has received a classification as a carcinogenic agent by the US Department of Health and Human Services with lung cancers being the most commonly associated with Cr(VI) intoxication (Cueller et al. 2022). Notably, high concentrations of Cr can substitute other metals in biological systems and have negative effects like cancer, kidney failure, neurodegenerative diseases (ND), and death (Monga et al. 2022a). On oral consumption, a part of Cr(VI) is extracellularly converted to Cr(III) as a protective mechanism (Proctor et al. 2002; De Flora et al. 2006). As soon as Cr(VI) enters the cells, it combines spontaneously with intracellular reducing substances such ascorbic acid, glutathione, cytochrome, etc. to produce short-lived intermediaries like Cr(V) and Cr(IV), free radicals, and ultimately Cr(III) (Costa 2003; Cheung and Gu 2007). The primary toxicity mechanism of Cr(VI) is explained by the fact that Cr(III) has a relatively low penetration, is largely trapped inside of cells, accumulates, and interacts with DNA (Zhitkovich 2011). Also, Cr(V) can undergo a redox cycle ton regenerate Cr(VI) along with generating ROS that could interact with DNA-protein multiplexes, creates oxidative stress and triggers multiple apoptosis signaling pathways compromising the cellular functions (De Flora et al. 2006; Wu et al. 2020). Then by producing ROS in excess and depleting physiological antioxidant molecules, Cr(VI) can change redox balance via Fenton reaction (Wang et al. 2007; Li et al. 2019a; Monga et al. 2022a). Cr and its related toxicity has been a point of contention of several decades now. Cr(VI) exposure can cause cellular injuries and dangerous health consequences in several ways. Genchi et al. (2021) have extensively reviewed Cr toxicity on human health. According to many investigations, long term exposure to Cr(VI) can lead to neurodegeration, renal damages, dermal sensitivities, genotoxicity, cytotoxicity and immune system disorders (Sun et al. 2015; Fu et al. 2020). Cr exposure triggers specific kinds of cellular responses in the vital organs of human bodies (Monga et al. 2022a) (Fig. 13.3) including epigenetic modifications, gene regulations, DNA modifications etc. For instance, one of the most active transcriptome responses to Cr(VI) in mouse lung cells was eukaryotic translation initiation factor (eIF2) signalling (Rager et al. 2019). EIF2 pathway is frequently up-regulated in tumor cells and is involved in cell proliferation and growth (Watkins and Norbury 2002; Sonenberg and Hinnebusch 2009). Also, Cr(VI) sensitive genes such as MLH1and RAD51 were down regulated displaying a decrease in DNA replication, recombination and repair (Rager et al. 2019). In liver cells, Cr(VI) interferes with mitochondrial functions: diminished my copy number, respiration and redox equilibrium and retarded my electron transport chain (Yang et al. 2020). Normal mt fusion and proliferation occur in a dynamic equilibrium, but Cr(VI) has the power to upset this balance and produce fission, which affects cellular homeostasis and leads to oxidative stress and cellular death (Li et al. 2019a; Monga et al. 2022a). In addition, Cr(VI) causes morphological and functional damage to the immune system's crucial organs, including the thymus, spleen, lymph nodes, and bone marrow (Hultman and Pollard 2022). Wistar rats were given Cr(VI) orally for 135 days, during which time Karaulov et al. (2019) observed morphological and functional changes in the lymphoid tissue, including lymphoreticular hyperplasia and plasma cystic macrophages. Studies in Cr-exposed individuals and laboratory animals, abnormal Cr(VI) deposition and ROS induced oxidative stress in brain tissue and motor function impairment (Travacio et al. 2000). Brain cells are much more prone to oxidative stress damages as compared to other cells due to: (1) they require aerobic respiration and use a lot of oxygen, (2) their cellular membranes contain relatively high quantities of Polly saturated fatty acids, and (3) their levels of antioxidant enzymes like GSH are quite low (Ferreira et al. 2015). They are therefore more vulnerable to the oxidation of proteins, lipids, and membrane pores, which reduces MMP and causes neuronal death (Zhao et al. 2019). For instance, exposure to Cr(VI) resulted in abnormal behaviors and symptoms such as increase in surfing and darting movements and impaired locomotion in Fish and Drosophila (Singh and Chowdhuri 2017). Evidence for developmental toxicity of Cr(VI) was found in a study on pregnant rats by Pribluda (1963). Rats given 1 mg/kg Cr demonstrated poor bone formation in their embryos. When compared to the control group, the group that received 2 mg/kg of Cr(VI) also demonstrated the lack of the sacral vertebrae (Marouani et al. 2017).

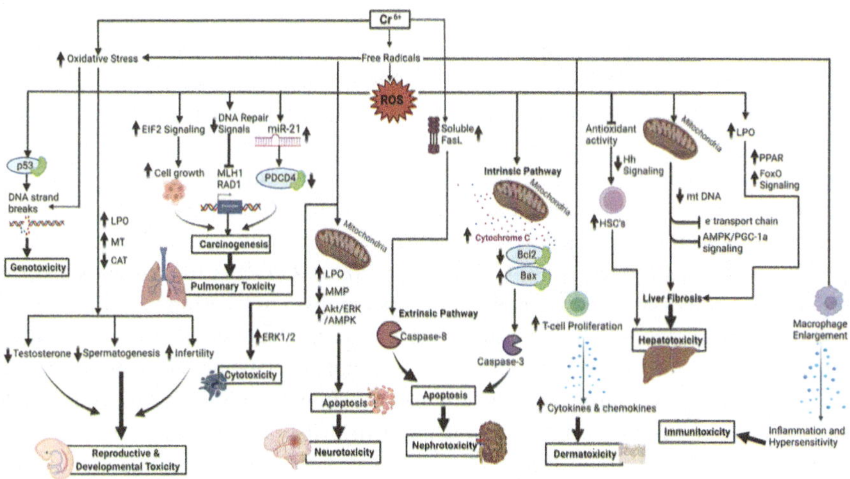

Fig. 13.3 Specific cellular responses with Cr(VI) toxicity. Adapted from Monga et al. (2022a)

13.7.2 Plants

The solubility of Cr(III) is much lower than Cr(VI), preventing its mobility in leaching into groundwater, thus affecting its bioavailability and absorption by plants (Cervantes et al. 2001). $HCrO_4^-$ and CrO_4^{2-} are the two most common forms of Cr(VI) in soils are that are quite easily absorbed by plants and travel rapidly downwards into deeper layers of soil and groundwater (Elahi et al. 2020). Cr(VI) levels above 5 mg/kg in soils and 0.5 mg/L in solution can be extremely dangerous for plant growth and metabolism (Elahi et al. 2020; Ayele and Godeto 2021). Jobby et al. (2018) have extensively listed some of the major effects of Cr(VI) toxicity in plants such as reduced uptake of nutrients, stunted growth, necrosis, chlorosis, decline levels of physiological and metabolic pathways etc. (Jobby et al. 2018). Leaves are the main organs for photosynthesis in plants; increasing Cr(VI) concentrations in soil leads to reduction in leaf area and biomass, suppression of chlorophyll production, loss of Mg^{2+} ions from chlorophyll molecules, inhibition of photosynthetic electron transport chain and thus photosynthesis failure leading to leaf necrosis and chlorosis (Stambulska et al. 2018). The oxidative stress generated in the plant cells due to Cr(VI) leads to lipid peroxidation, DNA strand breaks and chromosomal aberrations leading to cell death (Guo et al. 2021).

13.7.3 Microorganisms

Similar to its effects in plants and animals, due to requirement of Cr as an essential nutrient in trace amounts (Monga et al. 2022a); microorganisms are sensitive to both deficiency and excessive levels of Cr ions (Mishra and Bhargava 2016). HMs can have a significant effect in shaping microbial community structures in various ecosystems according to some reports (di Cesare et al. 2020). For instance, sedimentary microbes are vital for nutrient cycling, energy flow and organic matter remineralization. Under the effect of pollutants, the composition, abundance, and function of these microbial communities may change due to susceptibility and lead to decoupling of biogeochemical processes (di Cesare et al. 2020). Cr(VI) stress in the sensitive population of microorganisms disturbs their metabolism by altering their nuclei acid structure, cell membrane disruption, inhibition of enzyme activities and oxidative phosphorylation leading to LPO and osmotic imbalance (Ayangbenro and Babalola 2020). It causes cell enlargement and elongation while it inhibits cell division which is necessary for cellular growth and metabolism (Mishra and Bhargava 2016). On the other hand, due to their brief life cycles and basic genetic organization, some native microbes have evolved to modify their genetic make-up, conferring them the ability to survive in polluted environments. Bacteria has evolved several mechanisms (discussed below) such as efflux, intracellular/extracellular reduction, biosorption, extracellular binding by EPS etc. in order to tolerate toxic levels of HMs (Bruins et al. 2000; di Cesare et al. 2020).

13.8 Cr Pollution Remediation Measures and Practices

13.8.1 Wastewater Treatment

Eradication of toxic HMs such as Cr from industrial and domestic wastewater's is very essential in order to protect and maintain the standards of water streams, aquatic systems, and groundwater aquifers. Several technologies have been developed over the last decades with the goal of successful treatment of wastewater contaminated with HMs, particularly Cr. High solubility, bioavailability, and toxicity of Cr(VI) necessitates its removal from wastewater before discharged into the environment (Ukhurebor et al. 2021). More conventionally, Cr removal technologies were based on physical and chemical treatments such as chemical reduction, precipitation, membrane separation (ultrafiltration, nanofiltration, reverse osmosis, ion exchange membranes), flotation, solvent extraction, electrochemical methods (electrolysis, electro coagulation, electrodialysis) and ion-exchange (Srivastava et al. 2016; Ukhurebor et al. 2021) while the latest methods are more biotechnology based using bacteria (living and dead biomass), fungi, agro-industrial waste materials etc. that create less toxic byproducts, are sustainable and economically viable (Cuellar et al. 2022). The two main processes used in these procedures are reduction, where Cr(VI) is changed to Cr(III) at an acidic pH, and precipitation, where Cr(III) is formed at an alkaline pH. Addition of iron can reduce this two-step process into one (Mitra et al. 2017; Ukhurebor et al. 2021):

$$\mathrm{CrO^{-2} + 8H^{+} + 3Fe + 2 \rightarrow Cr^{+3} + 3Fe^{+3} + 4H_2O}$$

According to Malaviya and Singh (2011), reduction and precipitation procedures are frequently used to remove Cr from wastewater, but they also utilize a lot of chemicals and produce too much secondary waste. On the other hand, membrane-based methods (ion exchange etc.) are better in a way that they don't produce secondary pollution but they are very expensive, consume high energy and ineffective at low Cr concentrations (Malaviya and Singh 2011; Ukhurebor et al. 2021). In a recent study, Liu et al. (2022) combined the flocculation and membrane separation processes to treat wastewater from a tannery containing Cr. They used flocculation ultrafiltration (UF) to pre-treat the wastewater before transferring the generated water directly into nanofiltration (NF) for concentration treatment. When the salt contents of the main and secondary freshwater were 200–500 mg/L and 800–1000 mg/L, respectively, the NF multistage treatment was utilised to control the freshwater recovery rate to 90%. Finally, the effluent was desalinated using electrodialysis (ED) (Liu et al. 2022). By modifying a standard polyacrylonitrile (PAN) UF membrane, Mantel et al. (2022) was able to combine UF and ion exchange into a combined filtration process. By using this technique, adsorptive dead-end filtering was used to remove particulate particles and dissolve Cr(VI) (Mantel et al. 2022). In addition to precipitation and reduction, adsorption has emerged as a cost-effective and a simpler method to treat Cr containing wastewater. Adsorbents before were manly composed of chemical materials such as activated carbon, chitosan, zeolites etc. (Owlad et al. 2009) with diverse adsorptive abilities and majority of the functioning at low pH. Carbamoyl chitosan, a derivatized form of chitosan, has however demonstrated remarkable results for the adsorption of Cr, with an adsorption efficiency as high as 438.8 mg/g (Chauhan et al. 2012). Moreover, carbon-based nano-materials such as graphene have shown good adsorption properties in its oxidized state (Agarwal and Singh 2017). As reviewed by Singh et al. (2020) several new derivatized nano materials like Polyaniline nanorods dotted on grapheme oxide, Polypyrrole/Fe_3O_4 Nanocomposite, Phosphonium-coated (MNPs) and carbon nano anions in recent years have shown promising results for Cr removal. Additionally, carbon nanotubes have become an effective adsorbent that may be used alone or in conjunction with any metal, such as FeO. Because of the larger surface area, this combination has been found to boost the adsorptive capacity for Cr. It also has the added benefit of enabling total metal removal by easy magnetic methods (Luo et al. 2013; Singh et al. 2020).

13.8.2 Soil Remediation

In all aquatic systems, the sediment is the part where dissolved constituents and contaminants tend to gather due to scavenger representatives and adsorptive components (Peng et al. 2009). Conventional remediation techniques like in-situ capping and relocation actions were widely practiced but are now becoming unsustainable due to various problems associated with land space, budget, contaminant conveyance paths and ecological compatibility. The majority of soil treatment technologies rely on physiochemical techniques like sediment washing (which involves dissolving metal contaminants in aqueous chelating agent solutions), electro-chemical treatment (which involves separating metal cations using an electro-magnetic field), and thermal treatment of the sediment (Akcil et al. 2015). Since Cr can be absorbed into plants from the soil, plants can be utilized for phytoremediation. Phytoremediation is a green technique that utilizes plants to remove non-degradable toxic metal ions from the soil (Anju 2017). It is a better technique as compared to conventional physical and chemical methods since it does not harm the ecosystem, in-situ treatment volume can be achieved since it involves both dissolved and sorbet pollutants (Genchi et al. 2021). Cr hyperaccumulation plants, including *Spartina argentinensis, Amaranthus dubius, Convolvulus arvensis*, and others, have already been reported in the literature (Guo et al. 2021). It is suggested that the hyperaccumulator grade requirement be set at 300 g/g given the extremely low Cr concentrations in plants, both in normal (1 g/g) and metalliferous (ultramafic) soils (50 g/g). *Pycnandra acuminata* exhibits leaves with a metal content that is at least 2–3-fold better than other plants that grow in typical soils and significantly higher than the plants that grow in soil that are metalliferous (Van der Ent et al. 2013). Mangrove afforestation zones at designated locations in The Vai River watershed of Vietnam were recently advised to be employed for their phytoremediation prospects (Nguyen et al. 2020; Monga et al. 2022a). Leeching, often referred to as soil washing, is a technique that treats contaminated soil using principles of physical separation, chemical extraction, or a combination of both. Particles made of soil differ physically from particles made of heavy metals. Physical separation therefore takes advantage of this distinction to concentrate these heavy metals into smaller amounts that may then be eliminated (Ukhurebor et al. 2021). Testing for the removal of Cr from soil using chemicals like acetic acid, ethylenediamine tetra acetic acid (EDTA), and HCl on samples of pond sludge demonstrated that the highest removal efficiency of Cr was achieved using 0.3 M of HCl with 82.69%, accompanied by EDTA at 72.52%, and the lowest efficiency was recorded by 3 M of acetic acid with 46.96%. (Abumaizar and Smith 1999). Oxalic acid (OA), citric acid (CA), and HCl had also been utilized in the elimination of Cr, with oxalic acid showing the highest potency in this regard (Sun et al. 2019b). It was concluded that while oxalic acid can be used to remediate soil for Cr, it must be careful not to leach vital minerals or reduce soil fertility (Ukhurebor et al. 2021).

13.9 Microbial Remediation Mechanisms and Technologies

Conventionally, ion-exchange, membrane filtration, and reduction-precipitation are the three most often used techniques for removing Cr(VI). However, operating these processes at a large-scale are very expensive (Cheung and Gu 2003), especially for developing countries. Due to their relatively high removal efficiency, low cost, and environmentally safe or sustainable practice, biological remediation approaches using microbial strains (bioremediation) or plant species (phytoremediation) have grown significantly in popularity as the preferred choice for chromium removal technologies (Nakkeeran et al. 2018; Lian et al. 2019; Prasad et al. 2021). Due to their incapacity to degrade, HMs can build up in the environment, causing a serious threat to human health and poisoning the food chain (Genchi et al. 2021). Thus, bioremediation has evolved as one of the safer and more effective alternatives for treating HM pollution as compared to conventional physical and chemical methods (Singh et al. 2020).

13.9.1 Bioremediation

The elimination and reduction of HMs from contaminated environments is possible with the help of the innovative technology known as bioremediation. Microbes are crucial to the bioremediation of metals. *Flavobacterium, Pseudomonas, Bacillus, Arthrobacter, Corynebacterium, Mycobacterium, Methanogens, Aspergilus niger, Rhizopus arrhizus, Azotobacter, Alcaligenes, Ganoderma applantus,* and others are among (Verma and Kuila 2019). Through bioremediation, a metal site can be rehabilitated to its prior state without compromising the ecosystems (Jobby et al. 2018).

Living organisms such as bacteria, fungi, yeast, algae, and plants have the ability to clean up after themselves, however bacteria and fungi have been demonstrated to be more proficient at it. These technologies have a number of benefits, including minimal energy requirements, low operating costs, no environmental or health risks, high efficiency, the potential for reuse, and metal recovery (Garbisu and Alkorta 2003). Metal is frequently used by microorganisms as a nutrition or energy source to meet their growth requirements (Tang et al. 2007) and metabolise these pollutants through enzymatic/or non-enzymatic mediated reactions into less toxic or harmless compounds such as CO_2 or CH_4, water and biomass (Vidali 2001; Jobby et al. 2018). Bioremediation is made of two terms: "bios" meaning life and "remediate" means to solve the issues. So 'bioremediation' refers to solving environmental issues with the use of living organisms. Bacteria, for instance can remove/accumulate/precipitate or reduce toxic pollutants into less toxic forms. Though mostly these processes need the right combination of nutrients, time, temperature, ph etc. for carrying out effective bioremediation of the contaminants. Bioremediation has emerged as a new sustainable technology for decontamination of polluted ecosystems (Nur-E-Alam et al. 2020).

13.9.2 Types of Bioremediations

There are primarily three kinds of bioremediation:

(1) **Biostimulation**: Chemicals or nutrients that activate microorganisms are used to stimulate them to start the cleanup process. Biostimulation was chosen as the treatment option in 1999 at the Ace Services Superfund Site, a chrome processing plant in Kansas (Jobby et al. 2018).
(2) **Bioaugmentation**: This procedure introduces bacteria to the surface of the contaminated area, where they are then allowed to proliferate. It is mostly used to remove soil contamination. Though Cr(VI) contaminated soils naturally contain organisms that have adapted to the environment and are therefore better bioremediators, bioaugmentation is typically not used as a method (Jobby et al. 2018).
(3) **Intrinsic bioremediation**: This technique uses the indigenous microorganisms to transform hazardous toxins into inert ones.
(4) **Mycoremediation**: This type of bioremediation uses fungus, not bacteria or other microorganisms, for remediation purposes. Effective bioremediation of soil depends on a number of parameters, including the elemental composition of the pollutant, the soil's moisture content and pH, the microbial comunities present at the contaminated site, and temperature (Asha and Sandeep 2013; Jobby et al. 2018). Here I t is essential to comprehend the precise mechanism of action for metal removal by microorganisms in order to develop an efficient microbial-based treatment approach since remediation activity is closely linked to microbial metabolism (Singh et al. 2020).

13.10 Fungal Bioremediation

Fungi can actively participate in the bioremediation of Cr(VI) due to their special ability to tolerate HM. Such fungi bioremediate Cr(VI) through a variety of processes, including biosorption, bioaccumulation, and bioreduction (Ghosh et al. 2021) (Table 13.3). These mechanisms depend on fungi genetics, metal ion and environmental factors (Hassen et al. 1998; García-Hernández et al. 2017). Shan et al. (2022) in a recent report, isolated a Cr(VI) reducing fungal strain, Fusaium proliferated S4 from polluted soils near a chemical plant in China. Additionally, they evaluated the diverse Cr(VI) removal capacities of distinct cellular components and listed the following cell components in order of strength: cytoplasmic, cellular secretions, and cell debris (Shan et al. 2022). Various fungal strains have been reported in the literature: Fusarium chlamydosporium (Sharma and Malaviya 2014); Aspergillus and Rhizopus sp. (Ahmad et al. 2005); Aspergillus flavus, Fusarium sp., Helminthosporium sp., Aspergillus niger, and Aspergillus versicolor (García-Hernández et al. 2017). According to reports, certain Aspergillus sp. are frequently utilised as biosorbents to remove and sequester Cr(VI). Galactosamines, chitin, glucan, and certain lipids and amino acids in the cytosol are polysaccharides that are significant in the fungal metabolism of Cr(VI).

13.11 Algal Bioremediation

Algae has been shown previously to prevent eutrophication in wastewaters. For bioremediation, algae utilises the mechanism of photochemical reduction (Table 13.3). Algae has potential for Cr(VI) biodegradation because it produces oxygen during photosynthesis that is used by heterotrophic bacteria to generate biomass (Ghosh et al. 2021). Algae specifically uses its secondary metabolites such as phytochelations, metallothioneins and its cell wall constituents such as glucuronic acid, alginates, and other cell wall functional groups like −OH, NH_2, SO_4^{2-}, −COH for the biosorption of Cr(VI) ions (Elahi et al. 2020; Ghosh et al. 2021). Cr(VI) then bioaccumulates in the algal cell wall as a result of this. After 27 days of incubation period, algae including *Euglena sp., Chlorella vugaris, Spirulina sp., Spirogyra sp., Scenedesmus sp., Cladophora sp., Ceranium sp., Selenastrum sp.*, and *Nosctoc linkia* demonstrated a Cr(VI) detoxification efficacy of 97% from the culture system (Ghosh et al. 2021). I In another article, authors proposed that thylakoid membrane of Chlorella vulgaris in the presence of sodium alginate (SA) hold a capacity to reduce Cr(VI)–Cr(III) with 70% effectiveness in 4 days of incubation (Lee et al. 2017; Ghosh et al. 2021). Transgenic algae perform better than raw algae, according to more recent research. Genetic engineering can be used to improve the genes that express metal-binding proteins on algal membrane surfaces (Cheng et al. 2019).

Table 13.3 List of microorganisms (Bacteria, fungi and algae) with Cr(VI) bioremediation potential along with their mechanism of removal and optimum experimental conditions. NB = Nutrient broth; BM = Broth medium; ZMB = Zobell marine broth; LB = Luria Bertani broth; PDB = Potato Dextrose broth

Microbial strain	Source of isolation	Mechanism studied	MIC (Mg/L)	Initial Cr(VI) conc^{n}	Media	Optimum pH; T (°C)	Reduction ability (%)	Incubation/ contact time	References
Bacteria									
Bacillus sp. AKVCRR04	Marine sediment, Versova creek, Mumbai	Bioflocculation/ biosorption	2000	400	NB	7; 37	96.08	48 h	Monga et al. (2022b)
Pseudomonas aeruginosa AKVCRR02	Marine sediments of Versova creek, Mumbai	Bioflocculation/ biosorption	2000	400	NB	7; 37	95.15	120 h	Monga et al. (2022b)
Lactobacillus plantarum MF042018	Marine samples from Alexandrian Mediterranean Seacoast, Egypt	Bioaccumulation	100	100	Broth medium		30.2 ± 0.5%		Ameen et al. (2020)
Marinobacter hydrocarbonoclasticus	Equatorial Indian Ocean and Arabian Sea coastalwaters	Siderophore production, exopolysacharides Aerobic/anaerobic reduction	–	55.15	NB	32 ± 2 °C	88% aerobically; 89% anaerobically	5 days	Vijayaraj et al. (2019)
Enterobacter cloaceae (AK-I-MB-71a)	Marine	Exopolysaccharide sequestration		25, 50, 100			60–70		Iyer et al. (2004)

(continued)

Table 13.3 (continued)

Microbial strain	Source of isolation	Mechanism studied	MIC (Mg/L)	Initial Cr(VI) conc^{n}	Media	Optimum pH; T (°C)	Reduction ability (%)	Incubation/contact time	References
Bacillus licheniformis	Marine samples from Tamil Nadu, India	Enzymatic reduction, extracellular surfactants interactions	1500	20	ZMB		100	24–72 h	Kavitha et al. (2011)
Bacillus sp. MTCC 5514	Marine samples from Tamil Nadu, India	Enzymatic reduction, extracellular surfactants interactions	2000	20–2000	ZMB		100	24–96 h	Gnanamani et al. (2010)
Exiguobacterium Indicom MW1	Marine water of Paradip fort, Odisha		1500	100	AMM and M9	8; 35	91% in AMM; 92% in M9	192 h	Mohapatra et al. (2017)
Brevibacillus laterosporus	Marine water of Paradip fort, Odisha		2100	100		8; 35	92	120 h	Mohapatra et al. (2017)
Bacillus subtilis SHBB	Sediment		1000	100		7; 37; 4% NaCl	98	72 h	Swapna et al. (2016)
Pseudochrobactrum sp. B5	Marine	Bioreduction	2000	1000	LB		100	96 h	Ge et al. (2013)
Proteus sp. H24	Marine	Bioreduction	1500	1000	LB		100	144 h	Ge et al. (2013)
Klebsiella pneumoniae	Soil and water samples from phillipines	Biosorption	550	100	LB	7; 30	87	7 days	Bennett et al. (2013)

(continued)

Table 13.3 (continued)

Microbial strain	Source of isolation	Mechanism studied	MIC (Mg/L)	Initial Cr(VI) conc^n	Media	Optimum pH; T (°C)	Reduction ability (%)	Incubation/ contact time	References
Bacillus firmus	Soil and water samples from Phillipines	Biosorption	550	100	LB	7; 30	96	7 days	Bennett et al. (2013)
Mycobacterium sp.	Soil and water samples from Phillipines	Biosorption	750	100	LB	7; 30	91	7 days	Bennett et al. (2013)
Bacillus Amyloliquefaciens	Sediment	Biosorption				7	82.10	60 min	Ramachandran et al. (2022)
Pseudomonas chengduensis PPSS-4	Marine sediment of Paradip fort, Odisha	Biosorption by EPS, biofilm	2000	10	LB	6; 37; 4% NaCl	72.29	4 h	Priyadarshanee and Das (2021)
Halomonas sp. TA04	Marine sediments, southern Italy	Bioreduction (DPC method)	4.0 mM	0.5 mM	YEPG-NaCl	7–8, 28	81.5	24 h	Focardi et al. (2012)
Sporocarcina saromensis M52	Sediments from intertidal zone in Xiamen, China	Bioreduction (DPC method)	500	100	Modified LB	7–8.5; 35	100	24 h	Ran et al. (2016)
Sphingopyxis macrogoltabida SUK2c	Marine water of Sukinda valley, Odisha	Biosorption and extracellular reduction		4	NB	1; 30	55%	2 h	Prabhakaran et al. (2019)

(continued)

Table 13.3 (continued)

Microbial strain	Source of isolation	Mechanism studied	MIC (Mg/L)	Initial Cr(VI) conc^{n}	Media	Optimum pH; T (°C)	Reduction ability (%)	Incubation/ contact time	References
Acinetobacter calcoaceticus	Marine samples	Aerobic reduction	500	100	LB	7; 30	67.14% at pH 7; 70.53 at pH 8	24 h	Mishra et al. (2010)
Bacillus sphaericus OT4b31	Marine samples	–		30		4; 30	25% with living biomass; 44.5% with dead biomass	24 h	Velásquez and Dussan (2009)
Exiguobacterium sp. GS1	Water	–		8	TSB	7–8; 35–40	91	8 h	
Fungi									
Penicillium janthinellum P1 living biomass	Marine sediments	Biosorption	–	250	PDB	1; 30	87	8	Chen et al. (2019)
Penicillium janthinellum P1 dead biomass	Marine sediments	Biosorption	–	100	PDB	1; 30	58.6	12	Chen et al. (2019)
Aspergillus sydowii	Marine sediments from Mandovi estuary	Reduction	1000	300	Liquid broth	5; 28	26	7 days	Lotlikar et al. (2018)
Aspergillus flavus	Marine seaweeed associated	Bioaccumulation	100	100	Potato dextrose (PDB)	RT	25%	15 days	Vala et al. (2004)

(continued)

Table 13.3 (continued)

Microbial strain	Source of isolation	Mechanism studied	MIC (Mg/L)	Initial Cr(VI) conc^{n}	Media	Optimum pH; T (°C)	Reduction ability (%)	Incubation/ contact time	References
Aspergillus niger	Marine seaweed associated	Bioaccumulation	100	100	Potato dextrose	RT	25%	15 days	Vala et al. (2004)
Trichoderma viride	Seawater from Egypt	Biosorption and bioaccumulation	1000	125	–	6	98%	45 min	El-Kansas and El-Taher et al. (2009)
Aspergillus flavus	Soil and water samples from Phillipines	Bioreduction	600	150	PDB	2; 30	98	7 days	Bennett et al. (2013)
Aspergillus sp.	Soil and water samples from Phillipines	Bioreduction	600	150	PDB	2; 30	99	7 days	Bennett et al. (2013)
Aspergillus niger	Soil and water samples from Phillipines	Bioreduction	600	150	PDB	2; 30	98	7 days	Bennett et al. (2013)
Algae									
P. tricornutum CCY0033 (microalgae)	North Sea beach, Netherlands	Biosorption to EPS	1 mg/L	1 mg/L	MDV	23	35	3 days	Hedayathkhah et al. (2018)
N. pelliculosa CCMP543/ CCY0399 (microalgae)	Oyster pond, Masacheusetts, USA	Biosorption to EPS	1 mg/L	1 mg/L	MDV	23	32	3 days	Hedayathkhah et al. (2018)
Dunaliella salina (Microalgae)	Sambhar salt lake, Rajasthan, India	Biosorption	–	25		8.6	66.4	120 h	Kaushik and Raza (2019)

13.12 Bacterial Bioremediation

Due to its toxicity, Cr contaminated soil comprises a lower microbial population as compared to the soil non-contaminated by Cr (Viti et al. 2003). The bacterial species that are present are the ones that are able to tolerate/resist chromium toxicity. As per Gadd (1992), "tolerance" is the "capacity of a microbes to survive metal toxicity by means of basic components and/or ecological modification of toxicities," whilst "resistance" is the "ability of microbes to survive toxic effects of metal exposure through a detoxifying process designed in direct reaction to the metal species involved." Various researches have reported bacteria with the ability to bioremediate Cr(VI) and investigated an array of mechanisms that these bacteria adopt for their own survival (Banerjee et al. 2019; Baldiris et al. 2018; Li et al. 2021; Elahi et al. 2022; John and Rajan 2022; Kookhaee et al. 2022; Yakasai et al. 2022). These bacteria could use single or a group of these strategies to counteract the toxicity of Cr(VI) (Bharagava and Mishra 2018). Most commonly these mechanisms include bioreduction/biotransformation/enzymatic reduction, biosorption, bioaccumulation, efflux, precipitation, cytosolic binding etc. (Banerjee et al. 2019), though microbes differ in their potential to utilize these strategies. Cr(VI) bioreduction and biotransformation have been extensively studied in bacterial system as compared to other microorganisms (Elahi et al. 2022) (Table 13.3). Numerous tolerance mechanisms for bacteria to cope with the HMs have been postulated. Examples of fundamental strategies used by bacteria to survive and thrive in metal-stressed environments include active efflux, intracellular sequestration, enzymatic transformation, and oxidation/reduction of harmful metal ions (Zeng et al. 2020). Kathiravan et al. (2011) isolated Bacillus sp. from tannery effluent contaminated site and studied bioremediation process in batch and continuous operations. Nine strains that could withstand chromium up to 700 mg/L were reported by Park et al. (2000). Camargo et al. (2003) reported the optimal pH of 7–9 and temperature of 30 °C for maximum chromium reduction activity by *Bacillus sp.* Megharaj et al. (2003) examined the potential for *Arthrobacter sp.* and *Bacillus sp.* to reduce Cr(VI) and discovered that *Arthrobacter sp.* could do so up to 50 g/ml while Bacillus sp. could only do so to the extent of 20 g/ml. Cr resistant bacterial train *Bacillus cereus S-6* was isolated from effluents of tannery with the ability to reduce Cr(VI) to less toxic Cr(III). The cytosol and membrane preparations of the bacteria could reduce upto 67 and 43% of Cr(VI) within 24 h of incubation. Turick et al. (1996) reported various bacteria from different soil types for Cr(VI) reduction potential. Cheung and Gu (2003) studied the reduction of Cr(VI) to less toxic Cr(III) using *Pseudomonas putida PRS2000* for which the chromate reductase activity was instituted to be linked with soluble protein rather than membrane fraction. According to Costa (2003), bacterial strains *CrT-11, CrT-12, Brevebacterium sp. CrT-13, CrT-14* isolated from tannery effluents could tolerate upto 40 mg/ml of K_2CrO_4 on nutrient agar.

13.13 Molecular Mechanisms for Bacterial Bioremediation of Cr(VI)

13.13.1 Adsorption by Functional Groups on the Surface of the Cell

The biosorption of the metal is mostly caused by the cell wall of bacteria because it is the first to come into contact with the metal ions (Wang and Chen 2009; Gutiérrez-Corona et al. 2016). This interaction mainly depends upon the functional groups present on the bacterial cell wall and the physiochemical conditions of the medium (Karthik et al. 2017). Bacterial cell surface comprises various functional groups like hydroxyl (−OH), carboxyl (RCOOH), carbonyl (−COOH), amide (CO–NH), sulfonate, phosphonate, phosphodiester etc. that form larger compounds like lipopolysaccharides (LPS) and peptidoglycans (Karthik et al. 2017). They interact with the metal ions through a chemical bond. Maximum Cr(VI) binding efficiency occurs at acidic pH because of the electrostatic attraction between the protonated (H^+) bacterial surface and Cr(VI) (as $HCrO_4^-$). However, with increase in pH, the $HCrO_4^-$ ions convert to CrO_4^{2-} and $Cr_2O_7^{2-}$ ions. With an increase in OH^- ions in the solution, the adsorption efficiency of Cr(VI) drops at basic pH (Yaashikaa et al. 2019; Pushkar et al. 2021). Next, the adsorbed Cr(VI) either bioprecipitates on the microbial cellular surface or is biotransformed into Cr(III) which is either enzymatically catalyzed by a chromate reductase or occurs spontaneously (Thatheyus and Ramya 2016; Jobby et al. 2018).

Under the LPS layer in Gram negative bacteria is a thin coating of peptidoglycan, and these two layers interact with HMs on the cell surface in significant ways. However, the cell membrane of Gram positive bacteria only has a substantial coating of peptidoglycan (Fang et al. 2018). Under Cr(VI) stress, gram-negative bacteria secrete more LPS, which acts as a metal chelator and facilitates Cr's attachment to the cell surface (Kiliç et al. 2010). In a research study between *E. coli* and *Staphylococcus epidermidis*, Quiton et al. (2018) found that Gram negative *E. coli* bacteria had a stronger biosorption capacity due to the negative charge of LPS structures on the cell wall. Nonetheless, in case of Gram positive bacteria, the presence of a high amounts of anionic polymers in the cell wall primarily made up of peptidoglycan teichoic or teichuronic acids, helped them perform Cr biosorption. Several cell surface ligands on Gram positive cell wall such as phosphoryl, carbonyl (COO−) etc. had a strong affinity towards metal ions such as Cr(VI) (Pushkar et al. 2021). Various analytical techniques, such as Fourier transformed infrared spectroscopy (FTIR) and scanning/transmission electron microscopy with energy X-ray spectroscopy (SEM-EDX/TEM-EDX), have been used to discover the functional groups on the cell walls of the bacteria participating in Cr(VI) metal-microbial interaction as well as the process of Cr(VI) adsorption, absorption, and reduction (Batool et al. 2014; Gutiérrez-Corona et al. 2016; Elahi et al. 2022; Li et al. 2021). EPS for instance, are functional high molecular weight organic polymers found on bacterial cell surfaces in either capsular or secreted forms (Kumar et al. 2019). They typically possess functional groups like phosphate, carboxyl, hydroxyl, amide etc. that are responsible for chelation and detoxification of metal ions (Mangwani et al. 2016). Li et al. (2021) isolated a novel bacterium *Stenotrophomonas acidaminiphila* 4-1that secreted EPS under Cr(VI) stress. They were the binding sites of adsorption of Cr(VI) on surface of cells as depicted by TEM. The adsorption was mostly mediated by electrostatic or complexing bonds as reported in various studies (Hussein et al. 2019). In a separate report, FTIR identified the amine, hydroxyl, and carboxyl chemical groups involved in Cr(VI) interaction in the cell walls of the bacteria Streptomyces werraensis LD22 (Latha et al. 2015).

FTIR analysis of untreated [without Cr(VI)] cells can reveal the possible functional groups involved in metal-microbe interaction (Elahi et al. 2022). However, under Cr(VI) stress significant shifts were observed in peaks of the FTIR spectra of the treated cells (Karthik et al. 2017; Banerjee et al. 2019; Elahi et al. 2022), validating the involvement of functional groups in Cr(VI) binding on bacterial cellular surface. Shifts in the FTIR peaks mostly involved amino, carboxyl, nitrogen, peptide (oxygen) that are all mostly C and O based (Banerjee et al. 2019).

In context to resistance to Cr(VI), a wider group of Gram negative bacteria have been documented in comparison to gram positive bacteria. Bacillus sp. Predominates among the Gram positive bacteria known to be resistant to Cr(VI) (Shaw and Dussan 2018; Pushkar et al. 2021). This observation was also reported by Satarupa and Amal (2010), in their study on chromate mine seepage water that showed prevalence of Gram negative bacteria. Despite the fact that Gram positive and negative bacteria differ from one another due to variations in the composition of their cell walls, they nevertheless have the samc gene for chromium resistance (Fig. 13.4). This is mostly due to the selection of Cr resistant bacteria over time and horizontal gene transfer among bacterial groups (Pushkar et al. 2021). This was also corroborated by another study by Patra et al. (2010). They showed >99% similarity between the test Gram positive bacterial strains namely *Arthrobacter aurescens* strain MM10, *Bacillus atrophaeus* strain MM20, and *Rhodococcus erythropolis* strain MM30 with the already documented Gram Positive *E. Coli* and *Shigella* sp. (Patra et al. 2010). Gram positive involves hydroxyl groups present on their surface during Cr(VI) biosorption at pH 1–4 (Prabhakaran and Subramanian 2017). Gram negative bacteria, on the other hand, can reduce Cr(VI) extracellularly due ton the presence of LPS, lipoproteins and phospholipids present in their outer membrane (Pushkar et al. 2021) (Fig. 13.4). According to Shaw and Dussan, lineages I and II can be used to group together the efflux pumps and regulators of both Gram-negative and Gram-positive bacteria. Aligning the amino acid sequences of these clusters revealed the presence of several amino acid signatures and conserved regions in the respective lineage (Shaw and Dussan 2018; Pushkar et al. 2021). The development of microbial methods for the reduction, elimination, and retrieval of metals from aqueous solution depends on our ability to understand how bacteria acquire metals. In the case of non-living biomass, the only mechanism is metal binding to cell walls and external surfaces that is independent of metabolism. Adsorption techniques including ionic, chemical, and physical adsorption are mostly used in metabolism-independent uptake (Ahluwalia and Goyal 2007). One of the responses and a self-defense strategy against Cr toxicity is the clumping of cells, which is caused when like charges on the bacterial surface tend to neutralise in the presence of Cr (Karthik et al. 2017). Insoluble Cr(III), a colloid form of Cr hydroxide that can form from Cr(VI), can also be absorbed on the surface of bacterial cells. This alters the total protein composition of the cell surface (Asatiani et al. 2004). Wang and Cui (2019) reported the formation of protrusions on the cell surface of the bacteria after Cr(VI) treatment (520 mg/L) which was due to absorption of Cr(III) and the changes it caused to the protein composition on the cellular surface. Additionally, due to their similar structure, sulfate transporters present on the surface of the cells also aids in the transport of chromate ions (CrO_4^{2}). Thus Cr(VI) competitively inhibits sulfate uptake, which is compensated by increasing the uptake of cysteine-containing molecules by cell (Gang et al. 2019). Thus, the bacterial cell surface plays a crucial role in Cr(VI) resistance and removal.

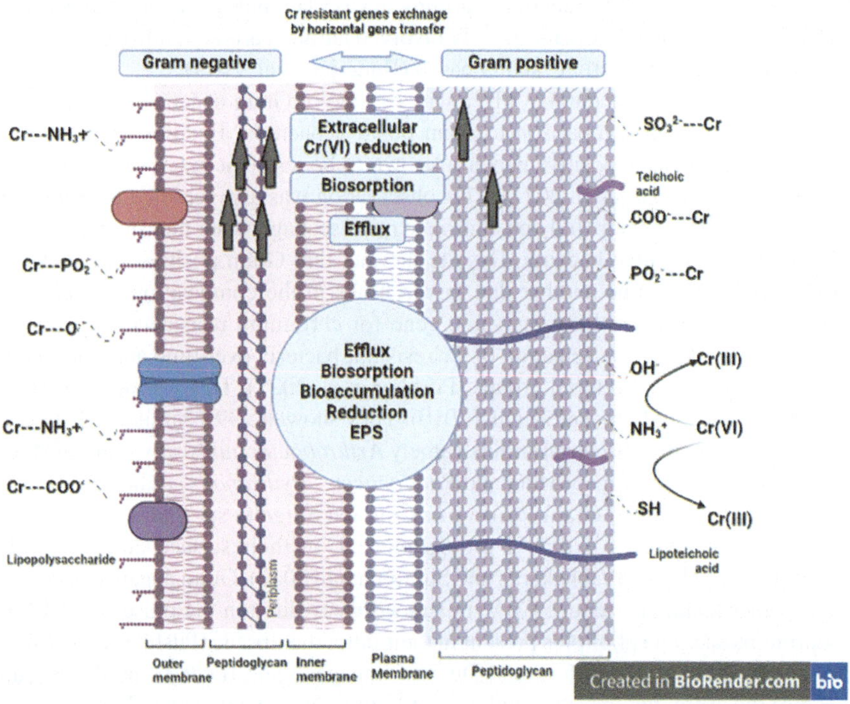

Fig. 13.4 A comparative diagram to depict Cr(VI) response of gram positive and gram negative bacteria

13.13.2 Extracellular Precipitation/Reduction: Role of Extracellular Biopolymers

Some microbial species have been observed to create extracellular biopolymers that facilitate flocculation (Monga et al. 2022b; Ayanbenro et al. 2019). Bioflocculation or biosorption is the process of any compound being absorbed by biological materials through metabolically independent or dependent absorption processes (Fourest and Roux 1992). Flocculants or surfactants are a class a amphipathic molecules that are eco friendly and aid in the HM removal from contaminated soil and sediments (Banerjee et al. 2015). They are also quite efficient at low HM concentrations because of which they are quite ideal for the treatment of effluents and wastewaters (Lin and Harichund 2011). Furthermore, in order to achieve a high bioflocculant yield at low costs and high flocculation activity efficiency, bioprospecting of strains with such capabilities is indispensable (Nwodo and Okoh 2013; Monga et al. 2022b). Numerous heavy-metal removal bacterial bioflocculants have been studied because they are non-toxic, environmentally safe, and biodegradable (Lin and Harichund 2011). Numerous bacteria, including *Bacillus, Pseudomonas, Acinetobacter*, and *Arthrobacter*, are well-known for producing bioflocculants (Banerjee et al. 2015). Though fewer studies on the biosorption of Cr(VI) with a bioflocculant have been reported. In a recent study conducted, a very effective bioflocculant, Na-Bsp, was successfully developed against kaolin particles utilising a tolerant Bacillus sp. strain with a high flocculant efficacy of 97.69 $\pm$ 0.61% and Fe^{3+} as a cofactor. On the surface of the bioflocculant, hydroxyl, carboxyl, and amine groups may have led to strong interactions with heavy metals. Cr(VI) has an adsorption capacity of 384.6 mg g^{-1} (Hua et al. 2021). According to a another recent study, the bioengineered strain F2-exoY-O recovered Cr(VI) more effectively than the wild strain because it produced more polysaccharide in its EPSs (Pi et al. 2021). An efficient and long-lasting method of removing chromium from contaminated habitats was provided via chelation between functional groups on EPS and Cr(VI), with a little conversion of Cr(VI) into Cr(III) on EPSs (Monga et al. 2022b). The bioflocculant synthesis of Bacillus sp. AKVCRR04 and AKVCRR05, which were isolated from the surface sediments of Mumbai's polluted Versova Creek, was also examined using the Kaolin clay assay. By the 5th day of incubation, the isolates displayed substantial flocculation activity (89.75 and 89.88%, respectively). Additionally, the growth profile of the isolates and the final pH in relation to the flocculation activity assay showed that the synthesis of bioflocculants peaked either in the late stationary phase or during the stationary phase (Monga et al. 2022b).

As mentioned above, for various molecules present in the extracellular environment, cell surface acts as the first line of defence and also transmits signals inside the cell. This is important in deciding the fate of the molecule as it enters the cellular environment (Pushkar et al. 2021). The EPS production is believed to be crucial protective strategy for bacteria to thrive and survive in environments polluted with HMs (Zeng et al. 2020). The EPS matrix mainly consists of negatively charged functional groups that aids in chelating metal ions and avoiding the direct contact of cells and the toxic pollutants (Wu et al. 2019; Zeng et al. 2020; Li et al. 2021). When Cr(VI) interacts with protonated biomass in an acidic pH environment, it is reduced to Cr(III) in the aqueous phase or in the biomass (Park et al. 2006). Using analytical methods including XPS, XAS, and SEM-EDX/TEM-EDX, this process has been established to occur in both bacteria and fungus (Park et al. 2007). Nevertheless, numerous studies have also shown that some bacterial species, including Cyanobacteria (Ozturk and Aslim 2008), Azotobacter (Joshi and Juwarkar 2009), Arthrobacter (Shuhong et al. 2014), Bacillus (Dogan et al. 2015), as well as fungal species, including Trichoderma and Schwanniomyces, produced EPS with the ability to remove Cr(VI) by an adsorption coupled reduction mechanism. Metal ions can be bioabsorbed by bacteria that naturally produce an EPS coating, keeping them from interfering with important biological components. These bacteria's EPS coating may offer sites where metal cations can be attached (Scott and Palmer 1990; Bruins et al. 2000). Several bacterial species, including Klebsiella aerogenes, Pseudomonas putida, and Arthrobacter viscosus, exhibit the capacity to bind metals extracellularly (Bruins et al. 2000).

Metal chelation by EPS is an interesting property that makes it important in the field of Cr(VI) remediation (Chug et al. 2016; Saba et al. 2019a, b). There are three different forms of EPS: soluble (S-EPS), loosely bound (L-EPS), and tightly bound (T-EPS). Briefly, the protein and the polysaccharide content of EPS act as electron donors in Cr(VI) reduction. The N and O groups on the LB and TB LBS transfer electron for this reduction. Cr(III) then immobilises onto the negatively charged groups on the EPS surface such as −OH, −ROOH (Pushkar et al. 2021). For non-enzymatic reduction, FTIR analysis revealed the role of −OH groups of polysaccharides and −NH groups of membrane proteins (Srinath et al. 2002). In *Bacillus sp.* MRP-3, functional groups of T-EPS played important role in Cr(VI) adsorption in comparison to LB-EPS in Cr(VI)/Cr(III) attachment (Shao et al. 2019). According to findings from a different study, Pseudochrobactrum saccharolyticum LY10 increases T-EPS expression when Cr(VI) levels increase (Long et al. 2019). As a cellular response mechanism, it has been observed that in activated sludge, bacteria produce more of the −N component of S-EPS in the presence of Cr(VI)/Cr(III) (Liu et al. 2020). Thus, EPS aids bacteria in removing chromium in a number of ways. The method of bioremediation of chromium and other heavy metals can be further improved with future study concentrating on the stimulation of EPS synthesis by bacterial cells (Pushkar et al. 2021).

13.13.3 Accumulation of Cr(VI) and Cr(III) on Cell Envelope: Biosorption and Bioaccumulation

Microorganisms adopts a variety of strategies and mechanisms to be able to survive in heavy metal polluted environment. Biosorption, bioaccumulation and biotransformation are frequently used by microbes to detoxify Cr(VI) to comparatively less harmful forms (Jobby et al. 2016). These techniques form a crucial part of bioremediation process. The superiority of biobased remediation methods over conventional physical and chemical methods are now well known (Jobby et al. 2018). There are two ways by which microorganisms incorporate metals into their cellular processes, first one is known as "passive uptake", more popularly known as "biosorption". This is a metabolism-independent uptake of metals that can take place in both living and dead microbial cell biomass. The other one is "active uptake" which involves energy and metabolism for the metal transport and can occur only in living biomass. These two modes of metal transport combined together are called "bioaccumulation" (Wang and Chen 2009; Gutiérrez-Corona et al. 2016). Biosorption being a passive process varies among different bacteria and depends largely on the cell wall composition, its physicochemical characteristics and that of the surrounding medium (Bharagava and Mishra 2018; Jobby et al. 2018). The physicochemical interaction between the metal pollutants and the surface of dead or live bacterial biomass that leads to adsorption of heavy metals on bacterial surface. It's a natural response of bacteria that is non-specific in nature and involves a formation of a chemical bond (Pushkar et al. 2021). Bioaccumulation on the other hand, utilizes the respiration energy of the metabolic pathway of bacteria to accumulate Cr(VI) within the cell wall (Wang et al. 1989; Jobby et al. 2018).

In first stage of biosorption, the physical adsorption may take place by forming a chemical bond, ion-exchange, adsorption, or precipitation on the surface of the bacteria. Depending on the type of bacterium and its environment, these physical adsorption processes of biosorption may function independently or in concert. It can also be accomplished by live and non-living bacterial cells because it is independent of bacterial metabolism (Jarosławiecka and Piotrowska-Seget 2014). Bioaccumulation is the second stage of biosorption; it is a slower procedure that involves active Cr transport into the bacterial cell that is regulated by metabolism. After bioaccumulation, Cr is released intracellularly by a number of pathways, such as localization to particular organelles, association with metallothionein, accumulation as a particulate HM, extracellular precipitation, or complexation (Srinath et al. 2002; Bharagava and Mishra 2018; Elahi and Rehman 2019a, b). According to Ma et al. (2019a, b) chromium is biosorbed either as Cr(VI) ions or as its reduced form, Cr(III) ions. Based on numerous biosorption studies on various bacteria that have been published in the literature, it can be assumed that biosorption is influenced by factors such as pH, temperature, biomass dosage, initial Cr concentration, contact time, etc. (Jobby et al. 2018). Cheng et al. (2021) used single-factor studies to understand the impact of Cr(VI) and RSM based on Box-Behnken design (BBD) to study the biosorption behavior of strain *Shewanella putrefaciens.* They could achieve Cr(VI) bio-removal with an efficiency of up to 85.68% under the optimized conditions of 16.57 h of contact time, a pH of 8, and 0.42 g/L of biomass (Cheng et al. 2021). The ability of *B. paraconglomeratum* ER41 to decrease Cr(VI) was demonstrated by Harboul et al. (2022). It could grow, totally biosorb, and bioreduce 100 mg/L of Cr(VI) in 48 h at pH 8 and 30 °C and demonstrated strong resistance to Cr(VI) (700 mg/L). The factors pH, temperature, chromium concentration, and contact time all play a significant role in the Cr(VI) reduction process. Additionally, the composition of bacteria's cell walls varies depending on the growth conditions and medium type. As a result, the same bacteria isolated from several sites exhibited diverse biosorption rates (Rizvi et al. 2020; El-Naggar et al. 2020). According to the research data, biosorption is a successful chromium bioremediation technique. The process of heavy metal bioremediation will benefit tremendously from additional study aimed at improving the biosorption capacity of microorganisms (Pushkar et al. 2021). Pulimi et al. (2012) employed statistical design tools such as Plackett-Burman design, Central Composite design etc. in order to optimise physical and chemical variables for Cr(VI) biosorption and biotransformation by strain *Acinetobacter junii* VITSUKMW2. A maximum of 99.95% of Cr(VI) removal was achieved in 12 h under optimised parameters of initial Cr(VI) levels (54 mg/L). El-Naggar et al. (2020) investigated biosorption based removal of Cr(VI) using *Pseudomonas* sp. NEWG-2. A statistical model based on RSM (response surface methodology) of face-centered central composite design was applied to the growth studies of P. alcaliphila NEWG-2 (FCCD). According to their FCCD test findings, the bacteria could proliferate and remove of 96.60% of 200 mg/L of Cr(VI) in the presence of yeast extract (5.6 g/l), glucose (4.9 g/l), and pH 7 for the duration of the 48 h incubation period. Furthermore, monolayer chromium ion adsorption on homogeneous sites on the bacterial surface was modelled as following both the pseudo-first-order model and the intraparticle diffusion model, demonstrating that the Langmuir model well explains chromium ion biosorption by *B. paraconglomeratum* (Harboul et al. 2022). In another report, the biomass of the metal tolerant *B. amyloliquefaciens isolated from a marine soil was optimized for biosorptions conditions.* Acidic pH and long contact times inhibited the effectiveness of biosorption. At pH 7 and 60 min of contact, the highest biosorption was 82.10% and 80.12%, respectively (Ramachandran et al. 2022). The authors further reported that the biosorption efficiency when declined at acidic pH and longer contact times. They identified the adsorption mechanism as monolayer and a favourable adsorption as indicated from the Freundlich model.

According to Srinath et al. (2002), *B. circulans, B. megaterium,* and *B. coagulans* were outstanding strains that were able to adsorb 34.5 mg, 32.0 mg and 39.9 mg Cr of dry weight respectively. The biosorption ability of the living and dead biomass of *Bacillus. coagulans* and *Bacillus. megaterium* was also evaluated by the authors, and they discovered that the dead cells were more efficient. *B. coagulans* dead cells absorbed 39.9 mg Cr g^{-1} dry weight while living cells only adsorbed 23.8 mg Cr g^{-1} dry weight. In the case of *B. megaterium,* similar outcomes were attained (15.7 and 30.7 mg Cr g^{-1} dry weight by living and dead cells, respectively). Inactive/dead cells perform better than active/living cells because they are more vulnerable to the harmful effects of metal ions, which can cause cell death during the metal removal process (Jobby et al. 2018). In comparison to other bioremediation methods, biosorption processes is reported to have various advantages because the metal is binding to the various multifunctional uniformly distributed binding sites on the cell surface; additional nutrients or chemicals are not required; simple and low cost to implement with high efficiency and re usability of the biosorbent. Additionally, the potential for metal recovery has drawn much research into the use of diverse biomass, including bacteria, fungus, and microalgae, for the removal of HMs, particularly Cr(VI) (Ayele and Godeto 2021). The following are additional benefits and drawbacks of non-living biomass (Modak et al. 1996; Ahluwalia and Goyal 2007):

- **Advantages**
  - Because dead biomass is growth independent, toxicity of cells and physiological constraints are not a limitation. So the problems related to nutritional requirements for optimised growth, aseptic conditions and disposal of by-products are not present. A wider range of operating conditions are possible in terms of pH, temperature, initial metal concentration etc.
  - Biomass in this case can be easily procured from several industries, biomass is essentially a waste for fermentation sectors.
  - Non-living biomass also behaves as an ion exchange, so the entire process is fast and efficient because of high metal loading capacity and desorption (recovery) abilities.
- **Disadvantages**

  Another drawback of employing dead biomass is that metal desorption is required before the biosorbent can be used again due to early saturation of the metal interaction sites.

  Because the cells are not metabolically active, any potential for biological process improvement like genetic engineering is limited.

  Also there is no biological control over the characteristic of the biosorbent or in altering the metal ionic state.

13.13.4 Biotransformation/Bioreduction

"Bioreduction" is a potential method for decreasing the level of Cr(VI) contamination. It involves conversion of Cr(VI) into Cr(III) using living systems (Wang et al. 1989; Jobby et al. 2018). Cr(VI) is 10–100 times more harmful than Cr(III) since it is a known carcinogen, a strong oxidant, and has a higher bioavailability in ecosystems (Costa 2003; Chang et al. 2019). The mechanism of biotransformation and reduction of Cr(VI) to lesser toxic Cr(III) has been thoroughly investigated in bacteria as compared to fungi, yeasts or actinomycetes (Elahi et al. 2022). As discussed previously, microbes are compelled to adopt a variety of strategies for their own survival under chromium stress. Cr tolerance/resistance and reduction are two independent phenomena employed by microbes in order to combat Cr(VI) stress. Strains that are able to resist Cr(VI) may not necessarily have the molecular capacity to reduce it also (Elahi et al. 2022). Bacterial strains with the ability to reduce Cr(VI) are popularly known as chromium-reducing bacteria (CRB) (Elahi and Rehman 2019b; Elahi et al. 2022). Various CRBs have been isolated from Cr(VI) contaminated soils (Karthik et al. 2017; Chang et al. 2019; Li et al. 2020, 2021), tannery sewage waste water (Elahi et al. 2022), and industrial effluents (Baldiris et al. 2018). Due cell membrane's impermeability to Cr(III) complexes, biotransformation mechanism of conversion of Cr(VI)–Cr(III) has been regarded as a potential solution for treatment of polluted wastes and reduce Cr toxicity in the environment (Karthik et al. 2017; Chang et al. 2019). Contrary to some metal ions in wastewater, such as Cu^{2+}, Cd^{2+}, Pb^{2+}, and Ni^{2+}, which can only be eliminated by biosorption, certain microorganisms can also detoxify Cr(VI) by reducing it to the less dangerous Cr(III) (Karthik et al. 2017; Chang et al. 2019). This method offers better prospects as a bioremediation process for Cr(VI) detoxification since Cr(VI) can be completely removed by microbial reluctant so after a certain operating time but still, the adsorption sites on these biosorbents are limited in number and prone to saturation (Vijayaraghavan and Yun 2008; Jobby et al. 2018; Chang et al. 2019). Until recently majority of studies had focused on Cr(VI) tolerance levels and basic bioreduction capabilities. Relatively, little is known with regards to the Cr binding sites involved in biosorption, intercellular accumulation and extracellular precipitation (Karthik et al. 2017). Moreover, these studies were mainly dependent on an indirect method of Cr(VI) elimination and did not consider Cr(III) compounds in their bioreduction evaluation. The validity of this methodology may be questioned in future (Baldiris et al. 2018). Because both reduction and adsorption can be used to eliminate Cr(VI), a basic examination of Cr(VI) elimination can scarcely confirm the true Cr(VI) reducing activity, and the valence state of the reduced Cr must be determined directly (Karthik et al. 2017; Baldiris et al. 2018). Cr(VI) tolerant Pseudomonas sp. DC-B3 isolated from a contaminated mine-soil demonstrated a strong ability to reduce Cr(VI) to less harmful (III) without any exogenous electron donor at pH 2. With increasing Cr(VI) concentration, both the reduction capacity and reduction rate increased linearly, achieving a reduction capacity of 32.0 mg Cr(VI)/g over a 75 h period at an initial concentration of 135.0 mg/L (Chang et al. 2019).

Following steps are involved in intracellular Cr(VI) reduction (Fig. 13.5):

1. **Cr(VI) biosorption on the surface of the cell**—As indicated in the previous sections, Cr(VI) ions form chemical bonds with bacteria's cell surface by making use of functional groups such amide, alkane, and amines.
2. **Cr(VI) transport**—Because there are no transport channels for Cr(VI) ions to enter the cells, they use SO_4^{2-} and phosphate channels instead due to their structural similarities (Mala et al. 2015; Elahi et al. 2022; Wang et al. 2017; Pushkar et al. 2021). Since Cr(VI) and SO_4^{2-} ions share a significant degree of similarity, Cr(VI) can pass through cellular membranes conveniently due to active sulphate transporters (Cervantes et al. 2001; Ayele and Godeto 2021). Once inside the cell, Cr(VI) is reduced to Cr(III), which is ultimately released from the cell (explained later). Cr(VI) intracellular reduction promotes chromate accumulation in the extracellular environment and ensures a low cytosolic concentration (Joutney et al. 2014).
3. **Cr(VI) reduction**—(1) Intracellular enzymatic reduction: The intracellular levels of Cr(VI) is reduced to insoluble Cr(III) by cytoplasmic molecules enzymatically or non-enzymatically (Thatoi et al. 2014; Gutiérrez-Corona et al. 2016; Singh et al. 2020); (2) Extracellular reduction by secreted enzymes: The cell benefits from this process because it saves energy by not having to carry Cr(VI)

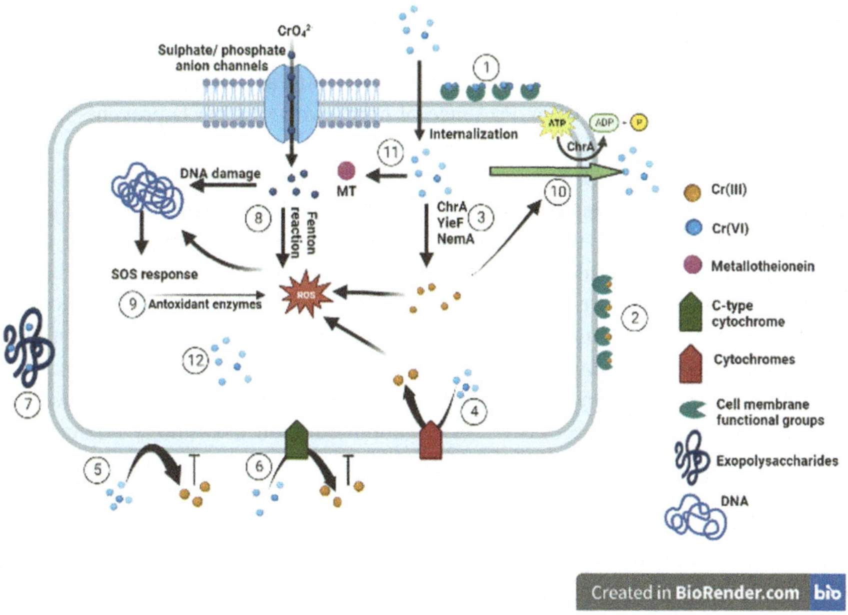

Fig. 13.5 Schematic diagram to illustrate various mechanisms employed by bacteria in response to Cr(VI) stress: (1) Biosorption of Cr(VI); (2) Biosorption of Cr(III); (3) Intracellular enzymatic reduction of Cr(VI)–Cr(III); (4) Reduction of Cr(VI)–Cr(III) by cytochrome in cytoplasm; (5) Cr(VI) reduction by cell membrane; (6) Cr(VI) reduction by c-type cytochrome under anaerobic conditions; (7) Cr(VI) entrapment by EPS; (8) Fenton reaction and SOS cellular response due to Cr(VI) stress with subsequent generation of ROS leading to cell damage; (9) Activation of antioxidants in as a defense response to ROS; (10) energy dependent efflux of Cr(VI) from cell; (11) complexation of Cr(VI) with metallotheioneins; (12) Bioaccumulation. Adapted from Pushkar et al. (2021)

into the cell and subsequently Cr(III) out of it. For instance, this mechanism for gram negative bacteria is predominated by NADH-dehydrogenase pathway under aerobic conditions. Since this an energy-intensive process, these enzymes are secreted only under Cr(VI) stress (Cheung and Gu 2007; Ayele and Godeto 2021). (3) Non- enzymatic extracellular reduction.

4. **Cr(VI) bioaccumulation**—The Cr(VI) that is bioreduced to Cr(III), then gets bioaccumulated in the cytoplasm (Karthik et al. 2017; Pushkar et al. 2021).

Several factors (directly or indirectly) determine the reduction potential of a bacterial strain including pH, initial concentration of chromate, presence of electron donors as well as co-existence of other metal ions in the samples (Mala et al. 2015; Singh et al. 2020; Elahi et al. 2022). The majority of the microorganisms identified have not been shown to be capable of reducing Cr concentrations by more than 60%. Additionally, the majority of the isolates perform poorly at high Cr loads (Singh et al. 2020). Under chromate stress, essential compounds produced by the bacteria during carbon oxidation for cellular growth gets utilized as electron donors for Cr(VI) reduction

(Karthik et al. 2017; Pushkar et al. 2021). In the presence of oxygen, the reduction process involves the generation of various transient toxic ions such as Cr(IV), Cr(V) and ROS that leads to oxidative stress and cellular damage (Bharagava and Mishra 2018; Elahi et al. 2022). Presence of high levels of Cr(VI) prevents normal cellular proliferation as the bacterial energy is mostly spent in reducing the Cr(VI) ions to lesser toxic forms (Parameswari et al. 2009). Bacteria in log-phase performed better Cr(VI) reduction than in any other phase mostly due to high number of active cells with maximum enzymatic activity (Ikegami et al. 2020). But the time taken form bioreduction would increase with Cr(VI) ion concentration which could be due to saturation of the enzyme (Jeyasingh and Philip 2005). Factors such as loss of microorganisms, toxicity to microorganisms, and uneven microbial growth at high Cr(VI) concentrations hamper the commercial applications of bioremediation. MIT (Microbial Immobilization Technology) is a popular research area today since has the potential to address the drawbacks of bioremediation technology (Jiang et al. 2022). Immobilization of microbial cultures has been shown to increase the stability and efficacy of organisms and to produce greater Cr adsorption/reduction than free organisms. For the purpose of immobilising Cr(VI) reducing or sorbing bacterial cells, matrices such as agar, agarose, polyethylene glycol, polyacrylamide, etc. have been investigated (Hora and Shetty 2016).

Several antioxidant enzymes such as GSH, GSSH and CAT are synthesized as a defensive mechanism to transform harmful ROS into safe compounds, thereby preventing metal generated ROS from altering the reduced environment within the cell (Elahi et al. 2022; Gu et al. 2020). Antioxidant profiling of *Bacillus cereus* b-525 k with or without 2 mM Cr(VI) stress depicted an increase in expression of all AOXs especially peroxidase (POX) with a 99% increase (Elahi et al. 2022). Because they can be produced in response to a variety of environmental stresses, including those caused by metal ions (Cd, Al, Zn, and Cu), drought, water, and gamma radiation, peroxidases are also classified as stress enzymes (Khalid and Jin 2013). The findings of Elahi et al. (2022) were well in line with those of Suthar et al. (2014), who also indicated that Cr(VI) stress causes a significant rise in all antioxidant enzymes. Elahi and Rehman (2019a) have previously emphasized on the importance of glutathione and non-protein thiols in reducing metal generated ROS toxicity. In reducing ROS toxicity, glutathione and non-protein thiols are crucial (Elahi and Rehman 2019a). Bacillithiol (BSH), a thiol molecule present in most Bacillus species, is likely involved in maintaining cellular redox balance and contributes to microbial resistance to several antibiotics, according to a 2009 study by Newton et al. (Elahi et al. 2022). Bacilli thiol (BSH) have been previously reported to in most Bacillus sp. Playing crucial roles in ROS toxicity and resistance to antibiotics (Newton et al. 2009). It was seen that *Pseudomonas brassicacearum* LZ-4 had potential to co-bioremediate naphthalene and chromate. Here, naphthalene was the sole carbon source that tremendously elevated the reduction capacity of the bacteria from 25 to 96.2%. Authors reported the upregulation of catabolic gene NahG gene in the presence of naphthalene that encodes for salicylate hydroxylase along with FAD as cofactor. FAD could be acting as the electron acceptor from NADH for subsequent Cr(VI) reduction (Huang et al. 2017).

13.13.4.1 Enzymatic Reduction

Chromate reduction in several microbial species depends upon the utilisation of Cr(VI) as a terminal acceptor catalysed by chromate reductases enzyme in their respiratory processes converting Cr(VI)–Cr(III) (Lovley and Philip 1994; Singh et al. 2020). Pseudomonas sp. was one of the earliest reported bacteria with Cr(VI) bioreduction abilities under anaerobic conditions (Singh et al. 2020). Later, E. coli was reported to biotransform Cr(VI)–Cr(III) aerobically (Shen and Wang 1993). There are two different kinds of Cr reductases: membrane-associated and intracellular, depending on where the reduction takes place. Numerous investigations have confirmed the existence of intracellular and membrane linked Cr reductase enzyme activities in the cellular membranes, cytoplasmic fractions and cell supernatants (Ilias et al. 2011). For the enzymatic Cr(VI) reduction mechanism in bacteria, either soluble cytosolic proteins or insoluble cell membrane enzymes are responsible for the catalysis. It has been widely documented that a variety of bacterial taxa, including *Pseudomonas, Bacillus*, and *Arthrobacter*, can reduce Cr(VI) enzymatically in either an aerobic or anaerobic environment, or even both (Ramírez-Díaz et al. 2008; Thatoi et al. 2014; Viti et al. 2014; Gutiérrez-Corona et al. 2016).

Indigenous species are frequently employed in the bioreduction process since they don't need extra nutrients to survive and proliferate when used in scale-up applications. This is a practical method that is affordable, secure, and generates no extra byproducts (secondary pollution). After the quick initial identification with 16S rRNA gene sequencing, the second stage in bioremediation studies is study of genomics to identify the enzymes that are involved in the reduction process (Baldiris et al. 2018). More recent reports have also determined the Cr(VI) reduction mechanism mediated by reductases present in microbial culture supernatant. Many chromate reductases such as ChrR, YieF, NemA and LpDH catalyse the bioreduction reaction by utilising electron donors like NAD(P)H, mediating the transfer of electrons to Cr(VI) and at the same time generating ROS (Reactive oxygen species) in two-step process known as Class I ("tight") and Class II ("semi-tight") (Thatoi et al. 2014). In comparison to membrane-associated chromate reductase enzymes, reductases that are soluble are better suited for protein engineering as they suit the environmental circumstances of contaminated sites. This makes them suitable for the development of biocatalysts for bioremediation (Thatoi et al. 2014; Baldiris et al. 2018). High extracellular chromate activity was reported for *Bacillus amyloliquefaciens* under optimised conditions by Rath et al. (2014), Gutiérrez-Corona et al. (2016). In a co-remediation study of pollutants chromate and pentachlorophenol, chromate reductase activity was reported in cytosolic fraction (48%) followed by culture supernatant (39.7%) and cell debris (12.3%) (Tripathi and Garg 2013; Gutiérrez-Corona et al. 2016). Baldiris et al. (2018) demonstrated the cytosolic nature of the chromate reductase responsible for chromate reduction in strain S. maltophilia. Their report was contradictory to Blake et al. (1993) who reported a membrane associated reductase responsible for chromate reduction in the same bacteria. Another study on *Bacillus sp.* TCL have reported the constitutive expression of membrane associated chromate reductase and loosely bound EPS as sites for Cr(VI) reduction or Cr(III) immobilisation respectively (Banerjee et al. 2019). Authors correlated the enhanced activity by membrane fractions with increased expression of membrane-bound reductases under chromate stress. The shifts in FTIR peaks to amino, carboxyl and nitrogen and oxygen of peptide bonds further suggested a protein (chromate reductase) mediated metal binding on cell surface under chromate stress. Numerous studies showed that EPSs and cytoplasmic extracts both contributed to the decrease of Cr(VI) by *Bacillus sp.* (Pan et al. 2014; Das et al. 2014; Li et al. 2019a, b).

The most crucial component for overcoming the difficulties in biodegradation and bioremediation of pollutants is the isolation of microorganisms from contaminated locations (Tang et al. 2021). Since the majority of bioreduction reactions are mediated by enzymes, variations in temperature and pH may have considerable effects on the rate of ionisation, the folding of proteins, and the activity of enzymes (Zhang and Li 2011). Although several research have focused on identifying the Cr(VI) reduction sites of various bacterial strains, the number of microorganisms and the intricate mechanisms involved in Cr(VI) reduction make this effort far still from sufficient. Li et al. (2019a, b) determined the Cr(VI) reduction sites of Cr tolerant *Bacillus sp.* M6 by comparing reduction rates in permeable cells (without phospholipid bilayer) and untreated cells as control. The permeable cells exhibited higher Cr(VI) reduction ability than intact cells, which implicated the involvement of cytoplasm of bacillus sp. M6 in the reduction process. Their results were in line with previous reports on *Planococcus maritimus* VITP21 and *Bacillus sp.* G1DM22 by Sangeetha et al. (2012) and Desai et al. (2008a, b) respectively. In both the studies, Tritonx-100 treatment was used to dissolve the phospholipid bilayer of the cell membranes, that subsequently released the reductive substances from the cytoplasm resulting in higher Cr(VI) reduction rate by the respective bacteria. In addition to cytoplasmic extracts, cell envelope extract also showed higher reduction rate when compared to untreated cells (Li et al. 2019a, b). Cell membrane and cytoplasm were both involved in Cr(VI) reduction and their potentials were similar. Cell membrane reductase involved the sulfate channels that transported Cr(VI) into the cytoplasm due to structural similarity of chromate and sulfate anion; and also reduced Cr(VI)–Cr(III) during the transport at the cell envelope (Li et al. 2019a, b). Interestingly, *Providencia sp.* reduced Cr(VI) mostly in the cell cytoplasm (Thacker et al. 2006), but *Thermus scotoductus* SA-01 reduced Cr(VI) primarily in the cell membrane (Opperman and Van Heerden 2007). Researchers have shown that Cr(VI) tolerance and reduction are two distinct mechanisms. Latter is detoxification of Cr(VI) and is usually not plasmid-related (Cervantes et al. 2001; Baldiris et al. 2018). The mechanism of Cr(VI) reduction varies among microbial strains depending upon their bio-geochemical activities and nutrient utilization patterns (Megharaj et al. 2003). According to Dhal et al. (2013) three most common reduction patterns are-(1) Aerobic reduction involving soluble chromate reductases that utilizes NADP or NADPH as cofactors (Park et al. 2000); (2) Anaerobic reduction that uses Cr (VI) as terminal electron acceptor in the electron transport cycle (Tebo and Obraztsova 1998) and (3) reductions brought about by chemical processes involving substances located within or extracellularly, such as glutathione, amino acids, nucleotides, vitamins, carbohydrates, or organic acids (Myers et al. 2000).

Aerobic Reduction of Cr(VI)

When oxygen is present, the bacterial Cr(VI) reduction process transforms into a two- or three-step process, primarily starting with the reduction of Cr(VI) species to the transient intermediates Cr(V) and/or Cr(IV) before being further reduced to Cr(III), which is known to be a thermodynamically stable end product. Here in this reduction procedure, the electron donors are NADH, NADPH and those from the endogenous reserves. The Cr(VI) reductase ChrR mentioned above reduces C(VI) to form Cr(V) followed by Cr(III) involves one-electron and a two-electron shuttle respectively (Lovley 1993). On the other hand, the enzyme YieF is unique in that it catalyses the direct reduction of Cr(VI)–Cr(III) through a four-electron transfer, where three electrons are used to reduce Cr(VI) and the fourth is transferred to oxygen (Ackerley et al. 2004). In the cytoplasm, typically aerobic reduction occurs. Aerobic Cr(VI) reduction is carried out by soluble enzymes such as flavoprotein, dehydrogenase, NADH-dependent nitroreductase, and azoreductase (Chai et al. 2019; Dong et al. 2018). *Pseudomonas sp.* GT7 was tested for four electron donors for Cr(VI) reduction (Zhang et al. 2016a, b). NADH and NADPH enhanced the Cr(VI) reduction by the soluble fraction of GT7. While NADH and NADPH portrayed similar stimulation levels; the effects were stronger as compared to citrate and succinate. Their results were in agreement with previous reports on the cytoplasmic fractions of others bacterial strains such as *Pseudomonas sp.* G1DM21 (Desai et al. 2008a, b) and T. scotoductus SA-01 (Opperman et al. 2008). Authors further suggested the to explore electron donors that are cheaper such as glucose and fructose as the ones used in their study were expensive (Zhang et al. 2016a, b).

The process of Cr(VI) resistance in *Ps. aeruginosa* has been attributed to increased or decreased efflux of Cr (VI) ions through the membrane (Nies and Silver 1995). A close relative of *Ps. Synxantha* with ability to reduce Cr (VI) was reported by Gopalan and Veeraman (1994). In contrast to the previously stated bacteria, which use reductases soluble in the cell cytoplasm, *Ps. Maltophilia* O-2 and *B. megaterium* TKW3 were shown to use reductases linked with membrane cell fractions. Several studies have reported the purification of Cr (VI) reductases from *Pseudomonads*. Also, Ishibashi et al. (1990), partially purified soluble Cr (VI) reductases from *Ps. Putida* PRS2000. Similar study was reported by Suzuki et al. (1992) from *Ps. ambigua* G1. Another soluble Cr (VI) reductase called ChrR was purified by Park et al. (2000) from *Ps. putida MK1* strain. On the basis of the amino acid sequence of purified ChrR protein, gene coding ChrR, *chrR* was recognized by Park et al. (2001). They also presented an open reading frame (ORF) *yieF* having high homology to *chrR*, ChrR showing most optimum reduction at 35 °C. ChrR was further acknowledged as a dimeric flavoprotein catalysing the reduction of Cr(VI) primarily at 70 °C (Ackerley et al. 2004). *Oceanobacillus oncorhynchi* W4 relied on biological reduction as the method of Cr(VI) removal than biosorption and the process was enhanced by addition of electron donors like glycerine followed by NADH, glucose and lactate (Zeng et al. 2019). A model strain of dissimilatory metal reduction bacteria, *Shewanella oneidensis* MR-1, has the ability to reduce Cr(VI)–Cr(III) either aerobically or anaerobically (Gang et al. 2019). This strain is reported to utilise endogenous electron donors such as NADH under aerobic conditions. When conditions are anaerobic, electrons are acquired from donors like membrane-associated cytochromes involved in the electron transport system (Gang et al. 2019). Supply of exogenous i.e. use of external donors such as glucose, lactose, sodium acetate, glycerol have also been reported to enhance the Cr(VI) reduction abilities of bacteria (Pushkar et al. 2021). Glycerol was reported to be an efficient electron donor for Cr(VI) reduction in *Bacillus sp.* M6 by Li et al. (2019a, b). However, under anoxic conditions, lactate acts as a spectacular electron donor for reduction (Huang et al. 2019). Furthermore, ChrT protein decreases Cr(VI) via using NAD(P)H, with a preference for NADPH > NADH > non-NAD(P)H (Gu et al. 2020). Although E. coli FACU displayed a decreased level of NADH (Mohamed et al. 2020). In both aerobic and anaerobic environments, Gram negative *Ps. aeruginosa, Serratia marescens, Alcaligenes faecal*, and *Klebsiella oxytoca* reduce Cr(VI) when Fe(II) and Fe(III) are present (Bansal et al. 2019). Humic acid or Anthraquinone-2,6-disulfonate (AQDS) improved the ability of *Aeromonas hydrophila* ATCC 7966 to reduce Cr(VI). As evidenced by the MtrC deleted mutant's inability to respond to AQDS dosage, the respiratory circuit played an important role in HM reductions (Huang et al. 2019). Thus, it concluded that NADPH, carbon sources, $K_2Cr_2O_7$ and K_2HPO_4 positively impacted the Cr reductase enzyme activity while it was negatively affected in the presence of NaCl, nitrogen sources, temperature and Ni (Banerjee et al. 2019; Ma et al. 2019a, b). As per several reports, after the translocation of Cr(VI) into the cell, bioreduction of Cr(VI)–Cr(III) can be depicted as following reactions: (Suzuki et al. 1992; Pushkar et al. 2021).

$$\mathrm{Cr(VI)} + 1\mathrm{e}^{-1} = \mathrm{Cr(V)} \tag{13.3}$$

$$\mathrm{Cr(V)} + 2\mathrm{e}^{-1} = \mathrm{Cr(III)} \tag{13.4}$$

Overall bio reduction of Cr(VI) to insoluble Cr(III) hydroxide reaction can be shown.

as:

$$\mathrm{CrO_4^{2-}} + 8\mathrm{H}^{+} + 3\mathrm{e}^{-1} = \mathrm{Cr(III)} + 4\mathrm{H_2O} \tag{13.5}$$

$$\mathrm{Cr(III)} + 4\mathrm{H_2O} = \mathrm{Cr(OH)_3} + 3\mathrm{H}^{+} + \mathrm{H_2O} \tag{13.6}$$

Anaerobic Reduction of Cr(VI)

The abundance of anaerobes with Cr(VI) reducing abilities offers excellent potential for in situ bioremediation of contaminated sediments. This only needs an additional supply of nutrients and some modifications to the existing physical environment in order to achieve the desired results (Romanenko and Koren'Kov 1977). When oxygen is absent, Cr(VI) acts as a terminal electron acceptor in the respiratory chain for a wide range of electron donors such as proteins, fats, carbohydrates, hydrogen, and NAD(P)H (Cheung and Gu 2007). The cytochrome families (like cyt b and c) were recurrently shown to be implicated in the enzymatic anaerobic Cr(VI) reduction. In anaerobic reduction process, the electrons generated by ubiquinone reduce cyt c (via cyt b) that subsequently gets oxidised to reduce Cr(VI)–Cr(III) extracellularly (Gang et al. 2019). Cr(III) then remains attached to the cell surface bound to various functional groups or is released in the surrounding environment. SRB have been extensively studied for Cr(VI) reduction. A number of Cr(VI) reducing anaerobes have identified including *B. Cereus, Ps. aeruginosa B. subtilis, Ps. ambigua, Ps. fluorescens, Micrococcus roseus, Desulfovibrio desulfuricans and D. vulgaris* (Cheung and Gu 2007). *Desulfovibrio vulgaris* was found to be involving soluble c3 cyt for Cr(VI) reduction (Turick et al. 1996). The process of Cr(VI) reduction under anaerobic situation was reported to be the result of both soluble and membrane-linked enzymes. Equation (13.5) below shows the total Cr(VI) reduction under anaerobic conditions using glucose as a carbon source. In an aqueous solution, the Cr(VI) is changed into the insoluble Cr(III) hydroxide. Other powerful electron donors can replace glucose, which makes the reduction process easier (Singh et al. 2011a).

$$C_6H_{12}O_6 + 8CrO_4^2 + 14H_2O + 3e^{-1} = Cr(OH)_3 + 10OH^- + 6CHO^- \quad (13.6)$$

13.13.4.2 Non-enzymatic Reduction

The non-enzymatic pathway for Cr(VI) reduction is carried out by chemical pathways in association with microbial metabolic compounds present in intra/extracellularly such as amino acids, sugars, antioxidants or nucleotides (Cervantes et al. 2001; Dhal et al. 2013; Gutiérrez-Corona et al. 2016). For instance, ascorbate can reduce Cr(VI). Also, riboflavin derivatives FAD and FMN are important coenzymes that can reduce chromate (Cervantes et al. 2001). *Microbacterium sp.* CR-07 bacterial supernatant was tested to reduce Cr(VI) which turned out to be was unaffected in the presence of 1% SDS solution, hot water, or pH value, showing that reaction was non-enzymatic. But glutathione was found in the supernatant in addition to the absence of reducing sugar showing that the reduction was caused by glutathione (Liu et al. 2012). Extracellular EPS caused highest Cr(VI) reduction rate in *Pseudochrobactrum saccharolyticum* LY10 (Long et al. 2019). Also, the microbial biomass in the form of other metal ion pollutants such as Fe(II) or H_2S, or organic molecules such as intracellular thiols or EPS can be utilized for non-enzymatic Cr(VI) reduction pathways

(Dwisandi et al. 2021). Moreover, Fe(II) and H_2S are the anaerobic byproducts of iron and sulfate reducing bacteria that can effectively reduce Cr(VI) individually or together under certain circumstances (Gutiérrez-Corona et al. 2016; Dwisandi et al. 2021). In *Pseudomonas stutzeri* KC, direct Cr(VI) reduction has been shown by means of siderophore pyridine-2,6-bis (thiocarboxylic acid) (pdtc) that could effectively reduce 86% of Cr(VI). The by-products of pdtc hydrolysis also reduce Cr(VI) (Zawadzka et al. 2007; Gutiérrez-Corona et al. 2016).

13.13.5 Transmembrane Efflux of Chromate

When Cr(VI) enters the cytoplasm, its interactions at the molecular level and how Cr exposure leads to cellular apoptosis, mutagenicity and carcinogenicity via oxidative stress pathways, DNA aberrations and epigenetic modifications have been extensively reviewed (Monga et al. 2022a). In order to overcome the stress, certain bacteria have evolved an active efflux mechanism as part of their cellular metabolic functions to expel harmful Cr ions into the periplasm or extracellular environment. Overexpression of these proteins thus is one of the ways of circumventing the Cr(VI) (Saba et al. 2019a, b; Mushtaq et al. 2022). Additionally, these efflux pumps are also used by bacteria to carry out other cellular functions such as in maintaining cell homeostasis, acquiring resistance to antibiotics, heavy metals, and salts, and surviving in harsh environments (Cánovas et al. 2003; Pal et al. 2020). Efflux of Cr(VI) has been documented as one of the key resistance mechanisms in various bacteria (Mushtaq et al. 2022) which is mostly conferred by ChrA protein (Ramírez-Díaz et al. 2008; Dong et al. 2018; Pushkar et al. 2021). ChrA proteins were first identified from *P. aeruginosa* and *Cupriavidus metallidurans* in relation to the efflux mechanisms. Afterwards, many of them were identified based on genome sequence analysis and grouped into a large Chromate ion transporter (CHR) family (Díaz-Pérez et al. 2007; Baaziz et al. 2017). They were further subdivided on the basis of protein lengths-short chain CHR (180 aa long) and long chain CHR (400 aa long). The several transmembrane regions of ChrA protein can be encoded by genes present on plasmid or chromosomes (Baaziz et al. 2017).

In *Pseudomonas aeruginosa*, ChrA is encoded by plasmids pUM505 (Cervantes et al. 1990; Ramírez-Díaz et al. 2008), comprises of 416 amino acid sequence and a structural configuration of 13 transmembrane segments (TMS) (Jiménez-Mejía et al. 2006; Ramírez-Díaz et al. 2008). It performs as an chemiosmotic pump to efflux Cr(VI) out of the cytoplasm with a proton motive force (Mushtaq et al. 2022). The efflux mechanism in Pseudomonas PAO1 was linked to NADH oxidation and ChrA had a Km of 0.12 mM Cr(VI), a Vmax of 0.5 nmol Cr(VI)/min per mg of protein. ChrA gene is usually found on plasmids or sometimes on chromosomes along with other genes such as ChrB, chrC, chrE, chrF in some bacterial species (Viti et al. 2014; Mushtaq et al. 2022). ChrA protein generates hydrogen ions leading to an electrochemical proton gradient across the cell membrane that expels the Cr(VI) from the cytoplasm (He et al. 2018). ChrC is involved in decreasing oxidative stress, whereas

ChrB is a regulator of Cr(VI) detection. Contrarily, ChrE has a role in facilitating the breakdown of some chromate-glutathione complexes (Viti et al. 2014; Mushtaq et al. 2022). The ChrB gene has a favourable regulatory effect by making the ChrAB protein more capable of metabolising chromate than the ChrA protein is, despite the fact that the ChrB protein cannot transport chromate from cells (He et al. 2018; Chen and Tian 2021). *Shewanella oneidensis* strain deleted of ChrA gene (genes belonging to large family of chromate ion transporters) showed lesser resistance to Cr(VI) than its wild type strain (Baaziz et al. 2017). Expression of ChrA in E.coli made the bacteria capable of resisting Cr(VI) stress and growing in the presence of high Cr(VI) concentrations. This has been reported in literature previously on plasmid encoded ChrA of *Shewanella sp.* and for multiple ChrA of *Burkholderia xenovorans* LB400 (Acosta-Navarrete et al. 2014; Baaziz et al. 2017). The pARI180 plasmid DNA carrying the respective gene was transformed into E. coli DH5 α strain that made the bacteria Cr(VI) resistant, but after the plasmid was lost E. Coli lost its resistance and became sensitive to Cr(VI) (Dhakephalkar et al. 1996; Chen and Tian 2021). Efflux of Cr(VI) can occur together with other harmful molecules. Many of the Cr resistant bacteria reported are also tetracycline resistant because both of them are transported of the cell using active efflux pumps (Pushkar et al. 2021). A Cr(VI) tolerant *Bacillus* strain TCL could effectively transport Cr(VI) and ethidium bromide (EtBr) out of the cell in order to reduce the intracellular stress (Banerjee et al. 2019; Pushkar et al. 2021). Additionally, the efflux mechanism is concentration dependent reaction since it is driven by an energy-dependent chemiosmotic homeostasis (Shaw and Dussan 2018; Pushkar et al. 2021).

13.14 Challenges in Developing Cr Bioremediation Technology

- Microbial remediation of Cr(VI) has been a subject of research for several decades now, and tremendous progress has been made so far. But still several questions needs answers especially in context of molecular cellular responses in the presence of the metal and their use in designing Cr bioremediation from the environment. For instance, *Ps. aeruginosa* PA01 has shown the involvement of oprE (responsible for outer membrane expression), rmlA (for cell LPS expression) and ftsK (cytoplasm) in tolerance to Cr (Rivera et al. 2008). But the roles of these genes are not yet documented for Cr resistance in detail. Similarly, other important genes with Cr reduction abilities but less explored are ChrT and YieF (Gu et al. 2020). Furthermore, as compared to gram negative, not many gram-positive bacteria has been reported with Cr bioremediation. There exists a clear gap in knowledge of the differences in the mechanisms that Gram positive and negative bacteria use for Cr resistance (Shaw and Dussan 2018).

- As the Cr concentrations increased, the rates of Cr(VI) elimination and reductions dropped, according to past researches on Cr(VI) bioremediation. This results from the strained metabolic pathways and impaired biological functions of the cells under severe Cr stress (Pan et al. 2014; Akkurt et al. 2022). At 200 and 300 mg/L Cr(VI) concentrations, *Bacillus sp.* CRB-B1 eliminated 86.15 and 43.1% of Cr(VI) respectively, although the reducing activity was impaired at concentrations higher than 300 ml/L. (Tan et al. 2020). Similar observations were reported by Huang et al. (2021) where *Sporosarcina saromensis* W5 strain which a novel facultative anaerobe, showed gradually decreased removal efficiency with increasing Cr(VI) levels. *Cellulosimicrobium sp.* isolated from leather industry wastewater could efficiently remove Cr(VI) up to 100 ml/L in 96 h. However, at higher concentrations of 200 and 300 mg/L the reduction ratio reduced drastically after the same amount of time (Bharagava and Mishra 2018).
- Due to lack of proper regulations, mixing of industrial wastes further complicates the problem. Varying pH affects the normal microbial growth cycle and makes the maintenance of an active and functional microbial population a challenge. In addition, the metabolic byproducts also may form complexes with the metal ions that may further complicate the desorption processes (Singh et al. 2020). Due to these complications researchers are now opting for non-living biomass, although research continues on re-usage and proper disposal methods for fully loaded adsorbents (Babangida et al. 2021).
- Biostimulation was suggested as a modification to stimulate the living biomass during bioremediation (Pradhan et al. 2017) by adding electron donors such as acetate, lactate etc. While their efficiency in Biostimulation also depended on the indigenous microbial community structure and physicochemical properties of the site; the continuous supply of nutrients was a challenge. Several reducing agents such as SO_2, H_2S, metallic Fe etc. that showed promising results in Biostimulation also created secondary pollution (Babangida et al. 2021).
- It has been emphasized in literature that HM stress will trigger bacterial EPS formation because EPSs act as the first defense of bacterial cell, preventing metals from the outer environment from interacting with essential cellular components. The potential of EPS to remove HMs from polluted environments has been comprehensively documented in the literature, with a primary focus on its biotechnological potential (Zeng et al. 2020). However, knowledge about the effects of HMs on EPS production as well as the correlation between EPS production and HM resistance in bacteria is still limited, particularly for exposure to different metals.
- The vulnerability of microorganisms to other toxins and environmental stressors present at the treatment site is one of the main bottlenecks in the bioremediation process. The majority of bioremediation research focuses on isolating Cr(VI) resistant bacteria and evaluating their bioremediation effectiveness in controlled lab settings. Therefore, future study must concentrate on the utilisation of microorganisms that can remove chromium from the actual polluted locations (Pushkar et al. 2021).

13.15 Strategies to Develop Cr Bioremediation Technology

13.15.1 Microbial Immobilization Technology (MIT)

In order to overcome issues in bioremediation treatment such as loss of microorganisms, loss of metabolic activity and toxicity at higher Cr concentrations and uneven microbial growth cycles can be overcome by MIT by chemically or physically confining the microbial cells or other biocatalysts in a specific area in the system and increase the microbial metabolism active for a longer time duration (Kathiravan et al. 2010; Jiang et al. 2022). Several carrier types has been devised over the years such as inorganic carriers (biochar, activated charcoal, diatomite etc.), organic carriers (alginate, agar, chitosan etc.), composite carriers (combination of polyvinyl alcohol and sodium alginate), and new carriers (modified carrier materials, nano materials) (Jiang et al. 2022). Biochar is an emerging material due to its large specific area, rich pore structure and functional groups for efficient absorption of microorganisms (Lehman and Joseph 2015). For instance, Zhu et al. (2021) compared the efficiencies of free (SRB6-2-1) and immobilized SRB IBXM700 using wheat straw biochar to treat Cr(VI) polluted wastewater. IBXM700 had a maximum removal efficiency of Cr(VI) of 286.54 mg g^{-1}, which was 166.3 and 30.8 mg g^{-1} greater than free SRB6-2-1 and XM700, respectively (Zhu et al. 2021; Jiang et al. 2022). Li et al. (2020) applied different formulations in the immobilization of strains *Bacillus cereus* D and *Bacillus cereus* 332 to compare the efficiencies of Cr(VI) reduction. The strongest reduction was achieved when sodium alginate (SA) was used to immobilise *B. cereus* D (66.9%) in 120 h. However, the immobilised beads of *B. cereus* 332 using SA with diatomite achieved a higher reduction rate of 88.9% in 72 h. Notably, the diatomite increased the hardness of the immobilised beads as compared to beads made with only SA that were not very hard and easily broke (Li et al. 2020). A similar study with SA was previously carried out by Samuel et al. (2013) to immobilise Acinetobacter johnsonii, E. coli, and B. subtilis for the removal of Cr(VI) from sewage water. The maximal Cr(VI) adsorption capacity (657 mg g^{-1}) was demonstrated by bacteria immobilised in the reactor using SA. The clearance rate of Cr (VI) after five adsorption and desorption tests was 74.22%. Also researchers frequently modify their composition and create a composite carrier to complement each other in order to further improve the effectiveness of bioremediation and circumvent the limitations with inorganic and organic biosorbent in a hostile environment in the actual treatment process. For instance, carbon nanotubes and SA were employed as a composite biosorbent for immobilisation of *Shewanella oneidensis* MR-1. They observed that the beads added with carbon nanotubes enhanced the Cr(VI) reduction by four times as compared to free/unfixed cells. The stability and reusability of the micro beads were both significantly enhanced by the inclusion of carbon nanotubes (Yan et al. 2013). More recently, in order to obtain a specific performance new carrier can be modified for the number of oxygen-containing functional sites, surface area, pore structure etc. to further improve the adsorption performance and immobilisation (Huang et al. 2015; Jiang et al. 2022). Modified biochar carrier materials has been recently reported using

Iron (Fe) and Zn (Jiang et al. 2022). The electrostatic interactions with the positively charged biochar enhanced the chemical adsorption capabilities of biochar, thereby improving the contact probability of microorganisms with Cr(VI) (Sun et al. 2019c). In order to devise a sustainable bioremediation technology, the recovery and regeneration of immobilized microorganisms and the carrier while removing Cr(VI) is crucial to avoid secondary pollution (Jiang et al. 2022). Researchers have tried using magnetically modified materials as immobilisation carriers and found excellent stability and reusability (Wang et al. 2021; He et al. 2020). Commercial technologies based on MIT approaches are few. AMT-BIOCLAIMTM is a commercially available product that contains immobilised *B. subtilis* cells on polyethyleneimine and glutaraldehyde. BioFIX is yet another method that has been developed. It uses a variety of biomasses like Sphagnum peat moss, algae, yeast, bacteria, and/or aquatic flora bound to a high density polysulfone. Elution cycles of more than 100 are achievable (Wang and Chen 2009; Singh et al. 2020).

13.15.2 Genetic Engineering

Most researches have focused on isolating indigenous/wild bacteria from contaminated sites for their potential use in the bioremediation processes as they are already adapted to complex environmental conditions. However, selective binding of metals and ability to remove them from polluted environments is lacking in these wild bacteria (Singh et al. 2011b; Ayangbenro and Babalola 2020). Now, it is widely accepted that molecular biology and genetic engineering of indigenous strains has a better potential application in designing bacteria for remediation tasks. They have been successfully demonstrated to have better remediative abilities, selective removal and metal binding capacity than wild type strains for degradation of pollutants under defined conditions (Singh et al. 2011b; Akkurt et al. 2022). Al Hasin et al. (2010) reported a genetically manipulated methanotroph *Methylococcus capsulatus* that could bioremediation Cr(VI) over a wide range of concentrations. In another study, Valls et al. (2000) in order to boost its affinity to metal ions, manipulated a mouse MT protein to be expressed onto the surface of the cell of a HM tolerant *Ralstonia eutropha* CH34 which was already adapted to survive in HM polluted soils. MTs are low-molecular-weight proteins that aid in detoxify HMs, protects cells from the oxidative damage by scavenging free radicals due to their high think content. Two human MT genes, MT2A and MT3, were recently cloned into E. Coli Jm109 by Akkurt et al. (2022). Due to the expression of the MT gene, which improved the reduction of Cr(VI) compared to wild type, these transformed strains were able to capture Cr ions inside the cells in addition to surface binding.

13.15.3 Enhancement of Bioremediation

13.15.3.1 Microbial Consortium for Bioremediation

Actual heavy metal polluted areas have much different environmental conditions as compared to laboratory conditions, thus the applications of pure cultures kept in sterile conditions in a laboratory are limited in the real world. A consortium of bacteria is more likely to sustain and survive in field conditions due to their competitiveness among each other (Tang et al. 2021). A mixed bacterial consortium attached to phosphate minerals and alginate improved the bioreduction efficiency of Cr(VI) and subsequent removal of Cr(III) (Ma et al. 2019a, b). Benazir et al. (2010) immobilised a consortium of *B. subtilis, Ps. aeruginosa* and *S. cerevisiae* using SA. The consortium demonstrated improved remediation efficiency and decreased Cr(VI) from initial concentration of 770–5.2–5.7 mg/L in the tannery effluent as when compared to individual cultures. In another study, Cr(VI) reducing bacteria *Morganella morganii* STB5 demonstrated improved reduction efficiencies of 70.41 and 68.27% when immobilised on electro spun polystyrene and polysulfone web respectively, beginning from an initial Cr(VI) level of 25 mg/L. These may be incorporated into setups for continuous treatment of Cr-contaminated discharge waters because they were reusable for at least five cycles (Sarioglu et al. 2016). The mixed microbial consortium of *Geotrichum sp.* and *Bacillus sp.* exhibits alternating growth and synergy. They have a significantly higher Cr(VI) bioremediation efficacy (Qu et al. 2018). Arshad et al. (2017) observed that the presence of 5% biochar and the microbial consortium reduced the toxicity of Cr to wheat plants (*Pseudomonas japonica* and *B. cereus*). The availability of hazardous Cr in the food chain was reduced due to the conversion of Cr(VI)–Cr(III), which limited its absorption by plants and resulted in a decrease in Cr toxicity. Such an amendment-based strategy might be useful in the case of significant Cr(VI) contamination of soil habitats. Additionally, it was discovered that this addition enhanced the physicochemical qualities of the soil. Therefore, using a combination of biochar and microorganisms to treat soils that are contaminated with Cr will result in soil conditioning in addition to Cr remediation.

13.15.3.2 Addition of Enhancer

The bioremediation activity can be enhanced by using a wide range of substrates, some of which may serve as nutrition sources or as co-donors of electrons. Also, Tang et al. (2021) have listed several minerals in Cr(VI) polluted sites that can contribute in enhancing the bioremediation under appropriate conditions. Magnetite can act as cytochrome OmcS to enhance extracellular electron transport. In both aerobic and anaerobic environments, Fe(II) could increase the removal efficiency of Cr(VI); on the other hand, Fe(III) displayed an inhibitory effect under anaerobic conditions and high concentrations during aerobic conditions (Bansal et al. 2019;

Tang et al. 2021). Phosphorus minerals were also suggested to be added to increase the removal of Cr(VI) because they could promote the formation of antioxidant enzymes and microbial resistance to Cr(VI). In addition to strengthening the genes involved in reducing Cr(VI), it also increases the capacity to absorb nutrients to lessen cell damage. The negatively charged and rough surface also aids in the removal of Cr(III) (Ma et al. 2020). Sulfur and its compounds can also be used as an electron donor for heterotrophic Cr(VI) reducers like *Desulfovibrio* and *Desulfuromonas*, and volatile fatty acids (VFAs) produced from abiotic sulphur oxidation can also be used to support bioreduction. This is due to the common coexistence of organic compounds and reduced sulphur compounds in groundwater aquifers (Zhang et al. 2020).

13.15.4 Bacterial Biofilm and Sequestration Through EPS for Cr(VI) Removal

A number of strategies have been developed by nature to counter the toxic levels of Cr(VI) in the environment, but nature was at its best when it manifested biofilms (Bhunia et al. 2022). The formation of biofilms, auto-aggregation in response to environmental conditions, and host colonisation are all facilitated by the various surface elements and cell surface derivatives expressed in bacteria, such as flagella, EPS, LPS, etc., in conjunction with various environmental signals, such as quorum sensing (Ghosh et al. 2021). In addition, high sorption capabilities, feasibility and low production cost are some of the main advantages of biofilms documented in HM removal. The positively charged cat ionic HMs accumulated in the environment adsorbs to the negative charges of the biofilms with electrostatic bonds. Thus, biofilms of various microbes can be regarded as efficient adsorbents for HMs (Priyadarshanee and Das 2021). EPS for instance, as mentioned before, acts as a resistance mechanism in HM toxicity such as Cr(VI). The two most commonly found forms of EPS are LMW (low molecular weight fractions) and HMW (high molecular weight fractions). While LMW is an inactive form of EPS produced under regular conditions, HMW form of EPS is produced under stress and the two combined gives rise to biofilms as a protective shield against Cr(VI). The biofilm acts by retarding the diffusion of Cr(VI) within the cellular membrane and helps the bacteria thrive b under stress (Ghosh et al. 2021). A marine bacterium strain called *Pseudomonas chengduensis* PPSS-4 was discovered by Priyadarshanee and Das (2021) from contaminated soils at Paradip Fort in Odisha, India. When compared to free planktonic cells, the bacteria demonstrated a significantly greater uptake of multiple-metals [Pb(II), Cr(VI), and Cd(II)] in biofilm mode. These findings were consistent with a prior work by Black et al. (2014), which found that a biofilm-forming bacterium eliminated Pb(II) at a rate of 83.7% compared to 72.6% with free cells. Contrary to this, in a recent report by Wadood et al. (2021), *Staphylococcus equorum* KS1 and *Staphylococcus equorum* KW1 isolated from contaminated soils and wastewater were more efficient in Cr(VI) reduction in their planktonic form (free cells) as compared to their biofilms in 24 h. Thus, for faster Cr(VI) reduction in wastewater planktonic cells are probably more suited than planktonic forms according to authors. Additionally, these bacteria were isolated from wastewater, and since flowing water and toxic pollutants are a common feature of wastewater environments where bacteria can live, these organisms have evolved to perform more efficiently in this type of environment (Elahi and Rehman 2019a; Wadood et al. 2021). On the other hand, S. equorum KS1 isolated from soil formed the firmest and thick biofilm in both presence and absence of Cr(VI) showing the biofilm forming character of soil-borne bacteria (Wadood et al. 2021).

13.15.5 Microbial Fuel Cells (MFCs) for Eco-Remediation of Cr(VI)

MFCs are another emerging technology to bioremediate Cr(VI) from soils and subsurfaces. A major problem faced in conventional bioremediation strategies is the generation of electron acceptor that reoxidizes the contaminant. Alternatives include inserting MFCs into soil sediments, where the anode will serve as the electron acceptor through biodegradation (Ghosh et al. 2021). MFCs are a unique technology of simultaneously generating renewable energy and Cr(VI) remediation making it an environmentally sustainable approach and has attracted many researchers (Ali et al. 2018; Yu et al. 2022). Dual-chamber of MFC was applied in a study by Tandukar et al. (2009) that completely removed Cr(VI) from polluted wastewater in operating time of 300 h and generated power of 55.5 mW/m^2 where Cr(VI) existed in hydroxide precipitated form. The organic and inorganic components of wastewater were decomposed by microbes in the anode. The electrons generated were transmitted to the cathode through an external circuit that reduced Cr(VI)–Cr(III) in a bio-electrochemical reaction and generated electricity (Tandukar et al. 2009). *Aeromonas, Pseudomomas* and *Thiuomonas* are some of the bacteria that could metabolise anoxic substrates to generate electrons thus, promoting catholic reduction of the target pollutants (Huang et al. 2008). However, still the effective mechanisms, optimizations of different influencing factors and the practical application of MFCs in different contaminated sites are still a long way. Yu et al. (2022) in a recent review have extensively the state-of-the-art experience of using MFCs for eco-remediation of Cr(VI), their performance and challenges associated with the technology. Proton exchange membranes (PEM) for instance, are ver important in maintaining anaerobic environment and in proton migration top cathode chamber. However, PEM needs to be cleaned every 6–7 months to avoid biological fouling which is a laborious process that makes the on-site and long-term application of MFCs a challenge (Xu et al. 2012; Yu et al. 2022).

13.15.6 Nano-Bioremediation (NBR) of Cr(VI): a Green Technology

When a substance is scaled down to the nanoscale, the surface area per unit mass ratio rises; as a result, more of that substance can interact with other particles and impact the levels of reactivity. Also, with NMs lesser activation energy is spent in chemical reactions or in other words, NMs show quantum effect. Another attribute of nanoparticles (NPs) is 'surface plasmid resonance' which can be used to detect toxic metal ions. Various metallic and non-metallic NMs of different shapes and sizes are available for custom environmental remediations: (1) NPs are capable of diffusing into contaminated areas where micro particles cannot; (2) they have stronger reactivity to redox amenable pollutants (Rizwan et al. 2014). Zero-valent iron NPs (nZVI) are one of the most commonly used and effective adsorbents NPs for Cr(VI) remediation from aqueous solutions (Mitra et al. 2017). Nano-bioremediation is a unique combination of nanotechnology and bioremediation. With the use of nanotechnology, this technology uses nano particles developed from prokaryotes (gram-negative rods, actinobacteria, etc.) and eukaryotes (fungi, algae, and plants) (Rajput et al. 2021; Hidangmayum et al. 2022). Le et al. (2015) reported dechlorination and biodegradation of biphenyls using Zn NPs and *Burkholderia xenovorans*. NPs derived from plant extracts such as, *Noaea mucronata* have been reported for the bioremediation of HMs from polluted water bodies (Mohsenzadeh and rad 2012). There have been several reports published on the toxicity of nZVI, however it is still unclear how nZVI might affect the ecosystem. Ravikumar et al. (2016) used biologically (BS-nZVI) and chemically (CS-nZVI) synthesised nZVI to test the cytotoxicity of five native isolated strains and their consortia. Cell membrane damage and a reduction in cell viability were observed. However, it was discovered that BS-nZVI had a less harmful impact on the consortium than CS-nZVI (Ravikumar et al. 2016). Fresh neem leaves (*Azadirachta indica*) extract was used to synthesise NPs in this study. Zhang et al. (2022) biosynthesized palladium nano particles with *Shewanella oneidensis* MR-1 (bio-Pd) under aerobic conditions for the subsequent bioreduction of Cr(VI). They could achieve the smallest average particle size of 6.33 ± 1.69 nm by maintaining a high cell: Pd ratio. The small size and uniform distribution of extracellular bio-Pds could completely reduce 200 mg/L of Cr(VI) within 10 min and also maintained high activity for five operating cycles much higher than commercial Pd/Cs. To overcome the slow electron, transfer rate in conventional wastewater treatment methods, Qian et al. (2022) evaluated the non-enzymatic Cr(VI) reduction mediated by SRB especially by speeding up the electron transfer by in-situ developed FeS- NPs. The Cr(VI) removal rate was one magnitude higher than without FeS NPs in addition to improved reduction efficiency via non-enzymatic reactions with sulfide. The bio-FeS NP@SRB functioned as an electronic bypass that improved the electron flux substantially and switched the reduction process from the cytosolic to extracellular environment, which had a greater detoxifying effect on microbes and eventually stimulated the electron transfer extracellularly and eventual Cr(VI) reduction (Qian et al. 2022).

Use of anti-oxidants as biomaterials for nano bioremediation: Several organic compounds possess strong potential for reducing Cr(VI) ions to less toxic Cr(III) in the presence of phenolic and −OH groups. Cr(VI) readily accepts a proton from the phenolic −OH groups and is converted to Cr(III) and the phenolic group to a quinone subsequently (Babangida et al. 2021). The challenges faced with conventional biosorption processes such as secondary pollution due to overloaded biosorbets are mentioned above. In contrast to the use of chemicals like zero-valent iron and others, the use of antioxidant compounds is currently the subject of intense research for its potential use in Cr(VI) detoxification. But physical state of the system or the bacteria can degrade natural antioxidants. Therefore, advancements in nanoscale technology are essential for protecting antioxidants from other hazardous co-contaminants, undesirable byproducts, and microbes. In addition, ti protection, NPs also provides stability and a controlled release of their contents for a long lasting efficacy (Babangida et al. 2021). Mystrioti (2014) reported on the application of green tea in the fabrication of Zn NPs for the reduction of chromium (VI) in a column design. They also looked into the efficiency of five plant juices and extracts, including red wine, *Mentha spicata, Syzygium aromaticum*, and *Camellia sinensis*, for producing suspensions of Fe NPs and using them to reduce the amount of Cr(VI) in the environment. Green tea, pomegranates, and red wine were discovered to be three of these plants that are more efficient at reducing Cr(VI) (Mystrioti et al. 2016). Additionally, it was demonstrated that palladium nanoparticles (PdNPs) serve an important role as a catalyst in the elimination of Cr(VI) utilizing formic acid (Omole et al. 2007). They observed that the reduction is sensitive to temperature, pH, PdNP concentrations, as well as formic acid levels and that it exhibits first-order kinetics with respect to the reactant. When polyamic acid was used as the reducing agent in a previous study by the same author, they discovered a promising potential. They came to the conclusion that the strategy offers a significant advantage over traditional approaches, which frequently take more time to achieve complete reduction. According to Sadik et al. (2014), PdNPs in soil resulted in a 93 4% conversion of chromium (VI) to (III), compared to a 15% conversion when formic acid was applied alone. Such findings support the majority of the rationales for immobilizing antioxidants as micro- and NPs, including safety from microbial action and prolonged release for improved efficacy. The function of −OH and ROOH groups in biodegradable polymers and metal ions in metal-based NPs is explained by the synergistic effect that the immobilization compounds have on the conversion of Cr(VI).

13.16 Conclusion and Future Prospects

This chapter extensively highlights on the issue of environmental chromium pollution that the world is facing and the current technologies that exist for Cr remediation. Microbial remediation or bioremediation offers several advantages over the conventional chemical and physical methods. Bacterial bioremediation mechanisms and technologies have been discussed in detailed in this chapter. This extensive analysis on this subject suggests the need of a better understanding of the microbial molecular mechanisms, responses and pathways in order to design an efficient bioremediation system for a particular contaminated site. Also, it is crucial to know the indigenous microbial community structure, their metabolic potential and the physicochemical conditions of the site in order to achieve a 'designer microbial approach'. Literature also indicates the lack of practical on-site use of bioremediation approaches that seem to be mainly limited to lab scale. Toxicity at higher metal concentrations, presence of co-contaminants in the system, compromised microbial growth, saturation of adsorption sites, secondary pollution etc. are some of the challenges faced while developing effective bioremediation technologies. In parallel, several researchers are working on developing new strategies such as immobilization, nano-bioremediation to achieve sustainable and efficient bioremediation. For instance, biochar and biosurfactants in combination with algae or duckweeds have emerged as attractive sorting agents that are not only sustainable but also aid in the abatement of global warming (Singh et al. 2020). When paired with other methods like phytoremediation and immobilization that can encourage the growth of the bacteria, bacterial bioremediation is faster, more economical, and much more sustainable (Pushkar et al. 2021; Singh et al. 2020). Nonetheless, the in-silico strategies to Cr(VI) remediation must continue analyzing novel genes, genomes, from cultured or uncultured novel strains to diversify the taxonomy and fill gaps in the existing literature (Bhunia et al. 2022).

References

Abumaizar RJ, Smith EH (1999) Heavy metal contaminants removal by soil washing. J Hazard Mater 70(1–2):71–86

Ackerley DF, Gonzalez CF, Keyhan M, Blake R, Matin A (2004) Mechanism of chromate reduction by the Escherichia coli protein, NfsA, and the role of different chromate reductases in minimizing oxidative stress during chromate reduction. Environ Microbiol 6(8):851–860

Acosta-Navarrete YM, Leon-Marquez YL, Salinas-Herrera K, Jacome-Galarza IE, Meza-Carmen V, Ramírez-Díaz MI, Cervantes C (2014) Expression of the six chromate ion transporter homologues of Burkholderia xenovorans LB400. Microbiology 160(2):287–295

Agarwal M, Singh K (2017) Heavy metal removal from wastewater using various adsorbents: a review. J Water Reuse Desalin 7(4):387–419

Aggarwal M, Anbukumar S, Vijaya Kumar T (2022) Heavy metals concentrations and risk assessment in the sediment of Ganga River between Kanpur and Prayagraj, UP, India. Sādhanā47(4):1–11

Ahluwalia SS, Goyal D (2007) Microbial and plant derived biomass for removal of heavy metals from wastewater. Biores Technol 98(12):2243–2257

Ahmad I, Zafar S, Ahmad F (2005) Heavy metal biosorption potential of Aspergillus and Rhizopus sp. isolated from wastewater treated soil

Akcil A, Erust C, Ozdemiroglu S, Fonti V, Beolchini F (2015) A review of approaches and techniques used in aquatic contaminated sediments: metal removal and stabilization by chemical and biotechnological processes. J Clean Prod 86:24–36

Akkurt Ş, Oğuz M, Alkan Uçkun A (2022) Bioreduction and bioremoval of hexavalent chromium by genetically engineered strains (Escherichia coli MT2A and Escherichia coli MT3). World J Microbiol Biotechnol 38(3):1–13

Al Hasin A, Gurman SJ, Murphy LM, Perry A, Smith TJ, Gardiner PH (2010) Remediation of chromium (VI) by a methane-oxidizing bacterium. Environ Sci Technol 44(1):400–405

Ali J, Sohail A, Wang L, Rizwan Haider M, Mulk S, Pan G (2018) Electro-microbiology as a promising approach towards renewable energy and environmental sustainability. Energies 11(7):1822

Ali I, Alharbi OM, ALOthman ZA, Alwarthan A, Al-Mohaimeed AM (2019) Preparation of a carboxymethylcellulose-iron composite for uptake of atorvastatin in water. Int J Biol Macromol 132:244–253

Ameen FA, Hamdan AM, El-Naggar MY (2020) Assessment of the heavy metal bioremediation efficiency of the novel marine lactic acid bacterium, Lactobacillus plantarum MF042018. Sci Rep 10(1):1–11

Anju M (2017) Biotechnological strategies for remediation of toxic metal (loid) s from environment. In: Plant biotechnology: recent advancements and developments. Springer, Singapore, pp 315–359

Arshad M, Khan AH, Hussain I, Anees M, Iqbal M, Soja G, Linde C et al (2017) The reduction of chromium (VI) phytotoxicity and phytoavailability to wheat (Triticum aestivum L.) using biochar and bacteria. Appl Soil Ecol 114:90–98

Asatiani NV, Abuladze MK, Kartvelishvili TM, Bakradze NG, Sapojnikova NA, Tsibakhashvili NY et al (2004) Effect of chromium (VI) action on Arthrobacter oxydans. Curr Microbiol 49(5):321–326

Asha LP, Sandeep RS (2013) Review on bioremediation-potential tool for removing environmental pollution. Int J Basic Appl Chem Sci 3(3):21–33

Ayangbenro AS, Babalola OO, Aremu OS (2019) Bioflocculant production and heavy metal sorption by metal resistant bacterial isolates from gold mining soil. Chemosphere 231:113–120

Ayangbenro AS, Babalola OO (2020) Genomic analysis of Bacillus cereus NWUAB01 and its heavy metal removal from polluted soil. Sci Rep 10(1):1–12

Ayele A, Godeto YG (2021) Bioremediation of chromium by microorganisms and its mechanisms related to functional groups. J Chem 2021

Baaziz H, Gambari C, Boyeldieu A, Ali Chaouche A, Alatou R, Méjean V, Jourlin-Castelli C et al (2017) ChrASO, the chromate efflux pump of Shewanella oneidensis, improves chromate survival and reduction. Plos One 12(11):e0188516

Babangida SK, Muhammad A, Garba A, Sallau AB (2021) Antioxidants in bioremediation of chromium (VI) by conventional and nanotechnological approaches: a review. Toxicol Environ Chem 103(2):162–183

Baldiris R, Acosta-Tapia N, Montes A, Hernández J, Vivas-Reyes R (2018) Reduction of hexavalent chromium and detection of chromate reductase (ChrR) in Stenotrophomonas maltophilia. Molecules 23(2):406

Banerjee G, Pandey S, Ray AK, Kumar R (2015) Bioremediation of heavy metals by a novel bacterial strain Enterobacter cloacae and its antioxidant enzyme activity, flocculant production, and protein expression in presence of lead, cadmium, and nickel. Wate Air Soil Pollut 226(4):1–9

Banerjee S, Misra A, Chaudhury S, Dam B (2019) A Bacillus strain TCL isolated from Jharia coalmine with remarkable stress responses, chromium reduction capability and bioremediation potential. J Hazard Mater 367:215–223

Bansal N, Coetzee JJ, Chirwa EM (2019) In situ bioremediation of hexavalent chromium in presence of iron by dried sludge bacteria exposed to high chromium concentration. Ecotoxicol Environ Saf 172:281–289

Barnhart J (1997) Occurrences, uses, and properties of chromium. Regul Toxicol Pharmacol 26(1):S3–S7

Bartlett RJ (1991) Chromium cycling in soils and water: links, gaps, and methods. Environ Health Perspect 92:17–24

Bartlett RJ, James BR (1988) Mobility and bioavailability of chromium in soils. Chromium Nat Hum Environ 20:571

Batool R, Yrjälä K, Hasnain S (2014) Impact of environmental stress on biochemical parameters of bacteria reducing chromium. Braz J Microbiol 45:573–583

Benazir JF, Suganthi R, Rajvel D, Pooja MP, Mathithumilan B (2010) Bioremediation of chromium in tannery effluent by microbial consortia. Afr J Biotech 9(21):3140–3143

Bennett RM, Cordero PRF, Bautista GS, Dedeles GR (2013) Reduction of hexavalent chromium using fungi and bacteria isolated from contaminated soil and water samples. Chem Ecol 29(4):320–328

Bharagava RN, Mishra S (2018) Hexavalent chromium reduction potential of Cellulosimicrobium sp. isolated from common effluent treatment plant of tannery industries. Ecotoxicol Environ Saf 147:102–109

Bhunia A, Lahiri D, Nag M, Upadhye V, Pandit S (2022) Bacterial biofilm mediated bioremediation of hexavalent chromium: a review. Biocatal Agric Biotechnol 102397

Black R, Sartaj M, Mohammadian A, Qiblawey HA (2014) Biosorption of Pb and Cu using fixed and suspended bacteria. J Environ Chem Eng 2(3):1663–1671

Blake RC, Choate DM, Bardhan S, Revis N, Barton LL, Zocco TG (1993) Chemical transformation of toxic metals by a Pseudomonas strain from a toxic waste site. Environ Toxicol Chem Int J 12(8):1365–1376

Bruins MR, Kapil S, Oehme FW (2000) Microbial resistance to metals in the environment. Ecotoxicol Environ Saf 45(3):198–207

Camargo FAO, Okeke BC, Bento FM, Frankenberger WT (2003) In vitro reduction of hexavalent chromium by a cell-free extract of Bacillus sp. ES 29 stimulated by Cu2+. Appl Microbiol Biotechnol 62(5–6):569–573

Cánovas D, Cases I, De Lorenzo V (2003) Heavy metal tolerance and metal homeostasis in Pseudomonas putida as revealed by complete genome analysis. Environ Microbiol 5(12):1242–1256

Cervantes CARLOS, Ohtake H, Chu LIEN, Misra TK, Silver S (1990) Cloning, nucleotide sequence, and expression of the chromate resistance determinant of Pseudomonas aeruginosa plasmid pUM505. J Bacteriol 172(1):287–291

Cervantes C, Campos-García J, Devars S, Gutiérrez-Corona F, Loza-Tavera H, Torres-Guzmán JC, Moreno-Sánchez R (2001) Interactions of chromium with microorganisms and plants. FEMS Microbiol Rev 25(3):335–347

Chai L, Ding C, Li J, Yang Z, Shi Y (2019) Multi-omics response of Pannonibacter phragmitetus BB to hexavalent chromium. Environ Pollut 249:63–73

Chang J, Deng S, Liang Y, Chen J (2019) Cr (VI) removal performance from aqueous solution by Pseudomonas sp. strain DC-B3 isolated from mine soil: characterization of both Cr (VI) bioreduction and total Cr biosorption processes. Environ Sci Pollut Res 26(27):28135–28145

Chauhan D, Jaiswal M, Sankararamakrishnan N (2012) Removal of cadmium and hexavalent chromium from electroplating waste water using thiocarbamoyl chitosan. Carbohyd Polym 88(2):670–675

Chen H, Lu Y, Yin P, Li X, Shan Y (2019) Exploring the mechanisms of biosorption of Cr(VI) by marine-derived Penicillium janthinellum P1. Int J Agric Biol 22(5):913–920

Chen J, Tian Y (2021) Hexavalent chromium reducing bacteria: mechanism of reduction and characteristics. Environ Sci Pollut Res 28(17):20981–20997

Cheng J, Gao J, Zhang J, Yuan W, Yan S, Zhou J et al (2021) Optimization of hexavalent chromium biosorption by Shewanella putrefaciens using the Box-Behnken design. Water Air Soil Pollut 232(3):1–14

Cheng SY, Show PL, Lau BF, Chang JS, Ling TC (2019) New prospects for modified algae in heavy metal adsorption. Trends Biotechnol 37(11):1255–1268

Cheung KH, Gu JD (2003) Reduction of chromate (CrO42−) by an enrichment consortium and an isolate of marine sulfate-reducing bacteria. Chemosphere 52(9):1523–1529

Cheung KH, Gu JD (2007) Mechanism of hexavalent chromium detoxification by microorganisms and bioremediation application potential: a review. Int Biodeterior Biodegradation 59(1):8–15

Chowdhary P, Yadav A, Kaithwas G, Bharagava RN (2017) Distillery wastewater: a major source of environmental pollution and its biological treatment for environmental safety. In: Green technologies and environmental sustainability. Springer, Cham, pp 409–435

Chowdhary P, Bharagava RN, Mishra S, Khan N (2020) Role of industries in water scarcity and its adverse effects on environment and human health. In: Environmental concerns and sustainable development. Springer, Singapore, pp 235–256

Chug R, Gour VS, Mathur S, Kothari SL (2016) Optimization of extracellular polymeric substances production using Azotobacter beijreinckii and Bacillus subtilis and its application in chromium (VI) removal. Bioresour Technol 214:604–608

Coetzee JJ, Bansal N, Chirwa E (2020) Chromium in environment, its toxic effect from chromite-mining and ferrochrome industries, and its possible bioremediation. Exposure Health 12(1):51–62

Costa M (2003) Potential hazards of hexavalent chromate in our drinking water. Toxicol Appl Pharmacol 188(1):1–5

Das P, Mishra S (2009) Hexavalent chromium [Cr (VI)]: yellow water pollution and its remediation. Sarovar Saurabh ENVIS Newsl Wetl Ecosyst 5(2):1–8

Das S, Mishra J, Das SK, Pandey S, Rao DS, Chakraborty A et al (2014) Investigation on mechanism of Cr (VI) reduction and removal by Bacillus amyloliquefaciens, a novel chromate tolerant bacterium isolated from chromite mine soil. Chemosphere 96:112–121

De Flora S, Iltcheva M, Balansky RM (2006) Oral chromium (VI) does not affect the frequency of micronuclei in hematopoietic cells of adult mice and of transplacentally exposed fetuses. Mutation Res/Genetic Toxicol Environ Mutagen 610(1–2):38–47

Desai C, Jain K, Madamwar D (2008a) Evaluation of in vitro Cr(VI) reduction potential in cytosolic extracts of three indigenous Bacillus sp. isolated from Cr(VI) polluted industrial landfill. Bioresour Technol 99(14):6059–6069

Desai C, Jain K, Madamwar D (2008b) Hexavalent chromate reductase activity in cytosolic fractions of Pseudomonas sp. G1DM21 isolated from Cr(VI) contaminated industrial landfill. Process Biochem 43(7):713–721

DesMarias TL, Costa M (2019) Mechanisms of chromium-induced toxicity. Curr Opin Toxicol 14:1–7

Dhakephalkar PK, Bhide JV, Paknikar KM (1996) Plasmid mediated chromate resistance and reduction in Pseudomonas mendocina MCM B-180. Biotech Lett 18(10):1119–1122

Dhal B, Thatoi HN, Das NN, Pandey BD (2013) Chemical and microbial remediation of hexavalent chromium from contaminated soil and mining/metallurgical solid waste: a review. J Hazard Mater 250:272–291

Di Cesare A, Pjevac P, Eckert E, Curkov N, Šparica MM, Corno G, Orlić S (2020) The role of metal contamination in shaping microbial communities in heavily polluted marine sediments. Environ Pollut 265:114823

Díaz-Pérez C, Cervantes C, Campos-García J, Julián-Sánchez A, Riveros-Rosas H (2007) Phylogenetic analysis of the chromate ion transporter (CHR) superfamily. FEBS J 274(23):6215–6227

Dogan NM, Doganli GA, Dogan G et al (2015) Characterization of extracellular polysaccharides (EPS) produced by thermal Bacil-lus and determination of environmental conditions affecting exopolysaccharide production. Int J Environ Res 9:1107–1116

Dong L, Zhou S, He Y, Jia Y, Bai Q, Deng P et al (2018) Analysis of the genome and chromium metabolism-related genes of Serratia sp. S2. Appl Biochem Biotechnol 185(1):140–152

Dwisandi RF, Mutiara F, Nurfauziah E, Meylani V (2021) Review effectiveness of indigenous local microorganisms in degrading hexavalent chromium (Cr (VI)) in Batik liquid waste. Biol Environ Pollut 1(1):19–29

Elahi A, Rehman A (2019a) Comparative behavior of two gram positive Cr6+ resistant bacterial strains *Bacillus aerius* S1 and Brevibacterium iodinum S2 under hexavalent chromium stress. Biotechnol Rep 21:e00307

Elahi A, Rehman A (2019b) Multiple metal resistance and Cr6+ reduction by bacterium, Staphylococcus sciuri A-HS1, isolated from untreated tannery effluent. J King Saud Univ-Sci 31(4):1005–1013

Elahi A, Arooj I, Bukhari DA, Rehman A (2020) Successive use of microorganisms to remove chromium from wastewater. Appl Microbiol Biotechnol 104(9):3729–3743

Elahi A, Rehman A, Hussain SZ, Zulfiqar S, Shakoori AR (2022) Isolation and characterization of a highly effective bacterium Bacillus cereus b-525k for hexavalent chromium detoxification. Saudi J Biol Sci 29(4):2878–2885

Elgarahy AM, Elwakeel KZ, Mohammad SH, Elshoubaky GA (2021) A critical review of biosorption of dyes, heavy metals and metalloids from wastewater as an efficient and green process. Clean Eng Technol 4:100209

El-Kassas HY, El-Taher EM (2009) Optimization of batch process parameters by response surface methodology for mycoremediation of chrome-VI by a chromium resistant strain of marine Trichoderma viride. Am-Eurasian J Agric Environ Sci 5(5):676–681

El-Naggar NEA, El-Khateeb AY, Ghoniem AA, El-Hersh MS, Saber WI (2020) Innovative low-cost biosorption process of Cr6+ by Pseudomonas alcaliphila NEWG-2. Sci Rep 10(1):1–18

Fang G, Li W, Shen X, Perez-Aguilar JM, Chong Y, Gao X et al (2018) Differential Pd-nanocrystal facets demonstrate distinct antibacterial activity against Gram-positive and Gram-negative bacteria. Nat Commun 9(1):1–9

Ferreira ME, de Vasconcelos AS, da Costa Vilhena T, da Silva TL, da Silva Barbosa A, Gomes AR et al (2015) Oxidative stress in Alzheimer's disease: should we keep trying antioxidant therapies? Cell Mol Neurobiol 35(5):595–614

Focardi S, Pepi M, Landi G, Gasperini S, Ruta M, Di Biasio P, Focardi SE (2012) Hexavalent chromium reduction by whole cells and cell free extract of the moderate halophilic bacterial strain Halomonas sp. TA-04. Int Biodeterior Biodegrad 66(1):63–70

Fourest E, Roux JC (1992) Heavy metal biosorption by fungal myce-lial by-products: mechanisms and influence of pH. App Microbiol Biotechnol 37(3):399–403

Fu SC, Liu JM, Lee KI, Tang FC, Fang KM, Yang CY et al (2020) Cr(VI) induces ROS-mediated mitochondrial-dependent apoptosis in neuronal cells via the activation of Akt/ERK/AMPK signaling pathway. Toxicol In Vitro 65:104795

Fulke AB, Kotian A, Giripunje MD (2020) Marine microbial response to heavy metals: mechanism, implications and future prospect. Bull Environ Contam Toxicol 105(2):182–197

Gadd GM (1992) Metals and microorganisms: a problem of definition. FEMS Microbiol Lett 100(1–3):197–203

Gang H, Xiao C, Xiao Y, Yan W, Bai R, Ding R et al (2019) Proteomic analysis of the reduction and resistance mechanisms of Shewanella oneidensis MR-1 under long-term hexavalent chromium stress. Environ Int 127:94–102

Garbisu C, Alkorta I (2003) Basic concepts on heavy metal soil bioremediation. EJMP & EP (Eur J Mineral Process Environ Protect), 3(1):58–66

García-Hernández MA, Villarreal-Chiu JF, Garza-González MT (2017) Metallophilic fungi research: an alternative for its use in the bioremediation of hexavalent chromium. Int J Environ Sci Technol 14(9):2023–2038

Ge S, Dong X, Zhou J, Ge S (2013) Comparative evaluations on bio-treatment of hexavalent chromate by resting cells of Pseudochrobactrum sp. and Proteus sp. in wastewater. J Environ Manag 126:7–12

Genchi G, Lauria G, Catalano A, Carocci A, Sinicropi MS (2021) The double face of metals: the intriguing case of chromium. Appl Sci 11(2):638

Ghosh S, Jasu A, Ray RR (2021) Hexavalent chromium bioremediation with insight into molecular aspect: an overview. Bioremediat J 25(3):225–251

Gnanamani A, Kavitha V, Radhakrishnan N, Rajakumar GS, Sekaran G, Mandal AB (2010) Microbial products (biosurfactant and extracellular chromate reductase) of marine microorganism are the potential agents reduce the oxidative stress induced by toxic heavy metals. Colloids Surf B 79(2):334–339

Godgul G, Sahu KC (1995) Chromium contamination from chromite mine. Environ Geol 25(4):251–257

Gopalan R, Veeramani H (1994) Studies on microbial chromate reduction by Pseudomonas sp. in aerobic continuous suspended growth cultures. Biotechnol Bioeng 43(6):471–476

Gorny J, Billon G, Noiriel C, Dumoulin D, Lesven L, Madé B (2016) Chromium behavior in aquatic environments: a review. Environ Rev 24(4):503–516

Gu R, Gao J, Dong L, Liu Y, Li X, Bai Q et al (2020) Chromium metabolism characteristics of coexpression of ChrA and ChrT gene. Ecotoxicol Environ Saf 204:111060

Guo S, Xiao C, Zhou N, Chi R (2021) Speciation, toxicity, microbial remediation and phytoremediation of soil chromium contamination. Environ Chem Lett 19(2):1413–1431

Gupta N, Yadav KK, Kumar V, Kumar S, Chadd RP, Kumar A (2019) Trace elements in soil-vegetables interface: translocation, bioaccumulation, toxicity and amelioration-a review. Sci Total Environ 651:2927–2942

Gutiérrez-Corona JF, Romo-Rodríguez P, Santos-Escobar F, Espino-Saldaña AE, Hernández-Escoto H (2016) Microbial interactions with chromium: basic biological processes and applications in environmental biotechnology. World J Microbiol Biotechnol 32(12):1–9

Harboul K, Alouiz I, Hammani K, El-Karkouri A (2022) Isotherm and kinetics modeling of biosorption and bioreduction of the Cr (VI) by Brachybacterium paraconglomeratum ER41. Extremophiles 26(3):1–14

Hassen A, Saidi N, Cherif M, Boudabous A (1998) Effects of heavy metals on Pseudomonas aeruginosa and Bacillus thuringiensis. Biores Technol 65(1–2):73–82

He J, Chen JP (2014) A comprehensive review on biosorption of heavy metals by algal biomass: materials, performances, chemistry, and modeling simulation tools. Biores Technol 160:67–78

He Y, Dong L, Zhou S, Jia Y, Gu R, Bai Q et al (2018) Chromium resistance characteristics of Cr (VI) resistance genes ChrA and ChrB in Serratia sp. S2. Ecotoxicol Environ Saf 157:417–423

He X, Li P (2020) Surface water pollution in the middle Chinese Loess Plateau with special focus on hexavalent chromium (Cr6+): occurrence, sources and health risks. Exposure Health 12(3):385–401

He C, Gu L, He H, Zhang Z, Wang X, Han F et al (2020) Dissolved organic matter modified magnetic carbon nanotubes enhance the bioremediation of azo dyes and Cr (VI). Environ Sci: Water Res Technol 6(7):1804–1815

Hedayatkhah A, Cretoiu MS, Emtiazi G, Stal LJ, Bolhuis H (2018) Bioremediation of chromium contaminated water by diatoms with concomitant lipid accumulation for biofuel production. J Environ Manag 227:313–320

Hidangmayum A, Debnath A, Guru A, Singh BN, Upadhyay SK, Dwivedi P (2022) Mechanistic and recent updates in nano-bioremediation for developing green technology to alleviate agricultural contaminants. Int J Environ Sci Technol 1–26

Hora A, Shetty VK (2016) Kinetics of bioreduction of hexavalent chromium by poly vinyl alcohol-alginate immobilized cells of Ochrobactrum sp. Cr-B4 and comparison with free cells. Desalin Water Treatment 57(19):8981–8989

Hua JQ, Zhang R, Chen RP et al (2021) Energy-saving preparation of a bioflocculant under high-salt condition by using strain Bacillus sp. and the interaction mechanism towards heavy metals. Che

Huang L, Zeng RJ, Angelidaki I (2008) Electricity production from xylose using a mediator-less microbial fuel cell. Bioresour Rechnol 99(10):4178–4184

Huang ZZ, Chen GQ, Zeng GM, Song ZX, Zuo YN, Guo Z, Tan Q (2015) Research progress of immobilized microorganism technology and its mechanisms in wastewater treatment. Environ Pollut Cont 37(10):77–85

Huang H, Tao X, Jiang Y, Khan A, Wu Q, Yu X et al (2017) The naphthalene catabolic protein NahG plays a key role in hexavalent chromium reduction in Pseudomonas brassicacearum LZ-4. Sci Rep 7(1):1–11

Huang XN, Min D, Liu DF, Cheng L, Qian C, Li WW, Yu HQ (2019) Formation mechanism of organo-chromium (III) complexes from bioreduction of chromium (VI) by Aeromonas hydrophila. Environ Int 129:86–94

Huang Y, Zeng Q, Hu L, Zhong H, He Z (2021) Bioreduction performances and mechanisms of Cr (VI) by Sporosarcina saromensis W5, a novel Cr (VI)-reducing facultative anaerobic bacteria. J Hazard Mater 413:125411

Hultman PA, Pollard KM (2022) Immunotoxicology of metals. In: Handbook on the toxicology of metals. Academic Press, pp 543–564

Hussein MH, Hamouda RA, Elhadary AMA, Abuelmagd MA, Ali S, Rizwan M (2019) Characterization and chromium biosorption potential of extruded polymeric substances from Synechococcus mundulus induced by acute dose of gamma irradiation. Environ Sci Pollut Res 26(31):31998–32012

Igiri BE, Okoduwa SI, Idoko GO, Akabuogu EP, Adeyi AO, Ejiogu IK (2018) Toxicity and bioremediation of heavy metals contaminated ecosystem from tannery wastewater: a review. J Toxicol 2018

Ikegami K, Hirose Y, Sakashita H, Maruyama R, Sugiyama T (2020) Role of polyphenol in sugarcane molasses as a nutrient for hexavalent chromium bioremediation using bacteria. Chemosphere 250:126267

Ilias M, Rafiqullah IM, Debnath BC, Mannan KSB, Hoq M (2011) Isolation and characterization of chromium (VI)-reducing bacteria from tannery effluents. Indian J Microbiol 51(1):76–81

Ishibashi Y, Cervantes C, Silver S (1990) Chromium reduction in Pseudomonas putida. Appl Environ Microbiol 56(7):2268–2270

Iyer A, Mody K, Jha B (2004) Accumulation of hexavalent chromium by an exopolysaccharide producing marine Enterobacter cloaceae. Mar Pollut Bull 49(11–12):974–977

Jamshed Z, VAmit P (2017) Review on heavy metal pollution in major lakes of India: remediation through plants. Afr J Environ Sci Technol 11(6):255–265

Jarosławiecka A, Piotrowska-Seget Z (2014) Lead resistance in micro-organisms. Microbiology 160(1):12–25

Jeyasingh J, Philip L (2005) Bioremediation of chromium contaminated soil: optimization of operating parameters under laboratory conditions. J Hazard Mater 118(1–3):113–120

Jeyasingh J, Somasundaram V, Philip L, Bhallamudi SM (2011) Pilot scale studies on the remediation of chromium contaminated aquifer using bio-barrier and reactive zone technologies. Chem Eng J 167(1):206–214

Jiang Y, Yang F, Dai M, Ali I, Shen X, Hou X et al (2022) Application of microbial immobilization technology for remediation of Cr (VI) contamination: a review. Chemosphere 286:131721

Jiménez-Mejía R, Campos-García J, Cervantes C (2006) Membrane topology of the chromate transporter ChrA of Pseudomonas aeruginosa. FEMS Microbiol Lett 262(2):178–184

Jobby R, Shah K, Shah R, Jha P, Desai N (2016) Differential expression of antioxidant enzymes under arsenic stress in Enterobacter sp. Environ Prog Sustain Energy 35(6):1642–1645

Jobby R, Jha P, Yadav AK, Desai N (2018) Biosorption and biotransformation of hexavalent chromium [Cr (VI)]: a comprehensive review. Chemosphere 207:255–266

John R, Rajan AP (2022) Pseudomonas putida APRRJVITS11 as a potent tool in chromium (VI) removal from effluent wastewater. Prep Biochem Biotechnol 52(2):163–170

Joshi PM, Juwarkar AA (2009) In vivo studies to elucidate the role of extracellular polymeric substances from Azotobacter in immo-bilization of heavy metals. Environ Sci Technol 43:5884–5889

Kapahi M, Sachdeva S (2019) Bioremediation options for heavy metal pollution. J Health Pollut 9(24)

Karaulov AV, Renieri EA, Smolyagin AI, Mikhaylova IV, Stadnikov AA, Begun DN et al (2019) Long-term effects of chromium on morphological and immunological parameters of Wistar rats. Food Chem Toxicol 133:110748

Karimi-Maleh H, Ayati A, Ghanbari S, Orooji Y, Tanhaei B, Karimi F et al (2021) Recent advances in removal techniques of Cr (VI) toxic ion from aqueous solution: a comprehensive review. J Mol Liq 329:115062

Karthik C, Ramkumar VS, Pugazhendhi A, Gopalakrishnan K, Arulselvi PI (2017) Biosorption and biotransformation of Cr (VI) by novel Cellulosimicrobium funkei strain AR6. J Taiwan Inst Chem Eng 70:282–290

Kathiravan MN, Rani RK, Karthick R, Muthukumar K (2010) Mass transfer studies on the reduction of Cr (VI) using calcium alginate immobilized Bacillus sp. in packed bed reactor. Bioresour Technol 101(3):853–858

Kathiravan MN, Karthick R, Muthukumar K (2011) Ex situ bioremediation of Cr(VI) contaminated soil by Bacillus sp.: batch and continuous studies. Chem Eng J 169(1–3):107–115

Kaushik G, Raza K (2019) Potential of novel Dunaliella salina from sambhar salt lake, India, for bioremediation of hexavalent chromium from aqueous effluents: an optimized green approach. Ecotoxicol Environ Saf 180:430–438

Kavitha V, Radhakrishnan N, Gnanamani A, Mandal AB (2011) Management of chromium induced oxidative stress by marine Bacillus licheniformis. Biol Med 3(2):16–26

Khalid AH, Jin HJ (2013) Heavy metal resistance of bacteria and its impact on the production of antioxidant enzymes. Afr J Microbiol Res 7(20):2288–2296

Kılıç NK, Stensballe A, Otzen DE, Dönmez G (2010) Proteomic changes in response to chromium (VI) toxicity in Pseudomonas aeruginosa. Biores Technol 101(7):2134–2140

Kookhaee F, Bafroee AST, Jabalameli L (2022) Isolation and characterization of chromium (VI) tolerant bacteria from tannery effluents. J Environ Health Sci Eng 1–16

Kotaś J, Stasicka ZJEP (2000) Chromium occurrence in the environment and methods of its speciation. Environ Pollut 107(3):263–283

Kumar H, Sinha SK, Goud VV, Das S (2019) Removal of Cr (VI) by magnetic iron oxide nanoparticles synthesized from extracellular polymeric substances of chromium resistant acid-tolerant bacterium Lysinibacillus sphaericus RTA-01. J Environ Health Sci Eng 17(2):1001–1016

Labied R, Benturki O, Eddine Hamitouche AY, Donnot A (2018) Adsorption of hexavalent chromium by activated carbon obtained from a waste lignocellulosic material (Ziziphus jujuba cores): Kinetic, equilibrium, and thermodynamic study. Adsorpt Sci Technol 36(3–4):1066–1099

Latha S, Vinothini G, Dhanasekaran D (2015) Chromium [Cr (VI)] biosorption property of the newly isolated actinobacterial probiont Streptomyces werraensis LD22. 3 Biotech 5(4):423–432

Le TT, Nguyen KH, Jeon JR, Francis AJ, Chang YS (2015) Nano/bio treatment of polychlorinated biphenyls with evaluation of comparative toxicity. J Hazard Mater 287:335–341

Lee L, Hsu CY, Yen HW (2017) The effects of hydraulic retention time (HRT) on chromium (VI) reduction using autotrophic cultivation of Chlorella vulgaris. Bioprocess Biosyst Eng 40(12):1725–1731

Lehmann J, Joseph S (2015) Biochar for environmental management: an introduction. In: Biochar for environmental management. Routledge, pp 1–13

Li S, Baiyun R, Lv Z, Li J, Han D, Zhao W et al (2019a) Exploring the kidney hazard of exposure to mercuric chloride in mice: disorder of mitochondrial dynamics induces oxidative stress and results in apoptosis. Chemosphere 234:822–829

Li M, He Z, Hu Y, Hu L, Zhong H (2019b) Both cell envelope and cytoplasm were the locations for chromium (VI) reduction by Bacillus sp. M6. Biores Technol 273:130–135

Li MH, Gao XY, Li C, Yang CL, Fu CA, Liu J et al (2020) Isolation and identification of chromium reducing Bacillus Cereus species from chromium-contaminated soil for the biological detoxification of chromium. Int J Environ Res Public Health 17(6):2118

Li L, Shang X, Sun X, Xiao X, Xue J, Gao Y, Gao H (2021) Bioremediation potential of hexavalent chromium by a novel bacterium Stenotrophomonas acidaminiphila 4-1. Environ Technol Innov 22:101409

Lian G, Wang B, Lee X, Li L, Liu T, Lyu W (2019) Enhanced removal of hexavalent chromium by engineered biochar composite fabricated from phosphogypsum and distillers grains. Sci Total Environ 697:134119

Lin J, Harichund C (2011) Isolation and characterization of heavy metal removing bacterial bioflocculants. Afr J Microbiol Res 5(6):599–607

Liu Z, Wu Y, Lei C, Liu P, Gao M (2012) Chromate reduction by a chromate-resistant bacterium, microbacterium sp. World J Microbiol Biotechnol 28(4):1585–1592

Liu T, Wang Z, Wu L, Guo M, Yang C, Cao X et al (2020) Acute impact of Hg2+, Cu2+, and Ag+ on the formation of biopolymers and nitrogenous soluble microbiological products in activated sludge for wastewater treatment. Environ Pollut 267:115388

Liu Z, Lei M, Chen G, Yuan J (2022) Treatment of Chromium removal wastewater from tanning by a new coupling technology. Processes 10(6):1134

Long D, Hashmi MZ, Su X, Pongpiachan S (2019) Cr (VI) reduction by an extracellular polymeric substance (EPS) produced from a strain of Pseudochrobactrum saccharolyticum. 3 Biotech 9(3):1–9. https://doi.org/10.1007/s13205-019-1641-8

Lotlikar NP, Damare SR, Meena RM, Linsy P, Mascarenhas B (2018) Potential of marine-derived fungi to remove hexavalent chromium pollutant from culture broth. Indian J Microbiol 58(2):182–192

Lovley DR (1993) Dissimilatory metal reduction. Annu Rev Microbiol 47(1):263–290

Lovley DR, Phillips EJ (1994) Reduction of chromate by Desulfovibrio vulgaris and its c 3 cytochrome. Appl Environ Microbiol 60(2):726–728

Luo C, Tian Z, Yang B, Zhang L, Yan S (2013) Manganese dioxide/iron oxide/acid oxidized multi-walled carbon nanotube magnetic nanocomposite for enhanced hexavalent chromium removal. Chem Eng J 234:256–265

Ma L, Chen N, Feng C, Hu Y, Li M, Liu T (2019a) Feasibility and mechanism of microbial-phosphorus minerals-alginate immobilized particles in bioreduction of hexavalent chromium and synchronous removal of trivalent chromium. Biores Technol 294:122213

Ma L, Xu J, Chen N, Li M, Feng C (2019b) Microbial reduction fate of chromium (Cr) in aqueous solution by mixed bacterial consortium. Ecotoxicol Environ Saf 170:763–770

Ma L, Chen N, Feng C, Li M, Gao Y, Hu Y (2020) Coupling enhancement of Chromium (VI) bioreduction in groundwater by phosphorus minerals. Chemosphere 240:124896

Mala JGS, Sujatha D, Rose C (2015) Inducible chromate reductase exhibiting extracellular activity in Bacillus methylotrophicus for chromium bioremediation. Microbiol Res 170:235–241

Malaviya P, Singh A (2011) Physicochemical technologies for remediation of chromium-containing waters and wastewaters. Crit Rev Environ Sci Technol 41(12):1111–1172

Mangwani N, Kumari S, Das S (2016) Bacterial biofilms and quorum sensing: fidelity in bioremediation technology. Biotechnol Genet Eng Rev 32(1–2):43–73

Mantel T, Glass S, Usman M, Lyberis A, Filiz V, Ernst M (2022) Adsorptive dead-end filtration for removal of Cr (VI) using novel amine modified polyacrylonitrile ultrafiltration membranes. Environ Sci: Water Res Technol

Marouani N, Tebourbi O, Hallègue D, Mokni M, Yacoubi MT, Sakly M (2017) Mechanisms of chromium hexavalent-induced apoptosis in rat testes. Toxicol Ind Health 33(2):97–106

McCartor A, Becker D (2010) World's worst pollution problems report 2010. Blacksmith Institute, New York, NY

Megharaj M, Avudainayagam S, Naidu R (2003) Toxicity of hexavalent chromium and its reduction by bacteria isolated from soil contaminated with tannery waste. Curr Microbiol 47(1):0051–0054

Mishra V, Samantaray DP, Dash SK, Mishra BB, Swain RK (2010) Study on hexavalent chromium reduction by chromium resistant bacterial isolates of Sukinda mining area. Our Nature 8(1):63–71

Mishra H, Sahu HB (2013) Environmental scenario of chromite mining at Sukinda Valley—A review. Int J Environ Eng Manag 4:287–292

Mishra S, Bharagava RN (2016) Toxic and genotoxic effects of hexavalent chromium in environment and its bioremediation strategies. J Environ Sci Health C 34(1):1–32

Mitra S, Sarkar A, Sen S (2017) Removal of chromium from industrial effluents using nanotechnology: a review. Nanotechnol Environ Eng 2(1):1–14

Modak JM, Natarajan KA, Saha B (1996) Biosorption of copper and zinc using waste Aspergillus niger biomass. Min Metall Explor 13(2):52–57

Moffat I, Martinova N, Seidel C, Thompson CM (2018) Hexavalent chromium in drinking water. J Am Water Works Ass 110(5):E22–E35

Mohamed AAR, El-Houseiny W, Abd Elhakeem EM, Ebraheim LL, Ahmed AI, Abd El-Hakim YM (2020) Effect of hexavalent chromium exposure on the liver and kidney tissues related to the expression of CYP450 and GST genes of Oreochromis niloticus fish: role of curcumin supplemented diet. Ecotoxicol Environ Saf 188:109890

Mohapatra RK, Parhi PK, Thatoi H, Panda CR (2017) Bioreduction of hexavalent chromium by Exiguobacterium indicum strain MW1 isolated from marine water of Paradip Port, Odisha, India. Chem Ecol 33(2):114–130

Mohsenzadeh F, Rad AC (2012) Bioremediation of heavy metal pollution by nano-particles of noaea mucronata. Int J Biosci Biochem Bioinform 2(2):85

Monga A, Fulke AB, Dasgupta D (2022a) Recent developments in essentiality of trivalent chromium and toxicity of hexavalent chromium: implications on human health and remediation strategies. J Hazard Mater Adv 100113

Monga A, Fulke AB, Gaud A, Sharma A, Ram A, Dasgupta D (2022b) Isolation and identification of novel chromium tolerant bacterial strains from a heavy metal polluted urban creek: an assessment of bioremediation efficiency and flocculant production. Thalassas: Int J Marine Sci 1–12

Mosphere 267:129324

Munjur HM, Hasan MN, Awual MR, Islam MM, Shenashen MA, Iqbal J (2020) Biodegradable natural carbohydrate polymeric sustainable adsorbents for efficient toxic dye removal from wastewater. J Mol Liq 319:114356

Mushtaq Z, Liaquat M, Nazir A, Liaqat R, Iftikhar H, Anwar W, Itrat N (2022) Potential of plant growth promoting rhizobacteria to mitigate chromium contamination. Environ Technol Innov 102826

Myers CR, Carstens BP, Antholine WE, Myers JM (2000) Chromium (VI) reductase activity is associated with the cytoplasmic membrane of anaerobically grown Shewanella putrefaciens MR-1. J Appl Microbiol 88(1):98–106

Mystrioti C (2014) Application of iron nanoparticles synthesized by green tea for the removal of hexavalent chromium in column tests. J Geosci Environ Protect 2(04):28

Mystrioti C, Xanthopoulou TD, Tsakiridis P, Papassiopi N, Xenidis A (2016) Comparative evaluation of five plant extracts and juices for nanoiron synthesis and application for hexavalent chromium reduction. Sci Total Environ 539:105–113

Nakkeeran E, Patra C, Shahnaz T, Rangabhashiyam S, Selvaraju NJBTR (2018) Continuous biosorption assessment for the removal of hexavalent chromium from aqueous solutions using Strychnos nux vomica fruit shell. Bioresour Technol Rep 3:256–260

Newton GL, Rawat M, Clair La JJ, Jothivasan VK, Budiarto T, Hamilton CJ et al (2009) Bacillithiol is an antioxidant thiol produced in Bacilli. Nat Chem Biol 5(9):625–627

Nguyen A, Le BV, Richter O (2020) The role of mangroves in the retention of heavy metal (Chromium): a simulation study in the Thi Vai River Catchment, Vietnam. Int J Environ Res Publ Health 17(16):5823

Nies DH, Silver S (1995) Ion efflux systems involved in bacterial metal resistances. J Ind Microbiol 14:186–199

Nur-E-Alam M, Mia MAS, Ahmad F, Rahman MM (2020) An overview of chromium removal techniques from tannery effluent. Appl Water Sci 10:205

Nwodo UU, Okoh AI (2013) Characterization and flocculation properties of biopolymeric flocculant (glycosaminoglycan) produced by Cellulomonas sp. Okoh J Appl Microbiol 114(5):1325–1337. https://doi.org/10.1111/jam.12095

Omole MA, K'Owino IO, Sadik OA (2007) Palladium nanoparticles for catalytic reduction of Cr (VI) using formic acid. Appl Catal B 76(1–2):158–167

Opperman DJ, Van Heerden E (2007) Aerobic Cr (VI) reduction by Thermus scotoductus strain SA-01. J Appl Microbiol 103(5):1907–1913

Opperman DJ, Piater LA, van Heerden E (2008) A novel chromate reductase from Thermus scotoductus SA-01 related to old yellow enzyme. J Bacteriol 190(8):3076–3082

Owlad M, Aroua MK, Daud WAW, Baroutian S (2009) Removal of hexavalent chromium-contaminated water and wastewater: a review. Water Air Soil Pollut 200(1):59–77
Ozden B, Guler E, Vaasma T, Horvath M, Kiisk M, Kovacs T (2018) Enrichment of naturally occurring radionuclides and trace elements in Yatagan and Yenikoy coal-fired thermal power plants, Turkey. J Environ Radioact 188:100–107
Oze C, Fendorf S, Bird DK, Coleman RG (2004) Chromium geochemistry of serpentine soils. Int Geol Rev 46(2):97–126
Oze C, Bird DK, Fendorf S (2007) Genesis of hexavalent chromium from natural sources in soil and groundwater. Proc Natl Acad Sci 104(16):6544–6549
Ozturk S, Aslim B (2008) Relationship between chromium(VI) resistance and extracellular polymeric substances (EPS) concentration by some cyanobacterial isolates. Environ Sci Pollut Res 15:478–480
Pal S, Misra A, Banerjee S, Dam B (2020) Adaptation of ethidium bromide fluorescence assay to monitor activity of efflux pumps in bacterial pure cultures or mixed population from environmental samples. J King Saud Univ-Sci 32(1):939–945
Pan X, Liu Z, Chen Z, Cheng Y, Pan D, Shao J et al (2014) Investigation of Cr (VI) reduction and Cr (III) immobilization mechanism by planktonic cells and biofilms of Bacillus subtilis ATCC-6633. Water Res 55:21–29
Pandey G, Madhuri S (2014) Heavy metals causing toxicity in animals and fishes. Res J Anim Vet Fishery Sci 2(2):17–23
Parameswari E, Lakshmanan A, Thilagavathi T (2009) Chromate resistance and reduction by bacterial isolates. Aust J Basic Appl Sci 3(2):1363–1368
Park CH, Keyhan M, Wielinga B, Fendorf S, Matin A (2000) Purification to homogeneity and characterization of a novel Pseudomonas putida chromate reductase. Appl Environ Microbiol 66(5):1788–1795
Park CH, Gonzalez C, Ackerley D, Keyhan M, Matin A (2001) Molecular engineering of soluble bacterial proteins with chromate reductase activity. In: Remediation and beneficial reuse of contaminated sediments, pp 103–112
Park D, Yun YS, Park JM (2006) Mechanisms of the removal of hexavalent chromium by biomaterials or biomaterial-based activated carbons. J Hazard Mater 137(2):1254–1257
Park D, Lim SR, Yun YS, Park JM (2007) Reliable evidences that the removal mechanism of hexavalent chromium by natural biomaterials is adsorption-coupled reduction. Chemosphere 70(2):298–305
Patra RC, Malik S, Beer M, Megharaj M, Naidu R (2010) Molecular characterization of chromium (VI) reducing potential in Gram positive bacteria isolated from contaminated sites. Soil Biol Biochem 42(10):1857–1863
Peng JF, Song YH, Yuan P, Cui XY, Qiu GL (2009) The remediation of heavy metals contaminated sediment. J Hazard Mater 161(2–3):633–640
Pi S, Li A, Qiu J et al (2021) Enhanced recovery of hexavalent chro- mium by remodeling extracellular polymeric substances through engineering Agrobacterium tumefaciens F2. J Cleaner Prod 279:123829
Prabhakaran DC, Subramanian S (2017) Studies on the bioremediation of chromium from aqueous solutions using C. paurometabolum. Trans Indian Inst Metals 70(2):497–509
Prabhakaran DC, Bolanos-Benitez V, Sivry Y, Gelabert A, Riotte J, Subramanian S (2019) Mechanistic studies on the bioremediation of Cr (VI) using Sphingopyxis macrogoltabida SUK2c, a Cr (VI) tolerant bacterial isolate. Biochem Eng J 150:107292
Pradhan D, Sukla LB, Sawyer M, Rahman PK (2017) Recent bioreduction of hexavalent chromium in wastewater treatment: a review. J Ind Eng Chem 55:1–20
Prasad S, Yadav KK, Kumar S, Gupta N, Cabral-Pinto MM, Rezania S et al (2021) Chromium contamination and effect on environmental health and its remediation: a sustainable approaches. J Environ Manag 285:112174
Pribluda LA (1963) Chromium content of the long bones of rats at different stages of pregnancy. Dokl. Akad. Nauk B. SSR 7:206–212

Priyadarshanee M, Das S (2021) Bioremediation potential of biofilm forming multi-metal resistant marine bacterium Pseudomonas chengduensis PPSS-4 isolated from contaminated site of Paradip Port, Odisha. J Earth Syst Sci 130(3):1–17

Proctor DM, Otani JM, Finley BL, Paustenbach DJ, Bland JA, Speizer N, Sargent EV (2002) Is hexavalent chromium carcinogenic via ingestion? A weight-of-evidence review. J Toxicol Environ Health A 65(10):701–746

Pulimi M, Jamwal S, Samuel J, Chandrasekaran N, Mukherjee A (2012) Enhancing the hexavalent chromium bioremediation potential of Acinetobacter junii VITSUKMW2 using statistical design experiments. J Microbiol Biotechnol 22(12):1767–1775

Pushkar B, Sevak P, Parab S, Nilkanth N (2021) Chromium pollution and its bioremediation mechanisms in bacteria: a review. J Environ Manag 287:112279

Qian D, Liu H, Hu F, Song S, Chen Y (2022). Extracellular electron transfer-dependent Cr (VI)/sulfate reduction mediated by iron sulfide nanoparticles. J Biosci Bioeng

Qu M, Chen J, Huang Q, Chen J, Xu Y, Luo J et al (2018) Bioremediation of hexavalent chromium contaminated soil by a bioleaching system with weak magnetic fields. Int Biodeterior Biodegradation 128:41–47

Quiton KG, Doma Jr B, Futalan CM, Wan MW (2018) Removal of chromium (VI) and zinc (II) from aqueous solution using kaolin-supported bacterial biofilms of Gram-negative E. coli and Gram-positive Staphylococcus epidermidis. Sustain Environ Res 28(5):206–213

Rager JE, Suh M, Chappell GA, Thompson CM, Proctor DM (2019) Review of transcriptomic responses to hexavalent chromium exposure in lung cells supports a role of epigenetic mediators in carcinogenesis. Toxicol Lett 305:40–50

Rajput VD, Singh A, Minkina T, Rawat S, Mandzhieva S, Sushkova S et al (2021) Nano-enabled products: challenges and opportunities for sustainable agriculture. Plants 10(12):2727

Ramachandran G, Chackaravarthi G, Rajivgandhi GN, Quero F, Maruthupandy M, Alharbi NS et al (2022) Biosorption and adsorption isotherm of chromium (VI) ions in aqueous solution using soil bacteria Bacillus amyloliquefaciens. Environ Res 212:113310

Ramírez-Díaz MI, Díaz-Pérez C, Vargas E, Riveros-Rosas H, Campos-García J, Cervantes C (2008) Mechanisms of bacterial resistance to chromium compounds. Biometals 21(3):321–332

Ran ZHAO, Bi WANG, Cai QT, Li XX, Min LIU, Dong HU et al (2016) Bioremediation of hexavalent chromium pollution by Sporosarcina saromensis M52 isolated from offshore sediments in Xiamen China. Biomed Environ Sci 29(2):127–136

Rath BP, Das S, Mohapatra PKD, Thatoi H (2014) Optimization of extracellular chromate reductase production by Bacillus amyloliquefaciens (CSB 9) isolated from chromite mine environment. Biocatal Agric Biotechnol 3(3):35–41

Ravikumar KVG, Kumar D, Kumar G, Mrudula P, Natarajan C, Mukherjee A (2016) Enhanced Cr (VI) removal by nanozerovalent iron-immobilized alginate beads in the presence of a biofilm in a continuous-flow reactor. Ind Eng Chem Res 55(20):5973–5982

Rivera SL, Vargas E, Ramírez-Díaz MI, Campos-García J, Cervantes C (2008) Genes related to chromate resistance by Pseudomonas aeruginosa PAO1. Antonie Van Leeuwenhoek 94(2):299–305

Rizvi A, Ahmed B, Zaidi A, Khan M (2020) Biosorption of heavy metals by dry biomass of metal tolerant bacterial biosorbents: an efficient metal clean-up strategy. Environ Monit Assess 192(12):1–21

Rizwan M, Singh M, Mitra CK, Morve RK (2014) Ecofriendly application of nanomaterials: nanobioremediation. J Nanopart 2014

Romanenko VI, Koren'Kov VN (1977) Pure culture of bacteria using chromates and bichromates as hydrogen acceptors during development under anaerobic conditions. Mikrobiologiia 46(3):414–417

Saba B, Khan M, Christy AD, Kjellerup BV (2019a) Microbial phyto-power systems—A sustainable integration of phytoremediation and microbial fuel cells. Bioelectrochemistry 127:1–11

Saba YR, Ahmed M, Sabri AN (2019b) Potential role of bacterial extracellular polymeric substances as biosorbent material for arsenic bioremediation. Ann Finance 23:72–81

Sadik OA, Noah NM, Okello VA, Sun Z (2014) Catalytic reduction of hexavalent chromium using palladium nanoparticles: an undergraduate nanotechnology laboratory. J Chem Educ 91(2):269–273

Samuel J, Pulimi M, Paul ML, Maurya A, Chandrasekaran N, Mukherjee A (2013) Batch and continuous flow studies of adsorptive removal of Cr (VI) by adapted bacterial consortia immobilized in alginate beads. Biores Technol 128:423–430

Sangeetha S, Silviya S, Gurunathan J (2012) Hexavalent chromium reduction by metal resistant and halotolerant Planococcus maritimus VITP21. Afr J Microbiol Res 6(47):7339–7349

Sanjay MS, Sudarsanam D, Raj GA, Baskar K (2017) Isolation and identification of chromium reducing bacteria fromtannery effluent

Sarioglu OF, Celebioglu A, Tekinay T, Uyar T (2016) Bacteria-immobilized electrospun fibrous polymeric webs for hexavalent chromium remediation in water. Int J Environ Sci Technol 13(8):2057–2066

Satarupa D, Amal KP (2010) Occurrence and evaluation of chromium reducing bacteria in seepage water from chromite mine quarries of Orissa, India. J Water Resour Protect 2010

Scott JA, Palmer SJ (1990) Sites of cadmiun uptake in bacteria used for biosorption. Appl Microbiol Biotechnol 33(2):221–225

Shan B, Hao R, Xu H, Zhang J, Li J, Li Y, Ye Y (2022) Hexavalent chromium reduction and bioremediation potential of Fusarium proliferatum S4 isolated from chromium-contaminated soil. Environ Sci Pollut Res 29(52):78292–78302

Shao W, Li M, Teng Z, Qiu B, Huo Y, Zhang K (2019) Effects of Pb(II) and Cr(VI) stress on phosphate-solubilizing bacteria (bacillus Sp. strain MRP-3): oxidative stress and bioaccumulation potential. Int J Environ Res Publ Health 16. https://doi.org/10.3390/ijerph16122172

Sharma S, Malaviya P (2014) Bioremediation of tannery wastewater by chromium resistant fungal isolate Fusarium chlamydosporium SPFS2-g. Curr World Environ 9(3):721

Shaw DR, Dussan J (2018) Transcriptional analysis and molecular dynamics simulations reveal the mechanism of toxic metals removal and efflux pumps in Lysinibacillus sphaericus OT4b. 31. Int Biodeter Biodegrad 127:46–61

Shen H, Wang YT (1993) Characterization of enzymatic reduction of hexavalent chromium by Escherichia coli ATCC 33456. Appl Environ Microbiol 59(11):3771–3777

Shuhong Y, Meiping Z, Hong Y et al (2014) Biosorption of Cu(2+), Pb(2+) and Cr(6+) by a novel exopolysaccharide from Arthrobacter ps-5. Carbohydr Polym 101:50–56

Singh R, Kumar A, Kirrolia A, Kumar R, Yadav N, Bishnoi NR, Lohchab RK (2011a) Removal of sulphate, COD and Cr (VI) in simulated and real wastewater by sulphate reducing bacteria enrichment in small bioreactor and FTIR study. Biores Technol 102(2):677–682

Singh JS, Abhilash PC, Singh HB, Singh RP, Singh DP (2011b) Genetically engineered bacteria: an emerging tool for environmental remediation and future research perspectives. Gene 480(1–2):1–9

Singh RK, Sanchan VK, Ansari MQ, Pandey DS, Kamyotra JS (2013) Central Pollution Control Board. Ground Water Pollution Due to Chromium Rich Hazardous Waste Disposal in Raina–Khanchandpur Area. Dist Kanpur Dehat(R), U.P., India: A Case study. http://www.upsbdb.org/pdf/Souvenir2013/ch-3.pdf. Accessed 2 Nov 2019

Singh P, Chowdhuri DK (2017) Environmental presence of hexavalent but not trivalent chromium causes neurotoxicity in exposed Drosophila melanogaster. Mol Neurobiol 54(5):3368–3387

Singh P, Itankar N, Patil Y (2020) Biomanagement of hexavalent chromium: current trends and promising perspectives. J Environ Manag 279:111547

Singh Sankhla M, Kumar R, Prasad L (2021) Seasonal variations of lead and chromium concentrations in the water samples from Yamuna River in Delhi, India. Iran J Toxicol 15(2):109–114

Sonenberg N, Hinnebusch AG (2009) Regulation of translation initiation in eukaryotes: mechanisms and biological targets. Cell 136(4):731–745

Srinath T, Verma T, Ramteke PW, Garg SK (2002) Chromium (VI) biosorption and bioaccumulation by chromate resistant bacteria. Chemosphere 48(4):427–435

Srivastava S, Agrawal SB, Mondal MK (2016) Characterization, isotherm and kinetic study of Phaseolus vulgaris husk as an innovative adsorbent for Cr (VI) removal. Korean J Chem Eng 33(2):567–575

Srivastav A, Yadav KK, Yadav S, Gupta N, Singh JK, Katiyar R, Kumar V (2018) Nano-phytoremediation of pollutants from contaminated soil environment: current scenario and future prospects. In: Phytoremediation. Springer, Cham, pp 383–401

Stambulska UY, Bayliak MM, Lushchak VI (2018) Chromium (VI) toxicity in legume plants: modulation effects of rhizobial symbiosis. BioMed Res Int 2018

Sun H, Brocato J, Costa M (2015) Oral chromium exposure and toxicity. Curr Environ Health Rep 2(3):295–303

Sun L, Guo D, Liu K, Meng H, Zheng Y, Yuan F, Zhu G (2019a) Levels, sources, and spatial distribution of heavy metals in soils from a typical coal industrial city of Tangshan, China. Catena 175:101–109

Sun Y, Guan F, Yang W, Wang F (2019b) Removal of chromium from a contaminated soil using oxalic acid, citric acid, and hydrochloric acid: Dynamics, mechanisms, and concomitant removal of non-targeted metals. Int J Environ Res Public Health 16(15):2771

Sun Y, Iris KM, Tsang DC, Cao X, Lin D, Wang L et al (2019c) Multifunctional iron-biochar composites for the removal of potentially toxic elements, inherent cations, and hetero-chloride from hydraulic fracturing wastewater. Environ Int 124:521–532

Suthar B, Pansuriya J, Mafatlal MK, Patel VR, Nataraj M (2014) Biochemical changes under chromium stress on germinating seedlings of Vigna radiata. Notulae Sci Biol 6(1):77–81

Suzuki TOHRU, Miyata N, Horitsu H, Kawai K, Takamizawa K, Tai Y, Okazaki M (1992) NAD (P) H-dependent chromium (VI) reductase of Pseudomonas ambigua G-1: a Cr(V) intermediate is formed during the reduction of Cr(VI)–Cr(III). J Bacteriol 174(16):5340–5345

Swapna TH, Papathoti NK, Khan MY, Reddy G, Hameeda B (2016) Bioreduction of Cr(VI) by biosurfactant producing marine bacterium Bacillus subtilis SHB 13

Tan H, Wang C, Zeng G, Luo Y, Li H, Xu H (2020) Bioreduction and biosorption of Cr(VI) by a novel Bacillus sp. CRB-B1 strain. J Hazard Mater 386:121628

Tandukar M, Huber SJ, Onodera T, Pavlostathis SG (2009) Biological chromium(VI) reduction in the cathode of a microbial fuel cell. Environ Sci Technol 43(21):8159–8165

Tang CY, Fu QS, Criddle CS, Leckie JO (2007) Effect of flux (transmembrane pressure) and membrane properties on fouling and rejection of reverse osmosis and nanofiltration membranes treating perfluorooctane sulfonate containing wastewater. Environ Sci Technol 41(6):2008–2014

Tang X, Huang Y, Li Y, Wang L, Pei X, Zhou D (2021) Study on detoxification and removal mechanisms of hexavalent chromium by microorganisms. Ecotoxicol Environ Saf 208:111699

Tebo BM, Obraztsova AY (1998) Sulfate-reducing bacterium grows with Cr(VI), U(VI), Mn(IV), and Fe(III) as electron acceptors. FEMS Microbiol Lett 162(1):193–199

Thacker U, Parikh R, Shouche Y, Madamwar D (2006) Hexavalent chromium reduction by Providencia sp. Process Biochem 41(6):1332–1337

Thatheyus AJ, Ramya D (2016) Biosorption of chromium using bacteria: an overview. Sci Int 4(2):74–79

Thatoi H, Das S, Mishra J, Rath BP, Das N (2014) Bacterial chromate reductase, a potential enzyme for bioremediation of hexavalent chromium: a review. J Environ Manag 146:383–399

Travacio M, Polo JM, Llesuy S (2000) Chromium (VI) induces oxidative stress in the mouse brain. Toxicology 150(1–3):137–146

Tripathi M, Garg SK (2013) Co-remediation of pentachlorophenol and Cr6+ by free and immobilized cells of native Bacillus cereus isolate: spectrometric characterization of PCP dechlorination products, bioreactor trial and chromate reductase activity. Process Biochem 48(3):496–509

Turick CE, Apel WA, Carmiol NS (1996) Isolation of hexavalent chromium-reducing anaerobes from hexavalent-chromium-contaminated and noncontaminated environments. Appl Microbiol Biotechnol 44(5):683–688

Ukhurebor KE, Aigbe UO, Onyancha RB, Nwankwo W, Osibote OA, Paumo HK (2021) Effect of hexavalent chromium on the environment and removal techniques: a review. J Environ Manag 280:111809

USEPA (2010) IRIS toxicological review of hexavalent chromium (external review draft). EPA/635/R-10/004A

USEPA (2014) Toxic and priority pollutants under the clean water act priority pollutant list

Vala AK, Anand N, Bhatt PN, Joshi HV (2004) Tolerance and accumulation of hexavalent chromium by two seaweed associated aspergilli. Mar Pollut Bull 48(9–10):983–985

Valls M, Atrian S, de Lorenzo V, Fernández LA (2000) Engineering a mouse metallothionein on the cell surface of Ralstonia eutropha CH34 for immobilization of heavy metals in soil. Nat Biotechnol 18(6):661–665

Van der Ent A, Baker AJ, Reeves RD, Pollard AJ, Schat H (2013) Hyperaccumulators of metal and metalloid trace elements: facts and fiction. Plant Soil 362(1):319–334

Vega Cuellar MÁ, Calderón Domínguez G, Perea Flores MDJ, Peña Barrientos A, Salgado Cruz MDLP, García Hernández AB, Dávila Ortiz G (2022) Use of microorganisms and agro-industrial wastes in the biosorption of chromium (VI): a review. Waste Biomass Valoriz 1–22

Velásquez L, Dussan J (2009) Biosorption and bioaccumulation of heavy metals on dead and living biomass of Bacillus sphaericus. J Hazard Mater 167(1–3):713–716

Verma S, Kuila A (2019) Bioremediation of heavy metals by microbial process. Environ Technol Innov 14:100369

Vidali M (2001). Bioremediation. An overview. Pure Appl Chem 73(7):1163–1172

Vijayaraghavan K, Yun YS (2008) Bacterial biosorbents and biosorption. Biotechnol Adv 26(3):266–291

Vijayaraj AS, Mohandass C, Joshi D (2019) Microremediation of tannery wastewater by siderophore producing marine bacteria. Environ Technol

Vincent JB (2017) New evidence against chromium as an essential trace element. J Nutr 147(12):2212–2219

Viti C, Pace A, Giovannetti L (2003) Characterization of Cr (VI)-resistant bacteria isolated from chromium-contaminated soil by tannery activity. Curr Microbiol 46(1):0001–0005

Viti C, Marchi E, Decorosi F, Giovannetti L (2014) Molecular mechanisms of Cr (VI) resistance in bacteria and fungi. FEMS Microbiol Rev 38(4):633–659

Wadood HZ, Latif A, Mukhtar H, Javed M, Mukhtar H, Rehman Y (2021) Planktonic cells of Staphylococcus and Bacillus species capable of faster chromium reduction in short incubation times as compared to their biofilms. Arab J Geosci 14(17):1–9

Wang PC, Mori T, Komori K, Sasatsu M, Toda K, Ohtake H (1989) Isolation and characterization of an Enterobacter cloacae strain that reduces hexavalent chromium under anaerobic conditions. Appl Environ Microbiol 55(7):1665–1669

Wang Y, Shi J, Wang H, Lin Q, Chen X, Chen Y (2007) The influence of soil heavy metals pollution on soil microbial biomass, enzyme activity, and community composition near a copper smelter. Ecotoxicol Environ Saf 67(1):75–81

Wang Y, Su H, Gu Y, Song X, Zhao J (2017) Carcinogenicity of chromium and chemoprevention: a brief update. OncoTargets Therapy 4065–4079

Wang J, Chen C (2009) Biosorbents for heavy metals removal and their future. Biotechnol Adv 27(2):195–226

Wang J, Zhang X, Yang Q, Zhang K, Zheng Y, Zhou G (2018) Pollution characteristics of atmospheric dustfall and heavy metals in a typical inland heavy industry city in China. J Environ Sci 71:283–291

Wang C, Cui Y (2019) Recognition of a new Cr(VI)-reducing strain and study of the potential capacity for reduction of Cr (VI) of the strain. BioMed Res Int 2019

Wang J, Zhao S, Ling Z, Zhou T, Liu P, Li X (2021) Enhanced removal of trivalent chromium from leather wastewater using engineered bacteria immobilized on magnetic pellets. Sci Total Environ 775:145647

Watkins SJ, Norbury CJ (2002) Translation initiation and its deregulation during tumorigenesis. Br J Cancer 86(7):1023–1027

Wu X, Wu X, Shen L, Li J, Yu R, Liu Y et al (2019) Whole genome sequencing and comparative genomics analyses of Pandoraea sp. XY-2, a new species capable of biodegrade tetracycline. Front Microbiol 10:33

Wu YH, Lin JC, Wang TY, Lin TJ, Yen MC, Liu YH et al (2020) Hexavalent chromium intoxication induces intrinsic and extrinsic apoptosis in human renal cells. Mol Med Rep 21(2):851–857

Xu J, Sheng GP, Luo HW, Li WW, Wang LF, Yu HQ (2012) Fouling of proton exchange membrane (PEM) deteriorates the performance of microbial fuel cell. Water Res 46(6):1817–1824

Xu X, Nie S, Ding H, Hou FF (2018) Environmental pollution and kidney diseases. Nat Rev Nephrol 14(5):313–324

Yaashikaa PR, Kumar PS, Babu VM, Durga RK, Manivasagan V, Saranya K, Saravanan A (2019) Modelling on the removal of Cr (VI) ions from aquatic system using mixed biosorbent (Pseudomonas stutzeri and acid treated Banyan tree bark). J Mol Liq 276:362–370

Yakasai HM, Muhammad F, Shehu U, Yusuf F, Ahmad FA (2022) Isolation and identification of chromium-reducing bacteria from challawa industrial area Kano State, Nigeria. J Adv Microbiol 15–23

Yan FF, Wu C, Cheng YY, He YR, Li WW, Yu HQ (2013) Carbon nanotubes promote Cr (VI) reduction by alginate-immobilized Shewanella oneidensis MR-1. Biochem Eng J 77:183–189

Yang Q, Han B, Xue J, Lv Y, Li S, Liu Y et al (2020) Hexavalent chromium induces mitochondrial dynamics disorder in rat liver by inhibiting AMPK/PGC-1α signaling pathway. Environ Pollut 265:114855

Yu Y, Ali J, Yang Y, Kuang P, Zhang W, Lu Y, Li Y (2022) Synchronous Cr (VI) remediation and energy production using microbial fuel cell from a subsurface environment: a review. Energies 15(6):1989

Zainith S, Purchase D, Saratale GD, Ferreira LFR, Bilal M, Bharagava RN (2019) Isolation and characterization of lignin-degrading bacterium Bacillus aryabhattai from pulp and paper mill wastewater and evaluation of its lignin-degrading potential. 3 Biotech 9(3):1–11

Zawadzka AM, Crawford RL, Paszczynski AJ (2007) Pyridine-2, 6-bis (thiocarboxylic acid) produced by Pseudomonas stutzeri KC reduces chromium (VI) and precipitates mercury, cadmium, lead and arsenic. Biometals 20(2):145–158

Zayed AM, Terry N (2003) Chromium in the environment: factors affecting biological remediation. Plant Soil 249(1):139–156

Zeng Q, Hu Y, Yang Y, Hu L, Zhong H, He Z (2019) Cell envelop is the key site for Cr(VI) reduction by Oceanobacillus oncorhynchi W4, a newly isolated Cr(VI) reducing bacterium. J Hazard Mater 368:149–155

Zeng W, Li F, Wu C, Yu R, Wu X, Shen L et al (2020) Role of extracellular polymeric substance (EPS) in toxicity response of soil bacteria Bacillus sp. S3 to multiple heavy metals. Bioprocess Biosyst Eng 43(1):153–167

Zhang K, Li F (2011). Isolation and characterization of a chromium-resistant bacterium Serratia sp. Cr-10 from a chromate-contaminated site. Appl Microbiol Biotechnol 90(3):1163–1169

Zhang JK, Wang ZH, Ye Y (2016a) Heavy metal resistances and chromium removal of a novel Cr (VI)-reducing Pseudomonad strain isolated from circulating cooling water of iron and steel plant. Appl Biochem Biotechnol 180(7):1328–1344

Zhang X, Zhao X, Wan C, Chen B, Bai F (2016b) Efficient biosorption of cadmium by the self-flocculating microalga Scenedesmus obliquus AS-6-1. Algal Res 16:427–433

Zhang B, Wang Z, Shi J, Dong H (2020) Sulfur-based mixotrophic bio-reduction for efficient removal of chromium (VI) in groundwater. Geochim Cosmochim Acta 268:296–309

Zhang Y, Zhao Q, Chen B (2022) Reduction and removal of Cr (VI) in water using biosynthesized palladium nanoparticles loaded Shewanella oneidensis MR-1. Sci Total Environ 805:150336

Zhao XY, Lu MH, Yuan DJ, Xu DE, Yao PP, Ji WL (2019) Mitochondrial dysfunction in neural injury. Front Neurosci 13:30

Zhitkovich A (2011) Chromium in drinking water: sources, metabolism, and cancer risks. Chem Res Toxicol 24(10):1617–1629

Zhu XL, Li X, Kou ZJ, Wang JQ, Shang XQ, Chen C (2021) The adsorption effects and mechanisms of biochar immobilized sulfate-reducing bacteria (SRB) on Cr (VI). J Agro-Environ Sci 40:866–875

Chapter 14
Effects and Responses of Chromium on Plants

Shipa Rani Dey, Monika Sharma, and Prasann Kumar

Abstract There are many ways that chromium is used in industry has resulted in its status as a serious environmental pollutant in the modern world. It is of recent concern that soil and water may be contaminated with chromium (Cr) due to its presence in the environment. When it comes to the toxicity of chromium, the level of toxicity is determined by the valence state of the element. In contrast to the valence state Cr(6+), which is very toxic and highly mobile, the valence state Cr(3+) is much less toxic and much less mobile. Chromium does not have a specific transport mechanism that enables it to be transported from plant to plant in a specialized manner. As a result, this element is taken up by plant components that are responsible for transporting essential ions throughout the plant. Chromium can have a toxic effect on the growth and development of plants by causing changes in the germination process. This is in addition to alterations in the growth of stems, roots, leaves, and other plant parts. Also, it is worth mentioning that these alterations may occur due to the toxic effects that Cr has on the growth and development of plants. In the physiological realm, it has been demonstrated that the presence of Cr in soil has a detrimental impact on the physiological processes that plants engage in, including photosynthesis, mineral nutrition, and the relationship between water and soil. As well as having the capacity to generate reactive oxygen species, plants have also been found to display a direct effect of Cr exposure on enzymes and other metabolites, which in turn can cause oxidative stress in plants, as a consequence of being exposed to Chromium. Because of their potential for bioremediation, the utilization of plants for the bioremediation of Chromium contamination that can accumulate or stabilise Chromium compounds has recently achieved a great deal of attention. This is due to the plants' potential for bioremediation.

Keywords Abiotic · Bioremediation · Chromium · Toxic · Mitigation · Physiological · Photosynthetic · Zero hunger · No poverty

---

S. R. Dey · M. Sharma · P. Kumar (✉)
Department of Agronomy, School of Agriculture, Lovely Professional University, Phagwara, Punjab 144411, India
e-mail: prasann0659@gmail.com

© The Author(s), under exclusive license to Springer Nature Switzerland AG 2023

N. Kumar et al. (eds.), *Chromium in Plants and Environment*, Environmental Science and Engineering, https://doi.org/10.1007/978-3-031-44029-8_14

14.1 Introduction

Chromium is found in a wide range of minerals and rocks due to its high reactivity. It is also found in freshwater, as it is easily soluble in water and can be taken up by aquatic organisms. It is also found in soil and sediment, as it is part of the Earth's crust and cycles through the environment. The two forms of chromium least prone to going through chemical changes are hexavalent chromium (also known as Cr-VI) and trivalent chromium (Cr-III) (Prasad et al. 2021). Hexavalent chromium is more prevalent in industrial areas due to its increased use as a corrosion inhibitor in metalworking and welding. Trivalent chromium is more prevalent in natural environments, as it is released from the weathering of rocks and minerals. It can also be found in certain foods, such as grains, fruits and vegetables, due to its uptake by crops. Hexavalent chromium has a strong oxidizing nature, which makes it a compound that can cause cancer and mutations in living organisms. Therefore, it is important for those living in industrial areas to be aware of the possible risks of hexavalent chromium exposure. Several nations have designated Cr as a high-priority pollutant, including the United States Environmental Protection Agency (USEPA) (Mushtaq et al. 2021). To ensure the safety of residents in these areas, it is important to understand the levels of hexavalent chromium that are present and to take necessary precautions to protect against its known harmful effects. It is thought that chromium's toxicity is due to its ability to pass through the intercellular membranes and produce intracellular Reactive Oxygen Species (ROS), which are extra-toxic. Therefore, it is essential to monitor the levels of hexavalent chromium in the environment and to take proactive measures to protect against its potential health risks. A thousand times more dangerous and one hundred times more mutagenic than Cr(3+), Cr(6+) is one hundred times more toxic than Cr(3+). This makes Cr(6+) a dangerous contaminant that must be monitored and managed to avoid potential health risks. Trivalent chromium, on the other hand, has been found to play an important role in the regulation of glucose, triglycerides, and cholesterol in humans. However, despite its potential health benefits, Cr(3+) must still be monitored and managed due to its toxicity, as even small amounts can cause serious health risks. Despite this, a higher concentration of Cr(3+) can inhibit the activity of metalloenzymes due to its ability to form complexes with organic compounds (Zhitkovich et al. 2001). Therefore, it is important to ensure that the concentration of Cr(3+) is carefully regulated to provide potential health benefits while avoiding its associated health risks. In soil, groundwater, and sediments, it ranks 2nd after arsenic (Kar et al. 2008; Ogundiran and Afolabi 2008). With this in mind, it is essential to maintain the concentration of Cr(3+) at optimal levels, to enable its desirable health benefits while avoiding the associated health risks. In the periodic table, Cr belongs to group VI-b (Mandich 1997). The oxidation state of chromium ranges from −2 to +6. Consequently, knowledge of the oxidation state of chromium is critical in determining the risk of exposure and the health implications associated with it. Smith et al. (2002) refer to chromium (6+) and chromium (3+) as the most significant oxidation states of Cr. Additionally, chromium (3+) is considered more toxic than chromium (6+), thus it is important to differentiate between the two-oxidation states for accurate assessment. The following is an excerpt from Shanker et al. (2005), which suggest that it is the element Cr which plays a greater role in plant growth than the other metals, but that this element overall is considered to be of lesser importance in the context of plant development. Furthermore, while chromium (6+) is typically considered less toxic than chromium (3+), it is important to take into account the element's oxidation state when evaluating its impact on plant growth. Evidence suggests that, although the element Cr has a positive effect on plant development, it is not as influential as other essential metals. The amount of elemental Cr that was released into the environment each year ranged from 2,000 to 3,200 tonnes in some Asian countries (Chandra et al.1997). According to Krishnamurthy and Wilkens, the groundwater and soil of those countries were found to contain a very high amount of Cr contamination, including 14,800 mg/l in groundwater and 25,900 mg/l in soil. This contamination was highly concerning, and urgent action was needed to reduce the amount of Cr released into the environment. It has become increasingly apparent that the accumulation of Cr in the soil is one of the most pressing environmental concerns on a global scale, because of its detrimental effects on both crop production and human health (Tiwari et al. 2009). Consequently, immediate attention must be given to developing strategies to reduce the amount of Cr released into the environment. The toxicity of Chromium to plants has been demonstrated by demonstrating the failure of plants to grow, the formation of chlorosis in the leaves, and the damage to the roots, as well as a decrease in grain yield (Scoccianti et al. 2006; Ali et al. 2013a). To minimize the impacts of Cr on the environment, it is essential to focus on interventions that prevent Cr from entering the soil, such as washing off industrial waste, ensuring proper disposal of hazardous waste, and monitoring activities that may increase Cr concentrations. Furthermore, plants that grow in places where there is a lot of chromium are more likely to produce reactive oxygen species (ROS) as a result of the presence of chromium, such as H_2O_2, OH–, and O_2. Additionally, Cr-resistant species may be used to reduce the levels of ROS in contaminated soil, thus providing a further safeguard against the environmental impacts of Cr. These reactive oxygen species have been recognized for their detrimental effects on the production of biomolecules and their ability to cause damage, as well as damage to membranes and electrolyte leakage (Ali et al. 2011, 2015a, b). To this end, the use of Cr-resistant species may provide an effective strategy to mitigate the potential environmental damage caused by ROS. Malondialdehyde, also known as MDA, is one of the final products of the peroxidation process. It is created when free radicals cause damage to lipids through the process of peroxidation, which results in the production of malondialdehyde. Therefore, the use of Cr-resistant species may offer a promising solution to reduce MDA levels and consequentially, the peroxidation of lipids. Aside from this, it also acts as a marker for the formation of free radicals and the resulting damage to the tissue those results from their presence, so it also acts as a warning indicator. Additionally, MDA can also be used as an indicator of the effectiveness of dietary antioxidants, as its levels will be inversely proportional to the number of free radicals present. In the process of seed germination, chromium can cause metabolic disorders because of its presence. In this way, MDA is an essential marker for monitoring the damage caused by free radicals, as well as an indicator of the effectiveness of antioxidants and the potential damage chromium can cause in the process of seed germination. As a result, it interferes with the process by which stored food is converted into energy to assist in the subsequent successful emergence and establishment of seedlings in the environment by interfering with the processes involving the transformation of food into energy. This disruption of energy production results in a decrease in stored energy reserves, thereby reducing the potential for successful seedling emergence and establishment. In a study conducted on cowpea seeds (Vigna sinensis (L.), Savi ex Hassk) containing various concentrations of Cr^{6+}. It has been demonstrated that a significant decrease in both the amylase activity in the seeds, as well as the total amount of sugar in the seeds, resulted from the exposure to $Cr^{6+.}$ Which in turn resulted in a depressing effect on germination characteristics (Nath et al. 2008). This suggests that Cr^{6+} had a negative impact on the germination process, leading to reduced seed viability. There was a correlation found between higher levels of chromium in its various valence states and a concurrent decrease in seed germination, as reported by Nagajyoti et al. (2010). The researchers studying the effects of a variety of trace metals on three distinct species of Veronica (Plantaginaceae), the researchers found a significant positive correlation between both the concentrations of iron and Chromium in the plant tissues (Zivkovic et al. 2012). Numerous studies have been conducted on the toxicity of Chromium in crop plants. The metabolism of plants, like maize (Zea mays) (Sharma and Pant 1994), barley (Hordeum vulgare) (Ali et al. 2004; Sharma et al. 1995a, b), and sorghum, is significantly impacted by chromium. For example, Riaz et al. discovered that the growth, yield, and biochemical parameters of chickpea (Cicer arietinum var NM-88) plants were severely inhibited when concentrations of the chromium salts $CrCl_3$ and K_2CrO_4 were at a level of 300 g/mL. This was the case even when the plants were grown under optimal conditions. Additionally, significant amounts of chromium were found in the roots at a concentration of 1.912 mg Cr/g of dry mass; however, even lower concentrations were also transferred into the shoots (0.086 mg Cr/g of dry mass) and the leaves (0.074 mg Cr/g of dry mass). Despite being used at a low concentration of 1.0 mM (Sharma et al. 1995a, b), the researchers reported that not a single seed could be produced at a concentration of this concentration. This illustrates the extremely low and ineffective efficiency of chromium in the plant's metabolism, making it clear that at such a low concentration, it fails to induce the necessary conditions for proper seed production. This concentration did not produce any viable seed, indicating that chromium could be toxic to the reproductive capacity of plants, even at relatively low concentrations.

14.2 Environment Concentrations and Sources of Chromium in the Air, Water, and Soil

The majority of chromium found in crustal rocks is derived from industrial sources, although naturally occurring chromium can also be found in crustal rocks. This is due to anthropogenic activities such as mining, smelting, and other industrial processes, as well as natural weathering and erosion processes. There are several different forms of ferrochrome ($Fe_2Cr_2O_4$) that are found in nature; the most common is ferrochrome ($Fe_2Cr_2O_4$) found in the earth's crust as well as other minerals. These ferrochrome deposits often contain high concentrations of chromium, making them attractive targets for extraction and use in industrial applications. Among the most significant environmental pollutants are the ones that are caused by human activity, specifically industrial processes that use chrome as a component of their components (primarily leather tanning, textile dyeing, textile pigment production, metallurgy, organic synthesis, cleaning agents, wood processing, anodizing aluminium, catalytic manufacture, alloy preparation, Cr plating, and wood preservation) (Alloway 2013). Chromium is released into the environment through emissions from these processes, and it can have a negative impact on air and water quality and can cause health issues for humans and animals. Sixty to seventy per cent of the world's total production of marketable chromite ore, which comes to a gross weight of 24,000_103 metric tonnes, is used up in the production of stainless steel and other alloys. The tanning of leather, electroplating, the production of pigments, and other chemical industrial processes use greater than 15% of the total energy (Papp and Lipin 2010). Currently, more than 4000 tanneries around the world participate in chrome tanning methods. In addition, chromite ore is also used for refractories, heaters, and bricks for the metal industry, further increasing the need for its production. Because of the tannery industry in India, an estimated 2,000–3,000 tonnes of elemental chromium are released into the environment every year because of pollution. Consequently, the effects of chromium pollution on the environment and human health have been a growing concern, with exposure to hazardous levels of chromium being linked to serious health conditions. The tanning agent chrome (Cr) is used in approximately 80–90% of the leather industry. The effluents from these tanneries contain approximately forty per cent of the Cr that is used in the form of Cr(6+) and Cr(3+) salts (Sundaramoorthy et al. 2010). As far as chromium concentrations are, concerned, freshwater concentrations of chromium can be as high as 0.1–0.5 mg/l, while salt water concentrations of chromium can be as low as 0.0016–0.05 mg/l (Kumar and Puri 2012). However, due to the high levels of chrome used in the tanning process, the effluent produced can cause environmental damage if not handled properly. The World Health Organization (WHO) suggests that the maximum possible limits for the discharge of Cr(6+) into inland surface water and drinking water should be 0.1 mg/l and 0.05 mg/l, respectively. There are a number of chemical concentrations that are expected to be permitted and these numbers refer to the maximum concentrations. Therefore, it is important to adhere to these maximum possible limits to ensure the safety and quality of drinking water and inland surface water. These are the maximum permissible limits. According to F€orstner and Wittmann (2012), Cr holds the position of the 21st most abundant element that can be found in the crust of the earth. A study by Polti et al. (2011) has shown that the amount of chromium in the soil can fluctuate between five and three thousand mg per gram, depending on the soil characteristics (Polti et al. 2011). Even though the amount of chromium in soil fluctuates, the maximum permissible limits for its concentration should be observed to ensure safety and health. There are many sources of chromium in addition to natural rocks, including solid wastes, industrial effluents, ferrochromium slag with chromium-based byproducts, leachates, and dust particles with concentrations of chromium that are significantly higher than those allowed in permissible limits. Moreover, such human activities can cause an exponential increase in the concentrations of Chromium in the environment, which can be a major cause of concern.

14.3 Chromium's Toxic Effects on Plants

The compounds that are made up of chromium are extremely toxic to plants and if they meet them, then their growth and development can be slowed or stopped. In addition, the compounds can act as a poison to animals that ingest them, leading to potentially severe health risks. The use of chrome compounds should be avoided at all costs. Therefore, the potential environmental and health risks associated with chrome compounds make them dangerous and should be avoided. In spite of the fact that certain crops are not affected by low levels of Chromium (3.8 10 4 AMS) (Huffman and Allaway 1973a, b), Chromium is toxic at a dry weight of 100 AMS per kg for the majority of higher plants (Davies et al. 2002). Consequently, it is important to be aware of the associated risks and take all necessary precautions to avoid the use of chrome compounds whenever possible. When chromium levels reach a certain threshold level, the element is no longer a necessary component of life and therefore toxic. Therefore, it is essential to take all the appropriate steps to limit exposure to chromium-based toxins, as their presence in the environment can be hazardous to our health. A plant's metabolism cannot be effected by the metal, which does not have any function within an ecosystem, even if it could play a role in the metabolic processes of a plant (Dixit et al. 2002). As such, it is important to ensure that chromium levels are strictly monitored, as any excess of the element can lead to hazardous levels of contamination and can have a detrimental effect on the environment. A plant's accumulation of Chromium can have a number of detrimental effects, including the loss of pigment content, stunted growth, the induction of chlorosis in young plants, ultrastructural modifications to the chloroplasts and cell membranes, mutated enzyme functions, and impairment of root cells. Chaudhury and Panda (2005) indicate that Chromium accumulates in plants in a negative way. As a result, it is important to ensure that the level of Chromium in soil is monitored to ensure that plants are not exposed to excessive levels of this element. Because of its toxicity, chromium can stop seeds from germinating and slow the development of radicles in plants as well as prevent seeds from germinating (Panda et al. 2002). Therefore, it is essential to closely monitor Chromium levels in soil to mitigate the potential impacts it can have on plant growth. Plants are susceptible to a variety of factors that can lead to a reduction in their size, including a reduction in their rate of cell division, which occurs because of the induction of chromosomal aberrations (Liu et al. 1993). To ensure optimum plant growth, it is critical to keep track of Chromium levels in order to reduce the chances of chromosomal aberrations. According to Yoon et al. (2006), there is a variation in the accumulation of metals in different species of plants, and they attribute this difference to the variation in the genetic code of the plants. To further reduce the chances of chromosomal abnormalities, it is important to use strategies to monitor and manage Chromium levels accurately, as this will ensure the highest potential for successful plant growth. Physiological and morphological traits of genotypes can be used as an indicator of genetic variation in a population (Ishikawa et al. 2006). To this end, it is imperative to monitor genetic variability in plants with physiological and morphological traits; this will provide valuable insight into maximizing potential growth (Ishikawa et al. 2006). It is important to understand that when the concentration of Chromium is in the micromole range, plant cells can exhibit severe symptoms of phytotoxicity. This knowledge is essential to assess the effect of Chromium on plant cells and to determine the threshold concentration of Chromium that is safe for the plants to grow in their optimal potential. A plant such as Lemna minor, Pistia sp., and Taxithelium nepalense have their unique taxonomy, particularly in terms of their ability to alter ultrastructure at the chloroplast level, which ultimately ends up inhibiting photosynthesis (Choudhury and Panda 2005). Furthermore, this understanding is key to improving our comprehension of how plants process and metabolize Chromium, allowing us to assess its full potential to be a beneficial component of soil health. There is evidence that higher concentrations of Cr can negatively affect the roots of plants, causing them to wilt and leading to plasmolysis in the root cells (McGrath 1995). However, it is important to note that a careful balance of Cr must be maintained in soil, as too much can be detrimental to the health of plants. The results of the study showed that hexavalent Chromium has the potential to cause severe phytotoxic effects at high concentrations (1 mM), such as the distortion of the chloroplast membrane and the severe disarrangement of the thylakoids at high concentrations (1 mM). Additionally, it was observed that the accumulation of Cr in the roots of plants significantly decreased at higher concentrations, indicating that the plants were unable to absorb the metal effectively. This suggests that as the concentration of Cr increases, the metal becomes increasingly toxic and can lead to reduced growth and development of the plant. Because thylakoids become seriously disorganized, these effects occur because of these conditions. As a result, plants are particularly sensitive to the presence of high concentrations of Cr, as it can greatly disrupt their growth and development. It has been shown that aquatic plants, such as Vallisneria spiralis, can store substantial quantities of chromium in their tissues, which results in a reduction for biomass of these plants (Vajpayee et al. 1999, 2000, 2001). This confirms that Cr is a critical factor in the growth and health of plants and that it is important to keep Cr levels within healthy limits. Because Chromium is capable of degrading aminolevulinic acid dehydratase, an enzyme that plays a significant role in the production of chlorophyll, it has a negative impact on the utilization of aminolevulinic acid. Therefore, managing Chromium levels is essential in order to ensure optimal plant growth and health. Carotenoids in plants are susceptible to being degraded when Cr is present (Rai et al. 1992). In solution culture, Cr is more dangerous. After all, it is in a soluble form, it is very simple for plants to take up. In soil, most of the Cr is no longer available because of adsorption, reduction, and precipitation (Zayed and Terry 2003). In the following sections, we will investigate several of the metabolic and physiological methods that are altered in plants due to the presence of Chromium. These changes can be observed in a variety of metabolic pathways and physiological processes.

14.4 Plant Chromium Uptake, Transport, and Distribution

In spite of the fact that we are still learning about the mechanisms involved in the absorption of Cr and the distribution of this nutrient throughout the plant's vegetative and reproductive parts, we still lack a comprehensive understanding of what causes this process. To gain a better insight into how this process works, further research is needed to explore the various mechanisms and pathways associated with Cr absorption and distribution. As Cr is not a vital element for the survival of plants, plants do not have specific mechanisms for it to be absorbed into their tissues through their roots. However, despite this, plants may still be able to accumulate Cr if it is available in the soil solution, allowing for a passive absorption process. Due to this, the carriers that are used during the uptake of heavy metals also serve as carriers during the uptake of heavy metals. Consequently, the effectiveness of Cr uptake by plants depends on its availability in the soil solution and its association with the soil carriers. Plants absorb Cr from the soil and carry it throughout their tissues in a manner that changes with time depending on the mechanism that they use to do so. Therefore, understanding the mechanism of Cr uptake by plants is critical for determining how much of it is available for use in the environment. The uptake process appears to be influenced by both active and passive transports; the former is more prevalent at lower concentrations, whereas the latter becomes more significant once the levels of the compound reach toxic levels (when membrane selectivity is lost). Moreover, proton-dependent transporters and anion channels facilitate the active uptake of Cr by plants, while passive transport mainly takes place through an anion-exchange mechanism. It is important to note that the effects of Cr contamination on plant physiology are determined by the metal speciation, which is responsible for the uptake, mobilisation, translocation and accumulation of chromium within the plant system, all of which contribute to the plant's vulnerability to harm. Therefore, it is essential to consider the chemical form of Cr in order to accurately assess its potential impacts on the plants. Several active mechanisms constitute the pathway for the transport of Cr(6+) that consists of ion transporters such as sulphate, one of the most important ion transporters (Cervantes et al. 2001). Consequently, understanding the chemical form of Cr is critical in order to accurately evaluate its effects on the plants.

Additionally, numerous active mechanisms facilitate the pathway of Cr(6+) through ion transporters, including sulphate, which is the most widely used ion transporter (Cervantes et al. 2001). As far as carrier binding goes, it is established that the metals chromium, iron, sulfur, and phosphorus are highly competitive for carrier binding with one another. Furthermore, these mechanisms enable Cr(6+) to bind to several other ions, such as iron, sulfur, and phosphorus, leading to intense competition for carrier binding. In order to establish tolerance to toxic metals by plants, it is believed that the plasma membrane of the root, which is the first functional structure to meet toxic metals, plays a very important role in the process by which these metals are tolerated by plants. Consequently, the plasma membrane is thought to possess special mechanisms that enable plants to effectively respond to and tolerate the presence of toxic metals in the soil. There was a reduction in the uptake of Cr(6+) with the use of metabolic inhibitors but there was no adverse effect on the uptake of Cr(3+) with the use of metabolic inhibitors. This suggests that plants have evolved specific mechanisms for dealing with different forms of chromium, allowing them to better survive in soils containing toxic metals. It is evident that the amount of metabolic energy available for the uptake of Cr(6+) is determined by the amount of energy available for the uptake of Cr(3+), but not the amount of energy available for the uptake of Cr(6+). This means that plants have developed ways to prioritize the uptake of Cr(3+) over Cr(6+) as Cr(3+) is less toxic than Cr(6+). They have also evolved ways to efficiently use the energy they have available to take up the less toxic form of chromium, allowing them to survive in soils with a higher concentration of toxic metals. According to Barcelo and Poshenrieder (1997), there are two types of uptake of Cr(3+): a non-active process, which is thought to occur in a passive manner, and an active process, which occurs in a more active manner. By actively taking up the less toxic form of chromium, the organism's survival rates are significantly increased, even in environments that are more toxic. Skeffington et al. found that the sulphate carrier is less effective at taking in Cr(VI), while Cr(3+) forms binuclear complexes by affixing itself firmly to the carboxyl group of amino acids found in proteins. It has been observed that the cells immediately convert Cr(6+) that has been taken into Cr(3+) after it has been taken in. In the interior of the cell, Cr(3+) can be found in the cytosol. Because of its low mobility and recalcitrant nature, chromium can remain in the soil for an extended period, which can the amount of the element that is absorbed by plants should be increased. Because root exudates contain organic acids, which can combine with chromium to form complexes, and, as a result, making chromium available for uptake by the root may be a significant factor in the increased accumulation of chromium in the root. According to Srivastava et al. (1999a, b), a higher rate of chromium absorption in the plant's roots of *Lycopercicum esculentum* could be attributed, in part, to the root contains carboxylic acid in addition to the amino acids that it also contains. The xylem of plants is where most of the movement of Cr takes place. The distribution of Cr within crops possessed a consistent quality that was independent of the characteristics of the soil, as well as the concentration of this component. The roots have always contained the highest quantity of the contaminant element, while the vegetative and reproductive organs have always contained the lowest quantities (Pulford et al. 2001). In the case of beans, the seeds were only found to contain 0.1% of the Chromium that had accumulated, while the roots were found to contain 98% of it (Huffman and Allaway 1973a, b). Chromium gets stuck in the vacuoles of the root cells, which makes it less toxic. The higher amount of Cr in the roots of plants could be a natural way for the plant to deal with the element's toxicity and limit its exposure to it (Shanker et al. 2004). Because Cr(6+) and Cr(3+) must travel through the endodermis via the symplast. Cr(6+) in cells can likely be readily reduced to Cr(3+). The retention of Cr(3+) in the root cortex cells in the presence of low concentrations of Cr(6+) is one of the factors that contribute to the decreased toxicity of Cr(3+). Even though Cr(6+) reducing enzymes have not been found in higher vascular plants, they have been found in many different types of bacteria and fungi (Cervantes et al. 2001). The amount of chromium that accumulated in *Vigna radiata* was relatively low, and as a result, only a trace amount was found in the shoot. In contrast, Vigna unguiculata showed the highest amount of chromium accumulation. In most plant species, According to Shanker et al. (2005), the chromium that is transported from the roots to the shoots occurs at a very slow rate as it moves from the roots to the shoots. However, Vigna unguiculata had significantly higher chromium accumulation than other species, demonstrating its ability to transport chromium from the roots to the shoots more quickly than other species. The level of chromium accumulation in the roots of both of these species of Vigna was significantly greater when compared to the accumulation in the stems of the plants. There is a possibility that the majority of the chromium is stored in the vacuoles of the root cells to render it non-toxic. There may be a reason for the poor translocation of chromium from the soil to the shoot due to this reason. This suggests that vacuolar sequestration of chromium in root cells may be an important mechanism to reduce its toxicity and limit translocation to the shoot. Possibly, this might be a way for the plant to protect itself from its natural toxicity by protecting itself in this way. Therefore, it is likely that the root of the plant is employing a protective strategy to limit the accumulation of chromium in other parts of the plant. Therefore, it can be concluded that Vigna species can accumulate chromium in their roots, primarily to reduce its toxicity and protect the plant from potential harm.

14.4.1 Growth and Development

Plant growth and development are essential for the continuation of life and the spread of species. Without these processes, our planet would be unable to sustain itself and many plants and animals would become extinct. To ensure the continuation of life, we must protect and foster healthy plant growth and development. Due to the ongoing nature of their activity, and the fact that they rely heavily on the resources found in the soil and air around them to survive, they can sustain themselves. Therefore, it is important to monitor and manage these processes to ensure the viability of our planet's ecosystems. Consequently, plants play an integral role in our global ecosystem, providing food, shelter, and oxygen while stabilizing temperatures and maintaining biodiversity. As such, it is necessary to protect and preserve our plant resources to safeguard the health of our environment. A major factor that contributes to the expression of growth is the genotype. This is both a function of the genotype and the environment, which includes both internal and external growth factors. Accordingly, it is essential to understand the genotype and its interplay with the environment to effectively conserve and protect our plant resources. Moreover, this knowledge can inform and support strategies to promote the health and sustainability of our global ecosystem. As a function of both the genotype and the environment, growth is primarily determined by both. Therefore, it is important to gain a comprehensive understanding of the genotype and its relationship with the environment to successfully preserve and safeguard our plant resources. Moreover, this information can be utilized to inform and facilitate the implementation of strategies that promote the wellbeing and durability of the entire planet. Ultimately, the growth of species is determined by both the genetic makeup and environmental conditions. There is a change in the sequence of plant growth and development in response to the presence of Chromium in the environment in which the plant is growing, which causes changes in the plant's growth and development. Chromium plays a significant role in altering the growth trajectory of plants, as it affects the genetic expression and environmental factors in tandem.

14.4.2 Germination of Seeds and Development of Seedlings

To understand how chromium affects physiological processes, one of the first things you need to know is that chromium affects the initial stages of the germination process in seeds. If a seed can germinate in an environment that already contains chromium, that means that that seed has a high tolerance for the presence of this metal (Peralta et al. 2001). According to Rout and Almeida, it took 200 AM Cr order to achieve a 25% reduction in the rate of seed germination in the weed Echinochloa colona. A study conducted in 1982 found that Phaseolus vulgaris exhibited a 48% lower germination rate when presented with high levels of hexavalent chromium (500 ppm) in the soil in comparison to a control plant when exposed to low levels (50 ppm) of hexavalent chromium. As a result of exposure to 40 parts per million of $Cr(6+)$ in a contaminated medium, the seeds of lucerne (Medicago sativa cv. Malone) proved to be less able to germinate and grow in an uncontaminated medium. These findings were published in 2001 by Peralta et al. (2001). This reduction was by 23%. There was a reduction in sugarcane bud germination of between 32 and 57% when 20 and 80 ppm of Cr were used, respectively (Jain et al. 2000). It has been speculated that the decreased germination of seeds due to chromium stress may be caused by the inhibitory effect of chromium on the interaction of amylases and the subsequent transport of sugars to the embryo axes when chromium is present (Zeid 2001). Alternatively, the increased protease activity that occurs as a result of the chromium treatment may also be a factor that may be responsible for the diminished germination of the Cr-treated seeds (Zeid 2001), as it probably increases the enzyme activity of the seed during the treatment process.

14.4.3 Root Growth

Heavy metals present in plants and crops have the effect of slowing down the rate at which roots grow. As a result, crops can be adversely affected, potentially leading to lower yields and reduced quality. It has been shown that this effect is observed in both trees and crops (Tang et al. 2001). To combat this, farmers must consider methods to reduce or remove heavy metals from the soil to maximize the health and productivity of their crops. It has been reported that Prasad et al. (2001) found that cadmium, chromium, and lead were the metals that caused the greatest damage to the new root primordia in Salix viminalis. To further reduce the harmful effects of heavy metals on crops, farmers must consider methods such as land application of compost, phytoremediation, and the use of nanoparticles to reduce soil concentrations of these pollutants. In contrast to the other heavy metals that were investigated in this study, Chromium caused significantly greater damage to the root length than any other heavy metal that was examined. To further mitigate the effects of Chromium, farmers must explore more effective strategies than the previously mentioned methods, such as the use of biochar or chemical binding agents. The Caesalpinia pulcherrima is one of the most significant arid-adapted trees, and its roots and dry weight were inhibited when a concentration of 100 ppm Cr was applied to it (Iqbal et al. 2001). However, it is yet to be established whether Caesalpinia pulcherrima could be a viable option for farmers dealing with Chromium contamination, and further research is needed to assess this potential. In the soil, when 20 mg of Cr(6+) kg^{-1} of $K_2Cr_2O_7$ were present as $K_2Cr_2O_7$ in the soil, both the total root weight of the wheat and the root length of the wheat were adversely affected. To explore the viability of Caesalpinia pulcherrima as a potential solution, further research is needed to understand how it may respond to, and counteract, the effects of Chromium contamination. The research conducted by Panda and Patra (2000) revealed that the presence of chromium caused a significant increase in the root length of seedlings when they were grown in the presence of nitrogen (N) nutrition levels resulting in the presence of chromium. To gain a better understanding of the viability of Caesalpinia pulcherrima, further research is needed to determine how it responds to, and combats, chromium contamination, beyond what was observed in the study of Panda and Patra (2000). As a result of increased concentrations of Chromium present in all of the nitrogen treatments, however, the roots grew shorter with increased concentrations of Chromium present. Moreover, additional analysis must be conducted to determine if Caesalpinia pulcherrima is capable of sustaining chromium pollution in the long-term, as the observed root growth inhibition suggests that it may not be a sustainable solution for long-term chromium contamination. It is interesting to note that Samantaray et al. (1999) conducted a study involving five different cultivars of mung bean and found that when the soil contaminated with chromite mine spoiled for 28 days after root emergence, root growth was severely affected. Furthermore, the root biomass of each cultivar was significantly reduced, suggesting the potential of mung bean as an indicator species for soil contamination. As part of our research with chromite mine spoil soil, we made the following observation during the course of our experiments. We observed that the root growth of each cultivar was severely stunted and the root biomass of each cultivar decreased as the contamination period increased. This finding highlights the potential of mung bean as a suitable indicator species for soil contamination. As revealed by studies conducted with a scanning electron microscope on roots that had been exposed to Cr, there was an increase in the number of root hairs that formed as well as an increase in the thickness of the pith and the cortical layers when compared with roots that had not been exposed to Cr (Suseela et al. 2002). This finding is indicative of mung bean's ability to adapt to soil contamination and absorb Cr, making it a promising indicator species for research. This could be due to the inhibition of root cell division or elongation or an extension of the root cycle in the roots as a result of chromium toxicity, causing the general response of decreased root growth. Furthermore, this adaptation could also be attributed to the plant's ability to form chelates and bind the Cr, preventing its accumulation in the root cells. Both of these possibilities are possible. When seedling roots come into direct contact with Cr in the medium, the roots collapse. Consequently, when high concentrations of both chromium species are present in combination, the roots cannot take up water from the medium, which is a possible cause of a reduction in root growth when high concentrations of both chromium species are present in combination. Therefore, it is important to ensure that only one species of chromium is present in the medium to avoid these negative effects on root growth.

14.4.4 Stem Growth

Several reports have indicated that the presence of Cr can have a negative impact on the development of the plant's height and the growth of its shoots (Rout et al. 1997). When chromium was added to nutrient solutions in sand cultures of oats in concentrations of 2, 10, and 25 parts per million, Anderson et al. (1972) found that the plant height decreased by 11%, 22%, and 41%, respectively, when chromium was added to nutrient solutions at a concentration of 2, 10, and 25 parts per million. The results were compared to those of the control group. Researchers believe that the presence of Cr(6+) resulted in shorter plant heights for several species of Curcuma sativus, Lactuca sativa, and Panicum miliaceum mainly due to the presence of Cr(6+). They conducted research in 1995 to determine the effects of Cr(6+). During their study on lucerne cultures, according to the findings of Barton et al. (2000), they discovered that the incorporation of Cr(3+) into the soil reduced the rate at which the shoots grew. In a glasshouse experiment conducted by Sharma and Sharma, they found that Wheat cv. UP 2003 that had been sown in sand containing 0.5 AM sodium dichromate experienced a significant reduction in plant height 32 and 96 days after planting. When Cr was applied to Sinapsis alba soil at rates of 200 or 400 mg kg^{-1}, along with nitrogen, phosphorus, potassium, and sulphur fertilizers, there was a discernible drop in the overall height of the plant (Hanus and Tomas 1993). A reduction in root growth may be the primary cause of the decrease in plant height. As a consequence, there may be a reduction in the transport of nutrients and water to the higher parts of the plant as a result. As such, the application of Cr may have negative impacts on plant growth and lead to an overall decrease in biomass. The movement of chromium from the part of the plant that is in the air to the part of the plant that is in the ground can also have a direct effect on the cellular metabolism of the leaves, which contributes to the shortening of the leaves. Furthermore, the decrease in leaf size could be caused by the lack of an adequate supply of chromium, which serves as a key nutrient for cell metabolism.

14.4.5 Leaf Growth

To determine the amount of crop harvest that is realized, the growth of the leaves, their expansion, and their total number all play a significant role in determining the amount of crop harvest that is achieved. The amount of Chromium that was added to the nutrient solution caused the number of leaves produced by each wheat plant to decrease by fifty per cent as result of the addition of Chromium. The results of a study conducted by Tripathi et al. (1999) showed that the leaf area and biomass of Albizia lebbek seedlings were significantly altered when exposed to a high concentration (200 ppm) of the contaminant Cr(6+). A study conducted by these scientists indicates that the characteristics of leaf growth could be used as bioindicators of contamination with heavy metals and to select species that are resistant to such contamination. Moreover, these findings demonstrate the importance of conducting studies which consider the impact of heavy metals on the growth characteristics of different species. It was observed that when bush bean plants were grown in a 1–10 Ag cm^3 Chromium medium, both the primary leaves as well as the trifoliate leaves lost a fair amount of leaf area. Furthermore, the results of this study highlight the need for further research which examines the effects of heavy metals on the growth of different plant species. Additionally, exposure to high concentrations of chromium can cause a significant reduction in leaf area. Compared to the primary leaves of the plants, the trifoliate leaves were more affected by the Chromium than the primary leaves. Therefore, it is clear that the degree of sensitivity to heavy metals varied between plant species and that the trifoliate leaves were more susceptible to Chromium than the primary leaves. When 100 ppm of Cr(6+) was introduced into the soil, researchers discovered that bush bean plants experienced a reduction in their dry leaf yield of up to 45%. The researchers Karunyal et al. (1994) conducted research to determine how tannery effluent affected the leaf area and biomass of the plant. A significant reduction was observed in the leaf area and leaf dry weight of Oryza sativa, Acacia holosericea, and Leucaena leucocephala regardless of the concentration of tannery effluent that was used for all three plants. The results of a study conducted by Singh (2001) which examines the effects of Cr(3+) and Cr(6+) on spinach, found that the application of Cr at levels of 60 mg kg^{-1} or higher resulted in burns on the tips or margins of the leaves, reduced leaf size, and slowed the growth rate of the leaves. A study conducted by Jain et al. (2000) found that leaves were affected by chlorosis when the Cr concentration in the leaves reached 40 ppm, whereas necrosis occurred at a Cr concentration of

80 ppm. According to the research conducted by Pedreno et al. (1997), using a range of heavy metals, it was found that chromium had the greatest impact on leaf growth in tomato plants and that this impact was greater on the younger leaves as opposed to the older leaves. Interestingly, Poschenrieder et al. (1993) found that a decrease in the amount of leaf biomass in P. vulgaris was correlated with an increase in the amount of chromium extractable from oxalate acid when leaf biomass decreased.

14.4.6 Total Dry Matter Production

The primary requirement for higher yields to be obtained from plants is the increase in the amount of dry matter produced by the biomass of the plants. It has been estimated that 80–90% of the dry matter produced by plants is composed of carbon compounds. Specifically, the presence of heavy metals and chromium stress in the environment is the basis for the accumulation of organic substances and the production of dry matter. According to Bishnoi et al. (1993a, b), the increase in the photosynthetic process and the increase in source size is what allowed the accumulation of organic substances and the production of dry matter. Vallisneria spiralis was the subject of a research project to investigate the effect of chromium accumulation and toxicity on biomass production. This was done as part of an investigation into the effects of chromium accumulation on biomass production. A significant negative impact was seen on the amount of dry matter that was produced as a result of the concentrations of Cr(6+) in the nutrient medium that was greater than 2.5 Ag ml^{-1} in the nutrient medium. According to Vajpayee et al. (2001), the purpose of the study was to determine the relationship between chromium accumulation and biomass production (Vajpayee et al. 2001). In a study conducted by Zurayk et al. (2001), it has been shown that the interaction between salinity and Cr(6+) reduced the amount of dry biomass accumulation caused by Portulaca oleracea. There was an interaction between these two factors that resulted in this outcome. When grown at a concentration of 0.5 mM Cr(6+), cauliflower of the cv. Maghi variety significantly reduced dry biomass (Chatterjee and Chatterjee 2000). The results of a study conducted by Kocik and Ilavsky (1994) on sunflower, maize, and Vicia faba indicated that the effect of chromium on the quality and quantity of biomass was not significantly affected by 200 mg kg^{-1} of Cr(6+). On the other hand, there was a positive correlation between the contents of the soil and the amount of chromium that was taken up by the plant tissue Kocik and Ilavsky's (1994). There was a discernible drop in the amount of dry biomass produced by S. alba during the flowering stage when Cr(VI) was added to the soil at rates of either 200 or 400 mg kg^{-1} in combination with N, P, K, and S fertilizers (Hanus and Tomas 1993). A higher dry weight (DW) of roots and leaves was observed in P. vulgaris and maize plants which had been exposed to 1 AM Cr(III) compared to control plants. This growth in DW was more noticeable in conditions where there was a shortage of Fe. It was shown that in the water cultured plants that were exposed to 10 ppm of Chromium, the dry weight of the whole plant was reduced by a significant amount, going from 88.4 g $plant^{-1}$ in the control group to 28.4 g $plant^{-1}$ in the group exposed to 10 ppm of Chromium.

14.4.7 Yield

There is no doubt that Cr has an equivalently negative impact on the yield and productivity of the crops, and as a direct result, it has had a profoundly negative impact on the vast majority of physiological and biochemical methods. The study conducted by Golovatyj et al. (1999) found that when 100 or 300 mg kg^{-1} of Cr was added to the soil, the yields of barley and maize decreased. It was found that the number of flowers produced by each wheat plant decreased by 50% at 0.05 AM chromium compared to the control, while the decline was even greater at 0.5 AM chromium than at 0.05 AM chromium. There was a 59% decrease in the number of grains produced by each plant when the control was in 0.05 AM chromium as compared to the control. Even though the control grain had the highest grain DW, an increase in Cr level resulted in a reduction of 58–92% of the grain DW of the control grain. It has been shown that a higher concentration of Cr leads to a decrease in tillering and an increase in seed deformities. Sharma and Mehrotra, in a study published in 1993, found that the amount of dry seed per plant was 2.11 g in the absence of Cr, 0.39 g in the presence of 20 ppm of Cr, and 0.16 g in the presence of 200 ppm of Cr according to their study. As a result of the effects that chromium has on the processes that take place within a plant during its early stages of growth and development, it is eventually responsible for a significant decline in yield as well as the amount of total dry matter within a plant. The reason for this is that chromium impairs the production, translocation, and partitioning of assimilates to the plant's economic components as they are absorbed into the plant. As chromium has a direct effect on plants, there was almost a complete decrease in the yield of the plants as well as in the amount of dry matter produced as a result of its presence. This could ultimately result in a lack of nutrients in the shoot of the plant, as it makes it harder for the plant to absorb minerals and water, leading to slow growth and development of the plant. There was an increase in the amount of Cr(+6) that could enter the roots passively as the plants grew, and when Cr(6+) was transferred from the roots to the shoots, it damaged the plants' photosynthetic and mitochondrial systems, which in turn resulted in a lack of growth. Moreover, oxidative damage may have resulted in the breakdown of the normally functioning mechanism for the selective uptake of inorganic nutrients. In this way, the roots would have been able to absorb higher concentrations of Cr(6+) as a result of this. On the other hand, ligand substitution does not affect the rate at which Cr(3+) reacts with the ligand. Therefore, it can form substitution-inert metalloprotein complexes in living organisms as a result of this property, which is a unique property that reduces the role it plays in the production of toxic symptoms in living organisms. There is an opinion that the toxicity of Cr(III) can be attributed to indirect effects, such as the change in pH or the stopping of ions from moving around, which are examples of indirect effects.

14.5 Effect on Photosynthetic Pigments and Carbon Assimilation

There have been several studies that have shown that plants that have been exposed to Cr(6+) stress have been shown to have reduced amounts of total chlorophyll, chlorophyll a, chlorophyll b, and carotenoids (Vajpayee et al. 2000; Appenroth et al. 2003; Rai et al. 2004; Paiva et al. 2009; Redondo-Gómez et al. 2011). As a result of 72 h of exposure to 100 M Cr(6+), Rai et al. (2004) observed that Ocimum tenuiflorum lost approximately 70%, 69%, 73%, and 87% of its chlorophyll, chlorophyll a, and chlorophyll b content, respectively, following 72 h of exposure to 100 M Cr(6+). It was done as a reaction to the harmful effects of Cr(6+) on the body. It was found that Chl a exhibited a higher degree of sensitivity when exposed to Cr(6+) stress than Chl b (Vajpayee et al. 2000; Paiva et al. 2009). It was found that in P. amarus, Rai and Mehrotra (2008) discovered that chlorophyll b was much more sensitive to the effects of Cr(6+) stress than chlorophyll a. In other words, Pandey et al. (2005) found that after exposure to Cr(6+), the amounts of chlorophyll pigments in B juncea increased after a period of fifteen days in a Cr(6+) controlled experimental condition. As a result of the constrained expansion of the leaves, they attributed this increase proportionately to the increased growth rate. There have been studies that have shown that Cr(III) has detrimental effects on the net photosynthetic rate of G americana seedlings when grown in water that is contaminated with Cr(III). The stomatal conductance of H. annuus as well as the net photosynthesis of the plant were both inhibited by Cr(3+) when it was present at a concentration of 100 M in the plant (Davies et al. 2002). A study conducted by Paiva et al. (2009) on Eichhornia crassipes showed that both Cr(3+) and Cr(6+) significantly decreased leaf gas exchange, Chlorophyll fluorescence parameters, and photosynthetic pigment levels, with Cr(6+) being the most toxic of the two compounds. The effect of exposure to Cr(6+) has also been observed to have an inhibiting effect on chlorophyll fluorescence emission spectrum, a decrease in the chlorophyll a/b ratio, as well as a reduction in PSII activity in Z. mays. It has been shown that Cr(6+) toxicity induced stomachal closure, reduced net photosynthetic rates, and reduced transpiration in O. sativa (Ahmad et al. 2011), *A. viridis* (Liu et al. 2008), L. perennial (Vernay et al. 2007), and a Cr-sensitive cultivar of *V. radiata*. It was determined that these changes were caused by a decrease in the net photosynthetic rate (Samantaray 2002). It has been suggested that the decreased chlorophyll content that has been observed as a response to chromium exposure is a result of the impaired activity of various enzymes responsible for chlorophyll biosynthesis. Two examples of such enzymes are protochlorophyllide reductase and aminolevulinic acid dehydratase. ALAD is also known as porphobilinogen synthase. The production of chlorophyll in plants is dependent on the presence of these two enzymes (Ganesh et al. 2008). According to Vasjpayee et al. (2000), it has been shown that Cr(VI) reduced chlorophyll levels in *Nymphaea alba* at a concentration of less than one mM, inhibited ALAD activity, and increased the amount of-aminolevulinic acid (ALA) in Nymphaea alba (Nymphaea alba). When there is a high concentration of Cr in the environment, ALAD activity is inhibited, which may contribute to the use of PBG (porphobilinogen) for the synthesis of chlorophyll that is limited in the presence of Cr toxicity. It has been suggested that the reduction of chlorophyll pigments that can be attributed to Cr(VI) stress may be connected to a reduction in the plant's ability to absorb magnesium and nitrogen, both of which are essential elements that are found in chlorophyll molecules. Additionally, magnesium and nitrogen deficiencies caused a precipitous drop in the amount of light that was absorbed from 500 to 2,600 nm as a result of a lack of magnesium and nitrogen. Upon the application of Cr, Sharma et al. (1995a, b) found that there was a significant reduction in the amount of Hill activity in T. aestivum (VI) under the influence of Cr. In addition to negatively affecting the assimilation of carbon dioxide, Cr(6+) also has a negative effect on negative photosynthetic rates (Vernay et al. 2007; Liu et al. 2008; Subrahmanyam 2008). In addition to obstructing the production of pigments that are involved in photosynthesis, the toxicity of Chromium causes a reduction in the amount of CO_2 that can be absorbed by plants as a result of a reduction in pigment production. The result of this is a decrease in the dry biomass of the plant. There is evidence to suggest that the detrimental effects of Cr can be traced back to its ability to cause changes in the activities of carbon fixation enzymes, as well as a disruption of Cr's effect on the electron transport chain of photosynthesis (Larcher 1995). As Joshi et al. (2003) demonstrated in a study on the leaves of Cyamopsis tetragonoloba, the activity of malate dehydrogenase and RuBP carboxylase in addition to the rate of photosynthesis were both inhibited by Cr(6+) when the concentration ranged from 2 to 6 parts per million. However, when the concentration of Cr(6+) was increased to 1 ppm, the opposite effect was observed concerning the observed effects. In response to the presence of Cr(6+), both Pueraria montana and Salvinia minima did not exhibit any discernible changes in their internal CO_2 concentrations. As a result, the amount of carbon that is readily available in the presence of Cr(6+) does not affect the amount of CO_2 that is uptaken (Nichols et al. 2000). Based on the results of Subrahmanyam (2008), it appears that the ratio of Fv/Fm in T. aestivum was not affected by Cr(VI) and that the photochemical processes occurring in PSII in T aestivum were not affected by Cr(VI). In contrast, it impeded the assimilation of CO_2 into the cells as demonstrated by the lower in vivo quantum yield of PSII, as well as a slower electron transport rate (Subrahmanyam 2008). In Salvinia natans plants exposed to Cr-rich wastewater, PSI and PSII activity increased, but the ratio of Fv to Fm did not change after they were exposed to Cr-rich wastewater. While RuBisCO (ribulose-1,5-biphosphate carboxylase–oxygenase) showed an increase in activity after the same exposure, RuBisCO activity decreased after the same exposure (Dhir et al. 2009). There has been evidence that Cr(VI) can cause a decrease in photochemical quenching while increasing non-photochemical quenching based on the work of Vernay et al. (2007), Liu et al. (2008), and Subrahmanyam (2008). In light of this, it can be assumed that ATP and NADPH levels have declined (Subrahmanyam 2008). Because plants have a diminished capacity for light absorption, this is evidence that Cr(6+) disturbs a mechanism that carries electrons during photosynthesis (Nichols et al. 2000). A study published by Dixit et al. (2002) suggests that Cr(6+) prevents uncoupled electron transport within a cell. It is suggested that Cr(6+) can be found to bind to many different sites along the electron transport chain in plants based on the findings of this study. As a result of these researchers' findings, they have concluded that the change may have been caused by a redox change in copper and iron carriers. As a result of their research, they proposed that Cr(6+) could have been transferred through the cytochrome system in the mitochondria to reduce the concentration of Cr(6+) in the mitochondria. A possible explanation for this could be that the reduced heme group of cytochrome served as a site of Cr(6+) binding, thereby preventing electron transport (Dixit et al. 2002). Apart from its ability to bind to the complex IV of cytochrome oxidase, Cr(6+) has also been shown to bind to cytochrome a3, which in turn reduces the activities of both of these molecules to a significant extent (Dixit et al. 2002). Cr(6+) ions may have a negative impact on photosynthesis and the transfer of excitation energy because of the abnormalities and ultrastructural changes that are caused by Cr(6+) ions. These abnormalities include undeveloped lamellar systems, widely spaced thylakoids, as well as a decreased number of grana in the chloroplasts (Paiva et al. 2009). A study conducted by Juarez et al. (2008) suggests that the pheophitinization of chlorophylls and the destruction of pigment–protein complexes observed in thylakoid membranes may be both outcomes of the production of reactive oxygen species (ROS) in response to the stress induced by Cr(6+). As a result of research involving Cr(6+) and PSII, researchers found that Cr(6+) significantly impacts the PSII performance index. This can be accomplished by reducing the number of active reaction centres produced by absorption, the yield of primary photochemistry, and how efficiently a trapped exciton can move electrons into the electron transport chain (Appenroth et al. 2001). The efficiency of primary photochemistry, as well as the efficiency with which a trapped exciton can move an electron into the electron transport chain, are all factors that have been considered. Accordingly, one of the primary targets of chromium toxicity was the reduction in the number of active reaction centres, as well as damage to the oxygen-evolving complex (Appenroth et al. 2001). As a part of their studies, Bishnoi et al. (1993b) investigated the effects of Cr(6+) on PSI and PSII activity in isolated chloroplasts of P. sativum which were exposed to the presence or absence of Cr(6+). It was found that the presence of Cr(6+) had a greater effect on the activity of PSI than the absence of Cr(6+). As a result of this study, it was discovered that PSI and the light-harvesting complex of PSII were less sensitive to the presence of Cr(6+) than PSI and the basic complex of PSII, as well as the connecting antenna of PSII. Nevertheless, a different study has found that PSII has a greater sensitivity to Cr(6+) than PSI, which suggests that they are both sensitive to Cr(6+) (Appenroth et al. 2003).

14.6 Changing the Balance of Nutrients as a Result

When plants are put under stress from chromium, both the uptake of nutrients and their biomass are impacted to a greater degree. Chromium interferes with the absorption of vital nutrients in a complex manner. Several studies have shown that both Cr(3+) and Cr(6+) interact with the consumption of macronutrients such as N, P, K, and Mg. These studies were done by Turner and Rust (1971), Sela et al., Biddappa and

Bopaiah, Moral et al., and Davies et al. (2002). It has been demonstrated that Cr(III) inhibits the uptake of vital mineral elements and lowers the amount of calcium that is present in the cells. In *H. annuus*, the concentrations of N, P, and K were reduced by 100 μM of Cr(3+), while concentrations of aluminium, iron and zink were increased (Davies et al. 2002). According to the findings of Liu et al. (2008), when copper, iron, and zinc were present in the environment of A viridis, their uptake was inhibited by the presence of Cr(6+). Citrullus vulgaris was exposed to Cr(6+), which resulted in an increase in the accumulation of phosphorus and zinc and a decrease in sulphur and copper in the plant as a result of the exposure (Dube et al. 2003). As far as the effect of Cr(6+) on the uptake of Mn is concerned, there have been contradictory reports concerning the effect that it has on the uptake of Mn. In C. vulgaris (Dube et al. 2003) and L. perenne, an increase in the uptake of manganese was caused by Cr(6+) (Vernay et al. 2007), but it caused a decrease in the absorption of manganese in Brassica oleracea (Chatterjee and Chatterjee 2000) and Amaranthus viridis (Liu et al. 2008). The levels of nitrogen, phosphorus, and potassium in the leaves of *Oryza sativa* were lower when Cr(6+) was present at 50–500 mg kg^{-1} in the soil (Ahmad et al. 2011). It is well established that Cr(6+) competes with both iron and phosphorus for surface root sites and binding sites (Chatterjee and Chatterjee 2000). As a result of Cr(6+) interfering with the absorption of iron, there is a reduction in the accumulation of iron, which is necessary for the biosynthesis of chlorophyll and heme. Additionally, there was a decrease in the levels of activity of the heme enzymes found in S. oleracea, which suggests that there was interference in the iron metabolism (Gopal et al. 2009). According to Turner and Rust (1971), the decreased amount of biomass in G. max that occurs after exposure to Cr(6+) is thought to be caused by direct interference of Chromium in the process of phosphorous metabolism, as well as there, is a limit placed on the amount of sulphur that can be incorporated into certain essential amino acids. In their study, Sundaramoorthy et al. (2010) illustrated that the primary reason for the decrease to determine the effects of Cr(6+) toxicity on Oryza sativa L., the total dry weight of the root and shoot was determined was a decrease in water uptake as well as an expansion of root cells. As a way of comparison, Han et al. (2004) found that impaired metabolic processes were responsible for 57% of the high-shoot dry mass in B. juncea when the concentration of Cr was 500 M. These processes restrict the extracting compounds that have been stored in the cotyledon and are essential for the continued expansion and development of the plant. At a concentration of 2 M, Cr(6+) has been shown to stop the roots of Z. mays from taking in K^+ and H^+. This suggests that it interferes with the transport activities of plasma membranes. When S. kali was exposed to a concentration of chromium between 5 and 20 mg l^{-1}, it took in less K, P, Mg, and Cu through its roots and less Ca, Fe, and Cu through its leaves; Nevertheless, the effect was not consistent regardless of the Cr speciation. In general, the conditions of Cr(3+) led to a reduction in the number of macronutrients and microelements that were absorbed in comparison to the conditions of Cr(6+). According to Redondo-Gómez et al. (2011), the uptake of essential nutrients was inhibited in cordgrass, *Spartina argentinensis*, when Cr(III) concentrations of 1.5 mg g^{-1} were present. During the research that was conducted by Barcelo and Poschenrieder (1997), it was determined that Cr(6+) is absorbed by higher plants through sulfate carriers, which are membrane transporters that are involved in the uptake of Cr(6+). According to Kleiman and Cogliatti (1997), after discovering an increase in Cr(6+) influx in sulfate-deprived T. aestivum plants, it is likely that sulfate plays a role in the transport of chromium in plants. The evidence suggests that the sulphate transporter (BjST1) is suppressed in roots of B juncea under Cr(6+) stress, resulting in reduced uptake of sulphate as well as a reduced expression of the sulphate transporter (BjST1). As a result, it can be concluded that sulphate carriers play a key role in Chromium transport. There is also evidence to suggest that the accumulation of Cr(6+) is greater in B juncea than in other species when the expression of the sulphate transporter (SHST1) is increased. The research study conducted by Kim et al. (2006) indicates that the uptake of Cr(6+) and the tolerance to Cr(6+) in transgenic tobacco are both controlled by an over expression of MSN1, a putative yeast transcriptional activator (Nicotiana tabacum).

14.7 Water Balance as a Result of the Effect

As discovered by Barcelo et al., the influence of Cr(6+) on water relations is dependent upon its concentration, and it also depends upon the type of leaves, which differs depending on the type of Cr(6+) applied. In primary leaves, the values of ψs and ψw were found to be lower at growth-inhibiting concentrations, while ψp was found to be higher. There was, however, a phenomenon in bifoliate leaves in which when the concentrations of all of the compounds did not have an inhibitory effect, w and p increased while s decreased, whereas trifoliate leaves, on the other hand, exhibited the opposite response. Gopal et al. (2009) illustrated that Cr(6+) lowers the physiological availability of water in S. oleracea leaves. There was a decrease in leaf water potential as well as an increase in diffusive resistance, both of which indicate that the plant is suffering from water stress.

14.8 Other Biochemical Effects

In addition, research has shown that the toxicity of Chromium harms the biochemical processes of plants in other ways. In the case of P. vulgaris, the occurrence of Cr(6+) stopped the production of ethylene from 1-aminocyclopropane-1-carboxylic acid made by the plant itself (ACC) (Poschenrieder et al. 1993). Even though Cr was responsible for the disintegration of the membrane that caused the inhibitory effect, the changes in metabolism that occurred as a result of Cr(6+) exposure were the cause of the inhibitory effect, as either inhibition of ACC synthase activity or the diversion of metabolic steps that occur before ACC-catalyzed reactions were to blame for this. These alterations were responsible for the inhibitory effect (Poschenrieder et al. 1993). It has been shown that Cr(6+) can disrupt the electron transport chain in mitochondria isolated from root cells of P. sativum, resulting in the chain's inactivation at concentrations of 20 and 200 M (Dixit et al. 2002). As a result of the study, it was revealed that both NADH: cytochrome c oxidoreductase and succinate: cytochrome c oxidoreductase activities have been significantly inhibited by the inhibitor, respectively, with cytochrome oxidase being the enzyme most sensitive to the inhibitor (Dixit et al. 2002). Because of Cr(VI; 150 mg l^{-1}), the amount of IAA and IBA found in the roots and shoots of *T. aestivum* decreased, while the amount found in the seeds increased (Zhang et al. 2009). Exposure to Cr(6+) in P. sativum root plasma membrane vesicles resulted in an increase in the NADPH-dependent superoxide production as well as the activity of NADPH oxidase, which led to the discovery that Cr(6+) was responsible for these changes, while there was a noticeable drop in the amount of activity displayed by NADH ferricyanide oxidoreductase. It is speculated that these findings might lead to the conclusion that Cr(6+) may disrupt the normal functioning of plasma membranes as a result of its effects (Pandey et al. 2009). There has been recent research suggesting that the presence of chromium in the earth's crust of metallurgical landfills (at a concentration of 1,346 mg kg^{-1} soil) can affect the ratio of saturated to unsaturated fatty acids found in the Lactuca serriola leaf. It has been demonstrated in previous studies that Cr(6+) inhibits the activity of plasma membrane H^+-ATPase (Shanker et al. 2005) and Na^+/K^+-dependent ATPase, whereas Cr(III) inhibits the activity of Ca^{2+}-dependent ATPase.

14.9 The Uptake and Translocation of Cr in Plants

There has been a significant amount of attention being paid all over the world to the uptake of Chromium by plants and its translocation to different parts of the plant as of late. Recent developments have been the focus of this attention. In addition to the fact that Chromium plays such an important role in human metabolism, it also plays a role in the development of carcinogenic effects on humans. As Cr does not participate in the metabolic processes that take place within plants, the mechanism by which plants take up Cr has not yet been fully understood because it does not participate in these metabolic processes. It has been shown that root systems in plants are the primary sources for chromium uptake, which varies from one plant type to another, as well as the type of chromium speciation, which may be Cr(3+) or Cr(6+) (Smith et al. 2002). Additionally, there is also evidence to suggest that the uptake of Cr from aqueous media is dependent on the pH, the concentration, and the salinity of the medium in addition to the presence of dissolved salts in the medium (Chatterjee and Chatterjee 2000). A study by Kocik and Ilavsky found that the formation of complexes between Cr and organic ligands appears to facilitate increased uptake of Cr by plant tissues by facilitating the formation of these complexes. To take in Cr(III), plants use a mechanism known as diffusion, the process of which is a passive one that is carried out at the cation exchange site of the plant cell wall. In an energy-dependent active process, Cr(VI) is taken up by the cell via the phosphate and sulphate transporters, because of its structural similarity to phosphate and sulphate (Chandra et al. 1997). There is evidence that ferric reductase enzymes are involved in the immediate change of Cr(VI)–Cr(III) in roots when Cr(VI) is transported by active transport (Biacs et al. 1995). The converted Cr(III) binds to the cell wall of the plant cells, thus preventing it from moving any further through the various tissues of the plant (Sharma and Mehrotra 1993). It has been shown that an increase in the expression of MSN1, a putative yeast transcriptional activator, in transgenic Nicotiana tabacum led to an increase in the tolerance and ability of the plant to absorb Cr and S (Smith et al. 2002). As a result of additional research carried out on N. tabacum, it was found that there was an increase in the expression of the gene for sulphate transporter one (NtST1) under the influence of Cr stress. The sulfate transporter seems to take up both S and Cr, which indicates that they are both taken up by the transporter. The ABC transporter is generally considered to be the mechanism used by prokaryotic organisms to transport sulphate from one place to another (Paiva et al. 2009). It has been discovered that there are some different sulphate transporters in eukaryotes that have varying degrees of affinity for the substrates in their environment. The plasma membrane of Chlamydomonas reinhardtii has been found to contain six different sulphate transporters and all of these sulphate transporters belong to the Na^+/SO_4^{2-} and H^+/SO_4^{2-} transporter families. A hypothesis has been put forth that these transporters may play a role in regulating the movement of Cr within plants (Redondo-Gómez et al. 2011). Chromium is the least mobile heavy metal among all the heavy metals found in plant roots among all the heavy metals. According to some studies conducted up to this point, the roots of plants contain the highest concentration of chromium when compared with other parts of the plant such as the leaves and stems (Smith et al. 2002). Earlier studies conducted by Zayed et al., for example, revealed that the formation of insoluble Cr compounds in roots led to an accumulation of chromium that was over a hundred times higher than that found in vegetable shoots as a result of the insoluble Cr compounds. Similarly, it was discovered that *P. sativum* and *S. oleracea* L. cv. Banarasi accumulated a greater quantity of Cr in their roots than in their leaves and stems. Notably, under conditions of Cr toxicity, bean plants accumulated 98% of the element in their roots, but only 0.1% of it was found in their seeds (Rai et al. 2004). When 0.50 mM of Cr(VI) was applied to Lolium perenne as part of another study, the results showed that there was a 10 times higher accumulation of Cr in the plant's roots than in its leaves. The cytoplasm and intracellular spaces of the rhizome of Iris pseudacorus were found to contain higher concentrations of Cr than those found in the root cell wall and the cytoplasm of Iris pseudacorus (Barcelo and Poshenrieder 1997). When compared to the shoots, the accumulation of Cr was greater in the plants' roots of *T. aestivum, A. sativa,* and *Sorghum bicolour* (Pandey et al. 2005). Although there is a restriction on the movement of Cr from the roots to the aerial parts of the plant, the chemical form of Cr can still affect the movement of Cr from the roots to the aerial parts of the plant. In particular, the application of exogenous EDTA (ethylenediaminetetraacetic acid) to the plants increases the uptake of Chromium and its translocation from the roots of the plants to their upper parts (Pandey et al. 2005).

14.10 Research Regarding the Effects of Cr Stress on Plants Has Recently Made Significant Advances

Even though many studies have been conducted regarding the effects of chromium stress on plants in the past, the exact molecular mechanisms involved in the effects of chromium phytotoxicity, plant defence against chromium exposure, as well as the translocation and accumulation of chromium in plants, in general, remain poorly understood (Dubey 2010). Despite this, due to the progress that has been made in recent years in the field of omics, investigations of this nature can now be conducted with a much-increased degree of precision, and a wider range of variables associated with physiological responses to Cr stress can now be considered. A significant amount of potential exists in the "omics" fields when it comes to studying the underlying mechanisms responsible for the toxicological effects of chemical pollutants, as well as the identification of new biomarkers of effect that will be generated as a consequence of this potential (Dowling and Sheehan 2006).

14.11 The Molecular Mechanisms by Which Cr is Detoxified in Plants

There are two main defence mechanisms that plants have developed to protect themselves from the potentially harmful effects of Chromium. There are two types of avoidance: avoidance and tolerance. Tolerance is believed to be caused by genes and proteins that play a role in the uptake and translocation of Chromium, the chelation process, and the sequestration of Chromium in the vacuoles. A chemical compound called chromium is capable of causing the death of cells in the body. The study published by Liu (2008) indicates that a reduction in S levels activates the detoxification of Cr(VI) or tolerance in wild-type Scenedesmus acutus strains through a decrease in the amount of Cr(VI) taken up by the cells during the detoxification process. In the absence of sulphate, the activation of "high-affinity sulphate transporters" led to a greater uptake of sulphur than of chromium, which led to a lower rate of chromium uptake (VI) when there was no sulphate present. In addition, S-starvation activated the S-uptake/assimilation pathway, which led to the production of S-containing molecules (GSH, PCs, or MTs) as a result of the process of S-starvation. It is these molecules that are ultimately responsible for the cells' ability to tolerate Cr(VI) in the long run.

14.11.1 Avoidance

As the first step in protecting plants from the harmful effects of chromium, it is necessary to prevent the roots from being able to absorb more of the element from the soil. Plant species that accumulate Cr, including aquatic plants and terrestrial plants, bind Cr ions to their cell walls, which are primarily composed of pectic sites, callose, and mucilage. In turn, this causes a reduction in the amount of Cr that is translocated into the cytosol in these plants (Vernay et al. 2007). Further, according to Ahmad et al. (2011), Cr ions are bound to the secondary cell wall by lignin, which also means that lignin plays a role in the binding of Cr ions to this cell wall. As demonstrated in a previous study (Samantaray 2002), the cell wall can act as a barrier to prevent Cr translocation through the cell, a function that is demonstrated by the accumulation of callose within the cell wall of Oryza *sativa* as well as the elevated expression of proteins related to the structure of the cell wall. The importance of the cell wall can be seen in both of these findings.

14.11.2 Antioxidant Response

As a result of Cr toxicity in plants, reactive oxygen species (ROS) are produced by the Fenton and Haber-Weiss reactions (Ganesh et al. 2008), which are then followed by the modulation of antioxidant enzyme activities. Plants are protected against reactive oxygen species (ROS) that are produced in response to calcium stress by the increased activity of antioxidant enzymes such as POD, catalase (CAT), APX, and SOD. Several antioxidant enzymes are involved in the interception of the chain reaction of free radicals, which either completely stops the oxidation process or significantly slows it down. Studies conducted on *Z. mays*, *Solanum lycopersicum*, and *B. oleracea* showed that Cr (VI) treatments increased glutathione (GSH) levels in both the roots and the leaves (Nath 2008). On the other hand, when *Jatropha curcas* was exposed to Cr, GSH activity was down (Joshi et al. 2003). A study conducted by Sharma et al. (1995a, b) found that in response to Cr stress, there is an increase in glutathione reductase activity (GR), one of the key enzymes in the Ascorbate-Glutathione pathway. Further to its role as a substrate for the biosynthesis of PCs, GR also serves as a metal chelator and a scavenger of oxygen radicals. This is in addition to its role as a substrate. As a result of a recent study on Miscanthus sinensis, it was found that upon exposure to 0.50 mM Cr, there was an overexpression of 36 proteins that are involved in oxidative stress, metabolism, molecular chaperone functions, among others (Subrahmanyam 2008).

14.11.3 Reduction of Cr(VI)–Cr(III)

Several chemical or enzymatic methods can be used to reduce Cr(VI)–Cr(III) so that there can be a reduction in the toxicity of Cr in plants. Plant cells can undergo this reduction chemically with the assistance of glutathione, cysteine, sulfite, and thiosulfate which are already present in the plant cell (Juarez 2008). It has been demonstrated that numerous bacteria associated with rhizospheric soils, such as

Bacillus species, Staphylococcus species, Ochrobacterium intermedium species, Pseudomonas species, Mesorhizobium species, and Cellulosimicrobium species, are capable of enzymatic reduction (Joshi 2003). In the electron transfer chain, these bacteria use chromate as the terminal electron acceptor of the electron transfer chain in the rhizosphere (Appenroth et al. 2001). These organisms possess both soluble and membrane-bound reductases.

14.11.4 An Approach to the Decontamination of Cr Using Phytoremediation

In the past few decades, scientists have studied tolerant and hyperaccumulator plants to learn more about how they work. In addition, they have also studied how they can be used as part of the phytoremediation process. As of now, nearly 500 plant species belonging to more than 45 plant families have been discovered. In the majority of the tolerant hyperaccumulator plants, the toxic metals were converted into forms that were less hazardous and could not be moved by the plant (Han et al. 2004). Most of the time, high-affinity ligands such as amino acids, peptides, and organic acids are what make the Cr hyper-accumulators work because they bind to receptors with high affinity. It is these ligands that chelate the metal ions and store them within the vacuoles of the cells. It is due to the increased mobilisation of metals from the rhizosphere by organic acids, their absorption by different families of transporters, and their movement into the shoot through xylem loading that Cr and other heavy metals accumulate too much in the plant. These factors are responsible for the excess accumulation of Cr and other heavy metals in the plant (Kleiman and Cogliatti 1997).

14.11.5 Phytoremediation by Hyperaccumulating Plants

To reduce the harmful effects of Cr exposure on the environment, phytoremediation could prove to be an approach that is both highly effective and relatively inexpensive. It uses plants to clean up contaminated soil and wastewater, which is both friendly to the environment and friendly to the environment as well. The results of numerous scientific studies have demonstrated that a variety of plant species can effectively remove Cr from polluted areas, which suggests that these plants could potentially be useful for the phytoremediation of polluted areas. As part of the phytoremediation process, there may be an opportunity to use a novel plant that has a long history of use in traditional Chinese medicine called Lonicera japonica Thunb, which has been used for centuries to accumulate Cr. According to Kim et al. (2006), one of the mechanisms explaining how these plants were able to tolerate Cr was their ability to produce anthocyanins, oxalic acid, and carotene in greater quantities. According to a study published in 2009 by Zhang et al., the macrophyte Callitriche cophocarpa can be used as an efficient biosorbent for removing Chromium from concentrated solutions, which are typical of industrial effluents (Zhang et al. 2009). A study published by Shanker et al. (2005) demonstrated that Vigna unguiculata has a significantly higher capacity to remove Cr than Arachis hypogea, making these plants a more promising candidate for the phytoremediation of soils contaminated with chromium. As opposed to accumulating Cr in their aerial parts, V. unguiculata accumulated it in their roots from the surrounding soil, instead of accumulating it in their aerial parts. It was found by Levizou et al. (2018) that Origanum vulgare has an exceptional capacity to bioaccumulate chromium both in the aerial part and in the roots when grown in chromium-contaminated soil, as evidenced by the results of an experiment that was carried out on both an indoor and an outdoor scale in a pot. Following Afonso et al. (2019) research, species of Solanum viarum Dunal accumulate high levels of chloride in their biomass when left to grow. The high level of bioavailability of these plants, means that they can be used for phytoremediation and are likely to be effective at treating areas that are contaminated with heavy metals such as chromium. In an empirical study conducted by Marieschi et al. (2015), the Cassia tora plant was found to be a potential phytoremediator of Cr from contaminated sites because of its high bioaccumulation activity, high tolerance, and transportation index. As a result of this, the plant was able to apply the phytostabilization program, a program designed to lessen the toxicity of Cr on the mining sites that have been overburdened with chromite. *A crop called Arundo donax L. can handle moderate to high levels of heavy metals and can store a significant amount of Cr, which makes it a promising crop for energy production* (Mangabeira et al. 2011). It was found that the highest concentrations of Cr were found in the roots of Diectomis fastigiata (2371 mg/kg dry matter) and the shoots of Vernonia cinerea (5500 mg/kg dry matter). As a result of this, it can be seen that these plants are capable of removing Cr from the environment through phytoremediation. As an additional demonstration of its exceptional hyperaccumulation properties towards Cr, Callitriche cophocarpa Sendtn was planted in the heavily polluted watershed with sediments to demonstrate its exceptional hyperaccumulation properties towards heavy metal viz. chromium (Zeng et al. 2014). As part of a research project, it was discovered that the common water hyacinth, Eichhornia crassipes, has an effective capacity for removing chromium from water. The researchers used a small-scale hydroponic experiment with varying concentrations of metal for a period of one month. When compared to the accumulation in the shoot, the level of Cr found in the roots was significantly higher (Costa et al. 2010). The plant known as Cirsium vulgare is an effective accumulator of Cr and has the potential to be used effectively for the phytoremediation of soils that have been contaminated with Cr (Yadav et al. 2010). There is a possibility that the aquatic macrophyte Ipomoea aquatica can effectively remove Cr from water bodies in a relatively short amount of time, and that the species grows rapidly, so it may be an ideal candidate for phytoremediation of water bodies that are contaminated in elements (Sinha et al. 2018).

14.12 Detoxification of Cr Using Anti-Oxidant Machinery and Other Innovative Strategies

In terms of wastewater treatment, Ipomoea aquatica has the potential to be used in the treatment of wastewater that has high levels of Cr contamination (Sharmin et al. 2012). Moreover, Scoccianti et al. (2006) also cited that the fast-growing, tolerant, hyperaccumulating aquatic plants possessed a bio-accumulation and translocation factor equal to or greater than one, as well as demonstrating the ability to phytoremediation Chromium through bioaccumulation. It has been shown in earlier studies of this plant that it can be used in the treatment of effluents to remove Cr(VI), indicating that this plant has a high potential for use in the treatment of effluents as a source of the spontaneous removal of Cr(VI). Additionally, the researchers were able to demonstrate that when the anti-oxidant machinery was treated with Chromium, only minor alterations were observed, compared to the control group (Shanker et al. 2004). There has been a development of a method for removing chromium from the environment in an innovative manner. The method utilized by this study involves the use of an iron-biochar nano-complex for immobilizing the bioavailable mobile fraction, the use of the hyperaccumulator Leersia hexandra for uptake, and the utilization of a microbial consortium to facilitate the plant's growth. When this system was used, L. hexandra was able to accumulate Cr at a rate ranging from 147.5 to 785.0 mg/kg biomass of plant tissue when this system was applied (Shahid et al. 2017). Remarkably, the leaves of *Salvia moorcroftiana* that have been chemically modified can be utilized as biomass for the biosorption detoxification of aqueous solutions (Huang et al. 2018). This is accomplished by the removal of Cr(IV) ions through endothermic and non-spontaneous thermodynamic processes. This could be a better alternative because it is cheap and has a high biosorption capacity. Besides being able to remove Cr(IV) from water, it can also be used in the removal of heavy metals such as Pb(II) (lead) and Cd(II) (cadmium). As a result of a greenhouse experiment, it was found that two free-floating macrophytes, Eichhornia sp. and Pistia sp., had the capability of increasing anti-oxidant activity and building up in water bodies that were contaminated with Cr(VI), meaning that they could be used to clean them up. The Gomphrena celosoides can accumulate a significant amount of chromium because of their elevated levels of proline and antioxidant enzyme activity (Whitacre 2010). Because of the high level of proline in Gomphrena celosoides, as well as the antioxidant enzyme activity that they produce, this plant can store a significant amount of chromium (Whitacre 2010). It is known that Calotropis procera accumulates a high concentration of Chromium, as well as increases the activities of the enzymes SOD, CAT, and GR in the presence of Chromium (Cervantes et al. 2001). Because of these properties, it can be used for phytoremediation of polluted arid soils that are contaminated with Chromium. There is a high level of Chromium accumulation in Calotropis procera.

14.13 Conclusions

There has been an increase in the amount of pollution caused by chromium in the environment. In addition, there is a growing recognition that chromium is a serious health risk for the biota. There is a need for more research into how plants protect themselves from the toxic effects of this metal. It was in this chapter that we explored various negative effects that being exposed to Chromium can have on plants. This was from both the perspective of their morphology and their physiological reactions. In addition to the detrimental effects caused by Cr on plants, several toxic effects can be induced by Cr on plants, including altering germination and the growth process of roots, stems, and leaves. Additionally, Cr can adversely affect the morphological and physiological processes of plants, including photosynthesis, water relations, mineral nutrition, germination, and stem growth. It should be noted that Cr can also cause oxidative stress in plant cells by disrupting the balance of redox within the cells on a molecular level, in addition to causing oxidative stress in plant cells. As mentioned in the chapter, plants have a variety of ways of protecting themselves from external threats. While many of these defences are still not well understood, recent advances in molecular and cellular biology, such as genomics, proteomics, and the newly created field of metallomics, are shedding more light on the complex strategies plants use to protect themselves against such threats. Furthermore, there is a dearth of research in this area because these spheres of expertise have only relatively recently begun to be applied to environmental problems and have resulted in a lack of research in this area. In the future, certain constraints, such as the need for advanced mass spectrometry equipment and its hyphenations in the case of proteomics and metallomics, may continue to prevent development in the field; however, these constraints are becoming, increasingly, becoming cheaper and more readily available, and additional research in this field is just around the corner, which is why this area continues to grow (Table 14.1).

Table 14.1 Efficient chromium accumulator plants: habitat, culture conditions and removal mechanism

Cr accumulator	Family	Habitat	Cr(VI) removal mechanism	Culture condition	Max % removal/ bioaccumulation capacity	Experimental period	Influent conc	References
Brachiaria mutica (Paragrass)	Poaceae	Perennial grass	NR	Soil field study (mine wastewater)	Transportation index (TI): 6.16 Total accumulation rate (TAR):8.2 mg/ kg/day	100 days	0.65 and 0.74 mg/L for Cr(VI)	Mohanty and Patra (2012)
Amaranthus viridis (Green amaranth)	Amaranthaceae	Perennial broadleaf herb	Increased activity of antioxidative enzymes	Hydroponic culture	Cr accumulation: Roots: 2624.39 mg/g Cr(VI) (dw) at 5.2 mg/L	20 days	0.052e5.2 mg/ L Cr(VI)	Liu et al. (2008)
Dicoma niccolifera	Asteraceae	Terrestrial	NR		Cr accumulation: >1000 mg/kg Cr			Banach et al. (2012)
Helianthus annuus (Sunflower)	Asteraceae	Annual forb	NR	Cr contaminated soil	70% chromium removal Cr accumulation: Roots (2730 mg Cr/kg dry tissue)	90 days	10 mg/L Cr(VI)	Ranieri et al. (2013)
Azolla (Water fern)	Salviniaceae	Aquatic fern	NR	Hydroponic culture	Cr accumulation: 356 and 964 mg/ kg dm Cr(VI) and Cr(III) at 1 mg/dm	12 days	1–20 mg/L Cr(VI)	Arora et al. (2006)

(continued)

Table 14.1 (continued)

Cr accumulator	Family	Habitat	Cr(VI) removal mechanism	Culture condition	Max % removal/ bioaccumulation capacity	Experimental period	Influent conc	References
Brassica juncea (Indian mustard)	Brassicaceae	Annual growing perennial herb	NR	Soil condition	Cr accumulation: 48 and 58 mg Cr per plant from Cr(III) and Cr(VI)-treated soils	69 days	Soil amended with 100 mg/kg of Cr(III or VI)	Bluskov et al. (2005)
Leersia hexandra (Southern cutgrass)	Poaceae	Perennial herb (grow in swamps)	Facilitates microbial growth	CWs (Lab-scale)	99.7%	120 days	5 mg/L Cr(VI)	Liu et al. (2015)
			Cr(VI) reduction and sequestration	Hydroponic culture	Highest bioaccumulation coefficients for leaves: 486.8 for Cr(III) and 72.1 for Cr(VI)	45 days	10 mg/L Cr(VI) and 60 mg/L Cr(III)	Zhang et al. (2007)

(continued)

Table 14.1 (continued)

Cr accumulator	Family	Habitat	Cr(VI) removal mechanism	Culture condition	Max % removal/ bioaccumulation capacity	Experimental period	Influent conc	References
Eichhornia crassipes (Water hyacinth)	Pontederiaceae	Free-floating perennial aquatic plant	Increased activity of antioxidative enzyme Cr(VI) reduction	Hydroponic culture	Maximum Cr accumulation: 2.52 _ 103 mg/g of water hyacinth in 20 mg/L Cr removal efficiency: 91%	42 days	3, 5, 7, 10 and 20 Cr(VI) mg/L	Zewge et al. (2011)
			Plants exposed to 520 mg/L Cr(VI) for 4 days did not survive	Hydroponic culture under greenhouse conditions	Maximum Cr accumulation: 1258 mg/kg (dw) 520 mg/L Cr(III) for 2 days	2–4 days	52 and 520 mg/L Cr(III) and Cr(VI)	Mangabeira et al. (2004)
Prosopis laevigata (Smooth Mesquite)	Fabaceae	Flowering tree	NR	Tissue culture conditions	Cr accumulation: roots: 8090 mg/kg Cr(VI) (dw)	50 days	0–176.8 mg/L Cr(VI)	Buendía-Gonzälez et al. (2010)
Bacopa monnieri (Smooth water hyssop)	Plantaginaceae	Perennial, creeping herb	NR	Hydroponic conditions	319.5 mg/kg DW for Cr at 10 mg/ml	8 weeks	0.01, 0.1, 1.0, 2.5, 5.0, 10 mg/L Cr	Shukla et al. (2007)

(continued)

Table 14.1 (continued)

Cr accumulator	Family	Habitat	Cr(VI) removal mechanism	Culture condition	Max % removal/ bioaccumulation capacity	Experimental period	Influent conc	References
Phalaris arundinacea (Reed canarygrass)	Poaceae	Perennial grass	NR	Horizontal subsurface flow CW	Cr accumulation: 14.7 mg/kg dry mass Roots: 18.5 mg/kg Cr	4 years	Municipal sewage with 0.5e4 mg/L Cr	Vymazal et al. (2007)
Hydroctyle umbellata (Marshpennywort)	Araliaceae	Anchored hydrophyte	NR	Hydroponic culture	Cr accumulation: 18,200 mg/kg	90 days	Semi-solid tannery (wet) sludge at 0, 20, 40, and 60% total Cr concentrations	Khilji (2008)
Convolvulus arvensis (Bindweed)	Convolvulaceae	Herbaceous perennial plant	NR	Tissue culture conditions	Cr accumulation: 3800 mg/kg Cr(VI) (dw)		20 mg/L Cr(VI)	Gardea-Torresdey et al. (2004)
Vetiveria zizanoides (Khas-khas)	Poaceae	Perennial grass	NR	Hydroponic culture CW	77–78% for Cr uptake ability	–	5–20 mg/L	Singh et al. (2015)

(continued)

Table 14.1 (continued)

Cr accumulator	Family	Habitat	Cr(VI) removal mechanism	Culture condition	Max % removal/ bioaccumulation capacity	Experimental period	Influent conc	References
					Max Cr accumulation: Stem (28.3 g/kg) 89.29% removal efficiency Cr accumulation: Roots 0.448 mg/kg (dw) Leaves 0.241 mg/kg (dw)	100 days	NR	Srisatit and Sengsai (2003)
Spirodela polyrrhiza (Giant duckweed)	Lemnaceae	Perennial aquatic plant	NR	Continuous flow pond system	Maximum Cr accumulation: 4.423 mg Cr/g was found in plants grown in the first chamber of pond operated at pH 4.0 at 5.0 mg Cr/L	21 days	0.25–5.0 mg/L Cr(VI)	Mishra and Tripathi (2008)
Genipa americana L (Genipap)	Rubiaceae	Wood plant	NR	Hydroponic conditions	–	5 months	0, 5, 10, 15, 20, 25 and 30 mg/L Cr(III)	Barbosa et al. (2007)
				Hydroponic conditions	Reduction of 79 and 90% for 15 and 30 mg/L of Cr(VI)	15 days	15 and 30 mg/L Cr(III) and Cr(VI)	Santana et al. (2012)

(continued)

Table 14.1 (continued)

Cr accumulator	Family	Habitat	Cr(VI) removal mechanism	Culture condition	Max % removal/ bioaccumulation capacity	Experimental period	Influent conc	References
Miscanthus sinensis (Chinese silver grass)	Poaceae	Herbaceous perennial plant	Altered vacuole sequestration, nitrogen metabolism and lipid peroxidation	Hydroponic culture	–	3 days	0, 2.6, 5.2, 10.4, 15.6, 26, 39 or 52 mg/L Cr(VI)	Sharmin et al. (2012)
Pteris vittata (Chinese brake)	Pteridaceae	Fern species	NR	Hydroponic system	Cr accumulation: Fronds 234 mg/kg (dw) Roots 12,630 mg/kg (dw) at 2.6 mg/L Cr(VI)	14 days	0, 2.6, 13 and 65 mg/L	de Oliveira et al. (2014)
Salvinia minima (Water spangles)	Salviniaceae	Aquatic macrophyte (Free *floating fern*)	Increased activity of antioxidative *enzymes*	Outdoor condition	Cr accumulation: Submerged leaves 3358 mg/g Cr(VI) (dw) Floating leaves 637 mg/g Cr(VI) (dw)	7 days	26–208 mg/L Cr(III) or Cr(VI)	Prado et al. (2012)
Leersia hexandra (Southern cutgrass)	Poaceae	Perennial herb (grow in swamps)	Facilitates microbial growth	CWs (Lab-scale)	99.7%	120 days	5 mg/L Cr(VI)	Liu et al. (2015)

(continued)

Table 14.1 (continued)

Cr accumulator	Family	Habitat	Cr(VI) removal mechanism	Culture condition	Max % removal/ bioaccumulation capacity	Experimental period	Influent conc	References
			Cr(VI) reduction and sequestration	Hydroponic culture	Highest bioaccumulation coefficients for leaves: 486.8 for Cr(III) and 72.1 for Cr(VI) Chromium accumulated in leaves was 4868 mg Cr (III)/g and 597 mg Cr (VI)/g	45 days	10 mg/L Cr (VI) and 60 mg/L Cr (III)	Zhang et al. (2007)
Callitricha cophocarpa (Water-starwort)	Callitrichaceae	Aquatic macrophyte	Cr(VI) reduction	Hydroponic culture Wetland	Cr accumulation: 1000 mg/kg (dw)	3 weeks	2.6e36.4 mg/L Cr(VI)	Augustynowicz et al. (2020)
					Cr(VI) storage vascular bundles	7 days	5.2 mg/L Cr(III) and Cr(VI)	Augustynowicz et al. (2014)
					Cr accumulation: Cr(III) 28,385 mg/kg (dw)	5 days	26–208 mg/L Cr(III)	
Typha latifolia (Cattails) and *Phragmites australia* (Common reed)	Typhaceae Poaceae	Aquatic grass Perennial grass	NR	Horizontal subsurface flow CW	Maximum removal efficiency of 73%	17 months	Synthetic tannery waste water	Calheiros et al. (2007)

(continued)

Table 14.1 (continued)

Cr accumulator	Family	Habitat	Cr(VI) removal mechanism	Culture condition	Max % removal/ bioaccumulation capacity	Experimental period	Influent conc	References
Nymphaea spontanea (Water lilies)	Nymphaeaceae	Aquatic rhizomatous perennial herbs	NR	Hydroponic conditions	Cr accumulation: 2.119 mg/g from a 10 mg/L	9 weeks	1, 2.5, 5 and 10 mg/L Cr(VI)	Choo et al. (2006)
Spartina argentinensis (Cordgrass)	Poaceae	Perennial grass	NR	Glasshouse experiment	Cr accumulation: 15.1 mg/g Cr (VI) (dw) at 1040 mg/L	15 days	0–1040 mg/L Cr(VI)	Redondo-Gómez et al. (2011)
Gynura pseudochina (Purple passion)	Asteraceae	Herb	Cr (VI) reduction	Hydroponic culture	Cr accumulation: Tubers: 823.1 mg/kg Cr(VI) (dw) Shoots: 787.9 mg/kg Cr(VI) (dw)	2 weeks	100 mg/L Cr(VI)	Mongkhonsin et al. (2011)
Lemna sp. (Duckweed)	Araceae	Free-floating aquatic plants	NR	Hydroponic culture	4.423 mg Cr(VI)/g	7 days	5.0 mg/L Cr(VI) pH 4.0	Uysal (2013)
Salsola kali (Russian thistle)	Chenopodiaceae	Annual saltwart	NR	Agar based media	Maximum Cr accumulation at 20 mg/L Cr(VI): Roots: 2900 mg/kg Cr(VI) (dw) Stems: 790 mg/kg Cr(VI) (dw) Leaves:600 mg/kg Cr(VI) (dw)	15 days	0, 5, 10, and 20 mg/L Cr(VI)	Gardea-Torresdey et al. (2005)

Acknowledgements For their consistent support throughout the writing process, the authors would like to extend their sincere thanks to the Department of Agronomy, School of Agriculture, Lovely Professional University, Phagwara, Punjab, India.

References

Afonso TF, Demarco CF, Pieniz S, Camargo FAO, Quadro MS, Andreazza R (2019) Potential of SolanumviarumDunal in use for phytoremediation of heavy metals to mining areas, Southern Brazil. Environ Sci Pollut Res 26:24132–24142

Ahmad M, Wahid A, Ahmad SS, Butt ZA, Tariq M (2011) Ecophysiological responses of rice (*Oryza sativa* L.) to hexavalent chromium. Pak J Bot 43:2853–2859

Ali NA, Ater M, Sunahara GL, Robidoux PY (2004) Phytotoxicity and bioacuumulation of copper and chromium using barley (*Hordeum vulgare* L.) in spiked artificial and natural forest soils. Ecotoxicol Environ Saf 57(3):363–374.

Ali S, Bai P, Zeng F, Cai S, Shamsi IH, Qiu B, Zhang G (2011) Theecotoxicological and interactive effects of chromium and aluminum on growth, oxidative damage and antioxidant enzymes on two barley genotypes differing in Al tolerance. Environ Exp Bot 70(2):185–191

Ali B, Huang CR, Qi ZY, Ali S, Daud MK, Geng XX, Zhou WJ (2013) 5- Amino levulinic acid ameliorates cadmium-induced morphological, biochemical, and ultra structural changes in seedlings of oil seedrape. Environ Sci Pollut Res 20(10):7256–7267

Ali S, Bharwana SA, Rizwan M, Farid M, Kanwal S, Ali Q, Khan MD (2015a) Fulvic acid mediates chromium (Cr) tolerance in wheat (Triticum aestivum L.) through lowering of Cr uptake and improved antioxidant defense system. Environ Sci Pollut Res 22(14):10601–10609

Ali S, Chaudhary A, Rizwan M, Anwar HT, Adrees M, Farid M, Anjum SA (2015b) Alleviation of chromium toxicity by glycine betaine is related to elevated antioxidant enzymes and suppressed chromium uptake and oxidative stress in wheat (Triticum aestivum L.). Environ Sci Pollut Res 22(14):10669–10678

Alloway B (2013) Introduction. Heavy metals in soils. B. J. Alloway. Springer Netherlands, vol 22, pp 3–9

Anderson AJ, Meyer DR, Mayer FK (1972) Heavy metal toxicities: levels of nickel, cobalt and chromium in the soil and plants associated with visual symptoms and variation in growth of an oat crop. Aust J Agric Res 24:557–571

Appenroth KJ, Stöckel J, Srivastava A, Strasser RJ (2001) Multiple effects of chromate on the photosynthetic apparatus of *Spirodela polyrhiza* as probed by OJIP chlorophyll a fluorescence measurements. Environ Pollut 115:49–64

Appenroth KJ, Keresztes A, Sárvári É, Jaglarz A, Fischer W (2003) Multiple effects of chromate on Spirodela polyrhiza: electron microscopy and biochemical investigations. Plant Biol 5:315–323

Augustynowicz J, Sitek E, Bryniarski T, Baran A, Ostachowicz B, Urbańska-Stopa M, Szklarczyk M (2020) The use of Callitrichecophocarpa Sendtn. for the reclamation of Cr-contaminated freshwater habitat: benefits and limitations. Environ Sci Pollut Res 27:25510–25522

Barcelo J, Poshenrieder CH (1997) Chromium in plants. In: Canali S, Tittarelli F, Sequi P (eds) Chromium environmental issues. Franco Angeli Publisher, Milano, pp 101–129

Barton LL, Johnson GV, O'Nan AG, Wagener BM (2000) Inhibition of ferric chelate reductase in alfalfa roots by cobalt, nickel, chromium, and copper. J Plant Nutr 23:1833–1845

Biacs PA, Daood HG, Kadar I (1995) Effect of Mo, Se, Zn, and Cr treatments on the yield, element concentration, and carotenoid content of carrot. J Agric Food Chem 43:589–591

Bishnoi NR, Chugh LK, Sawhney SK (1993a) Effect of chromium on photosynthesis, respiration and nitrogen fixation in pea (Pisum sativum L) seedlings. J Plant Physiol 142:25–30

Bishnoi NR, Dua A, Gupta VK, Sawhney SK (1993b) Effect of chromium on seed germination, seedling growth and yield of peas. Agric Ecosyst Environ 47:47–57

Cervantes C, Campos-García J, Devars S, Gutiérrez-Corona F, Loza- Tavera H, Torres-Guzmán JC, Moreno-Sánchez R (2001) Interactions of chromium with microorganisms and plants. FEMS Microbiol Rev 25:335–347

Chandra P, Sinha S, Rai UN (1997) Bioremediation of Cr from water and soil by vascular aquatic plants. In: Kruger EL, Anderson TA, Coats JR (eds) Phytoremediation of soil and water contaminants. ACS Symposium Series, vol 664. Washington, DC7 American Chemical Society, pp 274–82

Chatterjee J, Chatterjee C (2000) Phytotoxicity of cobalt, chromium and copper in cauliflower. Environ Pollut 109:69–74

Choudhury S, Panda SK (2005) Toxic effects, oxidative stress and ultrastuctural changes in moss Taxithelium nepalense (Schwaegr.) Broth under chromium and lead phytotoxicity. Water Air Soil Pollut 167:73–90

Costa RC, Moura FC, Oliveira PE, Magalhães F, Ardisson JD, Lago RM (2010) Controlled reduction of red mud waste to produce active systems for environmental applications: heterogeneous Fenton reaction and reduction of Cr(VI). Chemosphere 78:1116–1120

Davies FT, Puryear JD, Newton RJ, Egilla JN, Grossi JAS (2002) Mycorrhizal fungi increase chromium uptake by sunflower plants: influence on tissue mineral concentration, growth and gas exchange. J Plant Nutr 25:2389–2407

Dhir B, Sharmila P, Saradhi PP, Nasim SA (2009) Physiological and antioxidant responses of Salvinia natans exposed to chromium rich wastewater. Ecotoxicol Environ Saf 72:1790–1797

Dixit V, Pandey V, Shyam R (2002) Chromium ions inactivate electron transport and enhance superoxide generation in vivo in pea (Pisum sativum L.cv. Azad) root mitochondria. Plant Cell Environ 25:687–690

Dowling VA, Sheehan D (2006) Proteomics as a route to identification of toxicity targets in ecotoxicology. Proteomics 6:5597–5604

Dube BK, Tewari K, Chatterjee J, Chatterjee C (2003) Excess chromium alters uptake and translocation of certain nutrients in Citrullus. Chemosphere 53:1147–1153

Dubey S (2010) Transcriptomic and metabolomic shifts in rice roots in response to Cr(VI) stress. BMC Genom 11:648–667

Förstner U, Wittmann GT (2012) Metal pollution in the aquatic environment. Springer, Berlin, Heidelberg, New York

Ganesh KS, Baskaran L, Rajasekaran S, Sumathi K, Chidambaram ALA, Sundaramoorthy P (2008) Chromium stress induced alterations in biochemical and enzyme metabolism in aquatic and terrestrial plants. Colloids Surf B Biointerfaces 63:159–163

Golovatyj SE, Bogatyreva EN, Golovatyi SE (1999) Effect of levels of chromium content in a soil on its distribution in organs of corn plants. Soil Res Fert 197–204

Gopal R, Rizvi AH, Nautiyal N (2009) Chromium alters iron nutrition and water relations of spinach. J Plant Nutr 32:1551–1559

Han FX, Sridhar BBM, Monts DL, Su Y (2004) Phyto-availability and toxicity of trivalent and hexavalent chromium to Brassica juncea. New Phytol 162:489–499

Hanus J, Tomas J (1993) An investigation of chromium content and its uptake from soil in white mustard. Acta Fytotech 48:39–47

Huang W, Jiao J, Ru M, Bai Z, Yuan H, Bao Z, Liang Z (2018) Localization and Speciation of Chromium in *Coptis chinensis* Franch. using synchrotron radiation X-ray technology and laser ablation ICP-MS. Sci Rep 8:8603

Huffman EWD, Allaway HW (1973a) Chromium in plants: distribution in tissues, organelles, and extracts and availability of bean leaf Cr to animals. J Agric Food Chem 21:982–986

Huffman EWD, Allaway WH (1973b) Growth of plants in solution culture containing low levels of chromium. Plant Physiol 52:72–75

Iqbal MZ, Saeeda S, Shafiq M (2001) Effects of chromium on an important arid tree (Caesalpinia pulcherrima) of Karachi city, Pakistan. Ekol Bratislava 20:414–422

Ishikawa S, Ae N, Murakami M, Wagatsuma T (2006) Is Brassica juncea a suitable plant for phytoremediation of cadmium in soils with moderately low cadmium contamination?—Possibility of using other plant species for Cd phytoextraction. Soil Sci Plant Nutr 52:32–42

Jain R, Srivastava S, Madan VK, Jain R (2000) Influence of chromium on growth and cell division of sugarcane. Indian J Plant Physiol 5:228–231

Joshi UN, Rathore SS, Arora SK (2003) Chromium induced changes in carbon and nitrogen metabolism in guar. In: Henry A, Kumar D, Singh NB (eds) Advances in arid legumes research, pp 355–360

Juarez AB, Barsanti L, Passarelli V, Evangelista V, Vesentini N, Conforti V, Gualtieri P (2008) In vivo microspectroscopy monitoring of chromium effects on the photosynthetic and photoreceptive apparatus of Eudorina unicocca and Chlorella kessleri. J Environ Monit 10:1313–1318

Kar D, Sur P, Mandal SK, Saha T, Kole RK (2008) Assessment of heavy metal pollution in surface water. J Environ Agric Food Chem 9:444–457

Karunyal S, Renuga G, Paliwal K (1994) Effects of tannery effluent on seed germination, leaf area, biomass and mineral content of some plants. Bioresour Technol 47:215–218

Kim YJ, Kim JH, Lee CE, Mok YG, Choi JS, Shin HS, Hwanga S (2006) Expression of yeast transcriptional activator MSN1 promotes accumulation of chromium and sulfur by enhancing sulfate transporter level in plants. FEBS Lett 580:206–210

Kleiman ID, Cogliatti DH (1997) Uptake of chromate in sulfate-deprived wheat plants. Environ Pollut 1–2:131–135; Environ Chem Lett 11:229–254 249

Kocik K, Ilavsky J. Effect of Sr and Cr on the quantity and quality of the biomass of field crops. In: Production and utilization of agricultural and forest biomass for energy: proceedings of a seminar held at Zvolen, Slovakia, pp 168–178

Kumar M, Puri A (2012) A review of permissible limits of drinking water. Indian J Occup Environ Med 16(1):40

Levizou E, Zanni AA, Antoniadis V (2018) Varying concentrations of soil chromium (VI) for the exploration of tolerance thresholds and phytoremediation potential of the oregano (Origanum vulgare). Environ Sci Pollut Res 26:14–23

Liu DH, Jaing WS, Li MX (1993) Effect of chromium on root growth and cell division of Allium cepa. Isr J Plant Sci 42:235–243

Liu D, Zou J, Wang M, Jiang W (2008) Hexavalent chromium uptake and its effects on mineral uptake, antioxidant defence system and photosynthesis in Amaranthus viridis L. Biores Technol 99:2628–2636

Mandich NV (1997) Chemistry & theory of chromium deposition: Part 1–Chemistry. Plat Surf Finish 84(5):108–115

Mangabeira PA, Ferreira AS, de Almeida AAF, Fernandes VF, Lucena E, Souza VL, dos Santos-Júnior AJ, Oliveira AH, Grenier-Loustalot MF, Barbier F (2011) Compartmentalization and ultrastructural alterations induced by chromium in aquatic macrophytes. Biometals 24:1017–1026

Marieschi M, Gorbi G, Zanni C, Sardella A, Torelli A (2015) Increase of chromium tolerance in Scenedesmusacutus after sulfur starvation: chromium uptake and compartmentalization in two strains with different sensitivities to Cr(VI). Aquat Toxicol 167:124–133

McGrath SP (1995) Chromium and nickel. In: Alloway BJ (ed) Heavy metals in soils. Blackie Academic and Professional, London, pp 152–217

Mushtaq Z, Asghar HN, Zahir ZA (2021) Comparative growth analysis of okra (Abelmoschus esculentus) in the presence of PGPR and press mud in chromium contaminated soil. Chemosphere 262:127865

Nagajyoti PC, Lee KD, Sreekanth TVM (2010) Heavy metals, occurrence and toxicity for plants: a review. Environ Chem Lett 8(3):199–216

Nath K, Singh DH, Shyam S, Sharma YK (2008) Effect of chromium and tannery effluent toxicity on metabolism and growth in cowpea (Vigna sinensis L. SaviexHassk) seedling. Res Environ Life Sci 1(3):91–94

Nichols PB, Couch JD, Al-Hamdani SH (2000) Selected physiological responses of Salvinia minima to different chromium concentrations. Aquat Bot 68:313–319
Ogundiran O, Afolabi TA (2008) Assessment of the physicochemical parameters and heavy metal toxicity of leachates from municipal solid waste open dumpsite. Int J Environ Sci Technol 5:243–250
Paiva LB, de Oliveira JG, Azevedo RA, Ribeiro DR, da Silva MG, Vitória AP (2009) Ecophysiological responses of water hyacinth exposed to Cr3+ and Cr6+. Environ Exp Bot 65:403–409
Panda SK, Patra HK (2000) Nitrate and ammonium ions effect on the chromium toxicity in developing wheat seedlings. P Natl Acad Sci India B 70:75–80
Panda SK, Mahapatra S, Patra HK (2002) Chromium toxicity and water stress simulation effects in intact senescing leaves of green gram (Vigna radiata L. var Wilckzeck K851). In: Panda SK (ed) Advances in stress physiology of plants. Scientific Publishers, Jodhpur, India, pp 129–136
Pandey V, Dixit V, Shyam R (2005) Antioxidative responses in relation to growth of mustard (Brassica juncea cv. Pusa Jaikisan) plants exposed to hexavalent chromium. Chemosphere 61:40–47
Pandey V, Dixit V, Shyam R (2009) Chromium effect on ROS generation and detoxification in pea (Pisum sativum) leaf chloroplasts. Protoplasma 236(1–4):85–95
Papp JF, Lipin BR, Kirk-Othmer O (2010) Encyclopedia of chemical technology. Wiley. http://onlinelibrary.wiley.com
Pedreno NJI, Gomez R, Moral G, Palacios J, Mataix J (1997) Heavy metals and plant nutrition and development. Recent Res Dev Phytochem 1:173–179
Peralta JR, Gardea Torresdey JL, Tiemann KJ, Gomez E, Arteaga S, Rascon E (2001) Uptake and effects of five heavy metals on seed germination and plant growth in alfalfa (Medicago sativa) L. B Environ Contam Toxicol 66(6):727–734
Polti MA, Atjián MC, Amoroso MJ, Abate CM (2011) Soil chromium bioremediation: synergic activity of actinobacteria and plants. Int Biodeterior Biodegrad 65(8):1175–1181
Poschenrieder C, Gunse B, Barcelo J (1993) Chromium-induced inhibition of ethylene evolution in bean (Phaseolus vulgaris) leaves. Physiol Plant 89:404–408
Prasad MNV, Greger M, Landberg T (2001) Acacia nilotica L bark removes toxic elements from solution: corroboration from toxicity bioassay using Salix viminalis L in hydroponic system. Int J Phytoremed 3:289–300
Prasad S, Yadav KK, Kumar S, Gupta N, Cabral-Pinto MM, Rezania S, Radwan N, Alam J (2021) Chromium contamination and effect on environmental health and its remediation: a sustainable approaches. J Environ Manag 285:112174
Pulford ID, Watson C, McGregor SD (2001) Uptake of chromium by trees: prospects for phytoremediation. Environ Geochem Health 23:307–311
Rai UN, Tripathi RD, Kumar N (1992) Bioaccumulation of chromium and toxicity on growth, photosynthetic pigments, photosynthesis, in vivo nitrate reductase activity and protein content in chlorococcalear green alga, Glaucocystis nostochinearum Itzigsohn
Rai V, Vajpayee P, Singh SN, Mehrotra S (2004) Effect of chromium accumulation on photosynthetic pigments, oxidative stress defense system, nitrate reduction, proline level and eugenol content of Ocimum tenuiflorum L. Plant Sci 167:1159–1169
Rai V, Mehrotra S (2008) Chromium-induced changes in ultramorphology and secondary metabolites of Phyllanthus amarus Schum & Thonn—An hepatoprotective plant. Environ Monit Assess 147:307–315
Redondo-Gómez S, Mateos-Naranjo E, Vecino-Bueno I, Feldman SR (2011) Accumulation and tolerance characteristics of chromium in a cordgrass Cr-hyperaccumulator, Spartina argentinensis. J Hazard Mater 185:862–869
Rout GR, Samantaray S, Das P (1997) Differential chromium tolerance among eight mungbean cultivars grown in nutrient culture. J Plant Nutr 20:473–483
Samantaray S, Rout GR, Das P (1999) Studies on differential tolerance of mungbean cultivars to metalliferous minewastes. Agribiol Res 52:193–201

Samantaray S (2002) Biochemical responses of Cr-tolerant and Cr-sensitive mung bean cultivars grown on varying levels of chromium. Chemosphere 47:1065–1072
Scoccianti V, Crinelli R, Tirillini B, Mancinelli V, Speranza A (2006) Uptakeand toxicity of Cr(III) in celery seedlings. Chemosphere 64(10):1695–1703
Shahid M, Shamshad S, Rafiq M, Khalid S, Bibi I, Niazi NK, Dumat C, Rashid MI (2017) Chromium speciation, bioavailability, uptake, toxicity and detoxification in soil-plant system: a review. Chemosphere 178:513–533
Shanker AK, Djanaguiraman M, Sudhagar R, Chandrashekar CN, Pathmanabhan G (2004) Differential antioxidative response of ascorbate glutathione pathway enzymes and metabolites to chromium speciation stress in green gram (Vigna radiata (L) R Wilczek, cv CO4) roots. Plant Sci 166:1035–1043
Shanker AK, Cervantes C, Loza-Tavera H, Avudainayagam S (2005) Chromium toxicity in plants—A review. Environ Int 31:739–753
Sharma DC, Mehrotra SC (1993) Chromium toxicity effects on wheat (Triticum aestivum L cv HD 2204). Indian J Environ Health 35:330–332
Sharma DC, Pant RC (1994) Chromium uptake: its effects on certain plant nutrients in maize (Zea mays L. cv. Ganga 5). J Environ Sci Health 29:941–948
Sharma DC, Chatterjee C, Sharma CP (1995a) Chromium accumulation and its effects on wheat (Triticum aestivum L. cv. HD 2204) metabolism. Plant Sci 111:145–151
Sharma DC, Chatterjee C, Sharma CP (1995b) Chromium accumulation by barley seedlings (Hordeum vulgare L,). J Exp Bot 1995:25241–25251
Sharmin SA, Alam I, Kim KH, Kim YG, Kim PJ, Bahk JD, Lee BH (2012) Chromium-induced physiological and proteomic alterations in roots of Miscanthus sinensis. Plant Sci 187:113–126
Singh AK (2001) Effect of trivalent and hexavalent chromium on spinach (Spinacea oleracea L). Environ Ecol 19:807–810
Sinha V, Pakshirajan K, Chaturvedi R (2018) Chromium tolerance, bioaccumulation and localization in plants: an over-view. J Environ Manag 206:715–730
Smith WA, Apel WA, Petersen JN, Peyton BM (2002) Effects of carbon and energy source on bacterial chromate reduction. Bioremed J 6:205–215
Srivastava S, Nigam R, Prakash S, Srivastava MM (1999a) Mobilization of trivalent chromium in presence of organic acids: a hydroponic study of wheat plant (Triticum vulgare). Bull Environ Contam Toxicol 63:524–530
Srivastava S, Prakash S, Srivastava MM (1999b) Chromium mobilization and plant availability—The impact of organic complexing ligands. Plant Soil 212:203–208
Subrahmanyam D (2008) Effects of chromium toxicity on leaf photosynthetic characteristics and oxidative changes in wheat (Triticum aestivum L.). Photosynthetica 46:339–345
Sundaramoorthy P, Chidambaram A, Ganesh KS, Unnikannan P, Baskaran L (2010) Chromium stress in paddy: (i) nutrient status of paddy under chromium stress; (ii) phytoremediation of chromium by aquatic and terrestrial weeds. CR Biol 333:597–607
Suseela MR, Sinha S, Singh S, Saxena R (2002) Accumulation of chromium and scanning electron microscopic studies in *Scirpus lacustris* l treated with metal and tannery effluent. Bull Environ Contam Toxicol 68:540–548
Tang SR, Wilke BM, Brooks RR, Tang SR (2001) Heavy-metal uptake by metaltolerant Elsholtzia haichowensis and Commelina communis from China. Commun Soil Sci Plant Anal 32(5–6):895–905
Tiwari KK, Dwivedi S, Singh NK, Rai UN, Tripathi RD (2009) Chromium(VI) induced phytotoxicity and oxidative stress in pea (Pisum sativum L.): biochemical changes and translocation of essential nutrients. J Environ Biol 30:389–394
Tripathi AK, Tripathi S, Tripathi S (1999) Changes in some physiological and biochemical characters in Albizia lebbek as bio-indicators of heavy metal toxicity. J Environ Biol 20:93–98
Turner MA, Rust RH (1971) Effects of chromium on growth and mineral nutrition of soybeans. Soil Sci Soc Am Proc 35:755–758

Vajpayee P, Sharma SC, Tripathi RD, Rai UN, Yunus M (1999) Bioaccumulation of chromium and toxicity to photosynthetic pigments, nitrate reductase activity and protein content of Nelumbo nucifera Gaertn. Chemosphere 39:2159–2169

Vajpayee P, Tripathi RD, Rai UN, Ali MB, Singh SN (2000) Chromium (VI) accumulation reduces chlorophyll biosynthesis, nitrate reductase activity and protein content in Nymphaea alba L. Chemosphere 41:1075–1082

Vajpayee P, Rai UN, Ali MB, Tripathi RD, Yadav V, Sinha S, Singh SN (2001) Chromium induce physiologic changes in Vallisneria spiralis L. and its role in phytoremediation of tannery effluent. Bull Environ Contam Toxicol 67:246–256

Vernay P, Gauthier-Moussard C, Hitmi A (2007) Interaction of bioaccumulation of heavy metal chromium with water relation, mineral nutrition and photosynthesis in developed leaves of Lolium perenne L. Chemosphere 68:1563–1575

Whitacre DM (2010) In: Metzler JB (ed) Reviews of environmental contamination and toxicology, vol 202. Springer, New York, USA

Yadav SK, Dhote M, Kumar P, Sharma J, Chakrabarti T, Juwarkar AA (2010) Differential antioxidative enzyme responses of Jatrophacurcas L. to chromium stress. J Hazard Mater 180:609–615

Yoon J, Cao X, Zhou Q, Ma LQ (2006) Accumulation of Pb, Cu, and Zn in native plants growing on a contaminated Florida site. Sci Total Environ 368:456–464

Zayed AM, Terry N (2003) Chromium in the environment: factors affecting biological remediation. Plant Soil 249:139–156

Zeid IM (2001) Responses of Phaseolus vulgaris to chromium and cobalt treatments. Biol Plant 44:111–115

Zeng F, Wu X, Qiu B, Wu F, Jiang L, Zhang G (2014) Physiological and proteomic alterations in rice (Oryza sativa L.) seedlings under hexavalent chromium stress. Planta 240:291–308

Zhang D, Jiang L, Shao Y, Chai B, Li C (2009) Variations in germination and endogenous hormone contents of wheat cultivars under Cr stress. Chin J Appl Environ Biol 15:602–605

Zhitkovich A, Song Y, Quievryn G, Voitkun V (2001) Non-oxidative mechanisms are responsible for the induction of mutagenesis by reduction of Cr(VI) with cysteine: role of ternary DNA adducts in Cr(III)-dependent mutagenesis. Biochemistry 40(2):549–560

ZivKovic J, Razic S, Arsenijevic J, Maksimovic Z (2012) Heavy metal contents Zn Veronica species and soil from mountainous areas in Serbia. J Serb Chem Soc 77(7):959–970

Zurayk R, Sukkariyah B, Baalbaki R (2001) Common hydrophytes as bioindicators of nickel, chromium and cadmium pollution. Water Air Soil Pollut 127:373–388

Chapter 15
Chromium Profile in Soil, Fodders and Plasma of Crossbreed Cattle

B. Devasena and J. V. Ramana

Abstract Chromium as an essential nutrient in livestock nutrition has been reported in early 1960s. Earlier it has been known more as heavy material which is potential bio accumulative toxin of daily production systems. More frequently the Chromium is seen in the trivalent and hexavalent structures though other oxidation states (from −2 to +6) also exist. Among the different oxidation states in which chromium is found in living organisms, trivalent chromium (Cr^{3+}) is the most stable and is contemplated to be a highly safe form of chromium. In India there is very poor documentation available regarding the heavy metal toxicity of daily feeds. In the field study carried out a total of 142 soil, 142 water, 608 dairy ration components and 790 plasma samples of crossbred cattle were collected and screened for chromium content. The chromium concentration in soil was below the permissible limit. The dry forages like paddy straw and groundnut straw indicated Chromium to a considerable extent (<1.6 Mg/Kg). While all the green forages (except Sweet Sudan grass) did not contain chromium in detectable concentration. Highest chromium concentration was found in particularly mineral mixtures and to a lesser extent in concentrate ingredients such as rice bran, wheat bran homemade concentrate mixtures and compounded cattle feeds. Different categories of crossbred cattle under investigation indicated that, Plasma chromium concentration of anoestrous cows, pregnant cows and pregnant heifers, observed to contain higher ($P < 0.01$) levels, followed by anoestrous heifers while lowest in lactating cows. The study indicated that there are no toxic levels of chromium in soil and dairy feeds.

Keywords Concentrates · Forages · Minerals · Plasma

B. Devasena (✉)
Department of Animal Nutrition, College of Veterinary Science, Sri Venkateswara Veterinary University, Tirupati, Andhra Pradesh, India
e-mail: devasenabusineni@yahoo.com

J. V. Ramana
COE, Sri Venkateswara Veterinary University, Tirupati, Andhra Pradesh, India

© The Author(s), under exclusive license to Springer Nature Switzerland AG 2023
N. Kumar et al. (eds.), *Chromium in Plants and Environment*, Environmental Science and Engineering, https://doi.org/10.1007/978-3-031-44029-8_15

15.1 Introduction

Mineral status of the livestock can be assessed by the sampling of forages fed to the animals and the soil in which they are cultivated. In many developing countries (including India) soil, plant and animal systems were not thoroughly examined because of several intricacies. Acquiring the information on the mineral status of the livestock, understanding of the interrelationships of various inorganic minerals in the forenamed system is essential for obtaining augmented livestock productivity. Both essential and nonessential elements can be toxic when levels exceed a certain threshold. Excessive deposition of toxic heavy metals particularly in the soil is a consequence of human activities which is often referred as heavy metal contamination. When these heavy metals enter the food chain it is highly hazardous to the nature and organisms in the environment. The concentration of essential and contaminant heavy metals such as chromium are potential bio-accumulative toxins in production system of livestock because the soils serve as long term sinks for such toxic metals. They are primarily absorbed on to organic matter of soil, followed by different forms of humified natural organic matter which also receive a larger concentration. Apart from this, metal oxides especially iron and manganese oxides, clay minerals etc. also incorporated with them (Li et al. 2005). The presence of high concentrations of manganese in the soil, as well as alkaline pH levels, may affect oxidation processes. When the forages are grown in such polluted environment, toxic minerals are absorbed from the soil under specific soil conditions as well as from the metal which is deposited on the surfaces of the plants. In addition, excess use of fertilizers with heavy metals used in the agriculture is the additional sources for heavy metal accumulation. Heavy metals exhibit a different physical and chemical properties based on which their mobility pattern in the soil takes place. Generally, some low amount of losses takes place as a consequence of absorption by the crops grown in that area, soil erosion and leaching (Aldrich et al. 2002). The soil productivity is diminished when there is long term accumulation of heavy metals in agricultural soil, which in turn inhibit the populations of microbes and fauna and that may create a risk to animal, human and health of ecosystem.

Among the essential trace minerals Chromium (Cr) is also listed which having importance biological system. Similar to many other minerals, Cr also plays several important roles in the metabolism of living organisms (Tokalioglu et al. 2000; Tokaliglu and Kartal 2005). A variety of metabolism related roles are played by Cr in both animals as well as plants (McDowell 2003). In man and laboratory animals Chromium has been considered as an essential trace element (Anonymous 1997), that comes from anthropogenic sources. It is seen in different oxidative conditions, but the three and six oxidation states are most commonly found in nature. The hexavalent form of Cr is an oxidizing mineral which is employed in most of the processes related to industry such as electroplating, welding and chromate painting. Trivalent chromium is considered to be essential to carry out normal carbohydrate, lipid, protein metabolism and the maintenance of nucleic acid (NA) structural integrity. Biologically active form of Chromium is found as part of an oligopeptide—chromodulin—potentiating the effect of insulin by facilitating insulin binding to receptors at the cell surface (Amata 2013). In carbohydrate and lipid metabolism the trivalent form of Cr is considered to be the most essential (NRC 2001). The importance of Chromium in the organism is related to the functionality of insulin, as a cofactor (Cr^{3+} is a component of glucose tolerance factor-GTF) it enhances the activity of insulin (Pechova and Pavlata 2007). The important role played by Cr^{3+} ions in enhancing insulin activity is by activation of the insulin receptor tyrosine kinase, and also it potentiates it by three-fold.

Because of ubiquitous nature of Cr it exists in air, water, soil and biological materials, and hence the supplementation is generally not considered essential. Through limited studies available, Cr supplementation in swine production reported to have beneficial responses but in dairy cattle not that encouraging (European Commission 2003). In spite of the fact that in most of animal species Cr is relatively harmless but, ingestion of high dose (30–40 mg/kg of B. wt./day) of Cr in dairy calves has resulted in toxicity (European Commission 2003).

Despite of the fact that Cr plays substantial role of in various physiological processes in ruminants, information on evaluation of mineral status of grazing ruminants, is quite limited, especially pertaining to the availability of Cr in the biological system of animals. The Chromium concentration required in feeds of livestock ranges between 0.3 and 1.6 mg/kg (Anonymous 1997) which is generally higher than the available Cr to livestock through livestock feeds. When Chromium levels exceed than these range which are deemed to be not only toxic to livestock but also awkwardly influence the reproductive potential of ruminants (McDowell 2003; McDowell and Arthington 2005). In view of the importance of this element to livestock, the premier focus of the present investigation was set to authenticate the practicability of broad feeding pattern for free grazing ruminants rather than appraising the actual daily ingestion of Cr for animals.

In the assessment of mineral status of livestock, sampling of forages offered to livestock and soil used for cultivation these forage plants is the chief activity involved. The sample and sampling procedure have the greatest value which need to be collected from soil, plant and biological tissues of animal in question and also depends upon the mineral to be determined from these variables used for assessment (McDowell 1985). Present investigation was planned in Chittoor district of Andhra Pradesh in India, in order to determine the chromium levels in soil (grazing and cultivated), drinking water, concentrate feeds, fodders (green/dry) as well as in plasma of crossbred dairy cattle.

15.2 Materials and Methods

15.2.1 Samples Collection

The study was conducted at Chittoor district (13.2172°N, 79.1003°E) in Andhra Pradesh state (India), which is classified under southern agro- climatic zone of Andhra Pradesh state. The district comprises of four revenue divisions and 66 revenue mandals. One village from each mandal was selected, which were geographically located apart in direction and a total sixty-six mandals of the district in which sampling procedure was carried out.

A minimum of two soil (grazing and cultivated land) samples from each mandal were collected (total of 142 soil samples). From four different sites within the pasture/grazing land, samples of soil were collected randomly. From the surface of the soil a minimum 15–20 cm depth was considered to collect the soil samples. Initially in the sampling area, surface litter was removed from the surface of the sampling point before taking the soil sample. Zigzag sampling design was followed, using a spade a depth of 20 cm (i.e. from the surface of soil to 20 cm beneath the soil surface) was cut and a thin portion of soil with about 5 cm thick was collected. Then this segment of soil was placed in pail and mixed with hands. A sub-sample of 300–500 g from the bulk soil sample was collected. This sample was placed in a plastic sample bag which contained label on the outside. Soon after the soil samples arrive at the laboratory, air drying at room temperature (25–35 °C) was carried out. Once soil samples were air dried, they were sieved by using a 2 mm sieve. All samples of soil collected were dried in the oven at appropriate temperature (100 ± 5 °C) until the constant dry weights were obtained.

With regard to water sampling, a minimum of two water (bores, canals, wells and river streams etc.) samples were collected (total of 142 samples) by composite sample method into a clean sterile collection bottle. During the preplanned time/after prefixed flow, smaller samples were collected in succession and mixed in the same container. When the samples were obtained by compositing the integrated samples of equal volume across the cross section were taken from different places at different depths-with equal flow, so that the overall composition of a stream has been represented. In larger water sources, because of the requirement of large number of samples, it was determined by section of a stream, along with depth and cross section where the water constitution will be constant.

The forage/fodder samples were collected at random from four different locations within the grazing area and pasture land which were being used for feeding of cattle. All forage/fodder samples collected were dried in hot air oven at appropriate temperature (80 °C) until constant dry weights are observed. Different feeds and fodders (Green and dry) offered (total 528) to the crossbred dairy cattle, by the farmers at the household level were also collected from each farmer in the study area. After collection all the representative samples of feeds and fodders, were subjected for drying in forced draft oven (at 80 °C for 24 h), subsequently pulverized to pass through 2 mm size sieve and were stored plastic sample bags in moisture free condition, till further analysis.

From each village a minimum of 10 crossbred dairy cattle in diverse physiological phases (lactating cows, pregnant cows, pregnant heifers, cows in anoestrous condition and heifers in anoestrous condition) were considered for collection of blood samples. The jugular vein was punctured in an aseptic manner and blood was drawn and transferred in to heparinized blood collection vials (a total of 790 samples). After collection the blood samples were subjected for centrifugation process at 3000 rpm for about 20 min so that plasma was separated. After collection plasma samples were stored in deep freeze, (at −20 °C) for further analysis.

15.2.2 Processing of Samples

Soil sample (0.5 g) was dissolved in 0.1 N HCl and then a pinch of charcoal was added to prepare mineral extract. Then it was filtered by using no. 1 Whatman filter paper. Then diluted with double glass distilled water to get appropriate concentration for analysis (Singh et al. 1999). When water sample (45 ml) was digested, 5 ml of concentrated nitric acid was used. Similarly digesting 0.5 g of feed/fodder sample 15 ml of nitric acid was used. Whereas 2 ml of plasma sample digested with 15 ml of nitric acid. The micro wave sample digester (CEM Mars X-press) was used in the digestion process of all samples. All the digested samples were filtered through Whatman filter paper no. 1 to obtain mineral extract and dilution was carried out with double glass distilled water to get the suitable concentration. The Chromium was estimated in the mineral extract prepared after digestion of the samples, in an atomic absorption spectrophotometer (Perkin Elmer, Avanta PM-A-6287), with 0.1 ppm detection limits. For the determination of chromium monoelement hollow cathode lamp was employed. Initially the calibration of AAS was carried out by following the recommendations of instrument manufacturers. Absorbance reading was plotted on the Y-axis and that of concentration of different standard solutions of Cr mineral on X-axis for building up a standard curve was prepared by versus. Then, the AAS reading of sample of interest was plotted on the standard curve, and then concentration of Cr metal was calculated in the sample.

15.2.3 Statistical Analysis

The data generated on the Cr content of the samples (Viz. soil, water, feed/fodder and plasma) were subjected for statistical analysis with the help of SPSS software. Each variable was worked out with a one-way analysis of variance (ANOVA).

15.3 Results and Discussion

15.3.1 *Soil*

The results of soil analysis (Table 15.1) revealed the average chromium concentration was 4.08 ppm (observed the range from 2.4 to 6.2 ppm). The detected range of chromium in all the soil samples collected found to be less than the permissible ceiling of 100 ppm (Misra and Mani 1978) and within the reported range by Devasena et al. (2012). There was a variation in soil Cr content due to the season has been reported. Low amount of Cr was reported during December as compared to high concentration during January and during the entire study period which observed to be in the range of 0.006–0.007 mg/kg (Khan et al. 2010c). Contrary to this higher soil Cr content reported in some earlier studies (Hodgson 1990). It was opinioned that, soil Cr concentration was significantly influenced by the time of sampling. Elevated levels of Cr in soil samples was found during the October (1st sampling) and lower amount of soil Cr was found during the April (4th sampling) as reported by Danish et al. (2014). Besides, chromium found abundantly to a much greater degree in soil as compared to the crops (Underwood and Suttle 1999). In the present study the low Cr level was observed, for which the alkaline soil pH (7.30–7.63) could be a reason. Higher exchangeable Cr takes place in the acidic soil than in the alkaline soil which has been evidenced in several reports (Xu et al. 2020). Presence of ferrous oxide and manganese oxide in the soil aggravate the deposition of heavy metal (Li et al. 2005). In agricultural soils the major routes of heavy metal load predominantly include atmospheric deposition, sewage sludge, animal manures, agrochemicals and inorganic fertilizers. While the losses takes place due to the uptake by cultivation of crops or livestock products, soil leaching and erosion. (Nicholson et al. 1999).

Table 15.1 A pH and chromium content (ppm) of soil and water

	Chittor division		Tirupati division		Puttur division		Madanapalli division		District over all	
Soil	Mean	SE	Mean	SE	Mean	SE	Mean	SE	Mean	SE
	(n = 39)		**(n = 33)**		**(n = 35)**		**(n = 35)**		**(n = 142)**	
pH	7.63	0.12	7.37	0.16	7.49	0.14	7.30	0.09	7.45	0.07
Cr	3.5^{b}	0.42	6.2^{a}	0.38	4.2^{b}	0.82	2.4^{c}	0.62	4.08	0.22
Water	**(n = 39)**		**(n = 33)**		**(n = 35)**		**(n = 35)**		**(n = 142)**	
pH	7.42	0.08	7.33	0.08	7.69	0.10	7.40	0.06	7.46	0.08
Cr	ND	–	ND	–	ND	–	ND	–	ND	–

abc values bearing different superscripts in a row differ significantly ($P<0.01$)*

15.3.2 Water

The analysis of water for pH indicated a faintly alkaline range from 7.33 to 7.69. In all water samples (142) analyzed, chromium was not in the detectable concentration. Water quality in addition to availability is extremely important for animal health and productivity. Even low concentrations of water contaminants effect animal performance when consumed over a long period (NRC 2001). In nature Hexavalent form of chromium happens commonly. Water solubility is more for Hexavalent chromium compounds as compared to trivalent chromium compounds. During the process of disinfection of water oxidization of trivalent chromium to hexavalent chromium takes place. In drinking water the amount of acquired hexavalent chromium for which the primary source is from oxidation of naturally occurring chromium present in combustible geologic formations. Chromium compounds have been released to the various locations of the environment by means of effusion, defective storage and due to inappropriate industrial waste discarding practices. In water the Chromium compounds either trivalent or hexavalent chromium are very pertinacious (USEPA 2011).

Present results indicated that, there was no contamination of chromium from other sources into either soil or water. As per the guidelines of World Health Organization (WHO 2022) the maximum total permissible chromium content for drinking water is 0.05 mg/L. At present, there are no standard integrated regulations for chromium in drinking water for individual livestock species (WHO 2022). The United States Environmental Protection Agency (USEPA) has set a maximum contaminant level of 100 ppb or 0.1 mg/L (USEPA 2011) in drinking water and it is the regulatory authority for setting the maximum permissible level of total chromium in standards of drinking water.

15.3.3 Forages/Roughages

Paddy straw (0.16 ppm) and groundnut straw (0.49 ppm) contained chromium to a considerable extent among the dry roughages analyzed (Table 15.2). Out of dry forage samples (153) analyzed, paddy straw (88.3%) and groundnut straw (80.63%) samples contained chromium and in the others types of dry roughages it was not in the range that could be detected. In case of green forages (among the 175 samples), all samples of SSG-898 contained (0.03 to 0.29 ppm) chromium, while in other fodder (138) and tree foliage samples (20) analyzed the Cr concentration was not in the range that could be detected. The Cr concentration in the forage reported by Danish et al. (2014) ranged from 2.9–4.0 mg/kg and was greater than the range observed in the present study. It has been reported that different plant parts contain variable amount of Cr (Anderson et al. 1990). While relatively low chromium concentration (0.002–0.0025 mg/kg) of forages was reported by Khan et al. (2010c) which was not toxic for animals being reared. Chromium enters the food chain of living organisms when the cell walls of plants passively take up Cr^{3+} from the soil, through cation transport/exchange sites. The quality and availability of pastures in semi-arid areas is substantially affected in dry season due to scant rainfall patterns, when more rainfall is experienced improvements in availability and quality of grasses are seen. But forages are observed to be acutely deficient as compared to the requirement for growth and development of ruminant livestock which is suggesting the need for supplementation. The Cr concentration of fodder/forage samples in the present evaluation was found to be variable as compared to the reported values in literature but was not in the toxic range (Devasena et al. 2012).

Table 15.2 Chromium content (ppm) of feed and fodder samples

	n	Mean	SE	Detected	%detection
Dry roughages	**153**			**108**	**70.5**
Paddy straw	94	0.16 (0.04–0.20)	0.01	83	88.3
GN straw	31	0.49 (0.31–1.04)	0.08	25	80.6
Sorghum stover	23	ND	0.0	0	0
Ragi straw	5	ND	0.0	0	0
Green roughages	**175**			**17**	**9.7**
Sugar Cane tops	29	ND	0.0	0	0
SSG-898	17	0.19 (0.03.0.29)	0.03	17	100
Co-1	32	ND	0.0	0	0
Sorghum	11	ND	0.0	0	0
APBN	25	ND	0.0	0	0
Local Grass	41	ND	0.0	0	0
Tree foliage	20	ND	0.0	0	0
Concentrates	**200**			**88**	**44.0**
Rice Bran	70	0.22 (0.01–0.39)	0.02	56	80
Wheat Bran	23	0.30 (0.02–0.45)	0.01	10	44 ara>
Gram chuni	13	ND	0.0	0	0
Maize/Bajra grain	23	ND	0.0	0	0
Ground nut cake	80	ND	0.0	0	0
Soybean meal	11	ND	0.0	0	0
Sunflower meal	3	ND	0.0	0	0
Home made concentrate mixture	20	0.25 (0.18–0.45)	0.0	4	20
Compounded Cattle feed	30	0.51 (0.06–0.65)	0.02	17	57.0
Mineral Mixture	7	28.1 (18.4–49.4)	0.01	5	71.4

Values in the parentheses indicate range

15.3.4 Feed Ingredients

In case of concentrates a total of 200 samples analysed out of which 12% contained chromium in the detectable range. Among the rice bran (0.22 ppm) samples 80 and 44% of wheat bran (0.17 ppm) samples contained chromium to a considerable extent. But in gram chuni, cereal grains (Maize and Bajra grain), different protein supplements (Groundnut cake, Soybean meal and Sun flower meal, Homemade concentrate mixture) chromium was not detected. Compounded cattle feed which was commercially available in the market found to contain 0.51 ppm chromium and 57% of the samples were detected with chromium. Islam et al. (2016) reported higher levels of chromium (1.64–1.71 ppm) in commercial broiler feed. Considerably lower chromium concentration was observed than the NRC (2005) set permissible limit (500 ppm or μg/g). Whereas in the present investigation the Cr concentrations noticed are indicating lower concentration than the values reported by Li et al. (2005). However, higher Chromium concentration in protein meal feed (10–218 mg/Kg) was reported by Jothi et al. (2016). This therefore was suggestive that, regular screening for detection of heavy metals needed for protection from harmful effects.

In the present study 20% of the home made concentrate mixtures shown to contain the chromium in the range of 0.21–0.28 ppm. These compounded/homemade concentrate mixtures contained chromium because the brans often form the major component of the concentrate mixtures (prepared by the farmers at their home and commercially prepared compounded feeds which are marketed being used by the farmers in this area), since during the analysis brans were detected with considerably higher concentration of chromium could be the reason.

Accurately measuring naturally occurring Cr in feed ingredients is difficult because of Cr contamination that can occur during harvesting, processing, and collection of samples as well as laboratory analysis of Cr. Grinding whole grain samples, de-oiled meals, by products like brans and chunies in Wiley mill which pass through a stainless steel screen greatly increased analyzed Cr concentrations. Appreciable amount of Cr found in the feed ingredients emerge as a result of contamination from soil as well as due to metal proximity in the process of cultivation, harvesting of crops, procedure of feed ingredients preparation or combination of both (Spears et al. 2017).

15.3.5 Mineral Mixtures

In case of mineral mixtures 71.4% were containing chromium in the detectable level with average content of 28.1 ppm (18.4–49.4 ppm) which indicated moderately higher concentration of chromium as compared to the other feed ingredients analysed. Higher levels of Cr observed in the feed stuffs might be due to the mixing of these feed ingredients in processors and stainless steel containers (which particularly contain 18% Cr), may give rise to increased level of Cr in feed stuffs as a result of contamination. As per NRC (2001), acceptable normal level of chromium in the rations of livestock is 1.6 mg/Kg and that of upper limit for toxicity of Cr is 1000–3000 mg/Kg. Present research results suggested that, the Cr levels observed with regard to most of feed/fodder samples were quite lower than the typical values reported (Li et al. 2005). As per the research data available, maximum threshold limit of the Cr concentration was not available but a concentration range between 0.03 and 1.0 mg/kg was suggested in majority of studies. Chromium reported in the salt-range of Pakistan was in the range of 0.156–0.285 mg/g (Khan et al. 2010a, b). Whereas lower concentration (from 0.0003 to 0.0006 mg/L) was observed in central Punjab of Pakistan (Raj et al. 2006). Although there is no evidence of unfavorable effect of higher Cr concentration in forage and blood plasma, still the concentration can be decreased with the plants that absorb reduced quantity of chromium. Moreover, supplementation of area specific minerals and minerals mixture with specific combination in the livestock is suggested in view to conserve the environment and safeguard the human health.

The cycling of many heavy metals through the dairy food chain occurs which is possibly to be restricted by the soil–plant barrier in a closed system (Chaney 1990). The soil acts as barrier which limits transmission of metals in to the food chain by different chemical processes which might have been involved in the regulation of the bioavailability of the metals. The soil cation exchange capacity is the principle component, which is governed in turn, by other soil chemical properties such as pH, salinity, macro nutrients, micro nutrients like different minerals and their concentration (Alloway 1995). At times the soils are subjected for alterations, due to the addition of bio-solid wastes or manures because of theses specific properties in the modification the bio-availability of the metals will be diverged (Chaney 1990). Similarly even the certain barriers present in plant restricts transmission of certain heavy metals into the food chain, in view of the fact that metals are phytotoxic because of which considerable yield reduction occurs before the crop would results in risk of metal overdosing during the ingestion by live-stock in their lifetime.

15.3.6 Animals

Chromium concentration determined in the plasma of various categories of crossbred cattle summarized in Table 15.3. The outcome of the experimentation indicated that cows in pregnancy, pregnant heifers and cows in nonproductive condition were significantly ($P < 0.01$) higher (ranged from 0.31 to 0.33 ppm) in contrast to heifers in nonproductive stage (0.27 ppm) and cows which were lactating (0.15 ppm). The variations observed were remarkable ($P < 0.05$) in plasma concentration amongst the animals of diverse physiological categories and in between the animals belonging to four revenue divisions which can be attributed to differences in feeding practices adopted by the farmers (Devasena et al. 2012). The animals in the district were fed by paddy straw and groundnut straw to the major extent where as green fodder offered to a very limited extent. The supply of concentrate feeds was limited to those animals which are in productive condition (Lactating animals). In addition to this, when the animals are reared under grazing system, annual ingestion of about 600 kg of soil in dairy cows has been reported in many parts of the world (Healy 1974), which might be another contributing factor for the plasma levels of the minerals though the levels are considerably less in most of the feeds. The plasma Cr content varied between 3.5 and 4.3 mg/L. The variability in the concentration of Cr is presumed to be due to the divergent sources of blood collection (Young/growing animals, lactating dairy animals and non productive/dry animals) which might influence the amount of Cr level as well as there might be some synergy between time of collection in a year and source of blood collection which exemplify the concentration level.

There were reports which indicated that higher blood Cr levels were found during February and lower amounts during December (Danish et al. 2014). The higher blood Cr concentration was observed than the suggested value by different authors (Khan et al. 2010a, b; Raj et al. 2006). The required range for Cr concentration in livestock is 0.3–1.6 mg/kg (Khan et al. 2010a, b) and the variability do exist which might be there due to the type of condition of animals (young growing calves, lactating dairy cows and dry/pregnant cows) in which the blood was examined. In spite of the detrimental effect of higher Cr concentration in blood plasma, which is a result of higher Cr in forage that are fed to the livestock, yet it can be decreased by growing the plants that absorb less quantity of chromium. Limited information is available about the Cr concentrations in feeds and fodders fed to the livestock and even barely known regarding Cr bioavailability from feedstuffs which are commonly offered to livestock at field level (Bryan et al. 2004). Moreover, it was suggested that supplementation of area specific minerals mixture in rations of livestock would be beneficial which not only conserve the environment but also safeguard human health (Danish et al. 2014).

In spite of the fact that recycling of several heavy metals through the dairy food chain occurs which is possibly to be restricted by different variables such as cation exchange capacity of soil, salinity, pH, phytotoxicity of the metal etc. In dairy system it is essential that the sources of that particular metal should be recognized for effective prevention of such complications connected with accumulation in soil. The reproductive potential of ruminants may badly be affected due to the elevated concentrations which are toxic to livestock (Raj et al. 2006). Present study revealed that the analysis of samples of soil, feeds, fodders and biological samples of animals carried out to determine the chromium content revealed to contain lower than the permissible levels (NRC 2001) which suggested that, chances of cumulative toxicity are less.

15.4 Conclusion

At present times, alarming level of heavy metal pollution in livestock feeds and fodders is most obvious throughout the world. Present research results on determination of chromium reflected existence of nontoxic levels in soil, water and dairy feeds and fodders. The chromium concentration in plasma of crossbred cattle of

Table 15.3 Chromium content (ppm) in plasma of crossbred cattle

Category of animal	Chittoor division (n = 241)	Tirupati division (n = 197)	Puttur division (n = 180)	Madanapalli division (n = 172)	District overall* (n = 790)
Lactating cows (n = 131)	0.24 ± 0.001 (45)	0.24 ± 0.002 (32)	0.19 ± 0.001 (28)	0.17 ± 0.002 (26)	0.15^{c} ± 0.002
Pregnant cow (n = 106)	0.19 ± 0.002 (30)	0.16 ± 0.001 (27)	0.15 ± 0.002 (25)	0.20 ± 0.001 (24)	0.31^{a} ± 0.003
Pregnant Heifer (n = 122)	0.18 ± 0.001 (34)	0.22 ± 0.001 (31)	0.18 ± 0.002 (29)	0.26 ± 0.003 (28)	0.31^{a} ± 0.002
Anoestrous Cows (n = 302)	0.21 ± 0.003 (92)	0.33 ± 0.003 (75)	0.31 ± 0.002 (68)	0.25 ± 0.001 (67)	0.33^{a} ± 0.003
Anoestrous Heifer (n = 129)	0.20 ± 0.002 (40)	0.23 ± 0.002 (32)	0.18 ± 0.001 (30)	0.23 ± 0.003 (27)	0.24^{a} ± 0.002
Overall*	0.29^{a} ± 0.003	0.11^{b} ± 0.004	0.21^{b} ± 0.002	0.17^{b} ± 0.001	0.20^{b} ± 0.003

Values in the parentheses indicate percentage. [abc] Values bearing different superscripts in a column differ significantly ($P < 0.01$)*

different age and physiological groups was lower than the permissible levels (NRC 2001) which indicated that, chances of cumulative toxicity are less.

Acknowledgements The authors highly acknowledge the financial support provided by a grant from the ICAR, New Delhi, for this study. The authors wish to thank the Andhra Pradesh State Animal Husbandry Department officers for their help in collecting the samples during the survey and SVVU, Titupati, Andhra Pradesh, for the support rendered to execute this work.

References

Aldrich AP, Kistler D, Sigg L (2002) Specifications of Cu and Zn in drainage water from agricultural soils. Environ Sci Technol 36:4824–4830

Alloway BJ (1995) Soil processes and the behaviour of metals. 11–37 in Heavy metals in soils. 2nd edn. B. J. Alloway, ed. Kluwer Academic Publ., Dordrecht, The Netherlands

Amata IA (2013) Chromium in livestock nutrition: a review. Glob Adv Res J Agricult Sci 2(12):289–306 (Special Anniversary Review Issue). ISSN: 2315-5094. http://garj.org/garjas/index.htm

Anderson RA, Polansky MM, Bryden NA, Kanary JJ, Mertz W (1990) Seventh international symposium on trace elements in man and animals, Dubroobnik, Yugoslavia

Anonymous (1997) The role of chromium in animal nutrition. National Academy Press, Washington, DC

Bryan MA, Socha MT, Tomlinson DJ (2004) Supplementing intensively grazed late-gestation and early-lactation dairy cattle with chromium. J Dairy Sci 87:4269–4277

Chaney RL (1990) 20 years of land application research. Biocycle 31:54–59

Danish M, Ahmad N, Sharif I, Shahzad MN, Rizvi SS, Nazar MF (2014) Inter-relationshipof Soil-Forage-Plasma, and Milk Chromium: a case study in an arid Region of Pakistan. J Environ Anal Toxicol 4(3):21

Devasena B, Ramana JV, Prasad PE, Sudheer S, Prasad JR (2012) Chromium concentration in soil, feeds and plasma of animals in Chittoor district of Andhra Pradesh. Indian J Anim Nutr 29:384–387

European Commission (2003) Opinion of the scientific committee on animal nutrition on undesirable substances in feed. European Commission, Health and Consumer Protection Directorate, Brussels, Belgium

Healy WB (1974) Ingested soil as source of elements to grazing animals. In: Hockstra WG, Suttle JW, Gouther HE, Mertz W (eds) Trace element metabolism in animals-2. University Park Press, Baltimore, USA, pp 448–450

Hodgson J (1990) Grazing management: science into practice. Longman Scientific and Technical, Wiley, New York

Islam MM, Kabir SML, Sarker YA, Sikder MMH, Islam SKS, Akhter AHMT, Hossain MM (2016) Risk assessment of chromium levels in broiler feeds and meats from selected farms of Bangladesh. Bangladesh J Vet Med 14(2):131–134

Jothi JS, Yeasmin N, Anka IZ, Hashem S (2016) Chromium and lead contamination in commercial poultry feeds of Bangladesh. Int J Agril Res Innov Tech 6(2):57–60

Khan ZI, Ahmad K, Raza N, Al-Qurainy F, Ashraf M (2010a) Assessment of chromium concentrations in soil-plant-animal continuum: possible risk for grazing cattle. Pak J Bot 42:3409–3414

Khan ZI, Ashraf M, Mukhtar MK, Raza N, Ahmad K, Akram NA (2010b) A study on the transfer of iron in soil–plant–animal continuum under semi-arid environmental conditions in Sargodha, Pakistan. Biol Trace Element Res. https://doi.org/10.1007/s12011-010-8799-6

Khan ZI, Ahmad K, Raza N, Al-Qurainy F, Ashraf M, Hussain A (2010c) Assessment of chromium concentrations in soil-plant-animal continuum: possible risk for grazing cattle. Pak J Bot 42(5):3409–3414

Li Y, Mc Crory DF, Powell JM, Saam H, Jackson-Smith D (2005) A survey of selected heavy metal concentrations in dairy feeds. J Dairy Sci 88:2911–2922

McDowell LR (1985) Nutrition of grazing ruminants in warm climates. Academic Press, San Diego

McDowell LR (2003) Minerals in animal and human nutrition, 2nd edn. Amsterdam. Univ, Florida

McDowell LR, Arthington D (2005) Minerals for grazing ruminants in tropical regions. Extension bulletin. Animal Science Department, University of Florida, USA

Misra SG, Mani D (1978) Soil pollution. Ashish Publishing House, New Delhi

Nicholson FA, Chambers BJ, Williams JR, Unwin RJ (1999) Heavy metal contents of livestock feeds and animal manures in England and Wales. Biores Technol 70:23–31

NRC (2001) Nutrient requirements of dairy cattle. National Academy press, Washington, DC

NRC (2005) Mineral tolerance of domestic animals, National Research Council 2005. National Academy press, Washington, DC

Pechova A, Pavlata L (2007) Chromium as an essential nutrient: a review. Vet Med 52(1):1–18

Raj BG, Patnaik MC, Babu SP, Kalakumar B, Singh MV (2006) Heavy metal contaminants in water-soil-plant-animal continuum due to pollution of Musi river around Hyderabad in India. Indian J Anim Sci 76:131–133

Singh D, Chhonkar PK, Pandey RN (1999) Soil, plant, water analysis. A methods manual. ICAR, New Delhi

Spears JW, Lioyd KE, Krafka K (2017) Chromium concentrations in ruminant feed ingredients. J Dairy Sci 100(5):3584–3590

Tokalioglu S, Kart S, Gunes S (2000) Determinations of heavy metals in soil extract and plant tissues at around of a zinc smelter. Int J Environ Anal Chem 80:2010–2017

Tokaliglu S, Kartal S (2005) Determination of Cu, Pb, Cd, Ni, Cr, Co, Mn, Fe, and Zn in algae and vegetable samples using wet and dry ashing procedures. Trace Elem Electr 22:169–173

Underwood EJ, Suttle NF (1999) The mineral nutrition of livestock, 3rd edn. Midlothian, UK, pp 283–392

USEPA (2011) Basic information about chromium in drinking water. http://water.epa.gov/drink/contaminants/basicinformation/chromium.cfm

WHO (2022) Guidelines for drinking-water quality. World Health Organization, Geneva

Xu T, Nan F, Jiang X, Tang Y, Zeng Y, Zhang W, Shi B (2020) Effect of soil pH on the transport, fractionation, and oxidation of chromium (III). Ecotoxicol Environ Saf 195:110459

Chapter 16
Molecular Mechanisms of Chromium Tolerance in Plants: A Key Role of Antioxidant Defense

Viktor Husak and Maria Bayliak

Abstract Accumulative contamination of soil with chromium-containing compounds generates many adverse effects in plants. The toxic effect is manifested in the retardation of plant development, lower green biomass, morphological defects, and poorer flower quality and crop yield. Stimulation of reactive oxygen species (ROS) synthesis followed by oxidative stress development is considered to underlie the harmful effects of both Cr^{6+} and Cr^{3+}. Since a high intensity oxidative stress may cause the inhibition of photosynthetic mechanisms and oxidative modifications of cellular molecules, plants up-regulate various antioxidant and non-antioxidant defense mechanisms to avoid oxidation, maintain low stable-state ROS levels and repair damages that have already occurred in response to Cr(IV) exposure. These mechanisms include: (i) the antioxidant enzymes that directly scavenge ROS; (ii) the non-enzymatic antioxidant low molecular mass molecules (vitamins, terpenoids, phenolics, etc.) which are able to scavenge ROS and repair oxidized molecules, and some of them (ascorbic acid, glutathione) are used as substrates by antioxidant enzymes, in particular in so-called ascorbate-glutathione pathway; (iii) thioredoxins and glutaredoxins to restore oxidized proteins; (iv) metal-binding proteins such as phytochelatins and metallothioneins which sequestrate toxic metals in the specific plant compartments (vacuoles). In this review, we analyze current data on the involvement of redox regulators, particularly Rap2.4a and NPR1, in the regulation of adaptive antioxidant response of plants under Cr exposure.

Keywords Chromium · ROS · Oxidative stress · Antioxidant defense · Plant

V. Husak (✉) · M. Bayliak (✉)
Department of Biochemistry and Biotechnology, Vasyl Stefanyk Precarpathian National University, Ivano-Frankivsk, Ukraine
e-mail: viktor.husak@pnu.edu.ua

M. Bayliak
e-mail: maria.bayliak@pnu.edu.ua

© The Author(s), under exclusive license to Springer Nature Switzerland AG 2023
N. Kumar et al. (eds.), *Chromium in Plants and Environment*, Environmental Science and Engineering, https://doi.org/10.1007/978-3-031-44029-8_16

Abbreviations

APX	Ascorbate peroxidase
AsA	Ascorbic acid
CAT	Catalase
Grx	Glutaredoxin
HM	Heavy metal
HSP	Heat shock protein
MT	Metallothionein
NPR1	Non-expressor of the pathogenesis-related gene
PC	Phytochelatin
RNS	Reactive nitrogen species
ROS	Reactive oxygen species
SA	Salicylic acid
TF	Transcription factor
TGA	TGACG sequence-specific binding protein
Trx	Thioredoxin

16.1 Introduction

Metals with variable valence, such as chromium, iron, copper, cobalt, nickel, etc., have a wide practical application in various branches of industry. At the same time, they are one of the most widespread pollutants of the environment, because they are intensively released with industrial emissions and domestic waste (Uddin 2017). Metal retention in soil poses a serious danger to soil organisms, including agricultural plants, reducing their quality and crop yield (Alengebawy et al. 2021; Saud et al. 2022; Wyszkowska et al. 2023). Chromium(Cr) is a particularly dangerous metal pollutant for all organisms, from microorganisms to humans (Kubrak et al. 2010; Vasylkiv et al. 2010; Stambulska et al. 2018; Saud et al. 2022). Chromium can exist in eleven valence states, but the main valence states of Cr in the soil are Cr^{3+} and Cr^{6+}. Hexavalent Cr compounds are widely used in industry. In particular, Cr^{6+} compounds are used in the metallurgy (e.g., for the production of different steels, metal materials with high resistance to physical and chemical factors) as well as in the pigment production, leather tanning, etc. (Pellerin and Booker 2000; Shahid et al. 2017). Hexavalent chromium emitted into the atmosphere is airborne for some time in the form of particles or dust. As sediments, it contaminates soil and water. Rain precipitates Cr^{6+} particles from the atmosphere, depositing them in the soil, or they can be spread by wind for long distances (Yu et al. 2014; Tumolo et al. 2020). Emissions from diesel vehicles are the main source of Cr air pollution, as Cr is a component of diesel fuel (Wang et al. 2003; Yu et al. 2014). It is generally accepted that Cr^{6+} is more toxic than Cr^{3+}. However, many studies showed that both Cr^{6+} and Cr^{3+} can increase dramatically the amount of so-called reactive oxygen species

(ROS), triggering oxidative stress in different organisms with various harmful effects (Lushchak et al. 2009; Vasylkiv et al. 2010; Sawicka et al. 2021). Mutagenic and genotoxic effects of Cr^{6+} and Cr^{3+} ions are largely a result of ROS-induced damage to DNA (Figgitt et al. 2010; Novotnik et al. 2016; Sawicka et al. 2021).

Plants are able to absorb chromium from the soil and, as a result, excessive accumulation of chromium is detrimental for plants, causing defects in growth and development up to plant death. Chromium toxicity in plants is associated with its inhibitory effect on a number of metabolic processes, in particular photosynthesis, chlorophyll, and protein biosynthesis, as well as with DNA damage (López-Bucio et al. 2015; Stambulska and Bayliak 2020; Wakeel et al. 2020; Kumar et al. 2022; Saud et al. 2022; Wyszkowska et al. 2023).

Unlike animals, which can leave the contaminated area, plants are immobile, so they have evolved powerful defense systems to withstand pollution. In response to chromium exposure, many plants up-regulate a battery of enzymatic and non-enzymatic defenses to sequester chromium, keeping redox balance, preventing oxidative damage and repair damaged biomolecules (Anjum et al. 2017; Stambulska et al. 2018; Stambulska and Bayliak 2020; Kumar et al. 2022). If the capacity of defense systems is not enough to overcome the increase in ROS levels and treat the damage, chromium-induced damage to biomolecules can be significantly enhanced, which is manifested in severe oxidative stress with the corresponding detrimental consequences for plants. It should be noted that chromium toxicity is reduced in the presence of associative or symbiotic microorganisms, e.g. nitrogen-fixing rhizobia, which are able to accumulate heavy and transition metals and thus reduce metal uptake by plants (Stambulska et al. 2018; Sharma et al. 2022).

This chapter summarizes the available information on the mechanisms of toxicity of Cr^{6+} ions in plants, focusing on their involvement in the generation of ROS and cellular and tissue oxidative injuries. Studies point out that plants with more powerful antioxidant systems are more resistant to soil chromium contamination. Therefore, we have also analyzed in detail the plant antioxidant compounds contributing to prevention and neutralization of Cr toxic effects. The main molecular players of plant responses to redox perturbations are also discussed.

16.2 Reactive Oxygen Species and Reactive Nitrogen Species as Components of Plant Immunity and Energy Metabolism

In living organisms, reactive oxygen (ROS) and nitrogen (RNS) species are products of incomplete reduction of oxygen and nitrogen and are generated as primary products or either byproducts of basic metabolism. In plants, ROS and RNS as side products of metabolism are generated in plastids, mitochondria, peroxisomes, apoplast, and cytosol (Gupta and Kaiser 2010; Luis and Río 2013; Del Rio and Lopez-Huertas 2016; Mittler 2017; Qi et al. 2017). Intracellular ROS/RNS homeostasis is rigorously managed to sustain proper cellular functions by a wide range of non-enzymatic (vitamins (A, E, K) organic acids, glutathione, ionic metals (Fe^{2+}, Cu^{2+}, Zn^{2+}, and Mn^{2+}), phenolic and terpenoid compounds), and enzymatic scavengers (superoxide dismutase (SOD), catalase (CAT), enzymes engaged in the reduction-oxidation of non-enzymatic antioxidants (glutathione and ascorbic acid), amine oxidase, etc.) (Chaudiere and Ferrari-Iliou 1999, Apel and Hirt 2004; Choudhury et al. 2017; Mur et al. 2013; Cadet and Davies 2017; Samardzic and Rodgers 2017).

The main ROS formed by the plant cells are the typical for aerobic organisms and include hydrogen peroxide (H_2O_2), hydroxyl radicals ($OH^{\cdot}$), superoxide anions ($O_2^{\cdot-}$), and singlet oxygen ($^1O_2$). They can be signaling molecules in plants, controlling various physiological processes—differentiation, senescence and death of the cells, pathogen defense, acclimation, and abiotic stress survival (Foyer and Noctor 2005; Gechev et al. 2006; Di Meo et al. 2016; Dunnill et al. 2017).

In photosynthesizing chloroplasts of green organs, three main ROS-generated processes can be distinguished (Fig. 16.1). First one is ROS generation in photosystem II via overexcitation of chlorophyll (Fryer et al. 2003). The Mehler reaction in photosystem I, in terms of which O_2 accepts one electron converting to $O_2^{\cdot-}$, is a second process leading to ROS generation (Badger et al. 2000; Ort and Baker 2002). The third process is related to the Rubisco by the substantial production of H_2O_2 by glycolate oxidase (Douce and Neuburger 1999). Glycolytic acid also undergoes oxidation in peroxisomes, and this is accompanied by H_2O_2 release. Even under favorable conditions, a certain amount of oxygen formed in chloroplasts is metabolized into ROS (Asada 1999). One of the major producers for H_2O_2 are peroxisomes (Noctor et al. 2002). Peroxisomal membrane NADPH oxidase and peroxisomal matrix xanthine oxidase produce $O_2^{\cdot-}$ radicals that are quickly reduced to H_2O_2 (Del Rio and Lopez-Huertas 2016). Enzymes, such as superoxide dismutase and different oxidases (sarcosine, flavin, sulfite copper amine ones) can produce H_2O_2 in peroxisomes (Gilroy et al. 2016; Hasanuzzaman et al. 2020a, b). While in animals, mitochondria are the main producer of ROS, in plants chloroplasts do this function. The rate of chloroplast ROS formation is 20 times higher than mitochondrial one (Foyer and Noctor 2003). At low levels, ROS are responsible for the coordinated functioning of cellular organelles, cellular homeostasis, and adaptability to rapidly changing external conditions (Gechev et al. 2006). Mitochondrial respiratory complexes I and III are the major sites for the emergence of ROS, superoxide, and H_2O_2, but the factors that lead to the intensification of ROS production in mitochondria are mostly unknown (Rhoads et al. 2006). In apoplast, oxalate oxidase, amine oxidase, and peroxidases can produce ROS (You and Chan 2015; Mittler 2017; Qi et al. 2017).

In addition, the excessive production of ROS is an effective component of the plant immunity under pathogenic infections. In plant–pathogen interactions, the major site of elevated ROS levels is the apoplast (Qi et al. 2017). NADPH oxidase produces of ROS ($O_2^{\cdot-}$ and H_2O_2) and thus provides innate plant immunity. Also, the redox networks of cells (peroxiredoxins, thioredoxins, glutaredoxins) affect very strongly ROS levels. The enhanced ROS production reduces bacterial and fungal growth (Suzuki et al. 2011; Mittler et al. 2011). The increasing concentration of ROS can

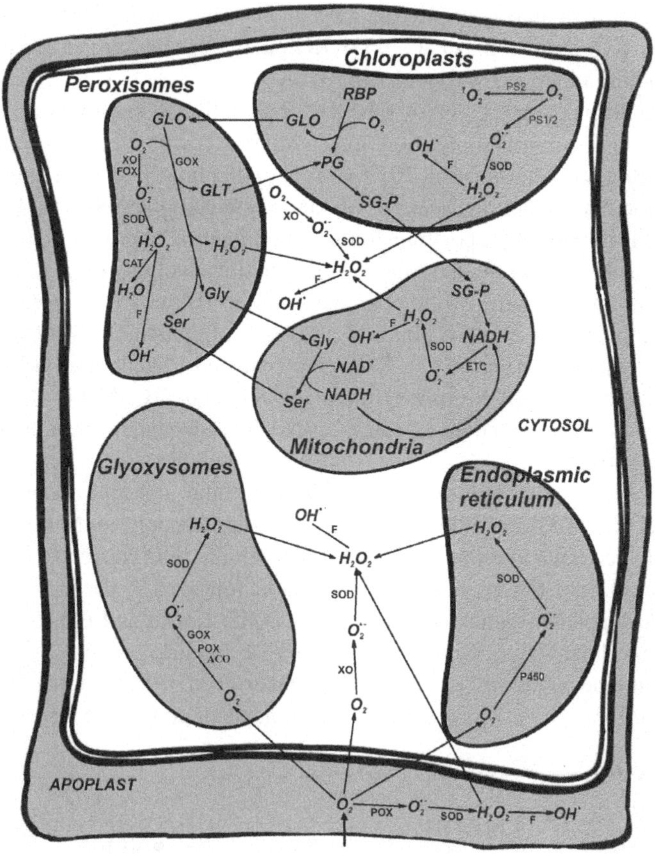

Fig. 16.1 Schematic overview of the ROS generating sites in the plant cell and their activity in various cellular components. The model is based on results described by various authors (Karuppanapandian et al. 2011; Hasanuzzaman et al. 2020a, b; Mandal et al. 2022). In chloroplasts, during photosynthesis, sunlight energy is captured and transferred to photosystem I (PS1) and photosystem II (PS2), where superoxide radical ($O_2^{\cdot-}$) can be formed. Then $O_2^{\cdot-}$ is converted to hydrogen peroxide (H_2O_2) by superoxide dismutase (SOD). Under excess light conditions, PS2 is able to generate singlet oxygen ($^1O_2$). The glycolate (CLO), produced in chloroplasts, moves to peroxisomes, where it is oxidized by glycolate oxidase (GOX) with formation of H_2O_2. In peroxisomes, H_2O_2 can also be formed from O_2 by xanthine and flavin oxidases (XO and FOX, respectively) coupled to SOD. In the mitochondrial electron transport chain, $O_2^{\cdot-}$ production is likely to occur in complex I—the flavoprotein region of NADH dehydrogenase segment reduces O_2 into superoxide $O_2^{\cdot-}$. Acyl-CoA oxidase (ACO), GOX, peroxidases (POX), and SOD are the primary enzymes responsible for $O_2^{\cdot-}$ and H_2O_2 generation in glyoxysomes. The endoplasmic reticulum-mediated ROS generation takes place by means of cytochrome P_{450}. Cell-wall-associated peroxidase and SOD are the main sources of $O_2^{\cdot-}$ and H_2O_2 in apoplast of plant cells. In the presence of redox-active metals, the extremely reactive hydroxyl radical ($^{\cdot}OH$) can be formed from H_2O_2 through the Fenton reaction (F). Other abbreviations used: CAT, catalase; GLO, glycolate; GLT, glycerate; Gly, glycine; PG, phosphoglycolate; RBP, ribulose 1,5-bisphosphate; Ser, serine; SG-P, sugar-phosphates

cause increase in synthesis of phytoalexins and other secondary metabolites which have defensive roles (Thoma et al. 2003). ROS may limit the spread of pathogens or play a signaling function (Mur et al. 2008).

Reactive nitrogen species (˙NO, $ONOO^-$, $˙NO_2$), like ROS, are signal molecules in plants. S-nitrosation and nitration are the main RNS-modifying mechanisms of plant cell behavior (Kharma et al. 2019; Corpas and Palma 2020; Corpas et al. 2021). S-Nitrosation is the reaction between ˙NO and cysteine, giving S–N=O bound. This mechanism is important for posttranslational modification of proteins and ˙NO signal transmission. As a result of alterations in the functional activity of proteins, changes also occur in the proteome of the cells (Corpas and Palma 2020). For example, S-nitrosoglutathione GS–N=O) is obtained as a result of glutathione S-nitrosation. The irreversible transformation of GS–N=O into oxidized glutathione (GSSG) occurs with the participation of S-nitrosoglutathione reductase (GSNOR) (Corpas et al. 2021). Then glutathione reductase catalyzes NADPH-dependent reduction of GSSG to GSH (Fig. 16.2).

S-nitrosoglutathione reductase determines cellular concentrations of GSNO (Feechan et al. 2005), since this metabolite is a stable supply of ˙NO in plant cells. Nitric oxide performs immune functions, since cellular ˙NO concentrations enhance salicylic acid levels (SA, plant immune activator) followed by the increased expression of SA-dependent genes (Feechan et al. 2005; Loake and Grant 2007). Salicylic acid modulates the activity of proteins (SA binding protein 3, SABP3) that perform immune functions in plants (Slaymaker et al. 2002; Kumar and Klessig 2003). Therefore, ˙NO is a key regulator of SA-dependent immune responses against pathogens.

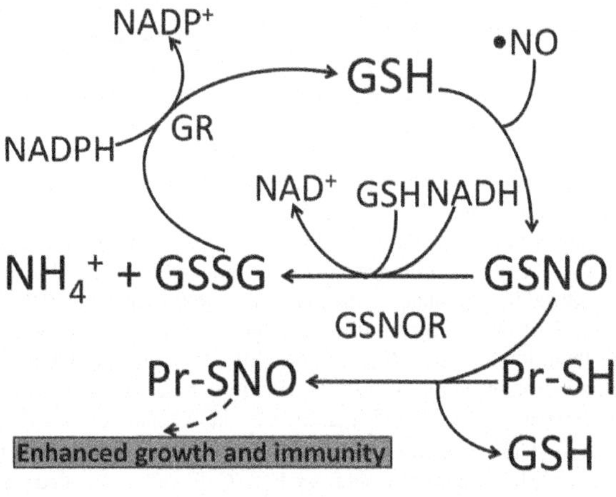

Fig. 16.2 S-nitrosylation of reduced form of glutathione (GSH) in plant cells. More details in the text. GSNO, S-nitrosoglutathione; GSNOR, S-nitrosoglutathione reductase; GSSG, oxidized form of glutathione; GR, glutathione reductase; Pr-SH, thiol-containing proteins; Pr-SNO, S-nitroso-containing proteins; NADH, nicotinamide adenine dinucleotide (reduced form); NAD^+, nicotinamide adenine dinucleotide (oxidized form); NADPH, nicotinamide adenine dinucleotide phosphate (reduced form); $NADP^+$, nicotinamide adenine dinucleotide phosphate (oxidized form)

Nitration is an addition of a nitro group ($^{-}NO_2$) to different biomolecules, especially proteins. The result of this type of protein nitration is a formation of nitrated amino acid residues (e.g., tyrosine, cysteine, tryptophan, methionine). These chemical modifications entail a loss of functional activity of the entire protein molecule (Corpas and Palma 2020). In addition, the interaction of metal ions with $^{\cdot}NO$ (metal nitrosylation) can occur in metalloproteins. Such an interaction can also lead to changes in protein metabolic activity (Corpas and Palma 2020). Reactive nitrogen species react also with fatty acids. Nitro-fatty acids act as messenger molecules, that modulate gene expression during developmental processes and stress events (Di Fino et al. 2021; Kolbert et al. 2019; Medrano-Macías et al. 2022).

16.3 Antioxidant System of Plants: Overview

As described above, ROS/RNS and other reactive species are continually being formed in plants as by-products of photophosphorylation and respiration. Reactive species are highly damaging molecules that rapidly oxidize protein and lipid molecules, nitrogen bases in DNA, exerting a strong damaging effect on individual organelles, cells, organs and, the plant organism as a whole. Therefore, cells have evolved sophisticated defense mechanisms to neutralize ROS/RNS. At the same time, complete neutralization of ROS/RNS does not occur, as they perform signaling functions within a single cell and at a distance. ROS/RNS regulate some cell functions and processes such as cell division, regeneration, organogenesis, cell death, immune responses, resistance to UV radiation, etc. Temporary increase of ROS/RNS levels is a component of plant immune defense against disease agents (viruses, microbes, fungi, parasites), mechanical injuries, and physical stressors (water deficit, temperature changes, etc.) (Hasanuzzaman et al. 2020a, b; Mandal et al. 2022). Thus, the cell strictly regulates the synthesis and degradation of ROS in accordance with its needs, so that the level of reactive forms allows for the effective performance of their regulatory functions and the amount of oxidative damage caused by them is minimized. Maintenance of low stationary ROS/RNS concentrations is achieved through the regulation of systems that produce reactive species, as well as systems that scavenge reactive species and restore /degrade oxidatively damaged biomolecules.

In plant cells, the system that protects against damaging effects of ROS/RNS is multilevel and includes: (1) antioxidant enzymes that are able directly to neutralize ROS/RNS—superoxide dismutases and various peroxidases; (2) ascorbate–glutathione cycle enzymes, destroying H_2O_2 and utilizing NADPH as a donor to reduce oxidized glutathione; (3) enzymes involved in NADPH generation (e.g., ferredoxin-NADP-reductase); (4) non-enzymatic low molecular mass antioxidants, such as redox-active vitamins, glutathione, phenolic compounds, terpenoids etc.; (5) enzymes that reduce peroxides and thiol-containing compounds (glutathione reductase, thioredoxin reductase, etc.); (6) enzymes involved in detoxification of xenobiotics (e.g., glutathione-S-transferase); (7) peptides/proteins capable of binding metals via SH-groups (phytohelatins and metallothioneins); (8) heat shock proteins

(HSPs) involved in formation of the native protein structure (folding) and restoring native structure of denatured proteins (Cobbett and Goldsbrough 2002; Michalak 2006; Anjum et al. 2017; Kapoor et al. 2019; Jedelská et al. 2020; Stambulska and Bayliak 2020; Fedenko et al. 2022). Since ROS and RNS are synthesized in different organelles and cytosol of the plant cell, a sophisticated network of defense mechanisms is present not only in the cytosol but also in the organelles of the plant cell, especially in mitochondria, chloroplasts, and peroxisomes (Geigenberger et al. 2017; Noctor et al. 2018; Martí et al. 2020). Also, the composition of the antioxidant system is slightly distinct in different parts of the plant (roots, flowers, fruits, leaves).

16.4 Low-Molecular Mass Antioxidants

16.4.1 Tocopherols

Tocopherols (collectively called vitamin E) are a lipophilic compounds that are synthesized in thylakoid membranes of plastids; therefore, they are found mostly in the green plant organs. There are four isomers of tocopherol (α, β, γ and δ) (Szewczyk et al. 2021). According to their antioxidant activity, the α isomer is the most active and δ one has the lowest antioxidant activity. The antioxidant activity is related to degree of methylation (Dumanović et al. 2021). Chromanol ring of tocopherols possesses the ability easily to donate elections to reduce oxidized molecules that determines the antioxidative effects of this vitamin. Due to their structure, tocopherols and tocotrienols are incorporated into the phospholipid bilayer of cell membranes, where they perform both antioxidant and structural functions (Szewczyk et al. 2021). The membrane of chloroplasts contains predominantly α-tocopherol; therefore, green parts of the plant are rich in α-tocopherol. At the same time, seeds contain mainly γ-tocopherol (Dumanović et al. 2021). In chloroplasts, tocopherols protect membrane lipids against oxidation by trapping singlet oxygen $^1O_2$, thus ensuring the normal functioning of photosystem II. The $^1O_2$ scavenging activity of tocopherols is extremely effective: in vitro it was calculated that more than 100 molecules of $^1O_2$ can be neutralized by one molecule of α-tocopherol until the latter degrades (Gill and Tuteja 2010). Alpha-tocopherol also interrupts lipid peroxidation in plasma and intracellular membranes, reacting with the lipid radicals (LO·, L· and LOO·), which arise from the oxidation of polyunsaturated fatty acids at the membrane-water surface (Dumanović et al. 2021). By giving protons to a radical molecule, tocopherol is converting to a tocopheroxyl radical, which is not reactive enough to initiate membrane peroxidation by itself (Das and Roychoudhury 2014). In model plant *Arabidopsis thaliana,* inactivation of genes, encoding enzymes of tocopherol biosynthesis, led to a strong oxidative state and lowered content of photosynthetic pigments in green plant parts (Semchuk et al. 2009, 2011). In plants, alpha-tocopherol was also found to perform protective functions contributing to the tolerance to salt stress (Semchuk et al. 2012), high/low temperature (Bergmüller et al. 2003), and xenobiotics (Semchuk et al. 2011).

16.4.2 Ascorbic Acid

Ascorbic acid, or ascorbate, is a common compound found in all cellular compartments. The concentration of ascorbate (AsA) is the highest in cytosol but is also substantial in chloroplasts (Smirnoff and Wheeler 2000). Under physiological conditions, AsA is implicated in photosynthesis serving a cofactor of enzymes (including those involved in the synthesis of phytotohormones and phenolic compounds) and in the control of the cell growth. Some fruits (e.g. black currant, pepper) show a high concentration of ascorbic acid, which may be the result of its compartmentalization and stabilization in acidic vacuoles (Smirnoff and Wheeler 2000).

Glucose, synthetized in chloroplasts during photosynthesis, is a precursor of ascorbic acid. Metabolic conversion of glucose occurs mostly via formation of GDP-mannose first and then L-galactose, which is oxidized to L-galactono-1.4-lactone with conversion of the latter to AsA by specific dehydrogenase located in mitochondrial membrane (Siendones et al. 1999). Plant mitochondria are involved in both de novo synthesis of AsA and the regeneration of oxidized forms of AsA such as dehydroascorbic acid (Gill and Tuteja 2010). Ascorbate is considered to be the most powerful ROS scavenger due to its ability to easily give electrons and protons. It is able to directly react with $O_2^{\cdot -}$, H_2O_2 and $^{\cdot}OH$, converting them to non-radical molecules, thus contributing to maintenance of membrane integrity. In addition, ascorbate regenerates α-tocopherol by reducing its oxidized form—tocopheroxyl radical. Moreover, ascorbate is a component of the ascorbate-glutathione cycle, described in more detail below. In reduced form, AsA maintains the activity of enzymes containing transition metal ions as prosthetic groups, preventing the oxidation of these metal ions (Gill and Tuteja 2010). Important function of ascorbate is serving as cofactor for several peroxidases and other enzymes like methionine sulfoxide reductases (Noctor et al. 2018). AsA-dependent enzymes are called ascorbate peroxidases (APXs). They are polygenic group of proteins. In particular, there was identified eight genes encoding APXs in *Arabidopsis* (Narendra et al. 2006) and 11 APX genes in poplar tree (Leng et al. 2021).

Plants with higher ascorbate content demonstrated the enhanced tolerance to stressors whose action leads to oxidative stress (Ding et al. 2009; Akram et al. 2017). Exposure to heavy metals was found to decrease ascorbate levels with increasing content of its oxidized forms (dehydroascorbic acid) (Gill and Tuteja 2010) In particular, Cd exposure decreased ascorbate levels in *Cucumis sativus* chloroplasts (Zhang et al. 2002), leaves of *A. thaliana* (Skórzyńska-Polit et al. 2003) and pea (Rodríguez-Serrano et al. 2006), roots and nodules of *Glycine max* (Balestrasse et al. 2008). Exposure to hexavalent chromium decreased ascorbate levels in tomato roots, while the addition of exogenous ascorbate improved tolerance of tomatoes to Gr (Al-Huqail et al. 2020). Decrease in ascorbate levels under Cr and Cd exposure was shown to be connected with up-regulation of APXs which use AsA a substrate to counteract a toxicant-stimulated ROS production (Sinha et al. 2018).

16.4.3 Glutathione

Glutathione (GSH) is a tripeptide (L-γ-glutamyl-L-cysteinylglycine) found in all living organisms, including plants. It is synthesized by γ-glutamylcysteine synthetase and glutathione synthetase (Lushchak 2012). In plants, glutathione synthesis occurs in the chloroplasts and cytosol and then GSH can be recruited also in other organelles, including nucleus (Dumanović et al. 2021). In plant cells, 85–90% of GSH is cytosolic (Kosakivska et al. 2021). Glutathione has two forms in the cell, an reduced one (GSH) and an oxidized one formed by disulfide binding of two molecules of glutathione (GSSG) (Noctor et al. 2018). In the cell, glutathione is largely present as GSH, when GSSG consists of about 1% of the total content (Lushchak 2012). For the cell, it is important the GSH/GSSG ratio which determines whole redox status of the cell. In plants, excessive ROS production may cause oxidation of GSH and decease in GSH/GSSG levels indicating more oxidative state of the cell (Queval et al. 2011).

Glutathione has a variety of functions in the cell; in particular, as we mentioned above, it is a key maintainer of redox cellular state in the range of values essential for adequate functioning of redox-sensitive signaling pathways controlling processes connected with growth and aging (seed development, formation of organs and fruits, senescence) (Noctor et al. 2018). This is evidenced by growth retardation and higher vulnerability to oxidant compounds in *A. thaliana* mutants, defective in glutathione synthesis (Cobbett and Goldsbrough 2002). As an antioxidant, GSH can directly react with ROS such as hydroxyl radical ($HO^{\cdot}$) or peroxynitrite ($ONOO^{-}$) (Lushchak 2012) Also, GSH is involved in eliminating peroxides (H_2O_2, organic peroxides) with the help of appropriate enzymes and is involved in the regeneration of the oxidized form of AsA, which is formed during H_2O_2 reduction (Foyer and Halliwell 1976). In detail, ascorbate-glutathione cycle is described in the section below. Glutathione is also able to conjugate with molecules having free SH-groups with forming respective disulfides in a reaction called S-glutathionylation (Lushchak 2012). So, GSH reacts with cysteine-rich proteins. This process is reversible and affects the protein function. In particular, S-glutathionylation inhibited the activity of the chloroplastic glyceraldehyde-3-phosphate dehydrogenase, an enzyme of the Calvin-Benson cycle in plants (Zaffagnini et al. 2007). S-glutathionylation is accepted to be a cellular protective strategy to protect cysteine residues in proteins against irreversible oxidation (Lushchak 2012). GSH is a substrate of glutathione-S-transferases to conjugate and eliminate different xenobiotics (Lushchak 2012). Also, GSH can interact with certain metal ions, especially cadmium, copper, zinc, mercury, arsenic, and lead (Lushchak 2012; Kosakivska et al. 2021). Chelating Cu^{2+} ions by GSH prevented their participating in the Fenton and Haber-Weiss reactions. In addition, GSH seems to be involved in Cr^{6+} reduction to Cr^{3+} that is considered as a way to decrease chromium toxicity (Lushchak 2012). However, some studies proposed that CSH participates in detoxification of heavy metals not directly but as a precursor molecule for phytochelatins (Cobbett and Goldsbrough 2002; Kosakivska et al. 2021).

Short term treatment of certain plants (maize, tomato and cauliflower) with Cr^{6+} increased GSH levels in the leaves and roots (di Toppi et al. 2002). Enhanced GSH levels were also reported in sorghum (*Sorghum bicolor* L.) treated with Cr^{6+} (Kumar et al. 2019). Some studies report that supplementation with exogenous glutathione improved plant growth and tolerance to Cr toxicity (Zeng et al. 2012; Sharma et al. 2020; Wen et al. 2022). The addition of exogenous GSH improved the growth of soybean and rice exposed to chromium and increased antioxidant capacity in the treated plants (Zeng et al. 2012; Wen et al. 2022). Treatment with exogenous glutathione resulted in chromium accumulation in the roots of plants rather than in the green part. This indicates that glutathione compartmentalizes chromium ions in plants, thus reducing chromium toxicity (Zeng et al. 2012; Wen et al. 2022).

16.4.4 Carotenoids

Carotenoids are isoprenoid polymeric compounds with diverse roles in plant metabolism, including those related to confer resistance to oxidative stress. The known carotenoids include lutein, β-carotene, zeaxanthin, lycopene, etc. (Pérez-Gálvez et al. 2020). As photosynthetic pigments, carotenoids absorb light at 400–50 nm and transfer it to chlorophyll (Havaux et al. 1998). Secondly, they protect the photosynthetic apparatus by quenching singlet oxygen $^1O_2$ and $^{\cdot}OH$ formed during light stage of photosynthesis (antioxidant function) (Stahl and Sies 2003). Antioxidant function of carotenoids also include prevention of the formation of $^1O_2$ by reacting with excited chlorophyll (Chl*) and breakdown of the chain reactions of lipid peroxidation (Das and Roychoudhury 2014). Thirdly, they as lipid-soluble molecules are components of the photosystem I and provide stability of the proteins of the photosynthetic complex and the thylakoid membrane as a whole (structural role) (Gill and Tuteja 2010). In *Capsicum annuum*, exposure to Cr-containing compounds was reported to result in increased carotenoid levels (Sharma et al. 2020). This suggests that carotenoids can contribute to Cr tolerance in plants.

16.4.5 Phenolic Compounds

Phenol compounds are ubiquitous in plant kingdom. Up to now, there are identified over eight thousand phenolic compounds in plants. Depending on the structure, phenolic compounds are divided into several classes: simple phenols, phenolic acids (gallic acid, caffeic acid, ferulic acid. etc.) stilbenes (e.g., resveratrol), lignans, coumarins, and flavonoids (including simple flavonoids (e.g., quercetin, luteolin, apigenin, catechins) and condensed tannins (Dai and Mumper 2010). Phenols are secondary metabolites in plants and are synthetized from intermediates of carbohydrate metabolism. In the cell, phenols are mainly accumulated in vacuoles, as well as in apoplast. Phenolic compounds confer pigmentation of flowers and fruits, germination of pollen, defense against UV radiation and different pathogens (Das and Roychoudhury 2014).

Plant phenols are well-known as powerful antioxidants. Antioxidant properties of plant phenols include:

(i) direct scavenging of free radicals (superoxide radical anion $O_2^{\cdot-}$, hydroxyl radical $^{\cdot}OH$, peroxynitrite $ONOO^-$);
(ii) inhibition of lipid peroxidation through interaction with peroxide radicals ($RO^{\cdot}$, $ROO^{\cdot}$) and reduction of hydroperoxides (ROOH);
(iii) quenching of excited molecules (e.g., singlet oxygen $^1O_2$);
(iv) chelation of transition metal ions through aromatic hydroxyl groups (prevention of the Fenton reaction);
(v) reduction of oxidized low molecular weight antioxidants (e.g., tocopheroxyl radical);
(vi) direst binding followed modulation of pro-/antioxidant enzymes, as well as the impact on cellular signaling pathways at biochemical and molecular level (Rice-Evans et al. 1997; Michalak 2006; Dai and Mumper 2010; Bayliak et al. 2016b; Fedenko et al. 2022).

Similarly, to other non-enzymatic antioxidants (e.g., ascorbic acid and carotenoids), plant phenols may also demonstrate pro-oxidant properties, which depend on many external factors—pH, oxygen availability and presence of transition metals (Dai and Mumper 2010; Bayliak et al. 2016a, b). Exposure to heavy metals, including Cr^{6+}, leads to increased levels of phenolic compounds in the treated plants, suggesting a defensive role of phenolics against Cr toxicity (Michalak 2006; Stambulska and Bayliak 2020; Alsherif et al. 2022).

16.5 Protein Non-enzymatic Antioxidants

16.5.1 Metal-Binding Proteins

In animals and fungi, heavy metals (HM) induce the increase in content of metal-binding proteins so-called metallothioneins (Zhou and Goldsbrough 1994). Metallothioneins are small Cys-rich proteins that binding HM make them metabolically inactive, conferring heavy metal tolerance. Like animals and fungi, plants possess several metal-binding proteins, which function to sequestrate metals with variable valence (Cobbett and Goldsbrough 2002; Pal and Rai 2010). Among the plant metal-binding proteins, the phytochelatins (PCs) and metallothioneins (MTs) are the best described (Cobbett and Goldsbrough 2002). PCs are enzymatically synthesized peptides, whereas MTs are proteins. PCs are a polymer based on glutathione monomers. The main role of PCs is to detoxify xenobiotic metals (Ag, Cd, Hg) and maintain essential metal homeostasis (Cu, Zn, Se, Ni) (Cobbett and Goldsbrough 2002; Kosakivska et al. 2021). It should be noted that overexpression of phytochelatin synthase gene confers contrast tolerance to HM in transgenic *A. thaliana*, in particular, both enhanced cadmium (Cd) tolerance and Cd hypersensitivity was observed dependently on experimental conditions (Wang et al. 2012). Also, phytochelatin synthase had contrasting effects on cadmium and arsenic accumulation in rice grains dependently on experimental conditions (Uraguchi et al. 2017).

Plant MTs are low molecular mass proteins, rich in cysteine residues and are capable of binding mono-or divalent metal ions, in particular Cu ions (Domènech et al. 2006). Moreover, a correlation between MT gene expression and copper tolerance has been observed in *A. thaliana* (Murphy and Taiz 1995). Plants contain several types of MT which are distinguished by their structure and subcellular localization (MT1-MT4) (Kosakivska et al. 2021). In sorghum, Cr stress was shown to induce MT transcription but did not induce PC synthesis, suggesting a role of MT in Cr detoxification (Panda and Choudhury 2005). On contrary, another study showed that the PC content was significantly up-regulated in green parts of plants exposed to Cr^{6+}. Also, plants with higher PC content accumulated chromium predominantly in roots and cell walls, that was proposed to be a mechanism improving plant tolerance to Cr (Saud et al. 2022).

16.5.2 Heat Shock Proteins

Heat shock proteins (HSPs), termed also chaperons, are ubiquitous in plant and animal cells. They were originally identified as proteins whose levels increase in the response to heat stress (Vierling 1991; Park and Seo 2015). To date, it has been established that the functions of HSPs are much broader. The integrity of cells is maintained by heat shock proteins which participate in correct formatting of native structure of de novo synthesized proteins, restoration of structure of misfolded or denatured proteins, preventing aggregation of damaged proteins and directing to degradation of the proteins which cannot be repaired under different stresses (Vierling 1991; Park and Seo 2015; Waters and Vierling 2020). Increased HSP90 was found in rice seedlings which were exposed to Cr^{6+} (Zeng et al. 2014). It was shown that copper, cadmium and chromium induced the synthesis of HSPs in *A. thaliana* (Cui et al. 2019; Appiah et al. 2021).

16.6 Enzymatic Antioxidant Defense

16.6.1 Thioredoxins and Glutaredoxins

Thioredoxins (Trx) and glutaredoxins (Grx) belong to families of small oxidoreductases, containing SH groups in the active site, that enable the reduction of disulfide intramolecular bridges in a specific set of proteins (Meyer et al. 2012). Thus, like glutathione, these proteins help to maintaining the more reduced state of the cell. Proteomic approaches identified the putative Trx and Grx target proteins, which are components of many physiological processes and stress adaptive responses in plants (Meyer et al. 2012; Martí et al. 2020). Transgenic plants defective in Trx and Grx allow to identify that these redox proteins are components combating with pathogen and are induced by pathways mediated by phytohormones (Meyer et al. 2008, 2012).

In plants, the thioredoxin and glutaredoxin families include a large number of isoforms. Thus, at least 20 genes encoding thioredoxins and about 50 genes encoding glutaredoxins are identified in *A. thaliana* (Gelhaye et al. 2005; Meyer et al. 2008; Martí et al. 2020). Thioredoxins and glutaredoxins are classified depending on their primary structure and subcellular localization. By reducing intramolecular disulfide bonds in target proteins, Trx and Grx are oxidized themselves, forming intramolecular disulfides. Plant cells use different ways to reduce disulfide bounds to free thiol forms in the oxidized Trx/Grx. In chloroplasts, oxidized Trx can be reduced by a ferredoxin-thioredoxin reductase, whereas NADPH-dependent thioredoxin reductases catalyze reduction of Trx in mitochondria, nucleus, and cytoplasm (Smiri et al. 2010; Jedelská et al. 2020). Oxidized Grx can be reduced by glutathione (Meyer et al. 2008). Besides that, Grx is able to reduce oxidized Trx (Gelhaye et al. 2005). Diversity of metabolic pathways of regeneration of oxidized thioredoxins identified emphasizes the important role of these proteins in protection against stressors, especially oxidants (Gelhaye et al. 2005; Jedelská et al. 2020). Knock-out mutants in specific chloroplastic Trx showed defects in chloroplast structure and were more susceptible to stressors associated with increased ROS generation (Noctor et al. 2018). Heavy metals were found to affect thioredoxin reductase/thioreoxin system (Smiri et al. 2010, 2013). Moreover, it was shown that Trx possesses metal-chelating property, in particular regarding Cd^{2+} (Smiri and El Ghoul 2012).

16.6.2 Antioxidant Enzymes

Enzymatic antioxidant defense system includes of several antioxidant enzymes such as superoxide dismutases, catalases, and peroxidases (glutathione-ascorbate-dependent, GPx and APX, and non-specific for substrates ones, POX); enzymes of GSH and AsA redox metabolism—glutathione reductase (GR), glutathione S-transferase (GST), monodehydroascorbate reductase (MDHAR) and dehydroascorbate reductase (DHAR).

Superoxide dismutase (SOD) is one of the most effective enzymes for protecting plant cells from $O_2^{\cdot -}$ toxicity. Three main types of this metalloenzyme are described in plants—Cu, Zn-SOD, Mn-SOD, and Fe-SOD ones. The SOD isoforms differ in protein structure and subcellular distribution. Fe-SOD is located in plastids, Mn-SOD is in mitochondria, apoplast, and peroxisomes, and Cu, Zn-SOD occurs in peroxisomes, cytosol, and plastids (Pilon et al. 2011; Miller 2012). By catalyzing of $O_2^{\cdot -}$ dismutation to H_2O_2, SODs reduce the probability of oxidizing effects of the superoxide radical and the likelihood of its interaction with transition metal ions with the appearance of $^{\cdot}OH$ (Gill and Tuteja 2010).

Catalase is the heme-containing oligomeric enzyme, which catalyzes disproportionation of the product of the SOD reaction H_2O_2 to water and oxygen. In this reaction, one molecule of H_2O_2 is a donor of elections and the second is the electron acceptor. Catalase (CAT) activity has been detected in mitochondria, peroxisomes, and cytosol of different plants (Sofo et al. 2015). Catalase isozymes exhibit developmental stages and tissue specificity in plants. The one CAT molecule can convert 6 million H_2O_2 molecules per minute (Mehla et al. 2017). This is the maximum rate of substrate transformation into a product among antioxidant enzymes.

Hydrogen peroxide can be also scavenged by different peroxidases, which use H_2O_2 for reduction of different substrates. Phenol peroxidase (POX) can generate phenoxyl radical ($PhO^{\cdot}$) by oxidizing phenols with concomitant H_2O_2 reduction to water, and two molecules of H_2O are formed in this reaction. The polyphenol oxidase (PPO) found in the chloroplast thylakoid membrane can then interact with peroxidase, which helps to scavenge ROS. In PPO-catalyzed reaction, reduced phenols are oxidized to the quinone Q_A and H_2O (Boeckx et al. 2015).

Glutathione peroxidase (GPX) is the non-heme peroxidase family and protects cells from oxidative damage by reducing of the lipid hydroperoxides (LOOH) and H_2O_2 into non-toxic products or H_2O (Bela et al. 2015). GPX has been found in many cellular compartments (Koua et al. 2009). Currently, the functions of plant GPXs, which contain Cys residues in their active sites, are not completely understood. In stress conditions, GPXs can maintain H_2O_2 homeostasis, modify nuclear signaling proteins, and contribute to crosstalk between different signaling pathways (immune responses to pathogens, H_2O_2-mediated and the abscisic acid-mediated signaling). The plant GPXs also contribute to hormone-stimulated growth of main and lateral roots, plant regeneration, flower- and seed development, development of leaves (Bela et al. 2015).

Glutathione reductase (GR) is a protein that reduces of GSSG to the sulphydryl form GSH (Hasanuzzaman et al. 2017). In higher plants, three types of GR are predominantly localized in chloroplasts (about 80% of total CR activity), but its small amount is also found in peroxisomes, mitochondria, and cytosol (Edwards et al. 1990). In plants, maintaining a high ratio GSH/GSSG by chloroplastic GR is necessary important for normal oxygenic photosynthesis (Kornyeyev et al. 2003). Different GR types are activated by specific abiotic signals that ensure the specificity of the plant responses to stressors (Stevens et al. 1997).

Glutathione S-transferases (GST) are group of detoxification enzymes regulating different functions in plant cells such as the elimination of toxic substances, mitigation of oxidative stress, tyrosine degradation and hormone transport (Oakley 2011; Dong et al. 2016). Some GSTs can also scavenge toxic lipid hydroperoxides. GSTs catalyze conjugating of glutathione with metals/metalloids to reduce their toxicity (Dixon et al. 2011). In plant development, glutathione S-transferases also have functions, comprising formation of ion channels, modulation of regulatory kinases, and the glutathionylation of proteins (Dixon et al. 2010). GSTs are found in mitochondria, cytosol, and microsomes. In contrast to mitochondrial and cytosolic forms of GSTs, microsomal GSTs are identified as the integral membrane proteins that participate in eicosanoid and glutathione metabolism. Expression of GSTs was shown can be induced by fungal elicitors and pathogen attack, cold, drought, hormone treatments, phosphate starvation, heavy metals, salt, and wounding (Kim et al. 2011; Chen et al. 2013; Xu et al. 2015; Shukla et al. 2015; Jia et al. 2016).

Ascorbate peroxidase (APX) is a heme-containing peroxidase which scavenges H_2O_2 in the presence of ascorbate (Hasanuzzaman et al. 2019). APX has several isoforms with different cellular localization—cytosolic (cAPX), mitochondrial (mitAPX), chloroplastic (chAPX), and peroxisomal/glyoxisomal (mAPX) ones (Del Rio and Lopez-Huertas 2016). All APX isoforms participate in detoxifying H_2O_2 with oxidation of AsA to monodehydroascorbate (MDHA) and then to dehydroascorbate (DHA) (Fig. 16.3).

Monodehydroascorbate reductase (MDHAR) is a NADH/NADPH-dependent FAD-containing enzyme (Fig. 16.3). MDHAR is involved in AsA regeneration from

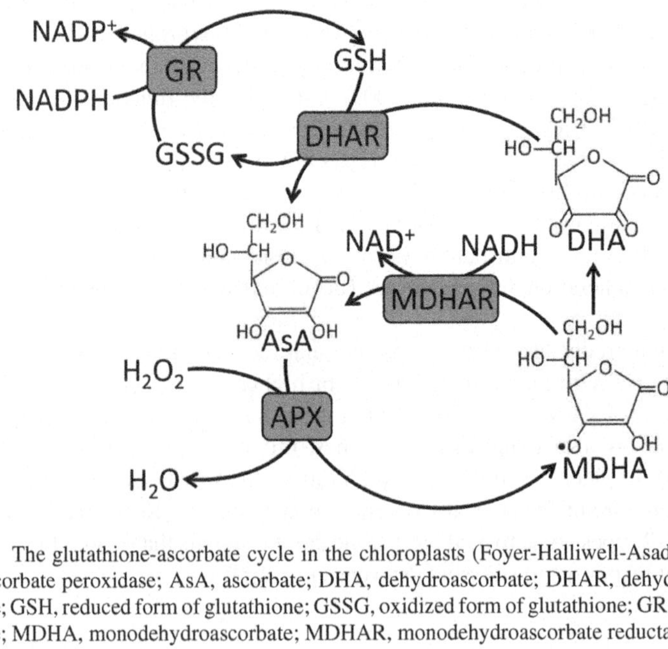

Fig. 16.3 The glutathione-ascorbate cycle in the chloroplasts (Foyer-Halliwell-Asada pathway). APX, ascorbate peroxidase; AsA, ascorbate; DHA, dehydroascorbate; DHAR, dehydroascorbate reductase; GSH, reduced form of glutathione; GSSG, oxidized form of glutathione; GR, glutathione reductase; MDHA, monodehydroascorbate; MDHAR, monodehydroascorbate reductase

MDHA and the phenoxyl radical reduction (Hasanuzzaman et al. 2019). Isoforms of this enzyme are in different plant cell components like chloroplasts, mitochondria, peroxisomes, glyoxysomes, and cytosol (Leterrier and Cagnac 2018).

Dehydroascorbate reductase (DHAR) can regenerate AsA by the oxidation of GSH to GSSG (Das et al. 2015). This reaction proceeds by a ping-pong mechanism, in which dehydroascorbate connects to the free reduced form of DHAR with the binding of GSH (Fig. 16.3).

DHAR can also be involved in the regulation of normal plant growth and development (Ding et al. 2020). Increased DHAR activity may be one of the mechanisms that regulate the regeneration of AsA from DHA under stress conditions accompanied by increased ROS formation (Locato et al. 2009). Therefore, increasing AsA content may restrict the detrimental effects of ROS accumulation (Wang et al. 2010).

16.7 Toxicity of Heavy Metals in Plants: Overview

Industrial waste, fertilizers, smelting, and wastewater discharges cause leaching of heavy metals (HMs) such as copper (Cu), manganese (Mn), lead (Pb), nickel (Ni), cadmium (Cd), cobalt (Co), iron (Fe), chromium (Cr), zinc (Zn), silver (Ag), arsenic (As), and platinum (Pt) into soil or groundwater (Gupta and Ali 2002; Dağhan and Ozturk 2015; Basheer 2018a). Plants need for their growth and development in small amounts some HMs (Co, Cu, Fe, Mn, Mo, Ni, V, and Zn). However, excessive amounts of these elements can become harmful to them. For example, Zn and Mn in excessive amounts impair the growth of plants and compete with Fe. The Fe^{2+} excess in cells participates in the Fenton reaction by forming $^{\cdot}OH$ (Shanmugam et al. 2011). HMs chelate by low molecular mass compounds (GSH, AsA, cysteine or others) and sequestrate in plant vacuoles. Other HMs (Pb, Cd, Hg, and Cr) are toxic for plants because of intensive protein denaturation and DNA damage. Most HMs are immobile, although some of them can be taken up by root system via endocytosis, diffusion, or through metal transporters (Krzesłowska 2011; Ali et al. 2015a, 2017; Basheer 2018b). The plant species accumulate HMs at different rates. Heavy metals impair virtually all molecular and biochemical processes in plants causing growth defects, root browning, chlorosis, and death (Gupta and Ali 2002; Dağhan and Ozturk 2015).

Heavy metals can bind to proteins, leading to inactivation or denaturation of the latter. The binding of HMs to membrane proteins disrupts the integrity of membranes, which is mainly reflected in the disruption of the transport of various substances. Ultimately, this leads to impaired photosynthesis and ion homeostasis (Hossain et al. 2012). The result of HMs stress is the overproduction of ROS including $^{\cdot}OH$, $O_2^{\cdot-}$, $OONO^-$, $^1O_2$, $^{\cdot}NO$, H_2O_2, hypochlorous acid (HOCl), and then resulting in oxidative stress (Rascio and Navari-Izzo 2011). This process is considered one of the primary mechanisms of HM toxicity in plants (Lushchak 2011; Sharma et al. 2019). Cu, Cd, Ni, Cr, and As ions exceeding permissible limits in plant cells lead to increased ROS amounts (Husak 2015; Shahzad et al. 2018). Depending on the bioactivity, HMs can be grouped into redox-active (Cr, Cu, Fe) and non-redox-active (Cd, Hg, Ni, Zn). Redox active metals may be oxidized in Haber-Weiss or Fenton reactions yielding highly dangerous free radicals having the capacity to produce oxidative injuries in plant cells. This results in excessive ROS generation and then disturbs the balance between prooxidant and antioxidant levels. Regarding redox-inactive metals, they form covalent bonds with sulfhydryl groups of protein.

All plant responses of HMs toxicity can be broadly classified as those that ensure the acquisition of tolerance to HMs or those reducing the uptake of metals (Krzesłowska 2011). The detoxification of HMs can occur by the chelation of metals and their transportation in the plant vacuoles. Under these conditions, the plant activates the synthesis of some stress-related proteins to prevent and cope with adverse effects of HMs (Ghori et al. 2019).

One of the most important pathways involved in plant abiotic stress is that mediated by the mitogen-activated protein kinases (MAPK) (Jonak et al. 2002). These phosphorylating cascades are also stimulated by hormones and control pathways of cell proliferation/differentiation, and expression of various stress-related genes. In plant cells, the MAPKs are diverse, and can interplay to synergetic action or inhibit the action of each other.

In the next sections, we discuss the toxic effects caused by chromium compounds and analyze the plants responses to the Cr stress.

16.8 Oxidative Stress as a Mechanism of Cr Toxicity

Chromium is used in different industries like electroplating, metallurgy, production of paints, pigments and paper, tanning, and manufacturing different chemical compounds (Ghani 2011). This large-scale application of Cr is a major cause of contamination of soil and water by its highly toxic valence state Cr^{6+} in the forms of chromate (CrO_4^{2-}), dichromate ($Cr_2O_7^{2-}$) and oxide (CrO_3). Toxic forms of Cr can generate H_2O_2, $O_2^{\cdot-}$, $^{\cdot}OH$, cause plant growth retardation, chlorosis, and roots destruction (Shanker et al. 2003a; Ozturk et al. 2015). The negative effects of Cr ions on plant photosynthetic systems, mineral and water metabolism have also shown (Stambulska et al. 2018).

In plant species, all toxic effects of Cr can be divided into four types. The first, Cr damages mitochondria and affects the synthesis of essential photosynthetic pigments, especially chlorophylls (Boonyapookana et al. 2002). The second mechanism is the increasing production of GSH and AsA or the increase in antioxidant enzyme activities as a direct response to Cr-induced ROS production to avoid significant oxidation of cellular components (Shanker et al. 2003a). Chromium ions can interact with iron-porphyrin in the active center of CAT, inactivating the enzyme (Sharma et al. 2003). A decrease in catalase activity may be one of the ways facilitating ROS increase in Cr-exposed plants and aggravating damaging effects of Cr. Cr can also inhibit other antioxidant enzymes such as GPX, GR, POX, and APX (Adrees et al. 2015; Ali et al. 2015b). The third, Cr induces the production of phytochelatins and histidine as secondary metabolites. These metabolites contribute toward Cr stress tolerance (Schmfger 2001). The fourth mechanism is the DNA damage (Shanker et al. 2003b). Most heavy metals interact directly with DNA, but Cr realizes its genotoxicity via intensive ROS production (Fang et al. 2014). Cr^{6+} can produce inter-DNA strand cross-links and nucleotide strand breaks (Fang et al. 2014; Meng et al. 2017). Chromium-DNA adducts are considered to be the primary cause of Cr^{6+}-induced mutagenicity (Fang et al. 2014).

Chromium toxicity involves the damage of cell membranes via the induction of lipid peroxidation, as well as the degradation of genetic material, and inactivation of enzymes, which results in the activation of programmed cell death (Zhang et al. 2016; Wakeel et al. 2019). An increase in malondialdehyde content was observed with a sharp increase in ROS upon chromium treatment (Ali et al. 2015a; Adrees et al. 2015). Increased malondialdehyde level is a marker of lipid oxidation; therefore, its accumulation leads to the deterioration of membrane permeability (Rahman et al. 2010).

High Cr concentration disturbs the photosynthetic process. Cr-induced chloroplast ultrastructural changes, alterations in auto-fluorescence of chloroplast, or binding of Cr to cytochrome groups were established (Shahid et al. 2017; Balasaraswathi et al. 2017). As a result, it disturbs CO_2 fixation and electron transport in photosystems I and 2, with inhibiting of respective enzymes (Shahid et al. 2017; Wakeel et al. 2020). Furthermore, high Cr levels mediate elevated ROS synthesis in chloroplasts, being involved in the suppression of photosynthesis (Rodriguez et al. 2011; Wakeel et al. 2020). Delta-aminolaevulinic acid dehydratase can be inhibited by the Cr accumulation in chloroplasts, which leads to a decrease in the content of chlorophyll (Yildiz et al. 2012; Paul 2016).

Chromium also affects the activities of many enzymes, since Cr is able to interact with their catalytic site, thus leading to the alteration in enzymatic activity (Yadav 2010). In plant roots, Cr inhibits such enzymes as nitrate reductase (Stambulska et al. 2018) and Fe-reductase (Barton et al. 2000). In the mitochondrial respiratory chain, Cr^{6+} can replace Cu and Fe ions in prosthetic groups of metal-containing proteins such as cytochrome oxidase (DalCorso 2012; Singh et al. 2013). The binding of Cr to complex IV may be the reason of the inhibition of cytochrome oxidase activity. It seems that Cr binds to cytochrome a3 (Dixit et al. 2002). Studies suggest that the plasma membrane NADPH oxidase (NOX) can produce ROS in association with Cr (Potocký et al. 2012; Weyemi and Dupuy 2012). Under these conditions, NOX, consuming cytosolic NADPH, produces $O_2^{\cdot -}$, which is quickly converted to H_2O_2 (Pourrut et al. 2008).

Chromium detoxification in plants occurs by certain mechanisms (Zayed and Terry 2003). The first is a reduction of Cr^{6+}–Cr^{3+} due to the formation of intermediate forms, Cr^{5+} and Cr^{4+} (Fig. 16.4). ROS can produce during this reduction and Fenton reaction. In the Fenton reaction, the catalytic power of Cr^{3+} is greater than iron, copper, or other metal ions (Shahid et al. 2017).

Cr^{4+} and Cr^{5+} ions which occur during radical oxidation of Cr^{3+} are catalytically active and can generate the hydroxyl radical (Fig. 16.4). However, the Cr involvement in this reaction is not well studied (Strile et al. 2003). Next mechanism is the Cr

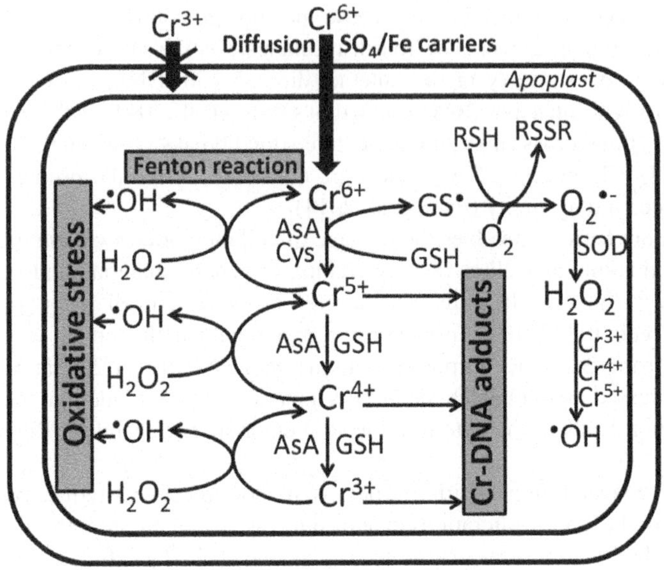

Fig. 16.4 Reactions of reduction and oxidation of chromium ions in plant cells. Cr^{6+} can actively enter the cells through channels for the transfer of the isoelectric andisostructural anions, such as SO_4^{2-} and HPO_4^{2-} channels. In the cell, glutathione (GSH), ascorbate (AsA) or cysteine (Cys) reduce Cr^{6+} to Cr^{5+}. The glutathione-derived thiyl radical ($GS^{\cdot}$) can further react with other thiol molecules (RSH) in oxygenated tissues to give the superoxide radical ($O_2^{\cdot -}$). Cr^{5+} can also be reduced by GSH or AsA to Cr^{4+} and then to Cr^{3+}. Once formed, Cr^{5+}, Cr^{4+} or Cr^{3+} species alter the DNA conformation followed by formation of Cr-DNA adducts. However, Cr^{3+}-DNA adducts play the dominant role in the chromium mutagenicity

immobilization in the cell wall and vacuoles (Han et al. 2004). Furthermore, Cr^{3+} can form highly stable complexes with glutathione, carbohydrates, NADH, $FADH_2$, and these complexes are stored in cell vacuole of plants (Lushchak 2012; Babula et al. 2008). Phytochelatins (PCs) and metallothioneins (MTs) are a very important for chromium detoxification in plants (Cobbett 2000; Cobbett and Goldsbrough 2002). However, their role in chromium detoxification in plants is not fully studied. Root exudates containing organic acids like citric, malic, aspartic, oxalic or carboxylic acids can form complexes with inorganic Cr, making them available for plant uptake (James and Bartlett 1983; Srivastava et al. 1999). However, organic acids in Cr detoxification in plants are studied not enough.

16.9 Redox-Sensitive Mechanisms of Plant Adaptive Response to Cr

As mentioned above, Cr toxicity is largely due to its ability to increase ROS production and facilitate non-enzymatic oxidation of cellular components. In response to the Cr-induced increase in ROS levels, plants up-regulate a protective battery of enzymatic and non-enzymatic molecules that help to combat with redox imbalance, repair existing oxidative damage and prevent new ones under stress. It was demonstrated that high antioxidant capacity predicts high HM tolerance (Alsherif et al. 2022). Studies indicate that in plants chromium treatment induces many antioxidant and related enzymes, non-enzymatic small antioxidants (glutathione, ascorbic acid, etc.), metal-binding enzymes, and repair components (Smiri et al. 2010; Stambulska et al. 2018; Saud et al. 2022). These biochemical markers were proposed to be indicators of resistance degree of plants to complex HM contamination (Alsherif et al. 2022). Molecular mechanisms of the up-regulation of defense mechanisms under Cr stress are studied scare, but it seems ROS are key participants in these signaling pathways.

Data suggest that molecular response to Cr is regulated at different levels—transcription of certain required genes, translation, and posttranslational modification. Posttranslational modifications ensure an immediately ("fast") response of the cell to the stimulus to stop initiation of adverse effects. These modifications include covalent changes in protein structure (e.g., phosphorylation-dephosphorylation, oxidation-reduction, association-dissociation of subunits) that affect protein function. In parallel, transcriptional ("slow") responses are triggered via multicomponent signaling cascades. ROS/RNS are proposed to be mediator of both "fast" and "slow" stress responses. Moreover, overlapping of two systems—"fast" and "slow" allows better adaptation of plants to stress challenges. Fast short-term signaling mechanisms are utilize the present set of proteins for defense, while a "slow" response includes the activation of expression of genes encoding defensive proteins (Lushchak 2011).

Various stresses, including heavy metals, stimulate processes that increase in different ROS levels. This increase is detected directly by redox-sensitive transcription factors (TFs) and specific sensors or act as mediators in a number of signaling pathways leading to global metabolic reprogramming (Foyer and Noctor 2013; He et al. 2018; Moore et al. 2016). Plant hormones, such as salicylic acid (SA) act through ROS-related mechanisms (Lushchak 2011; Foyer and Noctor 2013). Changes that will take place in the cell and the whole plant are determined by the type of ROS formed and the site of their subcellular generation (Foyer and Noctor 2013; Gadjev et al. 2006; Willems et al. 2016). Furthermore, elevated ROS specifically change the set of genes that are transcribed in a time-dependent manner. According to the conducted transcriptomic meta-analysis, the time of exposure to increased ROS might be a determinant in the pattern of gene expression. In a later period, the initial specific stress may cause more general changes in gene expression (Willems et al. 2016).

The direct effects of ROS, e.g. H_2O_2, on target proteins, involve the oxidation of sulfur-containing amino acids—cysteine and methionine. Thiol-groups of cysteine are first oxidized to sulfenic acid (Cys-SOH), which reacts rapidly with other thio-containing molecules, especially GSH. Such alterations protect protein cysteines from further irreversible oxidation to sulfuric (SO_2H) or sulfonic (SO_3H) acid (He et al. 2018; Martí et al. 2020). Among proteins, which undergo Cys-modification, are metabolic enzymes (e.g., chloroplastic glyceraldehyde-3-phosphate dehydrogenase), antioxidant oxidoreductases (thioredoxins, glutaredoxins), signaling proteins, such as protein kinases, phosphatases, and transcription factors (Lushchak 2012; He et al. 2018; Martí et al. 2020).

Under non-stressful conditions, many transcriptional factors (TFs) are in an inactive form in the cytoplasm, but oxidative stress activates them followed by their translocation to the nucleus. Here, we will review the effects of redox perturbations on gene expression in nucleus with a focus on two plant stress responses that are regulated by the redox-sensitive proteins NPR1/TGA and Rap2.4a.

The NPR1 protein (the non-expressor of the pathogenesis-related gene 1) is the main transcriptional coactivator for pathogenesis-related genes (PR) and other genes induced by salicylic acid. This protein also regulates SA-mediated suppression jasmonic acid signaling (Lushchak 2011; He et al. 2018). Under physiological conditions, NPR1 is the cytosolic protein in the form of oxidized oligomer with intermolecular disulfide bridges between Cys82 and Cys216 (Fig. 16.5a) (Mou et al. 2003). Nitric oxide (˙NO) may cause S-nitrosylation of Cys156 facilitating the formation of disulfide bonds between NPR1 monomers to form inactive oligomers. In response to pathogens or other stressors, plant hormone SA triggers thioredoxin-dependent S–S bond repair in oligomeric NPR1 and its dissociation into NPR1monomers, which are active form of the protein (Tada et al. 2008). The active NPR1 protein is translocated to the nucleus, where it interacts with the TGAs proteins, which are basic domain/leucine zipper TFs (Després et al. 2003). The complex NPR1/TGA binds to promotors of genes containing specific nucleotide sequence, namely TGACG sequence, leading to activation of transcription of stress-protective genes (Lushchak 2011; He et al. 2018). It should be noted that SA is able to directly bind to NRP1 leading to conformational changes in NPR1 (Wu et al. 2012). Plants with inactivated NPR1 showed poor systemic acquired resistance which is mediated by SA (Dong et al. 2001). Six TGA transcription factors were identified to cooperate with NPR1. The dominant role of TGA2, TGA3, TGA5, and TGA6 have been proved (Zhang et al. 2003; Kesarwani et al. 2007). It is proposed that HM due to H_2O_2-induced production may stimulate SA/NRP1 signaling to increase plant tolerance (Dutta et al. 2018).

Redox-changes in the structure can modulate the affinity of transcription factors towards DNA. This mechanism is well studied in transcriptional factor Rap2.4A, which is encoded by the At1g36060 gene. The Rap2.4A protein is able to bind the promoter element of the specific genes in redox-dependent manner (Fig. 16.5b) (He et al. 2018). In particular, it was found that the activation of the 2-Cys peroxiredoxin-A gene in *A. thaliana* is controlled by Rap2.4A. The use of Rap2.4a mutant lines demonstrates that this protein regulates tolerance to stressors by increasing photosynthetic ability and defense capacity of chloroplasts (Shaikhali et al. 2008; Rudnik et al.

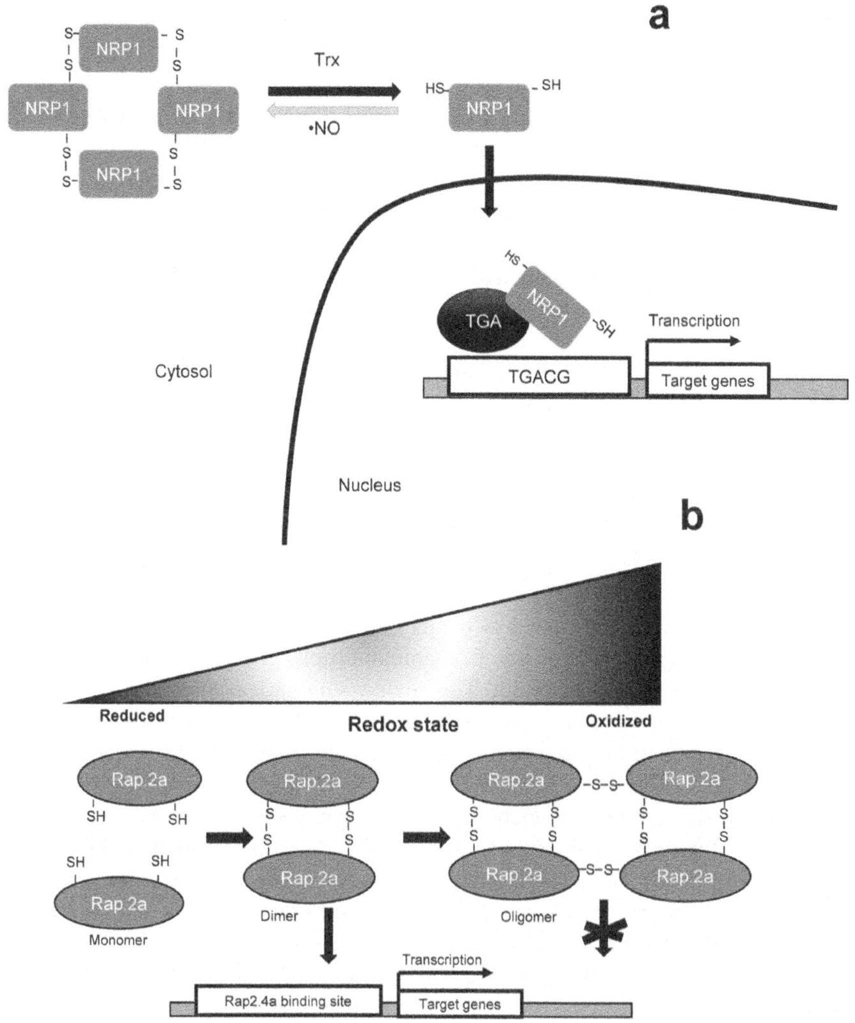

Fig. 16.5 Redox-regulation of NPR1 **a** and Rap2.4A, **b** transcriptional factors in adaptive response of plants to oxidative stress. Explanation in the text

2017). In particular, Rap2.4A controls expression of *ZAT10* gene, which encodes a transcription factor activating expression of cytosolic ascorbate peroxidase 2 APX 2 (Shaikhali et al. 2008). The ability of RAP2.4A to interact with DNA depends on the degree of oxidation Cys residues. Cys at positions 113, 286, and 302 was shown to determine the functional activity of Rap2.4a (Lushchak 2011). When these cysteine residues are oxidized, intermolecular disulfide bonds are formed, which results in the formation of a homodimer of Rap2.4A, which is its active form to induce transcription (He et al. 2018). The homodimer Rap2.4A moves to the nucleus and binds to target genes. It is assumed that before binding, the protein undergoes reduction by disulfide bridges, but the mechanism of this reduction is hypothetical. Only a moderate ROS increase leads to the Rap2.4a activation, when severe oxidative stress cause inactivation of this protein (Shaikhali et al. 2008). This is likely due to the fact that with increased ROS levels, there are formed aggregates of oxidized Rap2.4a protein that do not have DNA-binding activity. Thus, Rap2.4a serves as a highly sensitive and precisely controllable redox system that provides an adaptive response to small changes in the redox state, which can be a useful strategy to prevent the development of significant oxidative damage in the plant cell. It is not currently known whether Rap2.4a is involved in tolerance to heavy metals. However, its role as an inducer of antioxidant defense suggests that it may be involved in protection against metal toxicity. Detailed molecular redox mechanisms that provide chromium tolerance require further research.

16.10 Conclusions and Perspectives

Toxicity of heavy metals (HM) is the topic which is intensively studied. In this context, it is important to know how plants cope with and tolerate HMs, especially with regards to agricultural plants. Contamination of agricultural plants with HM has a direct effect on human health. This review summarizes available information on the relationship between chromium toxicity and ROS/RNS generation in plant cells. The analysis of literature data indicates that chromium causes disturbances in the redox balance through a number of mechanisms—directly inactivating protective proteins, disrupting the functioning of electron transport systems, destroying membrane integrity. In chloroplasts, Cr inhibits electron transport, reduces CO_2 fixation and glucose synthesis, causing drying, chlorosis, necrosis, and wilting of leaves. Plant cells up-regulate complex antioxidant non-enzymatic and enzymatic mechanisms that help the cells to adapt their metabolism to Cr stress. However, the mechanisms of Cr-induced toxicity at the biochemical and molecular levels still require deeper investigation. Molecular mechanism underling Cr tolerance is still poorly understood, and this is an avenue for future research.

Acknowledgements This work was supported by a grant from Ministry of Education and Science of Ukraine (#0122U000894).

References

Adrees M, Ali S, Iqbal M, Bharwana SA, Siddiqi Z, Farid M, Ali Q, Saeed R, Rizwan M (2015) Mannitol alleviates chromium toxicity in wheat plants in relation to growth, yield, stimulation of anti-oxidative enzymes, oxidative stress and Cr uptake in sand and soil media. Ecotoxicol Environ Saf 122:1–8. https://doi.org/10.1016/j.ecoenv.2015.07.003

Akram NA, Shafiq F, Ashraf M (2017) Ascorbic acid-a potential oxidant scavenger and its role in plant development and abiotic stress tolerance. Front Plant Sci 8:613. https://doi.org/10.3389/fpls.2017.00613

Alengebawy A, Abdelkhalek ST, Qureshi SR, Wang M-Q (2021) Heavy metals and pesticides toxicity in agricultural soil and plants: ecological risks and human health implications. Toxics 9:42. https://doi.org/10.3390/toxics9030042

Al-Huqail AA, Ali HM, Kushwaha BK, AL-Huqail AA, Singh VP, Siddiqui MH (2020) Ascorbic acid is essential for inducing chromium (VI) toxicity tolerance in tomato roots. J Biotechnol 322:66–73. https://doi.org/10.1016/j.jbiotec.2020.07.011

Ali S, Bharwana SA, Rizwan M, Farid M, Kanwal S, Ali Q, Ibrahim M, Gill RA, Khan MD (2015a) Fulvic acid mediates chromium (Cr) tolerance in wheat (*Triticum aestivum* L.) through lowering of Cr uptake and improved antioxidant defense system. Environ Sci Pollut Res 22:10601–10609. https://doi.org/10.1007/s11356-015-4271-7

Ali S, Chaudhary A, Rizwan M, Anwar HT, Adrees M, Farid M, Irshad MK, Hayat T, Anjum SA (2015b) Alleviation of chromium toxicity by glycinebetaine is related to elevated antioxidant enzymes and suppressed chromium uptake and oxidative stress in wheat (*Triticum aestivum* L.). Environ Sci Pollut Res 22:10669–10678. https://doi.org/10.1007/s11356-015-4193-4

Ali I, Suhail M, Basheer AA (2017) Advanced spiral periodic classification of the elements. Chem Int 3:220–224. https://doi.org/10.5281/zenodo.1473118

Alsherif EA, Al-Shaikh TM, Almaghrabi O, AbdElgawad H (2022) High redox status as the basis for heavy metal tolerance of *Sesuvium portulacastrum* L. inhabiting contaminated soil in Jeddah, Saudi Arabia. Antioxidants 11:19. https://doi.org/10.3390/antiox11010019

Anjum SA, Ashraf U, Khan I, Tanveer M, Shahid M, Shakoor A, Wang L (2017) Phyto-toxicity of chromium in maize: oxidative damage, osmolyte accumulation, anti-oxidative defense and chromium uptake. Pedosphere 27:262–273. https://doi.org/10.1016/S1002-0160(17)60315-1

Apel K, Hirt H (2004) Reactive oxygen species: metabolism, oxidative stress, and signal transduction. Annu Rev Plant Biol 55:373–399

Appiah C, Yang Z-F, He J, Wang Y, Zhou J, Xu W-Z, Nie G, Zhu Y-Q (2021) Genome-wide identification of Hsp90 gene family in perennial ryegrass and expression analysis under various abiotic stresses. Plants 10:2509. https://doi.org/10.3390/plants10112509

Asada K (1999) The water–water cycle in chloroplasts: scavenging of active oxygens and dissipation of excess photons. Annu Rev Plant Physiol Plant Mol Biol 50:601–639. https://doi.org/10.1146/annurev.arplant.50.1.601

Babula P, Adam V, Opatrilova R, Zehnalek J, Havel L, Kizek R (2008) Uncommon heavy metals, metalloids and their plant toxicity: a review. Environ Chem Lett 6(4):189–213. https://doi.org/10.1007/s10311-008-0159-9

Badger MR, von Caemmerer S, Ruuska S, Nakano H (2000) Electron flow to oxygen in higher plants and algae: rates and control of direct photoreduction (Mehler reaction) and rubsico oxygenase. Philos Trans R Soc Lond B: Biol Sci 355:1433–1445. https://doi.org/10.1098/rstb.2000.0704

Balasaraswathi K, Jayaveni S, Sridevi J, Sujatha D, Aaron KP, Rose C (2017) Cr-induced cellular injury and necrosis in *Glycine max* L.: biochemical mechanism of oxidative damage in chloroplast. Plant Physiol Biochem 118:653–666. https://doi.org/10.1016/j.plaphy.2017.08.001

Balestrasse KB, Yannarelli GG, Noriega GO, Batlle A, Tomaro ML (2008) Heme oxygenase and catalase gene expression in nodules and roots of soybean plants subjected to cadmium stress. Biometals. Int J Role Met Ions Biol Biochem Med 21:433–441. https://doi.org/10.1007/s10534-008-9132-0

Barton LL, Johnson GV, O'Nan AG, Wagener BM (2000) Inhibition of ferric chelate reductase in alfalfa roots by cobalt, nickel, chromium, and copper. J Plant Nutr 23(11–12):1833–1845. https://doi.org/10.1080/01904160009382146

Basheer AA (2018a) Chemical chiral pollution: impact on the society and science and need of the regulations in the 21st century. Chirality 30(4):402–406. https://doi.org/10.1002/chir.22808

Basheer AA (2018b) New generation nano-adsorbents for the removal of emerging contaminants in water. J Mol Liq 261:583–593. https://doi.org/10.1016/j.molliq.2018.04.021

Bayliak MM, Burdyliuk NI, Lushchak VI (2016a) Effects of pH on antioxidant and prooxidant properties of common medicinal herbs. Open Life Sci. 11:298–307. https://doi.org/10.1515/biol-2016-0040

Bayliak MM, Burdylyuk NI, Lushchak VI (2016b) Quercetin increases stress resistance in the yeast Saccharomyces cerevisiae not only as an antioxidant. Ann Microbiol 66:569–576. https://doi.org/10.1007/s13213-015-1136-8

Bela K, Horváth E, Gallé Á, Szabados L, Tari I, Csiszár J (2015) Plant glutathione peroxidases: emerging role of the antioxidant enzymes in plant development and stress responses. J Plant Physiol 176:192–201. https://doi.org/10.1016/j.jplph.2014.12.014

Bergmüller E, Porfirova S, Dörmann P (2003) Characterization of an Arabidopsis mutant deficient in gamma-tocopherol methyltransferase. Plant Mol Biol 52:1181–1190. https://doi.org/10.1023/b:plan.0000004307.62398.91

Boeckx T, Winters AL, Webb KJ, Kingston-Smith AH (2015) Polyphenol oxidase in leaves: is there any significance to the chloroplastic localization? J Exp Bot 66:3571–3579. https://doi.org/10.1093/jxb/erv141

Boonyapookana B, Upatham ES, Kruatrachue M, Pokethitiyook P, Singhakaew S (2002) Phytoaccumulation and phytotoxicity of cadmium and chromium in duckweed *Wolfa globosa*. Int J Phytoremed 4:87–100. https://doi.org/10.1080/15226510208500075

Cadet J, Davies KJA (2017) Oxidative DNA damage & repair: an introduction. Free Radic Biol Med 107:2–12. https://doi.org/10.1016/j.freeradbiomed.2017.03.030

Chaudiere J, Ferrari-Iliou R (1999) Intracellular antioxidants: from chemical to biochemical mechanisms. Food Chem Toxicol 37:949–962. https://doi.org/10.1016/s0278-6915(99)00090-3

Chen I, Chiu MH, Cheng SF, Hsu YH, Tsai CH (2013) The glutathione transferase of *Nicotiana benthamiana* NbGSTU4 plays a role in regulating the early replication of Bamboo mosaic virus. New Phytol 199:749–757. https://doi.org/10.1111/nph.12304

Choudhury FK, Rivero RM, Blumwald E, Mittler R (2017) Reactive oxygen species, abiotic stress and stress combination. Plant J 90:856–867. https://doi.org/10.1111/tpj.13299

Cobbett C (2000) Phytochelatins and their role in heavy metal detoxification. Plant Physiol 123:463–469. https://doi.org/10.1104/pp.123.3.825

Cobbett C, Goldsbrough P (2002) Phytochelatins and metallothioneins: roles in heavy metal detoxification and homeostasis. Annu Rev Plant Biol 53:159–182. https://doi.org/10.1146/annurev.arplant.53.100301.135154

Corpas FJ, Palma JM (2020) Assessing nitric oxide (NO) in higher plants: an outline. Nitrogen 1:12–20. https://doi.org/10.3390/nitrogen1010003

Corpas FJ, González-Gordo S, Palma JM (2021) Protein nitration: a connecting bridge between nitric oxide (NO) and plant stress. Plant Stress. 2:100026. https://doi.org/10.1016/j.stress.2021.100026

Cui Y, Wang M, Yin X, Xu G, Song S, Li M, Liu K, Xia X (2019) OsMSR3, a small heat shock protein, confers enhanced tolerance to copper stress in *Arabidopsis thaliana*. Int J Mol Sci 20:6096. https://doi.org/10.3390/ijms20236096

Dağhan H, Ozturk M (2015) Soil pollution in Turkey and remediation methods. In: Hakeem K, Sabir M, Ozturk M, Mermut A (eds) Soil remediation and plants: prospects and challenges. Elsevier, New York, pp 287–312

Dai J, Mumper RJ (2010) Plant phenolics: extraction, analysis and their antioxidant and anticancer properties. Molecules 15:7313–7352. https://doi.org/10.3390/molecules15107313

DalCorso G (2012) Heavy metal toxicity in plants. In: plants and heavy metals, springer, briefs in molecular science, Dordrecht, Netherlands, pp 1–25

Das K, Roychoudhury A (2014) Reactive oxygen species (ROS) and response of antioxidants as ROS-scavengers during environmental stress in plants. Front Environ Sci 2:53. https://doi.org/10.3389/fenvs.2014.00053

Das P, Nutan KK, Singla-Pareek SL, Pareek A (2015) Oxidative environment and redox homeostasis in plants: dissecting out significant contribution of major cellular organelles. Front Environ Sci 2:70. https://doi.org/10.3389/fenvs.2014.00070

Del Rio LA, Lopez-Huertas E (2016) ROS generation in peroxisomes and its role in cell signaling. Plant Cell Physiol 57:1364–1376. https://doi.org/10.1093/pcp/pcw076

Després C, Chubak C, Rochon A, Clark R, Bethune T, Desveaux D, Fobert PR (2003) The Arabidopsis NPR1 disease resistance protein is a novel cofactor that confers redox regulation of DNA binding activity to the basic domain/leucine zipper transcription factor TGA1. Plant Cell 15:2181–2191. https://doi.org/10.1105/tpc.012849

Di Fino L, Arruebarrena Di Palma A, Perk EA, García-Mata C, Schopfer FJ, Laxalt AM (2021) Nitro-fatty acids: electrophilic signaling molecules in plant physiology. Planta 254:120. https://doi.org/10.1007/s00425-021-03777-z

Di Meo S, Reed TT, Venditti P, Victor VM (2016) Role of ROS and RNS sources in physiological and pathological conditions. Oxidative Med Cell Longev 2016. https://doi.org/10.1155/2016/1245049

di Toppi LS, Fossati F, Musetti R, Mikerezi I, Favali MA (2002) Effects of Hexavalent Chromium on Maize, Tomato, and Cauliflower Plants. J Plant Nutr 25:701–717. https://doi.org/10.1081/PLN-120002953

Ding S, Lu Q, Zhang Y, Yang Z, Wen X, Zhang L, Lu C (2009) Enhanced sensitivity to oxidative stress in transgenic tobacco plants with decreased glutathione reductase activity leads to a decrease in ascorbate pool and ascorbate redox state. Plant Mol Biol 69:577–592. https://doi.org/10.1007/s11103-008-9440-3

Ding H, Wang B, Han Y, Li S (2020) The pivotal function of dehydroascorbate reductase in glutathione homeostasis in plants. J Exp Bot 71:3405–3416. https://doi.org/10.1093/jxb/eraa107

Dixit V, Pandey V, Shyam R (2002) Chromium ions inactivate electron transport and enhance superoxide generation in vivo in pea (*Pisum sativum* L. cv: Azad) root mitochondria. Plant Cell Env 25:687–693. https://doi.org/10.1046/j.1365-3040.2002.00843.x

Dixon DP, Sellars JD, Edwards R (2011) The *Arabidopsis* phi class glutathione transferase AtGSTF2: binding and regulation by biologically active heterocyclic ligands. Biochem J 438:63–70. https://doi.org/10.1042/BJ20101884

Dixon DP, Skipsey M, Edwards R (2010) Roles for glutathione transferases in plant secondary metabolism. Phytochemistry 71:338–350. https://doi.org/10.1016/j.phytochem.2009.12.012

Domènech J, Mir G, Huguet G, Capdevila M, Molinas M, Atrian S (2006) Plant metallothionein domains: functional insight into physiological metal binding and protein folding. Biochimie 88:583–593. https://doi.org/10.1016/j.biochi.2005.11.002

Dong X, Li X, Zhang Y, Fan W, Kinkema M, Clarke J (2001) Regulation of systemic acquired resistance by NPR1 and its partners. Novartis Found Symp 236:165–173; discussion 173–175. https://doi.org/10.1002/9780470515778.ch12

Dong Y, Li C, Zhang Y, He Q, Daud MK, Chen J, Zhu S (2016) Glutathione S-transferase gene family in *Gossypium raimondii* and *G. arboreum*: comparative genomic study and their expression under salt stress. Front Plant Sci 7:139. https://doi.org/10.3389/fpls.2016.00139

Douce R, Neuburger M (1999) Biochemical dissection of photorespiration. Curr Opin Plant Biol 2:214–222. https://doi.org/10.1016/S1369-5266(99)80038-7

Dumanović J, Nepovimova E, Natić M, Kuča K, Jaćević V (2021) The significance of reactive oxygen species and antioxidant defense system in plants: a concise overview. Front Plant Sci 11:552969. https://doi.org/10.3389/fpls.2020.552969

Dunnill C, Patton T, Brennan J, Barrett J, Dryden M, Cooke J, Georgopoulos NT (2017) Reactive oxygen species (ROS) and wound healing: the functional role of ROS and emerging ROS-modulating technologies for augmentation of the healing process. Int Wound J 14:89–96. https://doi.org/10.1111/iwj.12557

Dutta S, Mitra M, Agarwal P, Mahapatra K, De S, Sett U, Roy S (2018) Oxidative and genotoxic damages in plants in response to heavy metal stress and maintenance of genome stability. Plant Signal Behav 13:e1460048. https://doi.org/10.1080/15592324.2018.1460048

Edwards EA, Rawsthorne S, Mullineaux PM (1990) Subcellular distribution of multiple forms of glutathione reductase in leaves of pea (*Pisum sativum* L.). Planta 180:278–284. https://doi.org/10.1007/BF00194008

Fang Z, Zhao M, Zhen H, Chen L, Shi P, Huang Z (2014) Genotoxicity of tri- and hexavalent chromium compounds in vivo and their modes of action on DNA damage in vitro. Plos One 9:e103194. https://doi.org/10.1371/journal.pone.0103194

Fedenko VS, Landi M, Shemet SA (2022) Metallophenolomics: a novel integrated approach to study complexation of plant phenolics with metal/metalloid ions. Int J Mol Sci 23:11370. https://doi.org/10.3390/ijms231911370

Feechan A, Kwon E, Yun BW, Wang Y, Pallas JA, Loake GJ (2005) A central role for S-nitrosothiols in plant disease resistance. Proc Natl Acad Sci 102:8054–8059. https://doi.org/10.1073/pnas.0501456102

Figgitt M, Newson R, Leslie IJ, Fisher J, Ingham E, Case CP (2010) The genotoxicity of physiological concentrations of chromium (Cr(III) and Cr(VI)) and cobalt (Co(II)): an in vitro study. Mutat Res 688:53–61. https://doi.org/10.1016/j.mrfmmm.2010.03.008

Foyer CH, Halliwell B (1976) The presence of glutathione and glutathione reductase in chloroplasts: a proposed role in ascorbic acid metabolism. Planta 133:21–25. https://doi.org/10.1007/BF00386001

Foyer H, Noctor G (2003) Redox sensing and signaling associated with reactive oxygen in chloroplasts, peroxisomes and mitochondria. Physiol Plant 119:355–364. https://doi.org/10.1034/j.1399-3054.2003.00223.x

Foyer CH, Noctor G (2005) Oxidant and antioxidant signaling in plants: a reevaluation of the concept of oxidative stress in a physiological context. Plant Cell Environ 28:1056–1071. https://doi.org/10.1111/j.1365-3040.2005.01327.x

Foyer CH, Noctor G (2013) Redox signaling in plants. Antioxid Redox Signal 18:2087–2090. https://doi.org/10.1089/ars.2013.5278

Fryer MJ, Ball L, Oxborough K, Karpinski S, Mullineaux PM, Baker NR (2003) Control of ascorbate peroxidase 2 expression by hydrogen peroxide and leaf water status during excess light stress reveals a functional organisation of *Arabidopsis* leaves. Plant J 33:691–705. https://doi.org/10.1046/j.1365-313x.2003.01656.x

Gadjev I, Vanderauwera S, Gechev TS, Laloi C, Minkov IN, Shulaev V, Apel K, Inzé D, Mittler R, Van Breusegem F (2006) Transcriptomic footprints disclose specificity of reactive oxygen species signaling in *Arabidopsis*. Plant Physiol 141:436–445. https://doi.org/10.1104/pp.106.078717

Gechev TS, Van Breusegem F, Stone JM, Denev I, Laloi C (2006) Reactive oxygen species as signals that modulate plant stress responses and programmed cell death. BioEssays 28:1091–1101. https://doi.org/10.1002/bies.20493

Geigenberger P, Thormählen I, Daloso DM, Fernie AR (2017) The unprecedented versatility of the plant thioredoxin system. Trends Plant Sci 22:249–262. https://doi.org/10.1016/j.tplants.2016.12.008

Gelhaye E, Rouhier N, Navrot N, Jacquot JP (2005) The plant thioredoxin system. Cell Mol Life Sci CMLS 62:24–35. https://doi.org/10.1007/s00018-004-4296-4

Ghani A (2011) Effect of chromium toxicity on growth, chlorophyll and some mineral nutrients of *Brassica juncea* L. Egypt Acad J Biol Sci 2:9–15. https://doi.org/10.21608/eajbsh.2011.17007

Ghori N-H, Ghori T, Hayat MQ, Imadi SR, Gul A, Altay V, Ozturk M (2019) Heavy metal stress and responses in plants. Int J Environ Sci Technol 16(3):1807–1828. https://doi.org/10.1007/s13762-019-02215-8

Gill SS, Tuteja N (2010) Reactive oxygen species and antioxidant machinery in abiotic stress tolerance in crop plants. Plant Physiol Biochem 48:909–930. https://doi.org/10.1016/j.plaphy.2010.08.016

Gilroy S, Białasek M, Suzuki N, Górecka M, Devireddy AR, Karpinski S (2016) ROS, calcium, and electric signals: key mediators of rapid systemic signaling in plants. Plant Physiol 171:1606–1615. https://doi.org/10.1104/pp.16.00434

Gupta VK, Ali I (2002) Encyclopedia of surface and colloid science. Marcel Dekker, New York, pp 136–166

Gupta KJ, Kaiser WM (2010) Production and scavenging of nitric oxide by barley root mitochondria. Plant Cell Physiol 51:576–584. https://doi.org/10.1093/pcp/pcq022

Han FX, Sridhar BBM, Monts DL, Su Y (2004) Phytoavailability and toxicity of trivalent and hexavalent chromium to *Brassica juncea*. New Phytol 162(2):489–499. https://doi.org/10.1111/j.1469-8137.2004.01027.x

Hasanuzzaman M, Nahar K, Hossain MS, Mahmud JA, Rahman A, Inafuku M, Oku H, Fujita M (2017) Coordinated actions of glyoxalase and antioxidant defense systems in conferring abiotic stress tolerance in plants. Int J Mol Sci 18:200. https://doi.org/10.3390/ijms18010200

Hasanuzzaman M, Bhuyan MHMB, Anee TI, Parvin K, Nahar K, Mahmud JA, Fujita M (2019) Regulation of ascorbate-glutathione pathway in mitigating oxidative damage in plants under abiotic stress. Antioxidants 8:384. https://doi.org/10.3390/antiox8090384

Hasanuzzaman M, Bhuyan BMHM, Parvin K, Bhuiyan TF, Anee TI, Nahar K, Hossen MdSh, Zulfiqar F, Alam MdM, Fujita M (2020a) Regulation of ROS metabolism in plants under environmental stress: a review of recent experimental evidence. Int J Mol Sci 21:8695. https://doi.org/10.3390/ijms21228695

Hasanuzzaman M, Bhuyan MHMB, Zulfiqar F, Raza A, Mohsin SM, Mahmud JA, Fujita M, Fotopoulos V (2020b) Reactive oxygen species and antioxidant defense in plants under abiotic stress: revisiting the crucial role of a universal defense regulator. Antioxidants 9:681. https://doi.org/10.3390/antiox9080681

Havaux M, Tardy F, Lemoine Y (1998) Photosynthetic light-harvesting function of carotenoids in higher-plant leaves exposed to high light irradiances. Planta 205:242–250

He H, Van Breusegem F, Mhamdi A (2018) Redox-dependent control of nuclear transcription in plants. J Exp Bot 69:3359–3372. https://doi.org/10.1093/jxb/ery130

Hossain MA, Piyatida P, Silva JAT, Fujita M (2012) Molecular mechanism of heavy metal toxicity and tolerance in plants: central role of glutathione in detoxifcation of reactive oxygen species and methylglyoxal and in heavy metal chelation. J Bot 2012:1–37. https://doi.org/10.1155/2012/872875

Husak VV (2015) Copper and copper-containing pesticides: metabolism, toxicity and oxidative stress. J Vasyl Stefanyk Precarpathian Natl Univ 2(1):39–51. https://doi.org/10.15330/jpnu.2.1.38-50

James BR, Barlett RJ (1983) Behavior of chromium in soils. VI. Interactions between oxidation and reduction and organic complexation. J Environ Qual 12:173–176. https://doi.org/10.2134/jeq1983.00472425001200020004x

Jedelská T, Luhová L, Petřivalský M (2020) Thioredoxins: emerging players in the regulation of protein S-nitrosation in plants. Plants 9:1426. https://doi.org/10.3390/plants9111426

Jia B, Sun M, Sun X, Li R, Wang Z, Wu J, Wei Z, DuanMu H, Xiao J, Zhu Y (2016) Overexpression of GsGSTU13 and SCMRP in *Medicago sativa* confers increased salt–alkaline tolerance and methionine content. Physiol Plant 156:176–189. https://doi.org/10.1111/ppl.12350

Jonak C, Okresz L, Bogre L, Hirt H (2002) Complexity, cross talk and integration of plant MAP kinase signalling. Curr Opin Plant Biol 5:415–424. https://doi.org/10.1016/s1369-5266(02)00285-6

Kapoor D, Singh S, Kumar V, Romero R, Prasad R, Singh J (2019) Antioxidant enzymes regulation in plants in reference to reactive oxygen species (ROS) and reactive nitrogen species (RNS). Plant Gene 19:100182. https://doi.org/10.1016/j.plgene.2019.100182

Karuppanapandian T, Moon JC, Kim C, Manoharan K, Kim W (2011) Reactive oxygen species in plants: their generation, signal transduction, and scavenging mechanisms. AJCS 5(6):709–725

Kesarwani M, Yoo J, Dong X (2007) Genetic interactions of TGA transcription factors in the regulation of pathogenesis-related genes and disease resistance in *Arabidopsis*. Plant Physiol 144:336–346. https://doi.org/10.1104/pp.106.095299

Kharma A, Grman M, Misak A, Domínguez-Álvarez E, Nasim MJ, Ondrias K, Chovanec M, Jacob C (2019) Inorganic polysulfides and related reactive sulfur–selenium species from the perspective of chemistry. Molecules 24:1359. https://doi.org/10.3390/molecules24071359

Kim SI, Andaya VC, Tai TH (2011) Cold sensitivity in rice (*Oryza sativa* L.) is strongly correlated with a naturally occurring I99V mutation in the multifunctional glutathione transferase isoenzyme GSTZ2. Biochem J 435:373–380. https://doi.org/10.1042/BJ20101610

Kolbert Z, Barroso JB, Brouquisse R, Corpas FJ, Gupta KJ, Lindermayr C, Loake GJ, Palma JM, Petřivalský M, Wendehenne D, Hancock JT (2019) A Forty year journey: the generation and roles of NO in plants. Nitric Oxide 93:53–70. https://doi.org/10.1016/j.niox.2019.09.006

Kornyeyev D, Logan BA, Payton P, Allen RD, Holaday AS (2003) Elevated chloroplastic glutathione reductase activities decrease chilling-induced photoinhibition by increasing rates of photochemistry, but not thermal energy dissipation, in transgenic cotton. Funct Plant Biol 30:101–110. https://doi.org/10.1071/FP02144

Kosakivska IV, Babenko LM, Romanenko KO, Korotka IY, Potters G (2021) Molecular mechanisms of plant adaptive responses to heavy metals stress. Cell Biol Int 45:258–272. https://doi.org/10.1002/cbin.11503

Koua D, Cerutti L, Falquet L, Sigrist CJ, Theiler G, Hulo N, Dunand C (2009) PeroxiBase: a database with new tools for peroxidase family classification. Nucleic Acids Res 37:261–266. https://doi.org/10.1093/nar/gkn680

Krzesłowska M (2011) The cell wall in plant cell response to trace metals: polysaccharide remodeling and its role in defense strategy. Acta Physiol Plant 33(1):35–51. https://doi.org/10.1007/s11738-010-0581-z

Kubrak OI, Lushchak OV, Lushchak JV, Torous IM, Storey JM, Storey KB, Lushchak VI (2010) Chromium effects on free radical processes in goldfish tissues: comparison of Cr(III) and Cr(VI) exposures on oxidative stress markers, glutathione status and antioxidant enzymes. Comp Biochem Physiol C Toxicol Pharmacol 152:360–370. https://doi.org/10.1016/j.cbpc.2010.06.003

Kumar D, Klessig DF (2003) High-affinity salicylic acid-binding protein 2 is required for plant innate immunity and has salicylic acid-stimulated lipase activity. Proc Natl Acad Sci 100:16101–16106. https://doi.org/10.1073/pnas.0307162100

Kumar P, Tokas J, Singal HR (2019) Amelioration of chromium VI toxicity in Sorghum (*Sorghum bicolor* L.) using glycine betaine. Sci Rep 9:16020. https://doi.org/10.1038/s41598-019-52479-w

Kumar S, Wang M, Fahad S, Qayyum A, Chen Y, Zhu G (2022) Chromium induces toxicity at different phenotypic, physiological, biochemical, and ultrastructural levels in sweet potato (*Ipomoea batatas* L.) plants. Int J Mol Sci 23:13496. https://doi.org/10.3390/ijms232113496

Leng X, Wang H, Zhang S, Qu C, Yang C, Xu Z, Liu G (2021) Identification and Characterization of the APX Gene family and its expression pattern under phytohormone treatment and abiotic stress in *Populus trichocarpa*. Genes 12:334. https://doi.org/10.3390/genes12030334

Leterrier M, Cagnac O (2018) Function of the various MDAR isoforms in higher plants. In: Gupta D, Palma J, Corpas F (eds) Antioxidants and antioxidant enzymes in higher plants. Springer, Cham, Switzerland, pp 83–94

Loake G, Grant M (2007) Salicylic acid in plant defence—The players and protagonists. Curr Opin Plant Biol 10:466–472. https://doi.org/10.1016/j.pbi.2007.08.008

Locato V, de Pinto MC, De Gara L (2009) Different involvement of the mitocondiral, plastidal and cytosolic ascorbate-glutathione redox enzymes in heat shock responses. Phyiol Plantarum 135:296–306. https://doi.org/10.1111/j.1399-3054.2008.01195.x

López-Bucio J, Ortiz-Castro R, Ruíz-Herrera LF, Juárez CV, Hernández-Madrigal F, Carreón-Abud Y, Martínez-Trujillo M (2015) Chromate induces adventitious root formation via auxin signalling and SOLITARY-ROOT/IAA14 gene function *in Arabidopsis thaliana*. Biometals 28:353–365. https://doi.org/10.1007/s10534-015-9838-8

Luis A, Río D (2013) Peroxisomes and their key role in cellular signalling and metabolism. Springer, Dordrecht, The Netherlands

Lushchak OV, Kubrak OI, Torous IM, Nazarchuk TY, Storey KB, Lushchak VI (2009) Trivalent chromium induces oxidative stress in goldfish brain. Chemosphere 75:56–62. https://doi.org/10.1016/j.chemosphere.2008.11.052

Lushchak VI (2011) Adaptive response to oxidative stress: Bacteria, fungi, plants and animals. Comp Biochem Physiol C Toxicol Pharmacol 153:175–190. https://doi.org/10.1016/j.cbpc.2010.10.004

Lushchak VI (2012) Glutathione homeostasis and functions: potential targets for medical interventions. J Amino Acids 2012:736837. https://doi.org/10.1155/2012/736837

Mandal M, Sarkar M, Khan A, Biswas M, Masi A, Rakwal R, Agrawal GK, Srivastava A, Sarkar A (2022) Reactive oxygen species (ROS) and reactive nitrogen species (RNS) in plants–maintenance of structural individuality and functional blend. Adv Redox Res 5:100039. https://doi.org/10.1016/j.arres.2022.100039

Martí MC, Jiménez A, Sevilla F (2020) Thioredoxin network in plant mitochondria: cysteine s-posttranslational modifications and stress conditions. Front Plant Sci 11:571288. https://doi.org/10.3389/fpls.2020.571288

Medrano-Macías J, Flores-Gallegos AC, Nava-Reyna E, Morales I, Tortella G, Solís-Gaona S, Benavides-Mendoza A (2022) Reactive oxygen, nitrogen, and sulfur species (RONSS) as a metabolic cluster for signaling and biostimulation of plants: an overview. Plants 11:3203. https://doi.org/10.3390/plants11233203

Mehla N, Sindhi V, Josula D, Bisht P, Wani SH (2017) An introduction to antioxidants and their roles in plant stress tolerance. In: Khan MIR, Khan NA (eds) Reactive oxygen species and antioxidant systems in plants: role and regulation under abiotic stress. Springer, Singapore, pp 1–23

Meng X, Fang Z, Wei H, Yang Z, Huang Z (2017) Genotoxicity investigation of tri-and hexavalent chromium using a yeast SUP4-o mutation assay. Carcinog Teratog Mutagen 29:65–69. https://doi.org/10.3969/j.issn.1004-616x.2017.01.013

Meyer Y, Siala W, Bashandy T, Riondet C, Vignols F, Reichheld JP (2008) Glutaredoxins and thioredoxins in plants. Biochim Biophys Acta 1783:589–600. https://doi.org/10.1016/j.bbamcr.2007.10.017

Meyer Y, Belin C, Delorme-Hinoux V, Reichheld J-P, Riondet C (2012) Thioredoxin and glutaredoxin systems in plants: molecular mechanisms, crosstalks, and functional significance. Antioxid Redox Signal 17:1124–1160. https://doi.org/10.1089/ars.2011.4327

Michalak A (2006) Phenolic compounds and their antioxidant activity in plants growing under heavy metal stress. Pol J Environ Stud 15:523–530

Miller AF (2012) Superoxide dismutases: ancient enzymes and new insights. FEBS Lett 586:585–595. https://doi.org/10.1016/j.febslet.2011.10.048

Mittler R (2017) ROS are good. Trends Plant Sci 22:11–19. https://doi.org/10.1016/j.tplants.2016.08.002

Mittler R, Vanderauwera S, Suzuki N, Miller G, Tognetti BV, Vandepoele K, Gollery M, Shulaev V, Van Breusegem F (2011) ROS signaling: the new wave? Trends Plant Sci 16(6):300–309. https://doi.org/10.1016/j.tplants.2011.03.007

Moore M, Gossmann N, Dietz K-J (2016) Redox regulation of cytosolic translation in plants. Trends Plant Sci 21:388–397. https://doi.org/10.1016/j.tplants.2015.11.004

Mou Z, Fan W, Dong X (2003) Inducers of plant systemic acquired resistance regulate NPR1 function through redox changes. Cell 113:935–944. https://doi.org/10.1016/s0092-8674(03)00429-x

Mur LAJ, Kenton P, Lloyd AJ, Ougham H, Prats E (2008) The hypersensitive response; the centenary is upon us but how much do we know? J Exp Bot 59:501–520. https://doi.org/10.1093/jxb/erm239

Mur LA, Mandon J, Persijn S, Cristescu SM, Moshkov IE, Novikova GV, Hall MA, Harren FJ, Hebelstrup KH, Gupta KJ (2013) Nitric oxide in plants: an assessment of the current state of knowledge. AoB Plants 5:pls052. https://doi.org/10.1093/aobpla/pls052

Murphy A, Taiz L (1995) Comparison of metallothionein gene expression and nonprotein thiols in ten Arabidopsis ecotypes. Correlat Copper Toleran Plant Physiol 109:945–954. https://doi.org/10.1104/pp.109.3.945

Narendra S, Venkataramani S, Shen G, Wang J, Pasapula V, Lin Y, Kornyeyev D, Holaday AS, Zhang H (2006) The Arabidopsis ascorbate peroxidase 3 is a peroxisomal membrane-bound antioxidant enzyme and is dispensable for *Arabidopsis* growth and development. J Exp Bot 57:3033–3042. https://doi.org/10.1093/jxb/erl060

Noctor G, Veljovic-Jovanovic S, Driscoll S, Novitskaya L, Foyer C (2002) Drought and oxidative load in the leaves of C3 plants: a predominant role for photorespiration? Ann Bot 89:841–850. https://doi.org/10.1093/aob/mcf096

Noctor G, Reichheld J-P, Foyer CH (2018) ROS-related redox regulation and signaling in plants. In: Seminars in cell & developmental biology Redox signalling in development and regeneration, vol 80, pp 3–12. https://doi.org/10.1016/j.semcdb.2017.07.013

Novotnik B, Ščančar J, Milačič R, Filipič M, Žegura B (2016) Cytotoxic and genotoxic potential of Cr(VI), Cr(III)-nitrate and Cr(III)-EDTA complex in human hepatoma (HepG2) cells. Chemosphere 154:124–131. https://doi.org/10.1016/j.chemosphere.2016.03.118

Oakley AJ (2011) Glutathione transferases: a structural perspective. Drug Metab Rev 43:138–151. https://doi.org/10.3109/03602532.2011.558093

Ort DR, Baker N (2002) A photoprotective role for O_2 as an alternative electron sink in photosynthesis? Curr Opin Plant Biol 5:193–198. https://doi.org/10.1016/s1369-5266(02)00259-5

Ozturk M, Ashraf M, Aksoy A, Ahmad MSA (2015) Plants, pollutants and remediation. Springer, New York

Pal R, Rai JPN (2010) Phytochelatins: peptides involved in heavy metal detoxification. Appl Biochem Biotechnol 160:945–963. https://doi.org/10.1007/s12010-009-8565-4

Panda SK, Choudhury S (2005) Chromium stress in plants. Braz J Plant Physiol 17:95–102. https://doi.org/10.1590/S1677-04202005000100008

Park C-J, Seo Y-S (2015) Heat shock proteins: a review of the molecular chaperones for plant immunity. Plant Pathol J 31:323–333. https://doi.org/10.5423/PPJ.RW.08.2015.0150

Paul AK (2016) Evaluation of chromate reductase activity in the cell-free culture filtrate of *Arthrobacter sp.* SUK 1201 isolated from chromite mine overburden. Chemosphere 156:69–75. https://doi.org/10.1016/j.chemosphere.2016.04.101

Pellerin C, Booker SM (2000) Reflections on hexavalent chromium: health hazards of an industrial heavyweight. Environ Health Perspect 108:A402–A407

Pérez-Gálvez A, Viera I, Roca M (2020) Carotenoids and chlorophylls as antioxidants. Antioxidants 9:505. https://doi.org/10.3390/antiox9060505

Pilon M, Ravet K, Tapken W (2011) The biogenesis and physiological function of chloroplast superoxide dismutases. Biochim Biophys Acta 1807:989–998. https://doi.org/10.1016/j.bbabio.2010.11.002

Potocký M, Pejchar P, Gutkowska M, Jiménez-Quesada MJ, Potocká A, de Dios Alché J, Kost B, Žárský V (2012) NADPH oxidase activity in pollen tubes is affected by calcium ions, signaling phospholipidsand Rac/Rop GTPases. J Plant Physiol 169:1654–1663. https://doi.org/10.1016/j.jplph.2012.05.014

Pourrut B, Perchet G, Silvestre J, Cecchi M, Guiresse M, Pinelli E (2008) Potential role of NADPH-oxidasein early steps of lead-induced oxidative burst in *Vicia faba* roots. J Plant Physiol 165:571–579. https://doi.org/10.1016/j.jplph.2007.07.016

Qi J, Wang J, Gong Z, Zhou JM (2017) Apoplastic ROS signaling in plant immunity. Curr Opin Plant Biol 38:92–100. https://doi.org/10.1016/j.pbi.2017.04.022

Queval G, Jaillard D, Zechmann B, Noctor G (2011) Increased intracellular H_2O_2 availability preferentially drives glutathione accumulation in vacuoles and chloroplasts. Plant Cell Environ 34:21–32. https://doi.org/10.1111/j.1365-3040.2010.02222.x

Rahman MM, Rahman MM, Islam KS, Chongling Y (2010) Effect of chromium stress on antioxidative enzymes and malondialdehyde content activities in leaves and roots of mangrove seedlings *Kandelia candel* (L.) druce. J Environ Sci 26:171–179

Rascio N, Navari-Izzo F (2011) Heavy metal hyperaccumulating plants: how and why do they do it? And what makes them so interesting? Plant Sci 180(2):169–181. https://doi.org/10.1016/j.plantsci.2010.08.016

Rhoads DM, Umbach AL, Subbaiah CC, Siedow TN (2006) Mitochondrial reactive oxygen species. Contribution of oxidative stress and interorganellar signaling. Plant Physiol 141:357–366. https://doi.org/10.1104/pp.106.079129

Rice-Evans C, Miller N, Paganga G (1997) Antioxidant properties of phenolic compounds. Trends Plant Sci 2:152–159. https://doi.org/10.1016/S1360-1385(97)01018-2

Rodriguez E, Azevedo R, Fernandes P, Santos C (2011) Cr(VI) Induces DNA damage, cell cycle arrest and polyploidization: a how cytometric and comet assay Study in *Pisum sativum*. Chem Res Toxicol 24:1040–1047. https://doi.org/10.1021/tx2001465

Rodríguez-Serrano M, Romero-Puertas MC, Zabalza A, Corpas FJ, Gómez M, Del Río LA, Sandalio LM (2006) Cadmium effect on oxidative metabolism of pea (*Pisum sativum* L.) roots. Imaging of reactive oxygen species and nitric oxide accumulation in vivo. Plant Cell Environ 29:1532–1544. https://doi.org/10.1111/j.1365-3040.2006.01531.x

Rudnik R, Bulcha JT, Reifschneider E, Ellersiek U, Baier M (2017) Specificity versus redundancy in the RAP2.4 transcription factor family of *Arabidopsis thaliana*: transcriptional regulation of genes for chloroplast peroxidases. BMC Plant Biol. 17:144. https://doi.org/10.1186/s12870-017-1092-5

Samardzic K, Rodgers KJ (2017) Oxidised protein metabolism: recent insights. Biol Chem 398:1165–1175. https://doi.org/10.1515/hsz-2017-0124

Saud S, Wang D, Fahad S, Javed T, Jaremko M, Abdelsalam NR, Ghareeb RY (2022) The impact of chromium ion stress on plant growth, developmental physiology, and molecular regulation. Front Plant Sci 13:994785. https://doi.org/10.3389/fpls.2022.994785

Sawicka E, Jurkowska K, Piwowar A (2021) Chromium (III) and chromium (VI) as important players in the induction of genotoxicity—Current view. Ann Agric Environ Med AAEM 28:1–10. https://doi.org/10.26444/aaem/118228

Schmfger MEV (2001) Phytochelatins: complexation of metals and metalloids, studies on the phytochelatin synthase. Ph.D. thesis. Munich University of Technology (TUM), Munich

Semchuk NM, Lushchak OV, Falk J, Krupinska K, Lushchak VI (2009) Inactivation of genes, encoding tocopherol biosynthetic pathway enzymes, results in oxidative stress in outdoor grown *Arabidopsis thaliana*. Plant Physiol Biochem 47:384–390. https://doi.org/10.1016/j.plaphy.2009.01.009

Semchuk NM, Vasylyk IV, Kubrak OI, Lushchak VI (2011) Effect of sodium nitroprusside and S-nitrosoglutathione on pigment content and antioxidant system of tocopherol-deficient plants of *Arabidopsis thaliana*. Ukr. Biokhimichnyi Zhurnal 83:69–79

Semchuk NM, Vasylyk YV, Lushchak OV, Lushchak VI (2012) Effect of short-term salt stress on oxidative stress markers and antioxidant enzymes activity in tocopherol-deficient *Arabidopsis thaliana* plants. Ukr Biokhimichnyi Zhurnal 84:41–48

Shahid M, Shamshad S, Rafiq M, Khalid S, Bibi I, Niazi NK, Dumat C, Rashid MI (2017) Chromium speciation, bioavailability, uptake, toxicity and detoxification in soil-plant system: a review. Chemosphere 178:513–533. https://doi.org/10.1016/j.chemosphere.2017.03.074

Shahzad B, Tanveer M, Rehman A, Cheema SA, Fahad S, Rehman S, Sharma A (2018) Nickel: whether toxic or essential for plants and environment: a review. Plant Physiol Biochem 132:641–651. https://doi.org/10.1016/j.plaphy.2018.10.014

Shaikhali J, Heiber I, Seidel T, Ströher E, Hiltscher H, Birkmann S, Dietz K-J, Baier M (2008) The redox-sensitive transcription factor Rap2.4a controls nuclear expression of 2-Cys peroxiredoxin A and other chloroplast antioxidant enzymes. BMC Plant Biol 8:48. https://doi.org/10.1186/1471-2229-8-48

Shanker AK, Djanaguiraman M, Pathmanabhan G, Sudhagar R, Avudainayagam S (2003a) Uptake and phytoaccumulation of chromium by selected tree species. In: Proceedings of the international conference on water and environment held in Bhopal, India

Shanker AK, Sudhagar R, Pathmanabhan G (2003b) Growth phytochelatin SH and antioxidative response of sunflower as affected by chromium speciation. In: 2nd international congress of plant physiology on sustainable plant productivity under changing environment, New Delhi

Shanmugam V, Lo JC, Wu CL, Wang SL, Lai CC, Connolly EL, Huang J-L, Yeh K-C (2011) Differential expression and regulation of iron regulated metal transporters in *Arabidopsis halleri* and *Arabidopsis thaliana*—The role in zinc tolerance. New Phytol 190:125–137. https://doi.org/10.1111/j.1469-8137.2010.03606.x

Sharma D, Sharma C, Tripathi R (2003) Phytotoxic lesions of chromium in maize. Chemosphere 51:63–68. https://doi.org/10.1016/s0045-6535(01)00325-3

Sharma A, Kumar V, Shahzad B, Ramakrishnan M, Sidhu GPS, Bali AS, Handa N, Kapoor D, Yadav P, Khanna K (2019) Photosynthetic response of plants under different abiotic stresses: a review. J Plant Growth Regul 38:1–23. https://doi.org/10.1007/s00344-019-10018-x

Sharma A, Kapoor D, Wang J, Shahzad B, Kumar V, Bali AS, Jasrotia S, Zheng B, Yuan H, Yan D (2020) Chromium bioaccumulation and its impacts on plants: an overview. Plants 9:100. https://doi.org/10.3390/plants9010100

Sharma P, Chouhan R, Bakshi P, Gandhi SG, Kaur R, Sharma A, Bhardwaj R (2022) Amelioration of Chromium-induced oxidative stress by combined treatment of selected plant-growth-promoting rhizobacteria and earthworms via modulating the expression of genes related to reactive oxygen species metabolism in *Brassica juncea*. Front Microbiol 13:802512. https://doi.org/10.3389/fmicb.2022.802512

Shukla T, Kumar S, Khare R, Tripathi RD, Trivedi PK (2015) Natural variations in expression of regulatory and detoxification related genes under limiting phosphate and arsenate stress in *Arabidopsis thaliana*. Front Plant Sci 6:e898. https://doi.org/10.3389/fpls.2015.00898

Siendones E, González-Reyes JA, Santos-Ocaña C, Navas P, Córdoba F (1999) Biosynthesis of ascorbic acid in kidney bean l-galactono-γ-lactone dehydrogenase is an intrinsic protein located at the mitochondrial inner membrane. Plant Physiol 120:907–912

Singh HP, Mahajan P, Kaur S, Batish DR, Kohli RK (2013) Chromium toxicity and tolerance in plants. Environ Chem Lett 11:229–254. https://doi.org/10.1007/s10311-013-0407-5

Sinha V, Pakshirajan K, Chaturvedi R (2018) Chromium tolerance, bioaccumulation and localization in plants: an overview. J Environ Manag 206:715–730. https://doi.org/10.1016/j.jenvman.2017.10.033

Skórzyńska-Polit E, Dra̧żkiewicz M, Krupa Z (2003) The activity of the antioxidative system in cadmium-treated *Arabidopsis thaliana*. Biol Plant 47:71–78. https://doi.org/10.1023/A:1027332915500

Slaymaker DH, Navarre DA, Clark D, del Pozo O, Martin GB, Klessig DF (2002) The tobacco salicylic acid-binding protein 3 (SABP3) is the chloroplast carbonic anhydrase, which exhibits antioxidant activity and plays a role in the hypersensitive defense response. Proc Natl Acad Sci USA 99:11640–11645. https://doi.org/10.1073/pnas.182427699

Smiri M, Chaoui A, Ferjani EE (2010) Interaction between heavy metals and thiol-linked redox reactions in germination. Pak J Biol Sci 13:877–883. https://doi.org/10.3923/pjbs.2010.877.883

Smiri M, El Ghoul J (2012) Role for plant thioredoxin in Cd^{2+} chelation. Int J Veg Sci 18:93–105. https://doi.org/10.1080/19315260.2011.579692

Smiri M, Jelali N, El Ghoul J (2013) Cadmium affects the NADP-thioredoxin reductase/thioredoxin system in germinating pea seeds. J Plant Interact 8:125–133. https://doi.org/10.1080/17429145.2012.689865

Smirnoff N, Wheeler GL (2000) Ascorbic acid in plants: biosynthesis and function. Crit Rev Plant Sci 19:267–290. https://doi.org/10.1080/07352680091139231

Sofo A, Scopa A, Nuzzaci M, Vitti A (2015) Ascorbate peroxidase and catalase activities and their genetic regulation in plants subjected to drought and salinity stresses. Int J Mol Sci 16:13561–13578. https://doi.org/10.3390/ijms160613561

Srivastava S, Srivastava S, Prakash S, Srivastava MM (1999) Fate of trivalent chromium in presence of organic acids. Chem Spec Bioavail 10:147–150. https://doi.org/10.3184/095422998782775763

Stahl W, Sies H (2003) Antioxidant activity of carotenoids. Mol Aspects Med 24:345–351. https://doi.org/10.1016/s0098-2997(03)00030-x

Stambulska UY, Bayliak MM, Lushchak VI (2018) Chromium(VI) toxicity in legume plants: modulation effects of rhizobial symbiosis. BioMed Res Int 2018:8031213. https://doi.org/10.1155/2018/8031213

Stambulska UY, Bayliak MM (2020) Legume-Rhizobium symbiosis: secondary metabolites, free radical processes, and effects of heavy metals. In: Mérillon J-M, Ramawat KG (eds) Co-evolution of secondary metabolites, reference series in phytochemistry. Springer International Publishing, Cham, pp 291–322. https://doi.org/10.1007/978-3-319-96397-6_43

Stevens RG, Creissen GP, Mullineaux PM (1997) Cloning and characterization of a cytosolic glutathione reductase cDNA from pea (*Pisum sativum* L.) and its expression in response to stress. Plant Mol Biol 35:641–654. https://doi.org/10.1023/a:1005858120232

Strile M, Kolar J, Selih VS, Kocar D, Pihlar B (2003) A comparative study of several transition metals in Fenton like reaction system at circum neutral pH. Acta Chin Slov 50:619–632

Suzuki N, Miller G, Morales J, Shulaev V, Torres MA, Mittler R (2011) Respiratory burst oxidases: the engines of ROS signaling. Curr Opin Plant Biol 14(6):691–699. https://doi.org/10.1016/j.pbi.2011.07.014

Szewczyk K, Chojnacka A, Górnicka M (2021) Tocopherols and tocotrienols—Bioactive dietary compounds; what is certain, what is doubt? Int J Mol Sci 22:6222. https://doi.org/10.3390/ijms22126222

Tada Y, Spoel SH, Pajerowska-Mukhtar K, Mou Z, Song J, Wang C, Zuo J, Dong X (2008) Plant immunity requires conformational changes [corrected] of NPR1 via S-nitrosylation and thioredoxins. Science 321:952–956. https://doi.org/10.1126/science.1156970

Thoma I, Loeffler C, Sinha AK, Gupta M, Krischke M, Steffan B, Roitsch T, Mueller MJ (2003) Cyclopentenone isoprostanes induced by reactive oxygen species trigger defense gene activation and phytoalexin accumulation in plants. Plant J 34:363–375. https://doi.org/10.1046/j.1365-313x.2003.01730.x

Tumolo M, Ancona V, De Paola D, Losacco D, Campanale C, Massarelli C, Uricchio VF (2020) Chromium pollution in european water, sources, health risk, and remediation strategies: an overview. Int J Environ Res Public Health 17:5438. https://doi.org/10.3390/ijerph17155438

Uddin MK (2017) A review on the adsorption of heavy metals by clay minerals, with special focus on the past decade. Chem Eng J 308:438–462. https://doi.org/10.1016/j.cej.2016.09.029

Uraguchi S, Tanaka N, Hofmann C, Abiko K, Ohkama-Ohtsu N, Weber M, Kamiya T, Sone Y, Nakamura R, Takanezawa Y, Kiyono M, Fujiwara T, Clemens S (2017) Phytochelatin synthase has contrasting effects on cadmium and arsenic accumulation in rice grains. Plant Cell Physiol 58:1730–1742. https://doi.org/10.1093/pcp/pcx114

Vasylkiv OY, Kubrak OI, Storey KB, Lushchak VI (2010) Cytotoxicity of chromium ions may be connected with induction of oxidative stress. Chemosphere 80:1044–1049. https://doi.org/10.1016/j.chemosphere.2010.05.023

Vierling E (1991) The roles of heat shock proteins in plants. Annu Rev Plant Physiol Plant Mol Biol 42:579–620. https://doi.org/10.1146/annurev.pp.42.060191.003051

Wakeel A, Ali I, Wu M, Kkan AR, Jan M, Ali A, Liu Y, Ge S, Wu J, Gan Y (2019) Ethylene mediates dichromate-induced oxidative stress and regulation of the enzymatic antioxidant system-related transcriptome in *Arabidopsis thaliana*. Environ Exp Bot 161:166–179. https://doi.org/10.1016/j.envexpbot.2018.09.004

Wakeel A, Xu M, Gan Y (2020) Chromium-induced reactive oxygen species accumulation by altering the enzymatic antioxidant system and associated cytotoxic, genotoxic, ultrastructural, and photosynthetic changes in plants. Int J Mol Sci 21:728. https://doi.org/10.3390/ijms21030728

Wang Y-F, Huang K-L, Li C-T, Mi H-H, Luo J-H, Tsai P-J (2003) Emissions of fuel metals content from a diesel vehicle engine. Atmos Environ 37:4637–4643. https://doi.org/10.1016/j.atmosenv.2003.07.007

Wang Z, Zhang L, Xiao Y, Chen W, Tang K (2010) Increased vitamin c content accompanied by an enhanced recycling pathway confers oxidative stress tolerance in *Arabidopsis*. J Int Plant Biol 52(4):400–409. https://doi.org/10.1111/j.1744-7909.2010.00921.x

Wang F, Wang Z, Zhu C (2012) Heteroexpression of the wheat phytochelatin synthase gene (TaPCS1) in rice enhances cadmium sensitivity. Acta Biochim Biophys 44:886–893. https://doi.org/10.1093/abbs/gms073

Waters ER, Vierling E (2020) Plant small heat shock proteins—Evolutionary and functional diversity. New Phytol 227:24–37. https://doi.org/10.1111/nph.16536

Wen K, Li X, Huang R, Nian H (2022) Application of exogenous glutathione decreases chromium translocation and alleviates its toxicity in soybean (*Glycine max* L.). Ecotoxicol Environ Saf 234:113405. https://doi.org/10.1016/j.ecoenv.2022.113405

Weyemi U, Dupuy C (2012) The emerging role of ROS-generating NADPH oxidase NOX4 in DNA-damage responses. Mutat Res 751:77–81. https://doi.org/10.1016/j.mrrev.2012.04.002

Willems P, Mhamdi A, Stael S, Storme V, Kerchev P, Noctor G, Gevaert K, Van Breusegem F (2016) The ROS wheel: refining ROS transcriptional footprints. Plant Physiol 171:1720–1733. https://doi.org/10.1104/pp.16.00420

Wu Y, Zhang D, Chu JY, Boyle P, Wang Y, Brindle ID, De Luca V, Després C (2012) The *Arabidopsis* NPR1 protein is a receptor for the plant defense hormone salicylic acid. Cell Rep 1:639–647. https://doi.org/10.1016/j.celrep.2012.05.008

Wyszkowska J, Borowik A, Zaborowska M, Kucharski J (2023) Sensitivity of *Zea mays* and soil microorganisms to the toxic effect of chromium (VI). Int J Mol Sci 24:178. https://doi.org/10.3390/ijms24010178

Xu J, Xing X-J, Tian Y-S, Peng R-H, Xue Y, Zhao W, Yao Q-H (2015) Transgenic Arabidopsis plants expressing tomato glutathione S-transferase showed enhanced resistance to salt and drought stress. Plos One 10:e0136960. https://doi.org/10.1371/journal.pone.0136960

Yadav SK (2010) Heavy metals toxicity in plants: an overview on the role of glutathione and phytochelatins in heavy metal stress tolerance of plants. S Afr J Bot 76(2):167–179

Yildiz M, Terzi H, Cenkci S, Yildiz B (2012) Chromium(VI)-induced alterations in 2-D protein profiles and antioxidant defence systems of barley cultivars. Hacet J Biol Chem 40:257–265

You J, Chan Z (2015) ROS regulation during abiotic stress responses in crop plants. Front Plant Sci 6:1092. https://doi.org/10.3389/fpls.2015.01092

Yu CH, Huang L, Shin JY, Artigas F, Fan Z (Tina) (2014) Characterization of concentration, particle size distribution, and contributing factors to ambient hexavalent chromium in an area with multiple emission sources. Atmosp Environ (1994) 94:701–708

Zaffagnini M, Michelet L, Marchand C, Sparla F, Decottignies P, Le Maréchal P, Miginiac-Maslow M, Noctor G, Trost P, Lemaire SD (2007) The thioredoxin-independent isoform of chloroplastic glyceraldehyde-3-phosphate dehydrogenase is selectively regulated by glutathionylation. FEBS J 274:212–226. https://doi.org/10.1111/j.1742-4658.2006.05577.x

Zayed AM, Terry N (2003) Chromium in the environment: factors affecting biological remediation. Plant Soil 249(1):139–156. https://doi.org/10.1023/A:1022504826342

Zeng F, Qiu B, Wu X, Niu S, Wu F, Zhang G (2012) Glutathione-mediated alleviation of chromium toxicity in rice plants. Biol Trace Elem Res 148:255–263. https://doi.org/10.1007/s12011-012-9362-4

Zeng F, Wu X, Qiu B, Wu F, Jiang L, Zhang G (2014) Physiological and proteomic alterations in rice (*Oryza sativa* L.) seedlings under hexavalent chromium stress. Planta 240:291–308. https://doi.org/10.1007/s00425-014-2077-3

Zhang F, Shi W, Jin Z, Shen Z (2002) Response of antioxidative enzymes in cucumber chloroplasts to cadmium toxicity. J Plant Nutr 26:1779–1788

Zhang Y, Tessaro MJ, Lassner M, Li X (2003) Knockout analysis of Arabidopsis transcription factors TGA2, TGA5, and TGA6 reveals their redundant and essential roles in systemic acquired resistance. Plant Cell 15:2647–2653. https://doi.org/10.1105/tpc.014894

Zhang M, Smith JAC, Harberd NP, Jiang C (2016) The regulatory roles of ethylene and reactive oxygen species (ROS) in plant salt stress responses. Plant Mol Biol 91:651–659. https://doi.org/10.1007/s11103-016-0488-1

Zhou J, Goldsbrough PB (1994) Functional homologs of fungal metallothionein genes from Arabidopsis. Plant Cell 6:875–884. https://doi.org/10.1105/tpc.6.6.875

SPRINGER NATURE

GPSR Compliance

The European Union's (EU) General Product Safety Regulation (GPSR) is a set of rules that requires consumer products to be safe and our obligations to ensure this.

If you have any concerns about our products, you can contact us on ProductSafety@springernature.com

In case Publisher is established outside the EU, the EU authorized representative is:

Springer Nature Customer Service Center GmbH
Europaplatz 3
69115 Heidelberg, Germany

The manufacturer's authorised representative in the EU is Springer Nature Customer Service Centre GmbH, Europaplatz 3, 69115 Heidelberg, Germany. If you have any concerns regarding our products, please contact ProductSafety@springernature.com

Printed and bound by CPI Group (UK) Ltd, Croydon, CR0 4YY
30/09/2026
02225322-0018